Biology of Hawaiian Streams and Estuaries

Biology of Hawaiian Streams and Estuaries

Proceedings of the Symposium on the Biology of Hawaiian Streams and Estuaries

Hilo, Hawai'i
26–27 April 2005

Edited by
Neal L. Evenhuis & J. Michael Fitzsimons

Bishop Museum Bulletin in Cultural and Environmental Studies 3

Bishop Museum Press
Honolulu, 2007

Supported by a grant from the Sportfish Restoration Project of the United States Fish & Wildlife Service through the Division of Aquatic Resources, Department of Land & Natural Resources, State of Hawaii.

Available from:
Bishop Museum Press
1525 Bernice Street
Honolulu, Hawai'i 96817-2704

Division of Aquatic Resources
Department of Land & Natural Resources
1151 Punchbowl Street, Honolulu, Hawai'i 96813

Published by
Bishop Museum Press
1525 Bernice Street
Honolulu, Hawai'i 96817-2704, USA

Printed in the United States of America

ISBN 10: 1-58178-053-2
ISBN 13: 978-1-58178-053-6
ISSN 1548-9620

TABLE OF CONTENTS

Foreword

The representation of exceptional scientists, broad range of papers, and quality of research presented at this symposium are truly remarkable, and they underscore the progress made since the 1980s, when publication of anything related to Hawaiian streams was difficult because interest at all levels was minimal, and only a few pioneering researchers were willing to commit time to the subject.

The establishment of the Water Code in Hawai'i and the formation of the Commission on Water Resource Management under the Department of Land and Natural Resources encouraged the Division of Aquatic Resources (DAR), which is assigned responsibility by Statute for the management and protection of all living aquatic resources in Hawai'i, to focus on expanding the information base to assist in regulatory decision-making and lend credence to the development of instream flow standards. Fortunately, an existing collaboration between Dr. Robert Nishimoto (DAR) and Dr. J. Michael Fitzsimons (Louisiana State University) provided the seed to develop the program.

The resultant 1990 Symposium on Freshwater Stream Biology and Fisheries Management, entitled New Directions in Research, Management, and Conservation of Hawaiian Freshwater Stream Ecosystems, represented the first effort to organize existing state-of-the-art knowledge and bring varied interests together. It led to development of a longer range plan which with the current symposium has effectively been accomplished with reference to its principal objectives. The mark of good science, however, is that the advance of knowledge inevitably leads to more questions. The discussions during this symposium have posed an important set of new questions.

Application of the advancing knowledge to management has been lagging for institutional reasons more related to organization and influencing decision-makers than to the quality of the science involved, but that is no excuse for stepping back from new investigations to advance the science. In fact, as the institutional deficits are reduced, the demand for good science will increase rapidly. Effective steps to move the management component ahead have in fact been recommended in this symposium.

We are nevertheless at a crossroad. If there is not regulatory commitment to rational management, there is danger that this symposium will become an end point with a good scientific record indicative of what can be accomplished but insufficiently valued by decision-makers to justify the expense of continuation. Or, it can set the stage for implementation of a more comprehensive stream protection and instream flow management program that will mandate continuing expansion of scientific knowledge.

There will always be strong pressure for an end, in part because decision-makers and the public-at-large have little intrinsic familiarity with the biota supported by freshwater streams. As human population expands and development interests become overwhelming, water running out to sea may again in the prevailing view be regarded as wasted. If, however, the intellectual power inherent in the participants in this symposium is retained through future cooperation and unification, the pessimistic outcome can be overcome.

As understanding expands, it will become easier to justify the value of these resources. That is why your participation in this symposium was so important. And if you maintain your commitment, the symposium will indeed mark a step forward rather than an end point.

—William S. Devick, Administrator (Retired)
Division of Aquatic Resources, Department of Land and Natural Resources
State of Hawaii

Mililani Town, Hawai'i, USA
May 13, 2005

Biology of Hawaiian Streams and Estuaries. Edited by N.L. Evenhuis & J.M. Fitzsimons. *Bishop Museum Bulletin in Cultural and Environmental Studies* 3: 1–2 (2007).

Introduction

The stated purpose of the Symposium on the Biology of Hawaiian Streams and Estuaries is to provide "an overview of current knowledge about the animals and plants that live in the islands' fresh and brackish waters and [to] indicate key areas for future research". In the mid-1980s, this task could have been accomplished easily because so few studies had been published on the biology of Hawaiian streams. At that time, biologists working in Hawaiian streams and estuaries were limited mostly to Bob Kinzie and his graduate students at the University of Hawai'i, Bob Nishimoto and a few colleagues in the Division of Aquatic Resources, and Mike Fitzsimons and his graduate students from Louisiana State University. Our work was founded on the pioneering studies by Kenji Ego of the (then) Hawai'i Fish and Game, John Maciolek of the U.S. Fish & Wildlife Service, and Amadeo Timbol as a student at the University of Hawai'i and later as a faculty member at Kaua'i Community College.

However, two decades later there are by our rough count some 128 persons who have been active in biological studies of Hawaiian streams and estuaries or are doing relevant work on other high islands of the tropics. The symposium budget limited us to inviting only about 30 speakers to the conference in Hilo but many others expressed a desire to participate. It was encouraging to learn of the unanticipated interest in the symposium, but it was disappointing that we were not able to accommodate more speakers. In addition, we were joined by about 175 other individuals from campuses of the University of Hawai'i, county and state agencies, public and private schools, and other people, including those of Hawaiian ancestry, who are concerned with assuring the responsible use of flowing fresh waters in the islands.

Feedback from participants at the conference has been favorable and complimentary and, for this reason, we are additionally grateful to Peter Young, Chairperson of the Board of Land & Natural Resources and the Commission on Water Resource Management, and to Francis Oishi, Acting Administrator of the Division of Aquatic Resources, for giving us the opportunity to hold the conference. Their support and participation was central to the success of the meetings. The superb organizational skills of Gerri Kahili in the Hilo office of the Division of Aquatic Resources easily dispensed with any problems related to the logistics for the meetings. Gerri's work, which began months before the conference convened, was facilitated by Katharine Hind in the Division's Honolulu office, and, during the conference, Gerri received capable assistance from Van Dacanay, Laura Livnat, Wade Ishikawa, Andy Kahili, and Charlotte Fitzsimons.

This volume, developed from manuscripts prepared by conference participants, does not attempt to summarize all known information about the animals and plants that live in Hawaii's stream/estuarine ecosystems—now preferably and more accurately referred to as *ahupua'a*. Instead, it offers a broad range of topics that provide detailed insight into the variety and pervasiveness of scientific interest in the islands' fresh and brackish water environments.

When preparing for the last session of the conference during which areas for future study were to be identified, Mark McRae, Lori Benson McRae, Glenn Higashi, Bob Nishimoto, Ron Englund, and Mike Fitzsimons compiled a very long and (we thought) exhaustive list of ideas to prompt perhaps a few other suggestions from the audience. It is significant that the group enduring right to the end of the conference provided no less than 36 additional topics relevant to the management and preservation of Hawaii's streams and estuaries.

It is very easy to single out a single person who, far more than anyone else, has been responsible for making it possible to advance the knowledge of streams and estuaries in Hawai'i. During his tenure as Administrator of the Division of Aquatic Resources, William S. Devick's unflagging support and encouragement were the mainstays of survey and research efforts. Any successes in implementing science-based management and conservation of Hawaiian streams and estuaries were Bill's successes, and future successes are clearly the charge of those who follow him in office.

As we reflect on the papers presented at the conference and consider the manuscripts contributed for this volume, we are aware of the profound responsibility shared by researchers, local universities, state and federal agencies, and indeed anyone whose actions influence water-use decisions in the State of Hawai'i. For purely selfish reasons, the people of Hawai'i owe it to themselves to demand the very best science-based husbandry of the State's aquatic resources. We all also have a clear obligation to manage, conserve, and preserve these resources for the young people of Hawai'i while providing a model for people of all ages living among other islands of the tropical Pacific. Anyone attending the conference could not escape this conclusion after seeing and hearing the boys and girls of *Kanu 'o ka 'Āina* Public Charter School as they opened the symposium, set the tone for the entire conference, and made it readily apparent that the stewardship of Hawaiian *ahupua'a* is unmistakably <u>their</u> *kuleana*. However, until they reach the "full age of majority", the mandate to conserve *'o'opu*, *'ōpae*, *hīhīwai*, *āholehole*, and all the other indigenous animals and plants living in Hawaiian streams and estuaries must reside with the rest of us. It would be unconscionable to fail them.

Mike Fitzsimons
Baton Rouge

Bob Nishimoto
Hilo

Biology of Hawaiian Streams and Estuaries. Edited by N.L. Evenhuis & J.M. Fitzsimons. *Bishop Museum Bulletin in Cultural and Environmental Studies* 3: 3–9 (2007).

Hawaiian Stream Fishes: the Role of Amphidromy in History, Ecology, and Conservation Biology

ROBERT M. MCDOWALL
National Institute of Water and Atmospheric Research, P.O. Box 8602, Christchurch, New Zealand; email: r.mcdowall@niwa.co.nz

Abstract

Amphidromy, a pattern of migrations between fresh water and the ocean, known among some fish groups and especially the gobioids, is universal to all Hawaiian stream fishes and also to some gastropod molluscs and decapod crustaceans. A clear understanding of this pattern of migrations is profoundly important to diverse aspects of the biogeography, ecology, and conservation of the Hawaiian fauna, as it is for the faunas of islands widely across the tropics, subtropics, and extending into the northern and southern cool temperate zones.

Introduction

The small Hawaiian freshwater fish fauna, with only 5 gobioid species, represents a far eastern, oceanic outpost of the fauna of the Indo-West Pacific (Springer, 1982; Nishimoto & Kuamo'o, 1991, 1997; Fitzsimons & Nishimoto, 1995; Fitzsimons *et al.*, 2002). The fauna comprises four endemic species: *Lentipes concolor, Sicyopterus stimpsoni,* and *Stenogobius hawaiiensis* in the family Gobiidae and *Eleotris sandwicensis* in the related Eleotridae. In addition *Awaous guamensis* is a non-endemic Hawaiian gobiid that is found also in Guam in the Mariana Islands, New Caledonia, Vanuatu, and Fiji. All of these gobioid species are amphidromous (Ego, 1956; Tomihama, 1972; Maciolek, 1977; Radtke & Kinzie, 1991, 1996; Kinzie, 1993a, b; Ha & Kinzie, 1996). The stream fish fauna of Hawai'i resembles other aspects of the biota of Hawai'i in having an external dispersal origin (Wagner & Funk, 1995; Price & Wagner, 2004).

The nature and place of amphidromy

The presence of amphidromy in every one of these species is the core, fundamental driver of a whole series of biogeographical and ecological characteristics of the fauna, such that an understanding of amphidromy is crucial for understanding, managing, and conserving the fauna. Amphidromy is a specialised migratory behaviour that is widely present in the fish faunas of the islands of the globe, especially in the tropics and subtropics of the central and western Pacific, southeast Asia, and the Indian Ocean as far west as the Seychelles, Reunion, and the Comoro Islands (McDowall, 1988; Balon & Bruton, 1994). There are also amphidromous gobioids widely in the Caribbean islands and in Central America (Lyons & Schneider, 1990; Pringle, 1997; Fievet & Guennec, 1998). Amphidromous species are rarely found on continents.

Amphidromy involves reproduction in fresh water, an immediate migration of the newly hatched larvae to sea, a period of a few weeks to a few months feeding and growing at sea, and a return migration of small juveniles to fresh water—where subsequent feeding and growth to maturity take place, and so the life cycle is completed (Myers, 1949; McDowall, 1988, 1997a).

Implications of diadromy for Hawaiian streams

In the rest of this paper I outline the role of amphidromy in the historical biogeography, proximate ecology, and conservation biology of the fauna at a series of spatial and temporal scales.

***1) A dispersal fauna*:** Clearly, the presence of amphidromy in all members of the Hawaiian freshwater fish fauna reflects the role of life at sea in allowing freshwater fish to reach these remote and relatively youthful Hawaiian Islands in the first place (Polhemus, 1996; Kroenke, 1996; McDowall, 2003). Were there not amphidromous species elsewhere in the eastern/central Pacific (McDowall, 1988; Marquet & Laboute, 1992; Watson, 1992, 1995; Watson & Kottelat, 1994; Watson & Chen, 1998; Marquet *et al.*, 1996; Serét, 1997) it is possible that the Hawaiian Islands would have no freshwater fish at all. There is no hint that its fish fauna is derived from marine fishes in the seas around the islands (amphidromous gobioid species are not derivatives from local marine taxa across their range – McDowall, 1997b). And since five different genera, each known widely across the Indo-West Pacific, are represented in the fauna, there has plainly been rare, but repeated, dispersal of invading propagules, logically from the west, where other, and routinely amphidromous, species in all of these genera are to be found. The wider distribution of the only nonendemic Hawaiian freshwater fish, *A. guamensis* (Watson, 1992), implies that such eastward dispersions may still be going on, and there is no reason to think that it is not.

2) The absence of local island endemics: There are no local-island endemics in the Hawaiian freshwater fish fauna [unlike the widespread presence of local-island endemics on various Hawaiian islands in other biotic groups (Wagner & Funk, 1995)]; all five species are widespread across the Hawaiian archipelago, and this implies that either the range of each of the five species is relatively young in Hawai'i or, more likely, that there is continued dispersion by the five species across the archipelago. It seems likely that the amphidromous freshwater fish species are naturally spreading from older islands to younger islands, as the latter emerge, become ecologically stable, and acquire freshwater biotas [as is true of other biotic groups (Wagner & Funk, 1995)].

3) A lack of genetic structuring: Studies of genetic diversification using DNA sequencing show that there has been no local island diversification or structuring among the various Hawaiian Island populations of any of these gobioids, and this is consistent with the view that there is continuing dispersion of these gobies across the islands (Zink, 1991; Zink *et al.*, 1996; Fitzsimons *et al.*, 1990; Chubb *et al.*, 1998).

4) Recruitment into island stream fish populations: The fact that larvae of Hawaii's amphidromous gobioids spend part of their lives in the sea means that Hawaiian streams are continually subject to invasion of propagules from the sea (Ego, 1956). As a result, invasive juveniles are the mechanism for recruitment of new cohorts into the populations. An outcome of this is that ecological processes that take place at sea, beyond the streams of Hawai'i, affect recruitment processes. Amphidromy is therefore of crucial importance in determining the basic age structure and recruitment dynamics of freshwater fish populations in the islands' streams.

5) The role of marine ecological processes: Marine processes in the ecology and behaviour of these amphidromous species may be of crucial importance for recruitment success and population recovery of the faunas of steep islands—where there is a high likelihood of 'flash' floods sweeping newly hatched larvae out of the streams, and to sea.

6) Recruitment dynamics: Population structure and community dynamics are, to some extent, controlled by the processes of juvenile invasion into Hawaiian streams from the sea. Because all fish populations are governed by the continual immigration of juvenile gobioids, the fish communities present in Hawaiian streams result from differences among the species in their patterns and processes of invasion.

7) Differing patterns of invasion: It can be predicted that for some species, which invade only short distances up stream, all of the community ecological processes take place in the lower reaches of streams and make them vulnerable to anthropogenic impacts *(S. hawaiiensis and E. sandwicensis)*. For longer-distance migrators, it will be primarily the juveniles that are found in the lower reaches of streams and will be affected by such impacts.

8) Changing species richness: Because of differences in the distances that the various species penetrate upstream, there will be a downstream-upstream shift in species richness driven by invasion processes and dynamics.

9) Changing species composition: Parallel to that shift in species richness will be a change in species composition, in a downstream-upstream direction in streams, as one species after another drops out of the communities (McDowall, 1998) (This statement is applicable to the extent that young fish migrating upstream from the ocean and through the lower reaches of the streams, are counted along with resident adults).

10) Changing age and size structure of the populations: It is predictable that there will tend to be a shift in the age structure of the populations, with fewer small juveniles being found in the inland/upstream habitats (McDowall, 1988, 1990).

11) Changing trophic structures**:** Some Hawaiian gobioids are herbivorous and feed on periphyton taken from the substrate (Tomihama, 1972; Kido, 1996a, b; Fitzsimons *et al.*, 2003). There seems to be a downstream-upstream shift among species from carnivory to herbivory, probably parallel to a downstream-upstream increase in size/age that is associated with the duration of immigration. It is interesting that the species that penetrate least distances inland tend to be the predators (Fitzsimons & Nishimoto, 1995), and it is predictable that their prey may at times include the immigrating juveniles of other gobioids.

12) Climbing falls and finding food on the way: Some Hawaiian gobies are astonishing climbers of falls, with *Lentipes concolor* being a superb climber (Fitzsimons & Nishimoto, 1995; Englund & Filbert, 1997). It would be interesting to know whether this species is able to climb these heights, in part, as a result of its herbivory. This might give it an ability to feed and restore muscle energy by consuming algae (diatoms are algae) on the surfaces of the falls as it climbs. This would, in turn, mean that there is less urgency in getting to the tops of falls, compared with the energy constraints on climbing by carnivorous species for which food may be less easily available.

13) High risk island habitats**:** Freshwater habitats on small islands are likely to be ephemeral owing to the small size of stream catchments and because rainfall can be unreliable, especially on the leeward sides of islands. Moreover, some of the Hawaiian Islands are still active volcanoes. These various factors are likely to result in possibly repeated local extirpation of stream faunas, including their freshwater fishes.

14) Amphidromy and demography: Clearly, amphidromous migrations confer both risks and advantages on the populations, and these are presumably reflected in life history strategies. Amphidromous species run risks of dispersion and expatriation while at sea and, with that, difficulties in returning to freshwater habitats. These may be compensated for by small eggs, high fecundity, and the trophic advantages of life in the oceanic plankton where small food organisms may abound. Comparisons across the five Hawaiian species of fecundity, egg size, distribution, and duration of life at sea, and age and size at return to freshwater, would be of interest in clarifying demography.

15) Amphidromy and faunal restoration: Amphidromy provides a mechanism that facilitates the repopulation of streams that are affected by perturbations such as unreliable rainfall, or when streams are recovering their biotas after the impacts of active volcanism, or as new streams form once volcanic activity has settled down.

16) Amphidromy and the implications of habitat perturbation: Anthropogenic perturbation of the flows and biotas of Hawaiian streams is likely to have widespread impacts on fish communities, e.g.:

i) if stream mouths are blocked, as a result of flow fluctuations and water abstraction, entry into streams by juveniles from the sea will be prevented and effects on stream fish species profound;

ii) if stream reaches further inland are dewatered by water abstraction, even though entry to streams may still be possible, the immigrating larvae will be hindered in reaching suitable habitats further upstream, especially for those species that migrate long distances upstream. In addition to obstructing their upstream movement, hindrances to migration will make them more vulnerable to predation;

iii) if there are point-source polluting discharges into streams, a number of effects may ensue, including prevention of migration upstream past the discharge points and mortalities for resident fish populations downstream from the discharge point;

iv) and introduced species will rapidly discover the concentrated upstream migrations of juvenile gobioids as a rich source of food.

A broader perspective

What is discussed above for the freshwater fish fauna of Hawai'i needs to be seen in a broader context. Amphidromous gobies are important components of fish faunas of islands across the Indo-West Pacific, as well as widely in the Caribbean. Very similar issues to those in the Hawaiian fauna apply also to our understanding of the ecology and conservation of fish faunas broadly across the tropics and subtropics, where amphidromous gobies are widespread, although in few countries is the fauna so dominated by this group.

Conclusions: an overview of the role of amphidromy

Amphidromous migrations need to be seen as crucial elements in the ecology of Hawaiian streams:

- in facilitating overall distribution to and colonisation of the islands;
- of dispersal among the islands;
- in maintaining the species' community balance in the fish faunas;
- contributing to the nature (lack of) genetic structuring across the islands;
- juvenile recruitment into populations;
- the establishment of community structure;
- for restoration of fish communities following perturbation.

Thus amphidromy is of fundamental importance to the history and ecology of all indigenous freshwater fishes in Hawaiian streams, at a wide range of temporal and spatial scales, and should be viewed as a giant homeostatic mechanism that facilitates long term equilibrium in a fauna subject to the effects of geographical isolation and natural and anthropogenic perturbation (Fitzsimons & Nishimoto, 1995). The managers of Hawaiian waterways and those responsible for the conservation of their biotas need to be cognisant of:

— the patterns of fish migrations into streams from the sea;
— the importance of these migrations for maintaining the fish communities;
— the potential for diverse impacts both within and beyond the islands' streams to disrupt these communities.

I have focussed here only on fishes. There are decapod crustaceans and gastropod molluscs that are also amphidromous in the streams of Hawai'i and again more widely across the Indo-West Pacific (Ford & Kinzie, 1982; Schneider & Frost, 1986; Kano & Kase, 2003), and similar principles apply to how their ecologies can be affected by stream management practices. It is easy enough to imagine shrimps migrating into streams from the sea (Fievet *et al.*, 2001); migrating gastropod molluscs need a bit more imagination, but they, too, do migrate upstream into Hawaiian fresh waters from the sea.

Finally, what is described above for Hawaiian Islands applies equally to the freshwater biotas of islands throughout the tropics and subtropics, extending north to Japan and south to New Zealand, where there are similarly amphidromous fishes, decapod crustaceans, and gastropod molluscs. Across this broader spatial range the stream faunas are generally rather more diverse, and include other fish groups with alternative life history strategies.

Thus what is important for the conservation and management of the entire freshwater fauna of Hawaiian streams is almost equally important for island streams across a wide span of the globe.

Literature Cited

Balon, E.K. & M.N. Bruton. 1994. Fishes of the Tatinga River, Comoros, with comments on freshwater amphidromy in the goby *Sicyopterus lagocephalus*. *Ichthyological Exploration of Freshwaters* **5**: 25–40.

Chubb, A.L., R.M. Zink & M.J. Fitzsimons. 1998. Patterns of mtDNA variation in Hawaiian freshwater fishes: phylogeographic consequences of amphidromy. *Journal of Heredity* **89**: 8–16.

Ego, K. 1956. *Life history of freshwater gobies*. Freshwater Gamefish Management Research, Department of Land and Natural Resources, Territory of Hawai'i Project No. F-4-R: 1–24.

Englund, R. & R. Filbert. 1997. Discovery of the native stream goby, *Lentipes concolor*, above Hawaii's highest waterfall, Hi'ilawe Falls. *Bishop Museum Occasional Papers* **49**: 62–64.

Fievet, E. & B. Guennec. 1998. Migration de masse de *Sicydium* spp. (Gobiidae) dans les rivieres de Guadeloupe: implications pour le schema hydraulique des mini-centrales hydroelectriques au fil de l'eau. *Cybium* **22**: 293–296.

———., **S. Doledec & P. Lim.** 2001. Distribution of migratory fishes and shrimps along multivariate gradients in tropical island streams. *Journal of Fish Biology* **59**: 390–402.

Fitzsimons, J.M. & R.T. Nishimoto. 1995. Use of fish behaviour in assessing the effects of Hurricane Iniki on the Hawaiian island of Kaua'i. *Environmental Biology of Fishes* **43**: 39–50.

———., **J.E. Parham & R.T. Nishimoto.** 2002. Similarities in behavioural ecology among amphidromous and catadromous fishes on the oceanic islands of Hawai'i and Guam. *Environmental Biology of Fishes* **65**: 123–129.

———., **R. Zink & R.T. Nishimoto.** 1990. Genetic variation in the Hawaiian goby, *Lentipes concolor*. *Biochemical Systematics and Ecology* **18**: 81–83.

———., **M.G. McRae, H.L. Schoenfuss & R.T. Nishimoto.** 2003. Gardening behavior in the amphidromous Hawaiian fish *Sicyopterus stimpsoni* (Osteichthyes: Gobiidae). *Ichthyological Exploration of Freshwaters* **14**: 185–191.

Ford, J.I. & R.A. Kinzie. 1982. Life crawls upstream. *Natural History* (New York) **91**: 60–67.

Ha, P.Y. & R.A. Kinzie. 1996. Reproductive biology of *Awaous guamensis*, an amphidromous Hawaiian goby. *Environmental Biology of Fishes* **45**: 383–396.

Kano, Y. & T. Kase. 2003. Systematics of the *Neritilia rubida* complex (Gastropoda: Neritiliidae): three amphidromous species with overlapping distributions in the Indo-Pacific. *Journal of Molluscan Studies* **69**: 273–284.

Kido, M.H. 1996a. Diet and food selection in the endemic Hawaiian goby *Sicyopterus stimpsoni*. *Environmental Biology of Fishes* **45**: 199–209.

———. 1996b. Morphological variation in feeding traits of native Hawaiian stream fishes. *Pacific Science* **50**: 184–193.

Kinzie, R.A. 1993a. Evolution and life history patterns in freshwater gobies. *Micronesica* **20**: 27–40.

———. 1993b. Reproductive biology of an endemic, amphidromous goby *Lentipes concolor* in Hawaiian streams. *Environmental Biology of Fishes* **37**: 257–268.

Kroenke, L.W. 1996. Plate tectonic development of the western and southwestern Pacific: Mesozoic to the present. *In:* R.A. Keast & S.E. Miller (eds): *The origin and evolution of Pacific island biotas, New Guinea to eastern Polynesia: patterns and processes*. Amsterdam, SPB Academic. P. 19–34.

Lyons, J. & D.W. Schneider. 1990. Factors influencing fish distribution and community structure in a small coastal river in southwestern Costa Rica. *Hydrobiologia* **203**: 1–14.

Maciolek, J.A. 1977. Taxonomic status, biology and distribution of Hawaiian *Lentipes*, a diadromous goby. *Pacific Science* **31**: 355–362.

Marquet, G. & P. Laboute. 1992. Les Sicydiinae (Gobiidae) des eaux douces de la Polynesie francaise. Description de troi especies nouvelles. *Cybium* **16**: 213–231.

———., **B. Seret & R. Lecomte-Finiger**. 1996. Inventaires comparees des poissons des eaux interieures de trois iles oceanique tropicales de l'Indo-Pacifique (la Reunion, la Nouvelle-Caledonie et Tahiti). *Cybium* **21**(suppl): 27–34.

McDowall, R.M. 1988. *Diadromy in fishes: migrations between freshwater and marine environments*. London, Croom Helm. 308 p.

———. 1990. *New Zealand freshwater fishes: a natural history and guide*. Auckland, Reed. 553 p.

———. 1997a. Is there such a thing as amphidromy? *Micronesica* **30**: 3–14.

———. 1997b. The evolution of diadromy in fishes (revisited) and its place in phylogenetic analysis. *Reviews in Fish Biology and Fisheries* **7**: 443–462.

———. 1998. Fighting the flow: downstream-upstream linkages in the ecology of diadromous fishes in West Coast New Zealand rivers. *Freshwater Biology* **40**: 111–123.

———. 2003. Hawaiian biogeography and the islands' freshwater fish fauna. *Journal of Biogeography* **30**: 703–710.

Myers, G.S. 1949. Usage of anadromous, catadromous and allied terms for migratory fishes. *Copeia* **1949**: 89–97.

Nishimoto, R.T. & D.G.K. Kuamo'o. 1991. The occurrence and distribution of the native goby (*Lentipes concolor*) in Hawai'i Island streams with notes on the distribution of other native fish species. *New directions in research, management and conservation of Hawaiian freshwater stream ecosystems: proceedings of the 1990 symposium on freshwater stream biology and fisheries management*. Honolulu, Division of Aquatic Resources, Department of Land and Natural Resources. P. 77–95.

———. **& D.G.K. Kuamo'o**. 1997. Recruitment of goby postlarvae into Hakalau Stream, Hawai'i Island. *Micronesica* **30**: 41–49.

Polhemus, D.A. 1996. Island arcs and their influence on Indo-Pacific biogeography. *In:* R.A. Keast & S.E. Miller (eds): *The origin and evolution of Pacific island biotas, New Guinea to eastern Polynesia: patterns and processes*. Amsterdam, SPB Academic. P. 51–66.

Price, J.P. & W.L. Wagner. 2004. Speciation in Hawaiian angiosperm lineages: cause, consequence and mode. *Evolution* **58**: 2185–2200

Pringle, C.M. 1997. Exploring how disturbance is transmitted upstream: going against the flow. *Journal of the North American Benthological Society* **16**: 425–438.

Radtke, R. L. & R. A. Kinzie. 1991. Hawaiian amphidromous gobies: perspectives on recruitment processes and life history events. *In: New directions in research, management and conservation of Hawaiian freshwater stream ecosystems: proceedings of the 1990 symposium on freshwater stream biology and fisheries management*. Honolulu, Division of Aquatic Resources, Department of Land and Natural Resources. P. 125–141.

———. **& R.A. Kinzie**. 1996. Evidence of a marine larval stage in endemic Hawaiian stream gobies from isolated high-elevation locations. *Transactions of the American Fisheries Society* **125**: 613–621.

Schneider, D.W. & T.M. Frost. 1986. Massive upstream migrations by a tropical freshwater neritid snail. *Hydrobiologia* **137**: 153–157.

Serét, B. 1997. Les poissons d'eau de Nouvelle-Calédonie: Implications biogeographiques de recentes decouvertes. *Memoirs of the Museum National d'Histoire Naturelle* **171**: 369–378.

Springer, V.G. 1982. Pacific plate biogeography with special reference to shore fishes. *Smithsonian Contributions to Zoology* **367**: 1–182.

Tomihama, M.T. 1972. The biology of *Sicydium stimpsoni*: freshwater goby endemic to Hawai'i. B.Sc. (honors) thesis, University of Hawai'i, Honolulu. 123 p.

Wagner, W.L. & V.A. Funk. 1995. *Hawaiian biogeography: evolution on a hot-spot archipelago*. Washington, Smithsonian Institution Press.

Watson, R.E. 1992. A review of the gobiid fish genus *Awaous* from insular streams of the Pacific plate. *Ichthyological Exploration of Freshwaters* **3**: 161–176.

———. 1995. Gobies of the genus *Stiphodon* of French Polynesia, with descriptions of two new species (Teleostei: Gobiidae: Sicydiinae). *Ichthyological Exploration of Freshwaters* **6**: 33–48.

———. **& I.-S. Chen**. 1998. Freshwater gobies of the genus *Stiphodon* from Japan and Taiwan (Teleostei: Gobiidae: Sicydiini). *Journal of Ichthyology and Aquatic Biology* **3**: 55–68.

———. **& M. Kottelat**. 1994. *Lentipes whittenorum* and *Sicyopus auxilimentus*, two new species of freshwater gobies from the western Pacific (Teleostei: Gobiidae: Sicydiinae). *Ichthyological Exploration of Freshwaters* **5**: 351–364.

Zink, R.M. 1991. Genetic variation within and between populations of *Lentipes concolor* from Hawai'i and Kaua'i. *In*: *New directions in research, management and conservation of Hawaiian freshwater stream ecosystems: proceedings of the 1990 symposium on freshwater stream biology and fisheries management*. Honolulu, Division of Aquatic Resources, Department of Land and Natural Resources. P. 96–105.

———., **J.M. Fitzsimons, D.L. Dittman, D.R. Reynolds & R.T. Nishimoto**. 1996. Evolutionary genetics of Hawaiian freshwater fish. *Copeia* **1996**: 330–335.

Biology of Hawaiian Streams and Estuaries. Edited by N.L. Evenhuis & J.M. Fitzsimons. *Bishop Museum Bulletin in Cultural and Environmental Studies* 3: 11–21 (2007).

Behavioral Ecology of Indigenous Stream Fishes in Hawai'i

J. Michael Fitzsimons & Mark G. McRae

Museum of Natural Science, Louisiana State University, Baton Rouge, Louisiana 70803, USA; email: fitzsimons@lsu.edu; mmcrae@lsu.edu

Robert T. Nishimoto

Division of Aquatic Resources, 75 Aupuni Street, Hilo, Hawai'i 96720, USA; email: Robert.T.Nishimoto@hawaii.gov

Abstract

Five species of amphidromous fishes comprise the indigenous vertebrate fauna of Hawaiian streams. All have a marine larval phase, and, as adults, they live in freshwater and brackish environments marked by frequent flash floods. Although these fishes are closely related and live under similar conditions in the ocean and in streams, they are remarkably diverse in their behavior and ecology. They usually occupy species-typical sections of streams ranging from near the headwaters down to the mouth or estuary; species specificity is evident also in habitat selection by adults. Two species (*Eleotris sandwicensis* and *Stenogobius hawaiiensis*) are unable to climb waterfalls. Among the climbing species, two (*Awaous guamensis* and *Lentipes concolor*) use their pelvic disk and lateral fins for climbing, and the fifth species (*Sicyopterus stimpsoni*) uses the mouth and pelvic disk as holdfasts. The pattern of instream distribution coincides with the relative development of pelvic and oral suckers used in clinging to rocks and climbing waterfalls and with each species' station-holding ability in artificial streams, but the causal factors that prompt new recruits entering from the ocean to continue moving upstream to adult habitats is incompletely understood. One species is a carnivore, one is an herbivore, and the remaining three species are omnivorous. Males of all five species engage females in distinctive courtship behaviors that precede spawning, and all but one species provide some protection for the eggs by guarding the area in which one or more nests are included. Genetic studies for all species show sharp species-level differences among species, but there are no effects of isolation and subsequent differentiation among the five major islands. The data suggest that marine larvae disperse beyond their natal streams. In contrast, analyses of naturally occurring stable isotopes in new recruits indicate that they remain close to mouths of streams while at sea. Introduced freshwater fishes are the original source of potentially harmful parasites which have infected native stream fishes. Resident fishes differ in their susceptibility to parasites. The effect of other alien introductions on aquatic animals has not been determined. Adults in fresh water and free-living embryos and larvae in the ocean have a mix of characteristics within and among species attributable to r and K selection. Although Hawaii's stream fish species are diverse in behavior and ecology, their common amphidromous life cycle provides a basis for management and conservation. Hawaii's Division of Aquatic Resources has developed four levels of stream protection and a firm policy of "no net loss of habitat" to ensure the safety and continuance of the islands' indigenous fishes.

Introduction

No primary or secondary freshwater fishes occur naturally in Hawaiian streams and estuaries (Myers, 1938). However, over 40 species of fishes are known from the islands' fresh and brackish waters (Yamamoto & Tagawa, 2000). These include a few marine species that enter estuaries and lower stream reaches, a host of freshwater fishes introduced intentionally or by accident, and a small group of indigenous fishes that occur as adults in fresh and brackish water and have an obligate marine larval period in their amphidromous life cycle (Myers, 1949; McDowall, 2003). This latter group, derived from marine ancestors, includes two families of gobioid fishes, the Gobiidae and

Eleotridae, whose species are overwhelmingly predominant in island streams throughout the high islands of the tropical Pacific (Fitzsimons *et al.*, 2002). Among these islands, a resident stream fish is very likely to be a goby or an eleotrid, and it almost certainly will have an amphidromous life cycle. Among the amphidromous fishes in Hawai'i, the low species richness (five) and high degree of endemism (four out of five species or 80%) are in keeping with the unsurpassed isolation of the five major islands (Hawai'i, Maui, Moloka'i, O'ahu, and Kaua'i) in the archipelago that have sufficient elevation to produce the orographic rain needed to sustain perennial streams on windward slopes. Although species richness among Hawaii's amphidromous fishes is small, the degree of ecological and behavioral diversity among the five species is remarkable. Information from publications, field notes from 1984–2004, presentations from meetings, and conversations is used in this report to emphasize species similarities and differences in (1) distribution within a single stream and between streams, (2) food sources and feeding behavior, (3) reproductive behavior, (4) climbing ability, (5) genetics, (5) stable isotope analyses, (6) susceptibility to parasites, and (7) r and K selection. These fishes share an amphidromous life cycle and live under similar conditions in island streams where frequent and naturally occurring flash floods maintain ecological succession at an early stage of colonization in which the pioneering species function also as the climax species (Fitzsimons & Nishimoto, 1995). These similarities are the basis for well-established, effective procedures for conserving these animals throughout the Hawaiian high islands.

Discussion

Species distribution within and between streams

All five species of Hawaiian amphidromous fishes occur on each of the five major high islands, but all five species do not occur in every stream on these islands. Understanding the differences in the distribution of these animals between streams is simplified by understanding their distribution within individual streams.

Stream fishes in Hawai'i are species-specific in their selection of habitats along the length of a stream (Tate, 1997). During their amphidromous life cycle, all species enter the mouths of streams when their larvae move from the ocean toward adult habitats in brackish or fresh water. Two species, *Eleotris sandwicensis* (*'o'opu 'akupa*) and *Stenogobius hawaiiensis* (*'o'opu naniha*), remain in the estuary and lower stream reaches in areas of reduced current while the other species continue their migrations to sites farther upstream. The eleotrid lacks the fused pelvic fins which form a holdfast in gobies, and, although *S. hawaiiensis* has a pelvic disk, the structure is elongate, delicate, and better adapted for sitting on loose substrate (sand and gravel) than for adhering to rocks in swift water. The upstream extent of *E. sandwicensis* and *S. hawaiiensis* often is sharply bounded by the first waterfall or other obstruction a meter or more in height. *Awaous guamensis* (*'o'opu nākea*) and *Sicyopterus stimpsoni* (*'o'opu nōpili*) usually occur in the middle reaches of streams, but, because they differ in microhabitat selection, they seldom are close together (Kinzie, 1988). *Awaous guamensis* most often remains on or near the bottom in pools and behind boulders, and *S. stimpsoni* is common in more rapidly flowing runs and riffles and along the sides and upper surfaces of large rocks and boulders. Juveniles and adults of *Lentipes concolor* [*'o'opu hi'ukole* (males) and *'o'opu alamo'o* (females)] typically occur the farthest upstream well beyond the range of the other amphidromous fishes and are well known for their ability to climb high waterfalls (Englund & Filbert, 1997).

The pattern of linear instream distribution of amphidromous fish species in Hawaiian streams was mirrored precisely by each species' ability to "hold station" during experiments in artificial streams (Fitzsimons *et al.*, 1997). However, it is not known whether the distribution pattern is a simple matter of climbing ability, competition for space or food, or a reflection of "ghosts of competition past" (Connell, 1980) that occurred among the ancestors of Hawaiian stream gobies in the far western Pacific. In the latter regard, it is noteworthy that congeners and other close relatives of Hawaiian stream species living in Oceania have patterns of instream distribution that are strikingly similar to those of their Hawaiian kin (Fitzsimons *et al.*, 2002). As suggested for Micronesian stream gobies (Parham, 1995; Nelson *et al.*, 1997), the presence of predators in Hawaiian streams, primarily *E. sandwicensis* and marine-born *Kuhlia xenura* (*āholehole*) in estuaries and lower stream sections, may be a driving force that promotes upstream dispersal and species segregation. The latter hypothesis appears testable.

Topography determines which species occur in a given stream. A stream whose basin slopes gradually from the headwaters to the ocean is likely to have all five fish species. In a relatively long stream of this type, the species are likely to be separated, but, if short, there may be sites where the three upstream species (*A. guamensis, S. stimpsoni*, and *L. concolor*) occur together (Kinzie, 1988). A stream with a much steeper incline and few or no areas of reduced flow near the mouth probably will not include the two downstream species, *E. sandwicensis* and *S. hawaiiensis*. A stream ending in a waterfall that drops directly onto the beach or into the sea will not have these latter two species, and, depending on the height of the waterfall, *A. guamensis* and *S. stimpsoni* may be absent also. Streams terminating in a waterfall in excess of 30 m probably will have only *L. concolor* above the falls. Finally, a stream with a long estuary (such as Hanalei, Kahana, and Hi'ilawe streams on Kaua'i, O'ahu, and Hawai'i, respectively) will have few or no *L. concolor* near its headwaters. The relationship between stream topography and the dispersal capabilities of resident fishes is sufficiently well known that it is often possible to predict species presence and distribution in a stream merely by consulting a topographic map that includes contour intervals.

A lack of knowledge or intentional disregard of natural trends in fish distribution and the relationship between fish behavior and stream morphology has led to erroneous conclusions with implications for the conservation and management of Hawaiian stream fishes. Early stream surveyors apparently did not know that the range of adult *L. concolor* often begins well inland of the ranges of the other stream fishes. Surveys usually began at the mouth of a stream and proceeded upstream until animals were no longer observed. *Lentipes concolor* was recorded as absent in many streams where it was present farther inland. This misinformation led to a statewide classification of all streams as Lentipes Streams and Non-Lentipes Streams (Timbol *et al.*, 1980), designations which prompted a local conservationist group to sue the U.S. Fish and Wildlife Service to have *L. concolor* listed as Threatened or Endangered throughout the Hawaiian Islands. Even when advised that the basis for the listing was unfounded, the group persisted until the Hawai'i Division of Aquatic Resources (DAR) began surveys that located populations of these fish at remote sites on all islands, including O'ahu where the species was thought to have been extinct for many years (Higashi & Yamamoto, 1993). More recently, there have been several attempts (e.g., Kido & Smith, 1998) to standardize stream surveys in Hawai'i by using techniques arguably effective in streams of the mainland U.S. without regard to the connection between stream morphology and fish distribution, the implications of amphidromous life cycles, and the significance of the continuous cyclic recovery of streams from the frequent flash floods.

Food and feeding

Hawaiian amphidromous fishes run the gamut from carnivory (*E. sandwicensis*) to herbivory (*S. stimpsoni*) and omnivory (*S. hawaiiensis, A. guamensis*, and *L. concolor*).

Eleotris sandwicensis is a carnivore that feeds opportunistically on a wide variety of animal foods, including other fishes (even their own species), crustaceans, and aquatic snails. The fish's large head, protrusible upper jaw, underslung lower jaw extending beyond the tip of the upper jaw (prognathony), and conic teeth are adaptive for grabbing prey and swallowing it whole. The relatively short gut of *E. sandwicensis* in comparison to other gobioid fishes in Hawaiian streams (Kido, 1996a) is consistent with its carnivorous feeding habits. The fish makes short chases after small fishes, prawns, and shrimp, but usually sits motionless on the bottom until suitable prey comes within striking distance. *Eleotris sandwicensis* congregates in large numbers near the mouths of streams at times of peak recruitment of larval *'o'opu* from the ocean and often feeds until the belly is distended, but the environmental or behavioral cues prompting such opportune concentrations of these predators have not been investigated. A comparative study of feeding morphology and food habits among Hawaiian stream fishes (Kido, 1996b) revealed a large amount of algae in the gut of *E. sandwicensis* from Wainiha River, Kaua'i, but the author suggested that the plant material may have been ingested inadvertently while fish were feeding on animal material and noted that gut morphology and other features of these fish are indicative of carnivory.

Dentition, gut morphology, diet, and food selection (Kido, 1996a & b, 1997) distinguish *Sicyopterus stimpsoni* from other native freshwater fishes as an obligate herbivore. These fish feed preferentially on navicular, centric, and chain diatoms and pieces of filamentous green and blue-green algae which they scrape off rocks with brushing movements of the upper jaw armed with spatulate tricuspid teeth. Their feeding produces conspicuous scallop-edged patches or long trails on the sides and upper surfaces of rocks and boulders. Large patches prominently located on the surface of boulders near the center of the stream serve as the locus for the initiation of social behavior (territoriality and courtship) by usually large, conspicuously colored, dominant males (Fitzsimons *et al.*, 1993, 2003). While living offshore, young *S. stimpsoni* have a terminal mouth, but, upon entering fresh water, they are unable to feed (or climb) until the mouth migrates to a ventral (inferior) position. This extraordinary change in head morphology occurs within only 48 hours (Schoenfuss *et al.*, 1997).

The three omnivorous stream species occur in the estuary and lower stream reaches (*S.* hawaiiensis), the middle reaches (*Awaous guamensis*), and the upper sections of streams (*Lentipes concolor*). Although detailed studies are lacking, underwater observations suggest *S. hawaiiensis* are omnivores that ingest both plant and animal food by making stabbing bites at the upper surface of rocks or by swimming up into the water column to intercept items drifting downstream. *Awaous guamensis* feed on filamentous algae, diatoms, and small stream animals such as fly larvae and other benthic invertebrates (Kido, 1997). The fish usually bite off clumps of algae or take in a mouthful of sand or gravel from which algae and invertebrates are removed by comb-like gill rakers in the animal's "pharyngeal mill". Once cleaned, the particles of sand and gravel are spit out or dropped from under the lower edge of the gill cover. All five gobioid species have a filtering apparatus formed by gill arches and associated bony or soft projections, but the structure is best developed in *A. guamensis*, an omnivore, and the least in *S. stimpsoni*, an herbivore (Kido, 1996a). Although typically bottom feeders, *A. guamensis* occasionally swim up from the bottom and ingest pieces of material being washed downstream. *Lentipes concolor* feed on green algae and opportunistically on stream invertebrates (Kido, 1996a). In many streams, the most common food items are the larvae of the extremely abundant chironomid flies that lay their eggs on dampened parts of rocks near the water line. These fish also frequently dart up from the bottom and intercept particles carried downstream in manner reminiscent of the sallying behavior of insectivorous birds that fly up from a perch and capture insect prey. It has not been determined to what extent the omnivorous fishes are actually digesting plant material rather than receiving nutrition from the periphyton (*aufwuchs*) attached to it. Gut morphology (length and number of loops) among the omnivores *S. hawaiiensis*, *Awaous guamensis*, and *Lentipes concolor* is intermediate between that of the carnivore (*E. sandwicensis*) and the herbivore (*S. stimpsoni*) (Kido, 1996a).

Reproductive behavior

Hawaii's amphidromous stream fishes are egg layers. All but one species (*S. hawaiiensis*) are territorial and provide some degree of protection to egg masses by defending the larger area in which eggs are deposited. Eggs are extruded individually from the elongate urogenital papilla of females. Males fertilize the eggs shortly after they are deposited. Eggs are small (ca. 1/3 mm diameter) and encased in a gelatinous adhesive coat. The slightly buoyant eggs float up from the substrate where they have been attached, but the adhesive coat forms a stalk that anchors them and gives them the appearance of a tiny golf ball sitting on a tee. Eggs are deposited in one or more lines (*S. hawaiiensis* and small females of *E. sandwicensis, S. stimpsoni,* and *A. guamensis*) or in round and irregular patches (large females) sometimes reaching the diameter of a dinner plate (*A. guamensis*). Recently laid egg masses are white or light gray; patches of eggs where hatching has occurred are usually darker gray and have a ragged appearance. Eggs are deposited on the surface of rocks and boulders (*S. hawaiiensis* and occasionally *E. sandwicensis*), along the sides or underneath boulders (*E. sandwicensis, A. guamensis,* and *S. stimpsoni*), and in burrows usually located under piles of large rocks or under the downstream edge of large boulders (*L. concolor* and occasionally the other species).

All five species have an elaborate courtship ritual comprised of species-specific movements that precede pair-forming and spawning. Males approach females in or near their territories and begin a stereotypical sequence of display behavior that ends with the male leading the female to a spawning site. Pair forming is determined by female choice, but, in one species, *E. sandwicensis*, females have a mid-water display that solicits approach by males. Males of *S. hawaiiensis* initiate courtship with females, and, if the female is receptive, the male leads her to the nearest natural burrow or flat rock often near the margin of the stream where current is reduced. After a bout of spawning, the male may engage the same female in additional courtship with spawning at the same or different site. However, males usually swim along the margins of streams and pursue females encountered opportunistically. *Stenogobius hawaiiensis* differ from other stream fishes in that apparently neither the female, the eggs, nor the spawning site are defended against conspecific males.

Awaous guamensis may be the only amphidromous fish species in Hawai'i that has a spawning migration (Mainland in Titcomb, 1977; Yamamoto & Tagawa, 2000; Kido & Heacock, 1992). The fish are said to come downstream to sites near the mouths of streams in association with heavy rains during June or July. Streamside residents on Kaua'i are often familiar with the movements of these fish, and taro farmers have used the expressions *'o'opu* rain and *'o'opu* flood when describing the conditions associated with the downstream migration of *A. guamensis* in the Hanalei River (Yamamoto & Tagawa, 2000). We have observed large aggregations of spawning fish in the lower section of Wainiha River after being told by a local resident that the fish moved there a week earlier in response to a moderate flash flood. An old Hawaiian proverb refers to *A. guamensis* as *Ka i' a ka wai nui i lawe mai ai* ("the fish borne along by the flood"; Pukui, 1983 [1323]). Although accounts are anecdotal, their frequency and similarity afford them at least a tentative "ring of truth". The advantage of a downstream migration would be two-fold: The aggregation of adults at a common downstream spawning area would enhance the chances of an individual fish finding a receptive mate, and the newly hatched young would have a shorter distance to gain access to the sea. The latter would be of special significance in streams, such as Hanalei and Wainiha, which have long estuaries and adult fishes distributed farther inland than in most Hawaiian streams. In Japan, some amphidromous gobies (*Rhinogobius brunneus*) hatched far inland usually do not have sufficient yolk to sustain them during the long drift to the sea (Iguchi & Mizuno, 1999). The same may be true for gobies in certain Hawaiian streams.

Climbing

Hawaii's amphidromous fishes differ in their climbing ability. Two species, *E. sandwicensis* and *S. hawaiiensis,* apparently do not climb and are limited to estuaries and the lower stream reaches below the first waterfall of a meter or more in height. However, both have morphological and behavioral adaptations that allow them to maintain position in flowing water (Fitzsimons *et al.*, 1997). The head and body of *E. sandwicensis* is slightly flattened dorsoventrally, and the head is sloped slightly downward from the nape to the tip of the snout. Water flowing over the fish tends to press it into the substrate rather than forcing it up and away from the bottom. These fish are proficient in their ability to wedge themselves between rocks and boulders while using the pelvic and pectoral fins to maneuver and to push forward or backward. By hugging the bottom and working their way upstream behind rocks and boulders, the eleotrids are capable of progressing against currents that would displace other fishes of similar size and shape attempting to swim against the flow. *Stenogobius hawaiiensis* swim off the bottom, but these fish take advantage of slack water along the margins of streams and in the shelter of large rocks and boulders.

The three upstream species, *S. stimpsoni, A. guamensis,* and *L. concolor*, can climb waterfalls and adhere to the surface of rocks and boulders in swift water with a suction disk formed by the fusion of the pelvic (ventral) fins. Even among these three species, the mode of climbing and climbing ability differ. *Awaous guamensis* and *L. concolor* rely on the pelvic disk as a holdfast when climbing, but *S. stimpsoni* uses both the pelvic disk and the mouth alternately when "inching" forward in a strong current and when climbing a cascade or waterfall (Schoenfuss *et al.*, 1997; Schoenfuss & Blob, 2003; Nishimoto & Kuamo'o, 1997). The "powerburst climbers" (Schoenfuss & Blob, 2003), *Awaous guamensis* and *L. concolor,* push forward with the pectoral fins while detaching the pelvic disk and rapidly undulating the body from side to side. *Sicyopterus stimpsoni* and *A. guamensis* can negotiate waterfalls 10–20 m in height (Fitzsimons & Nishimoto, 1991), while adults of *L. concolor* reside in the pools above the highest waterfalls, such as Waipi'o Valley's Hi'ilawe Falls with a sheer drop of over 300 m (Englund & Filbert, 1997).

Genetics and isotope analysis

Surveys of genetic variation in Hawaiian stream fishes with protein electrophoresis (Fitzsimons *et al.*, 1990), mitochondrial DNA restriction site analysis (Zink *et al.*, 1996), and mtDNA sequencing (Chubb *et al.*, 1998) revealed no evidence of discernible geographic (island) structuring among the five species of amphidromous fishes. These data suggest that there is either a single offshore pool of larvae that recruit to each of the five major islands or that recruitment is from local offshore populations but with sufficient exchange of larvae between islands to prevent isolation and genetic differentiation.

A comparison of data from stable isotope analysis (carbon and nitrogen) of amphidromous fish recruits and fish species from marine habitats indicates that the source of nutrition for the stream species during their offshore larval period is near the mouths of streams (Hobson *et al.*, 2007). These findings support the idea that these larval fishes in the ocean remain close to freshwater plumes which function as nursery areas and suggest that amphidromous fish larvae may stay near their home island (or even natal stream?) rather than joining an offshore mass of larvae that contributes recruits to all the high islands. A finer focus with isotope analysis and genetics (such as microsatellite studies) may clarify details of the distribution, ecology, and population dynamics of amphidromous fishes during the marine period of their life cycle.

Parasites

The effects of alien species on indigenous aquatic animals are not well known. A striking exception is the investigation of parasites in native stream fishes associated with the introduction of alien fish species (Font, 1996, 1997a,b, 1998; Font & Tate, 1994). Fourteen species of helminth parasites have been found in stream fishes, but only three, a tapeworm (*Bothriocephalus acheilognathi*), leech (*Myzobdella lugubris*), and roundworm (*Camallanus cotti)*, which were introduced into the islands along with their alien fish hosts, are regarded as capable of causing disease in native fishes (Font, 2003). Hawaii's amphidromous fishes are diverse in their susceptibility to these potentially dangerous parasites. All three parasites are recorded by Font from *A. guamensis* and *E. sandwicensis*. *Lentipes concolor* has the roundworm, but not the other two parasite species. *Sicyopterus stimpsoni* often has leeches attached to the head and body but lacks both endoparasites. Finally, *Stenogobius hawaiiensis* has none of the three parasites.

r and K selection

Unbridled exponential growth is infrequent in natural populations. In the more universal logistic population growth, the rapid increase in number eventually peaks, slows, and stabilizes when the carrying capacity of the environment is reached. For this reason, population growth is more frequently represented by a sigmoid curve rather than a straight line. The equation for Logistic Population Growth is $I = rN (K - N / K)$, where I = the annual increase for the population, r = the annual growth rate, N = the population size, and K = the carrying capacity. The concepts of r and K selection originate from this equation. In a simplified definition, r selection is a form of natural selection that occurs when the environment has abundant resources that favor, among other things, a reproductive strategy of producing many offspring. Conversely, K selection is a type of selection in an environment that is at or near carrying capacity, and limited resources favor a reproductive strategy in which few offspring are produced. From that criterion alone, stream fishes in Hawai'i are clearly r-selected. However, there are other characteristics of r- and K-selected populations that can be considered: In environments where the availability of resources exceeds demands, the lack of competition promotes selection for early reproduction, fast growth, and short life spans (= r selection of Pianka, 1970). In environments where the demand for resources is about the same as the available supply, natural selection promotes competition among species that results in the evolution of slower growth, a delay in reproduction, and longer life spans [= K selection of Pianka (1970)]. In addition, extreme r-selected species are thought to be characteristic of unstable environments and density independent interactions, while extreme K-selected species are indicative of stable environments and density dependent interactions. These and other features of r- and K-selected species that appear most relevant to fishes in general are placed in Table 1 for a discussion of Hawaiian amphidromous fishes

Table 1. Characteristics of animals that are extreme r or K strategists. Loosely based on Campbell & Reece (2002); Gould (1997); Pianka (1970); Raven & Johnson (2002); and adapted for fishes.

r	K
unstable environment	stable environment
density independent interactions	density dependent interactions
small size	large size
energy used to make each individual is low (small size at birth)	energy used to make each individual is high (large size at birth)
weak, vulnerable	robust, well-protected
high mortality	low mortality
rapid growth	slow growth
early maturation	late maturation
short life expectancy	long life expectancy
each individual reproduces only once	individuals can reproduce more than once in their lifetime
little or no care for the offspring	much care for the offspring
variable population size	stable population size

with reference to the following questions: How do Hawaiian amphidromous fishes fit into the r-K selection continuum? Are they strongly associated with one extreme or do they show a blending or mosaic of the individual attributes of r and K selection?

To begin with, it can be argued that the equation for logistic population growth cannot be applied to indigenous stream fishes in Hawai'i because r (the annual growth rate) is based on reproduction by members of the population in essentially a closed system. Because of their amphidromous life cycle, every fish in every Hawaiian stream is a migrant and may or may not have begun life in the stream where it is observed, and there is no certainty that young animals are the offspring of adults in the same stream. The equation perhaps can be salvaged for Hawaiian stream fishes merely by changing r to R_s, the latter representing the site specific (s) annual rate of increase (R) attributable to recruitment, and rewriting it as $I = R_sN\ (\ K - N\ /\ K)$. The emphasis here is that population increases are caused by recruitment whether or not the recruits originate from the stream in question.

Although juveniles and adults of amphidromous species in Hawai'i live in an unstable environment because of recurring flash floods, it might be argued that their free-living embryos and larvae in the ocean live in a relatively more stable environment. If this assumption is correct, these fishes are r-selected in fresh water and K-selected in marine environments (Table 1). The situation in terms of r and K selection in the two environments is reversed when considering that density independent interactions, a feature of r-selected populations, are likely among the thousands of young fishes washed out of streams into the ocean, while the highly social adults are concentrated in finite sections of streams where interactions are density dependent, a feature of K-selected populations. Small size (r) characterizes the newly hatched free-living embryos (ca. 1.5 mm total length); adults are appreciably larger (K) even in comparison to the adults of most species of marine gobies and eleotrids. The energy used to make each individual, i.e., relative size at birth, is clearly low, an r-selected feature. Although the terms are subjective, amphidromous fishes offshore could be described as weak and vulnerable (r). In contrast, adults in streams, particularly the three upstream species, are much larger, robust, and mostly free from predators (K). Mortality rates are unknown for amphidromous fishes, but it is likely that there is higher mortality (r) among recently hatched fishes and lower mortality (K) among adult fishes once they are established in species-specific stream habitats. Evaluating growth as rapid or slow among amphidromous fishes is tentative because age-and-growth studies are lacking. Depending on the species, newly hatched embryos can increase in total length as much as 16 times (e.g., 1.5 mm TL for embryos vs almost 24 mm for recruiting *S. stimpsoni*; Nishimoto & Kuamo'o, 1997). Casual underwater observations suggest that males and females have indeterminant growth, but a change in size associated with the transition from juvenile to adult or with growth among adults is much less dramatic among stream-living fishes (K) than among conspecific young fishes in the ocean (r). Stream fishes mature late (K) when the lengthy larval period is added. Although there is no published information on age at maturity, the small size of spawning fishes observed underwater indicates they become reproductively competent in a short time (r), i.e., less than a year, after recruiting into streams. If *L. concolor* is typical of other stream fishes in Hawai'i, life expectancy would be considered long (K) in comparison to other similar sized fishes (Nishimoto & Fitzsimons, 1986). Stream gobies and eleotrids reproduce more than once in their lifetime (K); none die after spawning as do many species of anadromous fishes that also leave the marine environment and reproduce in fresh water. All but one species (*S. hawaiiensis*) are territorial at spawning sites and provide some amount of protection for the eggs (K). Population sizes are unknown for larval amphidromous fishes, but the potential for tracking changes in population size for adults in streams may be possible soon by using the freshwater database being developed by the Hawai'i Division of Aquatic Resources. In an island stream, population size reflects the balance between mortality and recruitment, but not reproduction, unless it can be demonstrated that recruitment involves significant homing to a natal stream. Recall that the equation proposed above for logistic population growth in Hawaiian streams substitutes recruitment for natural reproduction.

When the results of this discussion are tallied, larval fishes during the marine period of their life cycles appear to be mostly r-selected, while adults established in species-typical stream habitats are mostly K-selected.

Although this discussion is largely moot and academic, there are two possible conclusions that may have some practical merit. First, an amphidromous life cycle involving life in both fresh water and the sea makes it difficult, if not impossible, to apply to Hawaiian streams many analytical techniques basic to understanding the biology of streams and rivers on continents. Clearly, the biology of island amphidromous fishes is "strongly influenced by both these hydrologic systems" (Kinzie, 1988). Secondly, this discussion has proceeded mostly from educated guesses rather than hard data, and, instead of providing valuable reference material, it shows that even the most rudimentary baseline studies for Hawaiian stream fishes are either deficient or absent. The inevitable statement "more studies are needed" applies here.

Conservation

None of the five species of Hawaiian amphidromous fishes is currently threatened or endangered (Devick *et al.*, 1995). Because of the earth's rotation, trade winds continue to flow up the northeast sides of the high islands to cooler elevations where they produce the orographic rainfall that maintains forests, the perennial streams that run through them, and the animals that live in the clear, cold water. The larvae of amphidromous fishes, limpets, shrimp, and prawns continue to recruit from the ocean into brackish and fresh waters where they complete their life cycles. Periodic flash floods continue to clean out streams, and recruits from the ocean periodically "top off" populations of aquatic animals. This perpetual natural recycling maintains the dynamic conditions in Hawaiian streams that guarantee the persistence of a unique assemblage of organisms. These animals persist as living links to the evolutionary origins of both aquatic and terrestrial life on these islands, and they were vital to survival of the Polynesians who first colonized Hawai'i. The situation seems at once idyllic, perpetual, and maybe even unassailable. But there are catches.

In Hawai'i, freshwater organisms are confined to relatively small streams where they are easily vulnerable to human influences. Any material taken from a stream or dumped into it is likely to have an immediate effect. Any activity that blocks access to or from the sea shuts down the life cycles of the principal stream animals and eventually results in the loss of those species from the stream. Fresh water is usually abundant on each of the major islands. However, it is not limitless, and it seldom occurs naturally on the leeward sides of islands. The disparity in the distribution of rainfall on the islands in combination with human needs and wants – and these are not necessarily synonymous – automatically makes water a potentially scarce resource.

Until the late 1980s, the prevailing attitude was that water flowing into the sea was wasted (Devick, 1995). Indiscriminate diversion of water for agriculture (mainly sugar cane), development (often in arid areas), and other activities led to a wide-spread diminution of flow or a complete dewatering of many streams. The result was a loss of aquatic communities, a decline in taro production, and a reduction of productivity in the ocean near the coast where streams once emptied. Conditions began improving in 1987 with the establishment of a State Water Code and the formation of the Commission on Water Resource Management (Devick, 1995; Devick *et al.*, 1992); these two actions gave the State, rather than individual landowners, authority over the use of streams.

Hawai'i has been faulted for not joining other western states in the designation of minimum flow requirements to protect its stream animals. Critics do not understand the variable nature of flow in Hawaii's streams, know little or nothing of amphidromy, and have not experienced in their own states the flash floods that occur frequently in Hawai'i. In lieu of setting meaningless minimum flow requirements for all Hawaiian streams, William S. Devick, Administrator of the Division of Aquatic Resources, established a policy of "no net loss of habitat" in stream-use decisions (Devick, 1995). This approach has been successful, and, in retrospect, it is infinitely more logical and amenable to implementation than mainland procedures because the focus is on the requirements of stream animals rather than on a range of numbers describing the amount of water that flows over a section of stream during a certain period of time.

During Devick's service as DAR Administrator, he established four levels of stream protection (Devick *et al.*, 1992) that go well beyond conservation measures developed for streams in other states. Conservation zoning places designated streams into discrete Conservation Districts (1) which extend protection to entire stream corridors and incorporate all perennial streams. Interim Instream Flow Standards (2) for individual streams are based on existing flows, and activities affecting the stream beds, i.e., animal habitats, are closely regulated. The program of Stream Ecosystem Protection (3) halts further development of selected streams and their corridors while retaining existing uses. The development of a Natural Areas Reserve System (NARS) (4) preserves and restores designated areas that represent important freshwater ecosystems.

The Endangered Species Act is viewed sometimes as the first and sometimes as the last defense against the extinction of a species. A correlative idea is that protection of one species provides protection for all the other organisms in the area. This notion seems applicable to Hawaiian stream fishes because of their amphidromous life cycles, but, as emphasized throughout this paper, the five species of fishes are different in where they live in a stream and what they do for a living. A focused effort for restoring and protecting one species may or may not provide assistance to others. Instead, actions that provide protection of habitat by preserving entire ecosystems are preferable to a limited concentration on single species (Devick *et al.*, 1995). An ideal tactic is to maintain aquatic ecosystems so that invoking the Endangered Species Act never becomes necessary. The five species of fishes discussed in this paper possess remarkable diversity in their behavior and ecology, but the key to their protection lies with their common amphidromous life cycle. For an unaltered stream that flows briskly into the ocean along an open coast, the best protection may be to do nothing at all. Simply leave the stream unaltered. The animals themselves will take it from there.

Literature Cited

Campbell, N.E. & J.B. Reece. 2002. *Biology*. 6th ed. Benjamin Cummings, San Francisco. 1247 p.

Chubb, A.L., R.M. Zink & J.M. Fitzsimons. 1998. Patterns of mtDNA variation in Hawaiian freshwater fishes: The phylogenetic consequences of amphidromy. *Journal of Heredity* **89**: 8–16.

Connell, J.H. 1980. Diversity and coevolution of competitors, the ghost of competition past. *Oikos* **35**: 131–138.

Devick, W.S. 1995. The importance of stream beings. *'Elepaio* **55**: 45–48.

———., **J.M. Fitzsimons & R.T. Nishimoto**. 1992. *Conservation of Hawaiian freshwater fishes.* Division of Aquatic Resources, Department of Land and Natural Resources, State of Hawaii, Honolulu. 26 p.

———., **J.M. Fitzsimons & R.T. Nishimoto**. 1995. Threatened fishes of the world: *Lentipes concolor* Gill, 1860 (Gobiidae). *Environmental Biology of Fishes* **44**: 325–326.

Englund, R.A. & R. Filbert. 1997. Discovery of the native stream goby, *Lentipes concolor*, above Hawaii's highest waterfall, Hi'ilawe Falls. *Bishop Museum Occasional Paper* **49**: 62–24.

Fitzsimons, J.M. & R.T. Nishimoto. 1991. Behavior of gobioid fishes from Hawaiian fresh waters. 106–124. *In:* W.S. Devick (ed.), *New directions in research, management, and conservation of Hawaiian freshwater stream ecosystems.* Proceedings of the 1990 Symposium on Freshwater Stream Biology and Management, Hawai'i Division of Aquatic Resources.

———. **& R.T. Nishimoto**. 1995. Use of fish behavior in assessing the effects of Hurricane Iniki on the Hawaiian island of Kaua'i. *Environmental Biology of Fishes* **43**: 39–50.

———., **M.G. McRae, H.L. Schoenfuss & R.T. Nishimoto**. 2003. Gardening behavior in the amphidromous Hawaiian fish *Sicyopterus stimpsoni* (Osteichthyes: Gobiidae). *Ichthyological Exploration of Freshwaters* **14**: 185–191.

———., **R.T. Nishimoto & A.R. Yuen**. 1993. Courtship and territorial behavior in the native Hawaiian steam goby, *Sicyopterus stimpsoni. Ichthyological Exploration of Freshwaters* **4**: 1–10.

———., **J.E. Parham & R.T. Nishimoto**. 2002. Similarities in behavioral ecology among amphidromous and catadromous fishes on the oceanic islands of Hawai'i and Guam. *Environmental Biology of Fishes* **65**: 123–129.

———., **H.L. Schoenfuss & T.C. Schoenfuss**. 1997. Significance of unimpeded flows in limiting the transmission of parasites from exotics to Hawaiian stream fishes. *Micronesica* **30**: 117–125.

———., **R.M. Zink & R.T. Nishimoto**. 1990. Genetic variation in the Hawaiian stream goby, *Lentipes concolor. Biochemical Systematics and Ecology* **18**: 81–83.

Font, W.F. 1996. Parasites in paradise: exotic parasites in native Hawaiian gobioid stream fishes. *Newsletter, Introduced Fish Section, American Fisheries Society* **15**: 5–6.

———. 1997a. Distribution of helminth parasites of native and introduced streamfishes in Hawaii. *Bishop Museum Occasional Papers* **49**: 56–62

———. 1997b. Improbable colonists: helminth parasites of freshwater fishes on an oceanic island. *Micronesica* **30**: 105–115.

———. 1998. Parasites in paradise: patterns of helminth distribution in Hawaiian stream fishes. *Journal of Helminthology* **72**: 307–311.

———. 2003. The global spread of parasites: What do Hawaiian streams tell us? *BioScience* **53**: 1061–1067.

———. **& D.C. Tate**. 1994. Helminth parasites of native Hawaiian freshwater fishes: An example of extreme ecological isolation. *Journal of Parasitology* **80**: 682–688.

Gould, S.J. 1997. *Ontogeny and phylogeny.* Belknap Press, Cambridge, Massachusetts. 528 p.

Higashi, G.R. & M.N. Yamamoto. 1993. Rediscovery of 'extinct' *Lentipes concolor* (Pisces: Gobiidae) on the island of Oahu, Hawaii. *Pacific Science* **47**: 115–117.

Hobson, K.A., R.J.F. Smith & P. Sorensen. 2007. Applications of stable isotope analysis to tracing nutrient sources to Hawaiian gobioid fishes and other stream organisms. *In:* Evenhuis, N.L. & J.M. Fitzsimons (eds.), Biology of Hawaiian Streams and Estuaries. Proceedings of the Symposium on the Biology of Hawaiian Streams and Estuaries, Hilo, Hawai'i 26–27 April 2005. *Bishop Museum Bulletin in Culture and Environmental Studies* **3**: 99–111.

Iguchi, K. & N. Mizuno. 1999. Early starvation limits survival in amphidromous fishes. *Journal of Fish Biology* **54**: 705–712.

Kido, M.H. 1996a. Morphological variation in feeding traits of native Hawaiian stream fishes. *Pacific Science* **50**: 184–193.

———. 1996b. Diet and food selection in the endemic Hawaiian amphidromous goby, *Sicyopterus stimpsoni* (Pisces: Gobiidae). *Environmental Biology of Fishes* **45**: 199–209.

———. 1997. Food relations between coexisting native Hawaiian stream fishes. *Environmental Biology of Fishes* **49**: 481–494.

———. & D.E. Heacock. 1992. The spawning ecology of *'o'opu nakea* (*Awaous stamineus*) in Wainiha River and other selected north shore Kaua'i rivers, p. 142–157. *In*: W.S. Devick (ed.), *New directions in research, management, and conservation of Hawaiian freshwater stream ecosystems*. Proceedings of the 1990 Symposium on Freshwater Stream Biology and Management, Hawaii Division of Aquatic Resources.

———. **& G.C. Smith**. 1998. The Hawaii stream bioassessment protocol: a manual for biological monitoring and assessment of Hawaiian streams. Report to the Hawaii Department of Health, Honolulu.

Kinzie, R.A. 1988. Habitat utilization by Hawaiian stream fishes with reference to community structure in oceanic island streams. *Environmental Biology of Fishes* **22**: 179–192.

McDowall, R.M. 2003. Hawaiian biogeography and the islands' freshwater fish fauna. *Journal of Biogeography* **30**: 703–710.

Myers, G.S. 1938. Fresh-water fishes and West Indian zoogeography. *Annual Report of the Smithsonian Institution* **1937**: 339–364.

———. 1949. Usage of anadromous, catadromous, and allied terms for migratory fishes. *Copeia* **1949**: 89–97.

Nelson, S.G., J.E. Parham, R.B. Tibbatts, F.A. Camacho, T. Leberer & B.D. Smith. 1997. Distributions and microhabitats of the amphidromous gobies in streams of Micronesia. *Micronesica* **30**: 83–91.

Nishimoto, R.T. & J.M. Fitzsimons. 1986. Courtship, territoriality, and coloration in the endemic Hawaiian freshwater goby, *Lentipes concolor*, p. 811–817. *In:* Uyeno, T., R. Arai, T. Taniuchi, & K. Matsuura (eds.), *Indo-Pacific fish biology*. Proceedings of the second International Conference on Indo-Pacific fishes, Ichthyological Society of Japan, Tokyo.

———. **& D. G.K. Kuamo'o**. 1997. Recruitment of goby postlarvae into Hakalau Stream, Hawai'i Island. *Micronesica* **30**: 41–49.

Parham, J.E. 1995. Habitat use by an assemblage of tropical oceanic island streamfish. Unpublished M.S. Thesis, University of Guam, Mangilao. 118 p.

Pianka, E.R. 1970. On r and K selection. *American Naturalist* **104**: 592–597.

Pukui, M.K. 1983. *Olelo no'eau, Hawaiian proverbs and poetical sayings*. Bishop Museum Press, Honolulu. 351 p.

Raven, P.H. & G.B. Johnson. 2002. *Biology*. 6th edition. McGraw-Hill, Boston. 1,238 p.

Schoenfuss, H.L. & R.W. Blob. 2003. Kinematics of waterfall climbing in Hawaiian freshwater fishes (Gobiidae): vertical propulsion at the aquatic-terrestrial interface. *Journal of Zoology* (London) **261**: 191–205.

———., **T.A. Blanchard & D.G.K. Kuamo'o**. 1997. Metamorphosis in the cranium of postlarval *Sicyopterus stimpsoni*, an endemic Hawaiian stream goby. *Micronesica* **30**: 93–104.

Tate, D.C. 1997. The role of behavioral interactions of immature Hawaiian stream fishes (Pisces: Gobioidei) in population dispersal and distribution. *Micronesica* **30**: 51–70.

Timbol, A.S., A.J. Sutter & J.D. Parrish. 1980. Distribution, relative abundance, and stream environment of *Lentipes concolor* (Gill 1860), and associated fauna in Hawaiian streams. Water Resources Research Center, University of Hawaii at Mānoa. 64 p.

Titcomb, M. 1977. *Native use of fish in Hawaii*. University Press of Hawaii, Honolulu. 175 p.

Yamamoto, M.N. & A.W. Tagawa. 2000. *Hawaii's native and exotic freshwater animals*. Mutual Publishing, Honolulu. 200 p.

Zink, R.M., J.M. Fitzsimons, D.L. Dittman, D.R. Reynolds & R.T. Nishimoto. 1996. Evolutionary genetics of Hawaiian freshwater fish. *Copeia* **1996**: 330–335.

Biology of Hawaiian Streams and Estuaries. Edited by N.L. Evenhuis & J.M. Fitzsimons. *Bishop Museum Bulletin in Cultural and Environmental Studies* 3: 23–33 (2007).

The Hawaiian *Ahupua'a* Land Use System: Its Biological Resource Zones and the Challenge for Silvicultural Restoration

DIETER MUELLER-DOMBOIS
Department of Botany, University of Hawai'i at Mānoa, Honolulu, Hawai'i 96822, USA; email: amdhawaii@aol.com

Abstract

The land area of the major Hawaiian Islands was originally divided into districts called *moku*, and these were further subdivided into *ahupua'a*. The latter encompass landscape segments from the ocean to the mountain that served as the traditional human support systems. These life support systems were based on three to five biological resource zones. These were the upland/inland forest zone, or the *wao nahele*, the agricultural zone, or the *wao kanaka*, and the coastal zone, or the *kaha kai*. This latter zone included the strand area, fringing reefs, sea grass beds, lagoons, fish ponds, and estuaries, where present. Actually, estuaries, the *muliwai*, are mostly on the windward side of the islands and are part of a fourth biological resource zone, the *kaha wai* or freshwater ecosystems and streams. The ocean (*kai*), near the shore can be considered the fifth biological resource zone. Thus, the traditional land use was based on the vertical arrangement of a volcanic high island's natural ecosystems. This vertical arrangement allowed for maximizing the use of biodiversity over short distances and acknowledged the interactive influences of the biological resource and production zones. This interactive influence begins at the top, in the *wao nahele*. What happens there influences the three other production zones. Therefore, any *ahupua'a* restoration that aims at the reintroduction of adaptive and integrative management should start with silvicultural research at an operational scale. Silviculture is concerned with the care of forests. It is based on knowledge gained from research in forest ecology and should be a form of "low input management". With regard to the *ahupua'a* model, silviculture must focus on enhancing the natural processes associated with the function of the forested watershed and stream ecosystem. Silviculture should also aim at restoring a "Hawaiian sense of place" in those *ahupua'a* selected for stream restoration. This concept will be explained in some detail in this paper.

Introduction

The traditional land use in the Hawaiian Islands evolved from shifting cultivation into a stable form of agriculture around 1200 AD (Kirch, 2000). Stabilization required a new form of land use. This was the *ahupua'a* land use system, which consisted of vertical landscape segments from the mountains to the near-shore ocean environment, and into the ocean as deep as a person could stand in the water (Isabella Aiona Abbott, personal communication). The reason for converting from a shifting to a stabilized land use can be attributed to an increasing population pressure. Areas for cultivation are spatially more limited on islands as compared to continents. At the same time, also agricultural land use, to be stabilized in tropical environments, had to become more sophisticated than the traditional slash and burn practice of the initial colonizers, who are believed to have become settled in the windward valleys of O'ahu around 300 AD (Kirch, 2000).

In the term *ahupua'a*, the words *ahu* (stone altar or stone mound) and *pua'a* (pig), are combined. The *pua'a* was a carved wooden image of a pig head. These stone altars served as border markers and deposition places for offerings to the agricultural god *Lono* and a high chief (*ali'i nui*), who was the god's representative. Each *ahupua'a* in turn was ruled by a lower chief, or *ali'i 'ai*. He in turn appointed a headman, or *konohiki*. The *konohiki* served as general manager responsible for the use of an *ahupua'a* as a resource system. He in turn was assisted by specialists, or *luna*. For example, the *luna wai* was responsible for the fresh water flow and irrigation system (Kamehameha Schools, 1994).

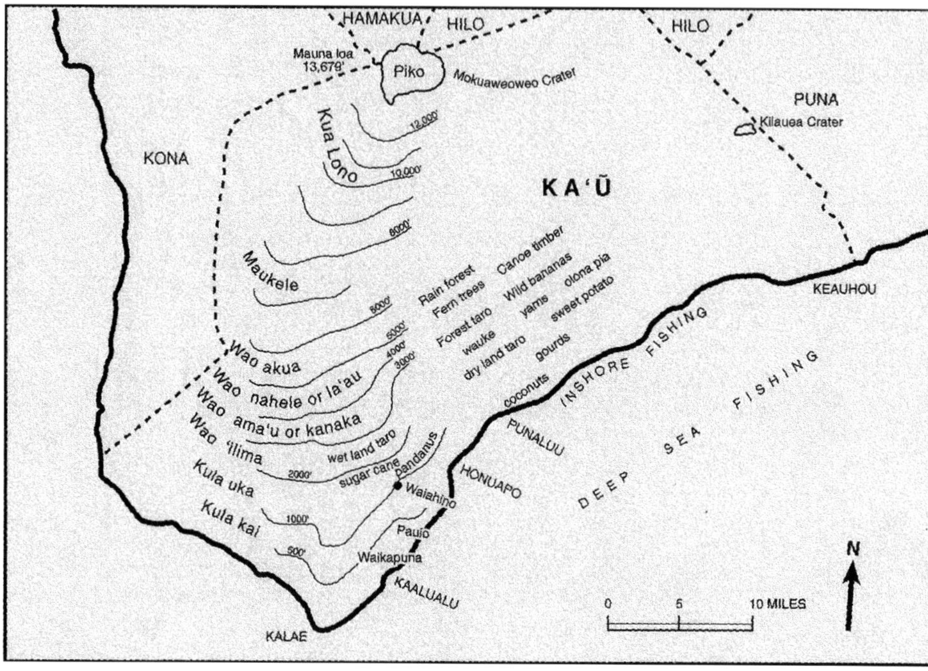

Figure 1. The vertical arrangement of Hawaiian ecological zones on the south slope of Mauna Loa, Hawai`i Island [after Handy & Handy (1972) with slight modifications].

Sophistication in the traditional Hawaiian land use practices becomes evident already from the way island areas were divided vertically, often in units of watersheds, and horizontally, in zones of ecosystem significance. Furthermore, the functionality of the individual zones was well understood as to their bioenvironmental potential. Wherever possible, the zones were modified by enhancing their natural ecosystem services.

In this paper I will first focus on the Hawaiian understanding of ecological zones and their uses. These are closely similar to an ecosystem model interpreted and used by an ecologist. I will then focus on the current need for taking care of the inland and upland forest as the protective cover in Hawaii's watersheds. From a professional viewpoint such care-taking is known as silviculture. Silvicultural research and management is now in great demand for restoring a Hawaiian sense of place and for introducing an adaptive and integrative form of management in selected Hawaiian *ahupua'a*.

The Hawaiian Ecological Zones

It was of particular interest to me as an ecologist to learn that the early Hawaiians had not only a great number of plants and animals identified by names, often by binomials, but that they also recognized a number of vegetation units (i.e., different land forms and plant communities) and types of ecosystems. The native Hawaiian author, David Malo, who lived in the early part of the 19th century, provided a rich record of indigenous environmental and ecological terms with brief definitions. His writings were translated by Dr. Nathaniel Emerson in 1898 and first published in 1903. A second edition was published by the Bishop Museum in 1951 with a number of reprinted versions of this classic record; the latest was published as Malo (1997). Definitions of Hawaiian environmental and ecological terms are best clarified when shown on a map. Such effort appeared in the book by

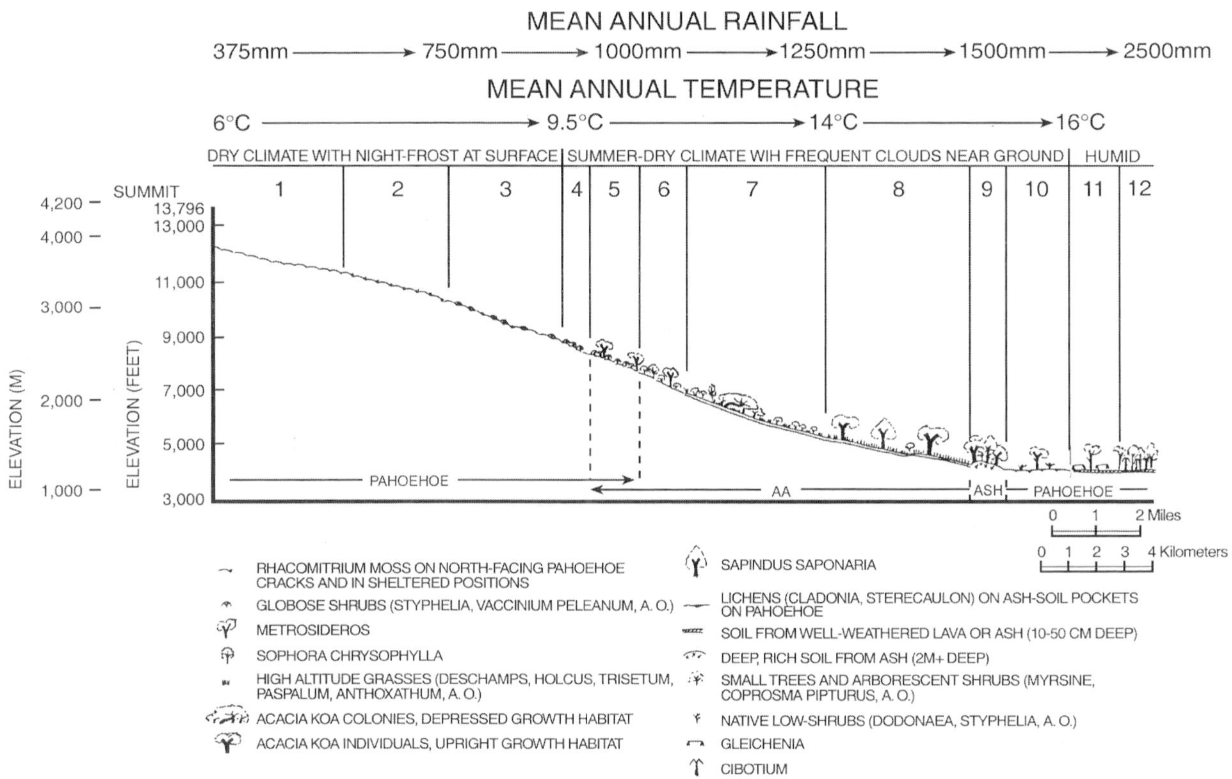

Figure 2. East-slope profile of ecological zones on Mauna Loa from 12,000 feet (3660 m) down to 4000 feet (1200 m) elevation near Kīlauea Caldera (after Mueller-Dombois et al. 1981). Profile segment 1 = Alpine stone desert, 2 = *Rhacomitrium* moss desert, 3 = Sparse alpine scrub, 4 = Alpine aggregate scrub, 5 = *Metrosideros* treeline ecosystem, 6 = Open subalpine *Metrosideros* scrub forest, 7 = Mountain parkland (formed by *Acacia koa* tree communities, *Styphelia* shrub communities, and grass communities), 8 = *Acacia koa-Sapindus* savanna, 9 = Closed kīpuka forest (segment 8 intergrades locally with segment 9), 10 = Open *Metrosideros* dryland forest (early successional stage), 11 = Open *Metrosideros* rainforest, 12 = Closed *Metrosideros* rainforest (near Thurston Lava Tube and Crater Rim Road).

Handy and Handy (1972), which is here reproduced with slight modifications as a map of the Ka'ū District on South Hawai'i (see Fig. 1).

Starting at the top of Mauna Loa, the crater is called *piko,* meaning navel and also naval string or umbilical cord (Pukui & Elbert, 1986). This Hawaiian term gives vivid reference to the crater's activity from which lava flows emerge and thereby create new land surfaces. The broad and barren summit area southward on the map diagram is called *kua lono,* which refers to a region near the mountain top (Pukui & Elbert, 1986). This area was classified as alpine stone desert and *Rhacomitrium* moss desert in Mueller-Dombois *et al.* (1981: 38 segments 1–2, Fig. 2).

The book of Malo (1997) gives additional *kua* zones. The word *kua* means "back", such as the back of a person. According to Malo, *kua-hiwi* refers to the backbone of the mountain implying spiny ridges, while *kua-lono* also refers to peaks or ridges and summits, but implies broader, plateau-like areas. *Kua-mauna* or simply *mauna* refers to the mountain side without any significant vegetation cover, perhaps including the sparse alpine scrub zone (segment 3 in Mueller-Dombois *et al.*, 1981: 38). Finally, the term *kua-hea* refers to the area below *kua-mauna,* where trees are stunted due to high altitude (Pukui & Elbert, 1986). This term refers to a zone of small trees, probably the open subalpine scrub forest (segment 6 in Mueller-Dombois *et al.*, 1981). It may also apply to the treeline ecosystem (segment 5), where forest gives way upslope to the alpine shrubland (segment 4). Alternatively, the term *kua-hea* may refer to the stunted trees on wind-swept ridges as found on the older, more dissected islands, which have similar stand structures. On the map diagram appears the term *maukele,* which simply means rainforest (*wao kele*, also spelled *ma'u kele* and *wao kele* in Pukui & Elbert, 1986). As rainforest, this area seems rather high in elevation on this map, shown here as from 6000 to 8000 feet (1830–2240 m). Malo points out that *wao-maukele* is also the area where "the monarchs of the forest grew". The monarchs were the tall (up to 30 m), endemic *Acacia koa* trees, which we know as having been assembled in mesic (less wet) higher elevation rainforests on Mauna Loa, Mauna Kea, and Haleakalā mountain.

The next four zones on Fig. 1 start with the term *wao,* which is a general term for inland region, usually forested (Pukui & Elbert, 1986). *Wao akua* literally means the wilderness of the gods and/or ghosts. In Malo's (1997) book, this zone is described as below *wao maukele,* which would fit the map diagram on Fig. 1. The *wao akua* is described as the zone in which smaller sized trees grew. The reduced tree size together with the distinction of this forest as being the realm of the gods (*akua*) and thus probably feared as a forest of ghosts, most likely points to *wao akua* implying what ecologists call "cloud forest". Next on the map is the *wao nahele* or *wao lā'au*, the general inland and upland forest region. This is followed down slope by the *wao 'ama'u* and *wao kanaka*. The latter refers to the zone where humans (the *kanaka*) work and cultivate the land. The former refers to the tall fern *'ama'u* (*Sadleria* spp.), which probably was a dominant plant in an open structured rainforest community, perhaps coexisting among cultivated, non-irrigated taro or *kalo* (*Colocasia esculenta*) and banana (*Musa* spp.) plantings interspersed with the endemic shrub *olonā* (*Touchardia latifolia*) used for cordage. Other cultivated plants were the Hawaiian introduced small tree *wauke* or paper mulberry (*Broussonetia papyrifera*) used for making *tapa* cloth, the sweet potato or *'uala* (*Ipomoea batatas*), and yams or *pi'a* (*Dioscorea pentaphylla*). These are vines with subterranean tubers, which were eaten by the Hawaiians cooked and when still warm (Neal, 1965). This species combination indicates that there was an agro-forest ecosystem present between 2000 and 4000 feet elevation (610–1220 m), which in the 20th Century was used for industrially grown sugar cane or *kō* (*Saccharum officinarum*). The fourth *wao* zone on the map diagram (Fig.1) is the *wao 'ilima*, so called because of the prevalence of the native shrub *Sida fallax*. That this shrub forms a zone below the *wao kanaka*, certainly indicates a reduction in mean annual rainfall below 3000 feet (915 m) on the *Ka'ū* map, which is definitely correct as shown in Mueller-Dombois *et al.* (1981: 30) on the climate diagram map of Hawai'i Island. Also wetland taro is indicated here, which in this drier zone can only refer to the cultivation of irrigated taro, since this is the same species as used for dryland taro cultivation in some rainforest environments such as the *wao ama'u* referred to above.

Two more zones are indicated below the *wao 'ilima*, namely *kula uka* and *kula kai*. The term *kula* refers to open, mostly grassland vegetation, and *uka* to an inland/upland location and *kai* to a seaward or coastal location. These *kula* zones were most likely the result of periodic fires. Burning was promoted periodically for *pili* grass (*Heteropogon contortus*) production as this grass played an important role as house thatching material (Egler, 1947; Mueller-Dombois & Fosberg, 1998).

The *Ahupua'a* Model

Figure 3 shows an *ahupua'a* model as designed by Luciano Minerbi (1999), with minor modifications. It portrays a typical land division on the windward side of an older Hawaiian island. Not being as high in elevation as the Island of Hawai'i, generally only up to around 3000 feet (915 m), it begins on the mountain side (*mauka*) with the upland/inland forest zone, the *wao nahele*. Note that the *wao nahele* is including a *wao akua* (cloud forest zone) around its upper fringe towards the summit, the *kua-hivi*. Below the *wao nahele*, through a transition zone (an ecotone), follows the agricultural zone, the *wao kanaka*. Continuing towards the ocean (*makai*) through another ecotone, lies the coastal zone, the *kahakai*. Note that in the *wao nahele* appear some springs (*pūnāwai*) that feed into the stream ecosystem, the *kahawai*. Near the seaward end of the *kahawai*, where the ocean tide brings saltwater into a mixing zone with the stream's fresh water is the estuary, known to the Hawaiians as the *muliwai*.

Since the *ahupua'a* served as the complete life support system for Hawaiian family groups (the *'ohana*) prior to European contact in 1778, there are many other important features noted on the diagram (Fig. 3). Besides habitation sites in the *kahakai* and *wao kanaka*, there were also transitional habitation sites near and in the *wao nahele*. Both the *wao kanaka* and *kahakai* included temples (*heiau*) and burial places (*hē*) as well as irrigated terraces (*lo'i kalo*) for taro cultivation. The *lo'i* were fed with fresh water from the main stream through a system of artificial ditches called *'auwai*. Water in these ditches had to be flowing and cool enough to prevent rotting of taro tubers. The *lo'i* were an important agricultural engineering invention for intensified crop production. Dryland taro and a suite of other crop plants including fruit trees, such as the bread fruit, *'ulu* (*Artocarpus altilis*) and mountain apple, *'ōhi'a 'ai* (*Syzygium malaccense*) were cultivated in tree gardens. Another important engineering invention were the Hawaiian fishponds (*loko i'a*), located in the ocean often next to the estuaries. They were large (hectare-sized) stonewalled enclosures with a sluice gate (*mākāhā*) at the seaward end, which allowed small fish to enter, but prevented big fish to escape. A 90 minute video production by Nālani Minton with anthropologist Marion Kelly (1992), gives a vivid account on the *lo'i kalo* and *loko i'a*. Further detail is provided by Mitchell (1992) and in the popular *ahupua'a* textbook of Kamehameha Schools (1994).

A Silvicultural Approach to *Ahupua'a* Restoration

Since an *ahupua'a* is typically a watershed valley, silvicultural research and management, based on forest ecological knowledge, should be a top priority. As such it will have an important complementary task in those *ahupua'a* selected for stream restoration activity.

In Webster's dictionary, silviculture is defined as "A branch of forestry dealing with the development and care of forests." Silviculture can also be understood as the practical application of forest science or forest ecological knowledge. Silviculture always has an applied research component and may involve experiments at an operational scale. When not applied to commercial forestry, silviculture can be considered a branch of applied conservation biology. Silvicultural approaches must be based on simulating and enhancing natural processes. In terms of labor and materials, they should be considered "low input management". As such, silviculture can be contrasted to horticulture.

Horticulture, by definition, is garden culture, which requires "high input management". In Webster's dictionary, horticulture is defined as "The art and science of growing fruits, vegetables, or ornamental plants". When applied to conservation of plant species, horticulture can also be considered a branch of applied conservation biology. But for restructuring or restoring native rainforests, silvicultural rather than horticultural techniques should be developed. Such silvicultural techniques should be based on ecological research as done in the Hawaiian rainforests.

Up to the mid-1960s, rainforest research in Hawai'i had been very limited. The most significant ecological research was that of Harold H. Lyon and a few of his contemporaries, who spent a decade

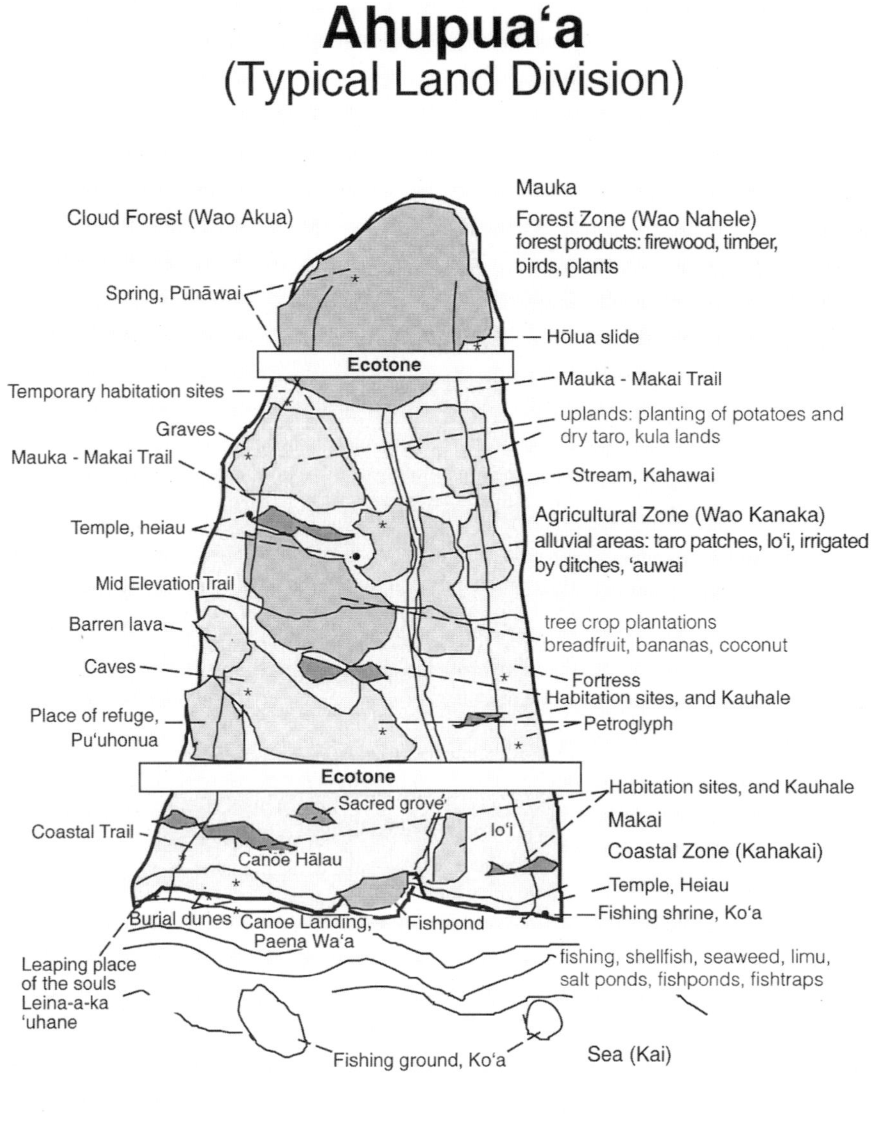

Figure 3. *Ahupua‘a* model after Luciano Minerbi 1999, slightly modified. Note the five biological resource and production zones from *mauka* to *makai*: *wao nahele*, *wao kanaka*, *kahakai*, *kai*, and the *kahawai* (stream ecosystem).

on researching the "Maui Forest Trouble" (Holt, 1983). This phase ended with Lyon's (1918) conclusion that (quote) "Our native forests are doomed".

Lyon's conclusion was based on his notion that the native *Metrosideros*-dominated rainforest was made up largely of pioneer species that could not adapt to aging soils. He thereafter postulated the idea that the missing climax species component has to be introduced from outside Hawai'i in order to save the Hawaiian watersheds. This was still the unwritten forest and watershed restoration policy in the state of Hawai‘i until about the mid-1970s.

Research under the Hawai'i IBP (International Biological Program) during the 1970s focused on the biological organization of selected native Hawaiian communities (Mueller-Dombois *et al.*, 1981). Among these was an 80 ha study plot in the Kīlauea rainforest on the Big Island. Subsequent research on the canopy dieback syndrome in the Hawaiian rainforests extended this research across the islands of Hawai'i, Maui, O'ahu, and Kaua'i and from there to the Pacific and Atlantic regions (Huettl & Mueller-Dombois, 1993).

For using a silvicultural approach to restoration, one needs to know first some of the keystone species that either stabilize or disrupt a specific rainforest community. Secondly, one needs to know something about their ecological properties and strategies. Such aspects will be discussed next. This will be followed by a set of silvicultural prescriptions for restoring Hawaiian rainforests.

Keystone Species

Among plants, keystone species are usually the dominants or the more robust ones in the community. In particular they are those whose population dynamics have a strong effect on the other species in the community. In the mature Hawaiian rainforest such species are the *'ōhi'a lehua* tree (*Metrosideros polymorpha*) and the *hāpu'u* tree ferns (*Cibotium* spp.). *'Ōhi'a lehua* dominates the canopy and the *hāpu'u* typically the sub-canopy. In less wet rainforests, the *koa* tree (*Acacia koa*) often joins the upper canopy as a second keystone species. Depending on habitat factors and geographic location, *koa* may even become an emergent tree reaching above the general canopy. Locally, other native tree, shrub, and vine species can be added as playing key roles. Among trees they include in upper Manoa Valley for example *'ahakea lau nui* (*Bobea elatior*), *hame* (*Antidesma platyphyllum*), *olomea* (*Perrotettia sandwicensis*), *lama* (*Diospyros* spp.), *kōpiko* (*Psychotria kaduana*), and *'ōlapa* (*Cheirodendron* spp.), among shrubs they include *'ohelo kau la'au* (*Vaccinium calycinum*), *ha'iwale* (*Cyrtandra* spp.), *ho'awa* (*Pittosporum glabrum*), *naupaka kuahiwi* (*Scaevola gaudichaudiana*), and *māmaki* (*Pipturus albidus*), among vines *'ie'ie* (*Freycinetia arborea*) and *maile* (*Alyxia oliviformis*). Many other robust native rainforest plants are listed by Stone & Pratt (1994: 173)

A number of alien invasives have now assumed the role of keystone species. Foremost among them is the feral pig (*Sus scrofa*). Pigs tend to destabilize the Hawaiian rainforest, in particular, because they seek out the native tree ferns, the *hāpu'u*, as a favored food item. They also promote locally the spread of strawberry guava (*Psidium cattleianum*), which is a key invasive tree in pig frequented sections of the Hawaiian rainforest. A shrub in this category is Koster's curse (*Clidemia hirta*). Locally in watershed forests on O'ahu, a particularly disturbing invasive keystone species is the often very tall (>30 m), canopy emergent albizia tree (*Falcataria moluccana*). Other recently spreading and penetrating trees are the introduced secondary and fast growing shoe button ardisia (*Ardisia elliptica*) and the octopus tree (*Schefflera actinophylla*). These secondary, fast growing trees form a new life-form group with several other alien species, which never really developed among the native species.

In the Hawaiian Islands, the primary rainforest has always renewed itself through the generational turnover of primary species without an intermediate successional phase that could be considered a "secondary forest". As is well known, a secondary forest is a typical phase in disturbed continental tropical rainforests, in which recovery of primary forest is considered a very long-term process.

Ecological Plant Properties and Strategies

For the purpose of this paper, only a few characteristics will be emphasized, which can be used for a silvicultural approach to forest restoration. During the IBP and canopy dieback studies, we surveyed many rainforest plots and transects. We enumerated all woody species by cover, density, and size. We also studied their substrate and found that most of the native rainforest species became established on decaying wood in developed mature forests. This stands in contrast to rainforest development on lava flows, where an assortment of hardy native pioneer species establish themselves in rock fissures without or with only very little organic matter (Smathers & Mueller-Dombois, 1974, 2007).

In mature rainforests we noted only three species that started commonly on mineral soil. These were the *hāpu'u* tree ferns, the *koa*, and *naio* (*Myoporum sandwicense*) trees. Most others had a significant log establishment index, meaning they started as seedlings on logs above the mineral surface (Cooray, 1974; Santiago, 2000). That means that most Hawaiian plants have an epiphytic beginning.

Such observation can be made easily in mature native rainforests, if one knows where to look for native fern sporophytes and tree seedlings. The first place to look, are the tree fern trunks. They often are the most favorable seed beds for *'ōhi'a lehua* germinants and small seedlings. If left alone, eventually one of them may succeed in becoming a sapling and thereafter a mature tree by extending its roots into the mineral soil. A precondition for this to happen is a canopy opening (Burton & Mueller-Dombois, 1984). This may occur naturally by loss of a tree fern frond or the decline of the tree fern itself after canopy opening. Many times one can observe stilt rooted *'ōhi'a lehua* trees that had an epiphytic start, either on a tree fern trunk or on a moss-covered dead tree trunk. For *'ōlapa* this seems to be the only mode of its natural establishment.

Silvicultural Restoration Tasks

Delimbing

Cutting off the limbs or big branches of the taller alien trees would be a useful first step in silvicultural restoration. This applies in particular to the huge albizia trees, which break easily and are a hazard for humans during strong winds. Delimbing should not be a clear-cut logging operation, but rather a carefully selected cutting and branch removal of selected alien trees to achieve partial opening of the forest canopy. Their limbs should be left on the ground and allowed to decompose *in situ*. To accelerate the decomposition process, the limbs, or thick branches, and in some situations the trunks of selected trees, may be cut into meter sections and split open. In mature and senescing Hawaiian rainforests, decaying logs, particularly when moss-covered, were found to be the favored micro-habitats for native fern gametophytes and woody plant seedlings to become established.

Fencing

Any section of rainforest considered for restoration needs to be fenced against pigs. Depending on financial resources one can begin with fencing of small enclosures, such as 100 m^2 plots. Of course, anything larger would always be preferable. The purpose is to create safe islands in *kīpuka* fashion within the larger forest infested by alien neophytes.

Reintroduction

From field research observations, it appears most efficient to begin with reintroducing the appropriate Hawaiian tree ferns into the fenced enclosures. On O'ahu Island this would preferably be *Cibotium chamissoi*, formerly named *C. splendens* (Palmer, 2003). But *C. menziesii* may also be considered. A natural hybrid of these two species was recently discovered in the Ko'olau mountains and called *Cibotium ×heleniae*. Such tree ferns are easily transplanted at any stage of their life cycle and/or raised in nurseries. Mature tree ferns are preferred. The reasons for reintroducing tree ferns are several. They can be planted directly into the mineral soil as they do not require a raised organic seedbed as do most of the other Hawaiian woody plants with exception of *Acacia koa* and *Myoporum sandwicense*. Tree ferns have a high value as watershed protectors in that they slow down the impact of heavy showers by forming a second canopy under the tree layer. They disperse the water away from their trunks in contrast to, for example, albizia trees. Albizia trees act as funnels for rain water due to their generally upward angled branch system. Because of this, they have a high rate of stem run-off, which is further accelerated due to their smooth bark. They are thus ill adapted as watershed tree cover in wet forests, where excess water is a problem. In contrast, tree ferns are expected to increase the rate of water percolation into the soil rather than contributing to run-off and erosion as do the alien albizia trees. A third major advantage is that tree fern trunks serve as epiphytic seed beds for many native ferns and woody plants. As mentioned before, many *Metrosideros* trees and almost all *Cheirodendron* trees start as seedlings epiphytically on tree fern trunks.

Weed control

In some situations, weed control may be the prerequisite prior to the introduction of native tree ferns into the *kīpuka*-type enclosures. Certainly, weed control may be considered an ongoing task until the tree ferns themselves become excluders of weeds on account of having developed a closed subcanopy in the *kīpuka*-type enclosures.

Inoculation

Wherever native woody plants and ferns are too far removed from the *kīpuka*-type enclosures, it may become necessary to inoculate the tree fern trunks and decaying coarse woody log segments on the ground with seeds and spores of selected native plants.

Monitoring

Another silvicultural research task involves monitoring the tree fern trunks and inoculated decaying wood segments for native plant establishment, growth, and survival. Monitoring will also be necessary in the *kīpuka*-type enclosures to keep weeds under control and the fencing in repair.

Soil scarification

In some of O'ahu's watershed forests, for example in the Kahana *ahupua'a*, it has been found that soil scarification will encourage germination of *koa* seeds. An abundance of koa seedlings has been observed there by Wirawan (1978), after removal of the *hala* litter associated with scarification of the surface mineral soil. Currently, there are only a few old senescing *Acacia koa* trees left in the canopy otherwise dominated by native *hala* (*Pandanus tectorius*) trees. Soil scarification in forest gaps will increase the *koa* component in the inland forest (the *wao nahele*) of the Kahana *ahupua'a*. It may also work in other *ahupua'a* where *koa* is in decline.

Removal of woody debris from streams

Hawaiian streams are known to be highly dynamic (Fitzsimons *et al.*, 2005). Their distances from *mauka* to *makai* are typically short, a few kilometers only, and their initial descents from the *wao nahele* are typically steep. Thus they can swell up quickly during rain storms. Woody debris that accumulates in the streams can be a cause of unexpected stream diversions as was the case in the damaging October 2005 flood in Mānoa Valley. Similarly in Kahana Valley the uncontrolled advance and overgrowing of the *hau* tree (*Hibiscus tiliaceus*) becomes a serious impediment for the dynamics of the Kahana stream which can result in flood damage in its *wao kanaka* and *kahakai*. It also interferes with the amphidromous native freshwater fauna (Fitzsimons *et al.*, 2005). Prevention of accumulation of woody debris in Hawaii's streams is a task falling into the realm of both silviculture and stream management which are in need of integration.

Conclusions

There are five major biological resource zones in an *ahupua'a*. These are the *wao nahele*, the *wao kanaka*, the *kahawai* the *kahakai*, and the near-shore *kai*, i.e., the ocean in front, including the coral reefs where present. Each of these five zones can be divided into subecosytems depending on their specific bioenvironmental settings. Since these ecosystems formed an integrated landscape unit that served as the life support system, they are of great cultural and ethno-ecological importance. They were managed cooperatively as integrated management units for several centuries prior to European contact. In any attempt towards their restoration, they should also be studied cooperatively by multidisciplinary teams, such as envisioned by the PABITRA (the Pacific-Asia Biodiversity Transect) Network. A manual of methods for such multidisciplinary island studies throughout the tropical Pacific was recently completed and put on the PABITRA web site (www.botany.hawaii.edu/pabitra/biodiversity). A recently published case study of the Kahana Valley *ahupua'a*, on windward O'ahu (Mueller-Dombois & Wirawan, 2005) provides further insights in form of a research synthesis of the valley's archeology, paleoecology, tenure-related management changes, contemporary botanical ecology, and the effects of past climate changes and human influences on the valley's geomorphology and vegetation. In the same issue of the journal *Pacific Science*, which deals with PABITRA methodology, is a biological assessment of Kahana Stream by Fitzsimons *et al.* (2005) as well as a paper on design considerations for island stream surveys by Parham (2005).

The eight silvicultural restoration tasks for Hawaiian rainforests discussed above may be considered a first set of prescriptions for adaptive management. It is suggested that these are applied in *kīpuka*-like fashion. This means that restoration should begin with fenced-in island-like nuclei of robust native plants. These comprise the ancient vegetation in usually a larger area of vegetation composed of neophytes. These native plant *kīpuka* may be small areas such as 10 by 10 m plots to begin with. They should be protected, monitored, and studied. Such native vegetation *kīpuka* will certainly provide a sense of Hawaiian place in our watershed forests. If they prove to have a reasonable survival value, they may eventually be expanded by silvicultural nurturing to become the vegetation matrix for reintroducing rare and endangered Hawaiian plants and animals. With further practical experiences gained from silvicultural experimentation at an operational scale, additional prescriptions will surely be developed.

Acknowledgments

This paper was inspired by my involvement in the Lyon Arboretum crisis of Fall 2004 and the search for a new vision of its place in the Hawaiian culture. I thank my wife, Annette Mueller-Dombois, for computer processing and editorial help and Professor Isabella Aiona Abott for reviewing the manuscript and offering appropriate suggestions.

Literature Cited

Burton, P.J. & D. Mueller-Dombois. 1984. Response of *Metrosideros polymorpha* seedlings to experimental canopy opening. *Ecology* **65**: 779–791.

Cooray, R.G. 1974. Stand structure of a montane rain forest on Mauna Loa. *Hawaii International Biological Program Technical Report* **44**, 98 p.

Egler, F.E. 1947. Arid southeast Oahu vegetation, Hawaii. *Ecological Monographs* **17**: 383–435.

Fitzsimons, J.M., J.E. Parham, L.K. Benson, M.G. McRae & R. T. Nishimoto. 2005. Biological assessments of Kahana Stream, Island of O'ahu, Hawai'i, with the use of procedures from the PABITRA Manual for Interactive Ecology and Management. *Pacific Science* **59**: 273–281.

Handy, E.S.C. & E.G. Handy. 1972. *Native planters in Old Hawaii: Their life, lore, and environment.* Bishop Museum Press, Honolulu. 233 p.

Holt, A.R. 1983. The Maui Forest Trouble: A literature review and proposal for research. *University of Hawaii Botanical Science Paper* **42**: 1–67. [http://www.botany.hawaii.edu/pabitra] (under Current and Planned Projects).

Huettl, F.R. & D. Mueller-Dombois, eds. 1993. *Forest decline in the Atlantic and Pacific regions.* Springer-Verlag. Berlin, London, New York. 366 p.

Kamehameha Schools. 1994. *The Ahupua'a Life in Early Hawai'i.* Third ed. Kamehameha Schools Press. Honolulu 59 p. plus colored poster.

Kelly, M. 1992. *Ahupua'a Fishponds and Lo'i: A film for our time.* Produced as VHS video by Nālani Minton and directed by Nā Maka o ka 'Āīna. [http://www.film-hawaii.com/NaMaka].

Kirch, P.V. 2000. *On the roads of the winds: An archeological history of the Pacific islands before European contact.* University of California Press. 424 p.

Lyon, H.L. 1918. The forests of Hawaii. *Hawaii Planter's Record* **20**: 276–279.

Malo, D. 1997. *Hawaiian antiquities, Mo'olelo Hawai'i.* Translated from the Hawaiian by Dr. Nathaniel B. Emerson 1898. Second ed. 1951. 7th reprint 1997. Bishop Museum Press. Honolulu. 278 p.

Minerbi, L. 1999. Indigenous management models and protection of the ahupua'a. *Social Process in Hawai'i* **39**: 208–225.

Mitchell, D.K. 1992. *Resource units in Hawaiian culture.* Fourth ed. Kamehameha Schools. Honolulu. 102 p.

Mueller-Dombois, **D**., **K**.**W**. **Bridges** & **H**.**L**. **Carson**, eds. 1981. *Island ecosystems: biological organization in selected Hawaiian communities*. United States/International Biological Program Synthesis Series 15. Hutchinson Ross Publishing Co., Stroudsburg & Woods Hole. 583 p.

———. & **F**.**R**. **Fosberg**. 1998. *Vegetation of the tropical Pacific islands*. Springer-Verlag, New York. 733 p.

———. & **N**. **Wirawan**. 2005. The Kahana Valley ahupua'a, a PABITRA study site on O'ahu, Hawaiian Islands. *Pacific Science* **59**: 293–314.

Neal, **M**.**C**. 1965. In gardens of Hawaii. *Bishop Museum Special Publication* **50**, 924 p.

Palmer, **D**.**D**. 2003. *Hawaii's ferns and fern allies*. University of Hawai'i Press, Honolulu. 324 p.

Parham, **J**.**E**. 2005. Aquatic survey techniques on oceanic islands: important design considerations. *Pacific Science* **59**: 283–291.

Pukui, **M**.**K**. & **S**.**H**. **Elbert**. 1986. *Hawaiian dictionary*. University of Hawai'i Press, Honolulu. 572 p.

Santiago, **L**.**S**. 2000. Use of coarse woody debris by the plant community of a Hawaiian montane cloud forest. *Biotropica* **32**(4a): 633–641.

Smathers, **G**.**A**. & **D**. **Mueller-Dombois**. 1974. *Invasion and recovery of vegetation after a volcanic eruption in Hawaii*. National Park Service Science Monograph Series **5**, 129 p.

———. & **D**. **Mueller-Dombois**. 2007. *Hawai'i, the fires of life*. Mutual Publishing, honolulu. 141 p.

Stone, **C**.**P**. & **L**.**W**. **Pratt**. 1994. *Hawai'i's plants and animals*. Illustrations by J.M. Yoshioka. University of Hawai'i Press, Honolulu. 399 p.

Wirawan, **N**. 1978. Vegetation and soil-water relations in a tropical rain forest valley on Oahu, Hawaiian Islands. Unpublished Ph.D. Dissertation, Botany Department, University of Hawai'i, Honolulu. 420 p.

Biology of Hawaiian Streams and Estuaries. Edited by N.L. Evenhuis & J.M. Fitzsimons. *Bishop Museum Bulletin in Cultural and Environmental Studies* 3: 35–61 (2007).

Opportunities in Stream Drift: Methods, Goby Larval Types, Temporal Cycles, *In situ* Mortality Estimation, and Conservation Implications

Kim N.I. Bell

Department of Ichthyology and Fisheries Science, Rhodes University, P.O. Box 94, Grahamstown 6140, South Africa; & Biology Department, Memorial University of Newfoundland, St. John's, NL, Canada, A1B 3X9; email: kbell@mun.ca

Abstract

Stream drift or rheoplankton study is relatively noninvasive. I describe methods and gear for quantitative stream drift sampling, partial self-sorting, goby larval identification, and analysis of cycles, and *in situ* mortality estimation. The rheoplankton of Dominica, W.I., includes abundant larvae of anadromous taxa (gobies, decapod crustaceans, and neritid gastropods), larval insects, acarines, calanoids, and nauplii. Other tropical volcanic coastal rivers tend to have similar assemblages. Goby larval types, of which one was confirmed as *Sicydium punctatum*, were distinguished by characters visible only in live larvae. This allowed separate analyses for each type. Abundances were analysed for temporal pattern on annual, lunar, and diurnal scales simultaneously. Twelve highly significant periodic regressions showed considerable diversity in temporal abundance patterns of different taxa or groups. Periodic regression can also enable rigorous inter-site, before/after, or impacted/unimpacted comparisons that are otherwise not reliable. A new approach to estimate *in situ* mortality was developed, and leads to important conservation and behavioural ecology implications.

Introduction

We still have much to learn about anadromous gobies, or, as some term them, amphidromous gobies. As fisheries, they are either nonexistent or by all the best accounts smaller than they used to be, and that is despite the fact that the value per kilo is likely higher per pound than any other fish locally available. Some species are listed as endangered. For fisheries and biodiversity conservation purposes, there is much we need to know. Why have the fisheries vanished or shrunk in so many places? Where are the oceanic habitats of larvae and postlarvae? How are the geographic groupings of sicydiines related; how did they disperse and where from? What larvae (free embryos) match what species? What are the main sources of mortality, or of recruitment variation? What reaches of a stream contribute the most to the larval pool? Do adult movements in-stream reflect this, as they should? Have growth rates changed? Why do the gobies in Hawai'i have much longer marine durations than those in Dominica? Are the abundances of adults in Dominican rivers typical, or are there fundamental reasons why some apparently suitable rivers have low abundances? For some of these, we have notions or even answers; but some questions lack even a background against which to pose them.

What do we know? Atwood (1791) is the first reference for sicydiine goby fisheries, and probably for sicydiine gobies. Manacop (1953) is the first reference, a milestone, for the anadromous life history, but the acceptance of his work may have suffered because it corrected a prevailing misunderstanding of the life history. The accepted knowledge that his work overturned was the following:

> "The Gobiidae are abundant in and about coral reefs, rivers, lakes, and mountain streams. A number of the small or minute kinds living in lakes or brooks are exclusively fresh-water fishes; but the vast majority, including all those of much economic importance, spawn in the sea, and the young ascend rivers and live in streams until mature. Indirect evidence is conclusive that those which survive the perilous journey to the sea return to their fresh-water haunts and continue to make the trips each way every year as long as they live" (Herre, 1927: 85c; elsewhere he lists *Sicyopterus lacrymosus* with others in the catadromous context).

Herre's (1927) assumption of catadromy was, of course, without evidence—nobody has ever reported adult sicydiines in the ocean, or re-entering rivers. His assumption of seasonally limited reproduction also could not have been supported by evidence, and I'm not sure whether it has yet been investigated; but in Dominica spawning is certainly pan-seasonal because larvae are obtainable in the stream drift year-round, and recruitment similarly is year-round (Bell & Brown, 1995; Bell *et al.*, 1995; Bell, 1997). Erdman properly conceded to logic with appropriate reservations, writing (1961) "In spite of 5 yrs turning over rocks in Puerto Rico, [I have] not yet found eggs", and (1986) "while circumstantial evidence of upstream spawning is strong, I have not yet found attached eggs or nests under … rocks up to about 20 inches in diameter".

Despite Manacop's remarkable work, the idea of catadromy persisted amongst his colleagues (e.g., Montilla, 1931; Blanco, 1956) and even according to Herre (1958), who included under "List A – Marine fishes returning to sea or to brackish water to spawn" a number of Gobiidae including the genera *Sicyopterus*, *Sicyopus*, *Chonophorus*, and others. The giant can perhaps be forgiven, especially as he begins "Most people like to go fishing … I first went fishing in 1878…", but it is unaccountably surprising and also lamentable that Manacop's impressive work made no impression on an ichthyologist of such prodigious enthusiasm, even 20 years previously when Manacop was involved in his fieldwork and Herre must surely have been well aware of it. Perhaps even scientists sometimes develop an investment, which is by definition unscientific, in their ideas. (Respectful differences over the term "anadromy" as opposed to "amphidromy" are not that kind of debate about biological fact, they are about utility and communication; both terms, as defined by their users, usually apply.) Manacop's work was explicitly motivated by the need to conserve the Philippine ipon fisheries based on, like tritri in Dominica, postlarvae of anadromous gobies, but the work was not subsequently put to good use. We can only guess the time and opportunities lost because of a failure to distinguish knowledge from hypothesis and provisional assumption.

We have much too sparse information on past fisheries (Jordan & Evermann, 1905; Titcomb, 1977) in Hawai'i, and elsewhere (summarised in Bell, 1999). We have good information on methods and social patterns involved with the Philippine fisheries (Montilla, 1931) which seem all but gone, though there are evidences some products still for sale (e.g., C. Chong, pers. comm.).

We know a fair bit about biology and fisheries in the Philippines (Manacop, 1953), biology and genetics in Hawai'i (Ego, 1956; Nishimoto & Fitzsimons, 1986; Radtke *et al.*, 1988; Fitzsimons & Nishimoto, 1990; Fitzsimons *et al.*, 1990; Kinzie, 1993) and early life history, recruitment dynamics, and fisheries in Dominica (Bell & Brown, 1995; Bell *et al.*, 1995; Bell, 1997).

Manacop (1953) is also the first rheoplankton (or stream drift) reference for anadromous goby larvae in the rheoplankton. (A later work by people who knew his work claimed an inability to duplicate his results with plankton nets – this speaks again to the notion of an accepted dogma protecting itself.) Drift of other fish larvae or eggs has been acknowledged or sampled (e.g., Cambray, 1985; Copp & Cellot, 1988; Flecker *et al.*, 1991; Pavlov, 1994). Rheoplankton has been explored in anadromous goby work (Iguchi & Mizuno, 1990, 1991), and of course from 1989 in my study in Dominica in the West Indies (Bell & Brown, 1990; Bell, 1994). Kinzie (1993) commented "A more realistic measure of reproductive output of the entire stream would be to use drift sampling to catch newly hatched free embryos", underscoring the fact that despite the long but sparse history of stream drift work with anadromous gobies, unexploited opportunities remain for quantitative stream-drift work.

I have been surprised by the rheoplankton data I collected in Dominica. Before I knew how to analyse it, it seemed to say little, and that was how I reported it. Given as well the reputation of plankton data for being noisy, the uninformative bivariate plot of abundance over time seemed a confirmation until I applied periodic regression to it (as I do to almost everything). So analysis is one of the themes of this paper.

This paper gives an overview of the methods and opportunities in stream drift or rheoplankton sampling. The primary methods are quantitative sampling, live sorting and counting, and larval identification. The secondary methods involve analysis of temporal patterns, and – to be reported elsewhere – *in situ* mortality estimation. These have research and conservation implications.

Single quantitative samples can be analysed by periodic regression for estimation of temporal patterns in abundances. Paired quantitative samples permit the *in situ* estimation of mortality in the stream drift; this does not seem to have been done before, except by Bell (1994).

Qualitative Sampling and Analysis

The previous literature has been criticised because "Virtually without exception, however, reports of drift lack statements of the precision of the estimate" (Allan & Russek, 1985). This is a clear reference to drift studies that are largely descriptive (with exceptions such as Kohler, 1985), and to data plotted as bars or line plots without objective analysis. Indeed, as this paper will show, it is very difficult to see temporal pattern on a bivariate basis, because pattern exists on multiple timescales that are superimposed. I hope here to provide a glimpse of the light at the end of the tunnel that will encourage a more adventurous application of stream drift study.

The desire for estimates of precision (e.g., Allan & Russek, 1985) can sometimes be satisfied not with replication at each time and place, but instead as the residual error of a regression. In fact I would suggest that effort is likely better spent on representation than replication, provided a regression is to result. I would go further than that, and say (supported by the results of this paper) that careful estimates of precision could be quite useless without establishing what the temporal and spatial patterns are. If replicated samples taken at two sites, one at 1045h on the third day after the new moon in September and the other at 0530h on the 12th day after the new moon in March, yield means that are "significantly" different, what does that mean? We know it is impossible to be in two places at once, therefore we cannot sample in two places at once. Can we say the difference is due to site? ... or due to before/after some environmental impact like hurricane or pollution? What if there is a pattern, and what if the difference found could be as well explained by the sample timing alone? We need some ways of characterising stream drift that take account of the pattern, or our work is nearly meaningless.

Rheoplankton, or stream drift, is rarely analysed for cycles in abundances. The expected natural cycles are the day, the tide, the lunar month, and the year (seasons). However, many studies are too brief to allow analysis of natural cycles longer than a day, and in many cases analyses that are possible are not attempted. This is possibly due to a lack of awareness of periodic regression (Bliss, 1958; Batschelet, 1981; Bell *et al.*, 1995; Bell, 2004) in biology.

Models like periodic regression bring many advantages. Obviously, they provide a description of patterns in abundance. Next, the mesor, which in the context of an ordinary regression is the intercept, gives a very robust estimate of central tendency that is effectively corrected for uneven sampling over the cycles—it is therefore much more meaningful than the simple mean of all data, and may be useful in comparing populations. Where periodic regressions explain a large amount of variation, they improve our ability to compare different populations because they reduce the amount of variation that is attributed to noise or error. When we have a temporal model of variation, observations can be put in the context of the pattern so that we avoid two possible errors: falsely attributing differences to situations or sites when they are really due to timing of samples; or falsely considering situations to be the same when this is only an accident of sample timing.

Sometimes more importantly, such models allow us to remove temporal trends and cycles from data so that we can analyse them with respect to other factors that are of interest, say, from the point of view of conservation. Conservation-related factors, such as the degree of development impact on sites, may not be capable of resolution with 'raw' data because the temporal/periodic signal overwhelms the factor signal.

Periodic regression is so readily manageable, and an understanding of cycles so important, that it should be a focal point for organising information in any studies where data can be affected by cycles.

Complex relationships cannot be fully represented on bivariate plots; indeed they can be quite difficult to visualize graphically. Any visualisation is therefore a crib, imperfect, but necessarily so, and a series of them can relate to a multiple regression. For these reasons, residuals become the chief avenue of diagnosis of the quality of an analysis that has more independent variables than can be graphically represented together.

Identification of larval goby larval types

Larvae of gobies can be distinguished using features that are visible when live. Identification from preserved larvae would have to rely on those features which reliably persist in the preserved state, and many of the features would not.

***In situ* mortality estimation**

Mortality rates in rheoplankton are often alluded to, but never have been estimated *in situ*. Only Bell (1994) provides a way to estimate drift mortality *in situ*. The approach and results will be briefly summarised here; the work has been extended and will be detailed elsewhere.

Why is it important? Considerable pressure is being exerted on natural systems by human systems (e.g., Brasher, 2003). Mortality estimated *in situ* is needed if we are to establish the losses that take place during the period of drift of hatchlings to the sea. Whatever the survival rate is, it is cumulative with time; this means that any increase in the drift distance that larvae must cover will result in an exponentially increased loss of larvae over that distance. Rates reported in Bell (1994) indicate survivals at or below 50% per hour (which is very close to per kilometer if the current speed is 0.3 m/s). At 50% mortality per hour only about 3% of larvae would survive a 5-hour drift. This is a very good reason for adults to select territories with the shortest drift times, i.e. nearest the sea. That in turn means the emphasis in conservation should be to preferentially preserve the habitats closer to the sea.

Materials and Methods

Primary Methods

Sampling

Sampling gear

Quantitative samples were taken using specially made small conical nets suspended at a settable depth below the water surface and with a mechanical integrating current meter outside of the net (Fig. 1). A variety of configurations were experimented with, including triple nets that produced three replicates at each set, but the preferred design is the one presented here.

The nets used were made from 81-µm Nitex™ mesh. The net sizes varied from 46 mm diameter at the mouth (a truly micro net), to 112 mm diameter at the mouth, used as a single net (tested also as a group of three but was not workable). The 112 mm diameter nets were ~1.0 m long (mesh portion of the cone). This made for a very high mesh/mouth area ratio, and therefore for a high filtering efficiency (FE). FE appeared to be close to 1.0, because particles drifting inside the net could not be seen to slow down, compared to particles outside the net, until they actually hit the mesh. Two practical considerations are: (1) overestimating FE reduces (does not increase) the calculated abundances; and (2) any constant error in FE has no effect on ratios of abundances. Calculations treated FE as equal to unity.

Nets had: [1] ballast chambers, made by gluing lids of film canisters onto the net collar, so that the canister could be snapped on with any desired air/water balance, to adjust buoyancy to near zero; [2] vanes (built onto the canisters) to induce rotation, and because of the attachment means the vane pitch could easily be adjusted; [3] an attachment for the current meter to be mounted on the side; and [4] a bridle with a swivel and clip for attachment to the deployment system. Rotation was in order to have the average position of the current meter equal to the center of the net mouth. There were other configurations but they were analogous.

Nets typically were attached to a bridle that clipped to a slidable element on a vertical cable between a float and a weight suspended from it. This allowed depth to be set (the target was usually about 1/3 of the water depth from the surface). A stayline and staypole was attached to the float. The gear was supportable either by a stayline alone (for sampling from bridges) or, for sampling from shore, the operator would hold the stayline taut on the upstream side and push out with the staypole on the downstream side to keep the gear in the desired position. All nets were made to be easily handled, rinsed, and emptied. A mechanical integrating current meter attached to the outside of the net provided an index of the amount total flow past the net (which, at FE = 1.0 is the same as

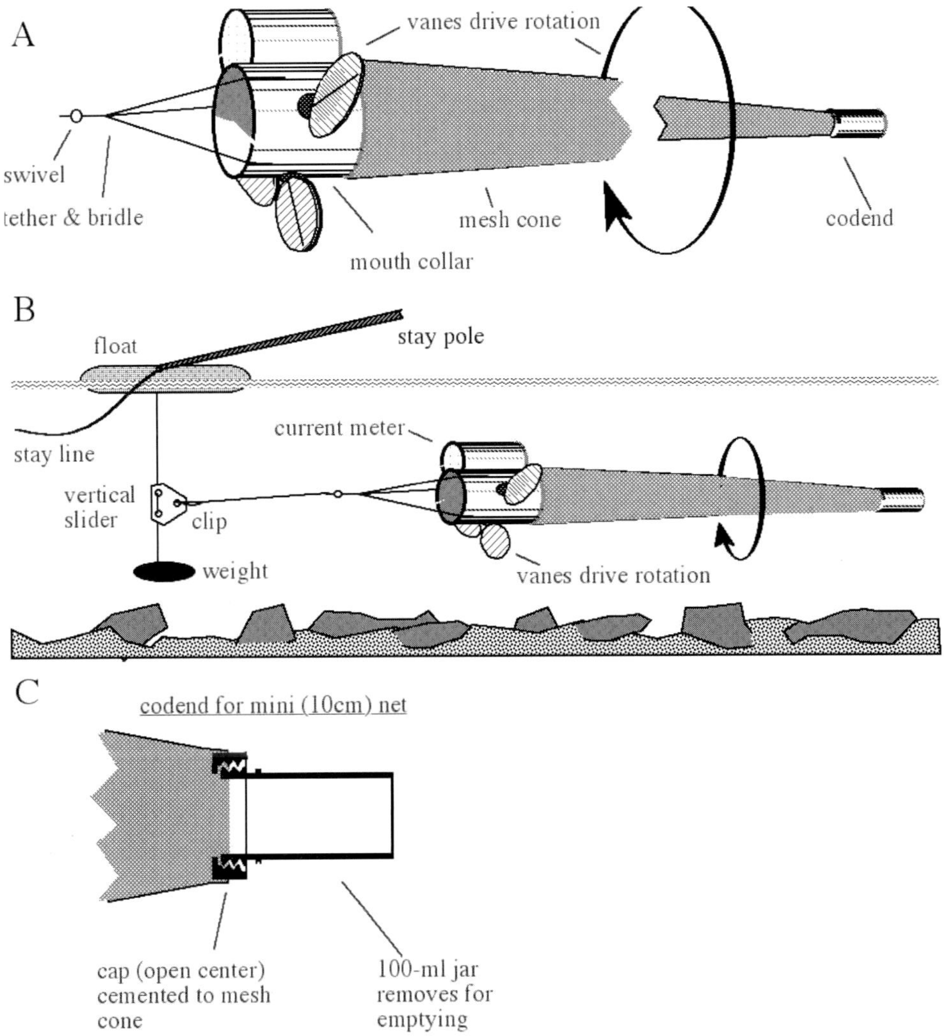

Figure 1. Plankton gear, semi-miniature for quantitative rheoplankton sampling in small rivers. **A**: Net, buoyancy chambers, vanes, attached current meter, and bridle and swivel attachment. Mouth diameter of net is ~11 cm diam. and the mesh cone is ~1 m long. Rotation caused by vanes allow the average position of the attached current meter to be the same as the position of the net. **B**: Deployment: float and weight system allows sample depth to be easily set; clip allows quick detachment. **C**: Threaded cod end for quick emptying.

flow inside the net). Calibration of the meter allowed this index to be converted to distance, which, multiplied by the mouth area of the net, estimated volume sampled and made the sample quantitative. A subtle point is that even after calibration (of virtually any meter), we have no estimate of what actually flows through the meter itself, we merely correlate a reading with a distance. That means we must avoid the a temptation to put the meter inside the net itself, even though it seems a neat thing to do, because that configuration cannot be calibrated by pulling the gear through a known distance.

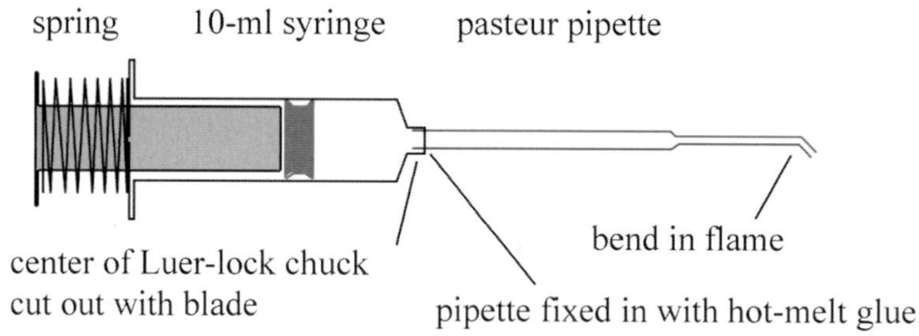

Figure 2. Spring-loaded suction pipette for counting live plankton out of a dish. Spring can be made by winding tempered wire around the syringe barrel; it will expand slightly when released. The plunger is pulled out, put through the spring, and replaced. Spring can be tensioned as needed. Needle end of syringe, inside the collar, is cut out with a blade. The Pasteur pipette is cut, bent as desired, and the end fire-polished (a butane lighter is hot enough); then it is glued with hot-melt glue into the collar. In use, syringe is gripped so that the thumb depresses the plunger at the start, and releases it in small jumps as plankton organisms are suctioned. This gives effective and easy control.

Sampling procedure

Each sample was accompanied by data for: location, date, time, water temperature, current meter reading, and elapsed time (typically 5–10 minutes, a sampled volume of about 1 m^3, which typically would yield several dozen larvae). Samples were rinsed from the net within a minute or two of ending the sample, into labeled carrying jars. Sufficient water was kept in order that the sample remained alive. Samples could remain alive for at least 24 hours, although counting was usually done before that. A check on this is the count of dead larvae; these were typically nil to < 5%, even after 24 hours.

Live counting of samples

Preservation can reduce a sample to a grey sludge of barely recognizable items stuck together, such that within normal limits of human patience they cannot be accurately counted. Live counting gives better results because live organisms are more readily detectable and identifiable. Live counting is necessary to identify larvae as types or species, because while black pigments may persist in preserved specimens, the green pigments certainly do not, and the yolk colours and textures would probably also be lost; there may also be distortion of larvae. Counting dishes used were small, ~20 mm diameter, so as to fit within the low-power field of the microscope.

Before counting, 'self-sorting' takes advantage of larval behaviour. Firstly, as the sample is allowed to stand for an hour in the collecting jar, debris and detritus settles to the bottom. The sample is then carefully decanted through a sieve (70 to 80 μm mesh) and the sample transferred to a counting dish while the water is returned to the sample. This allows the decanting procedure to be repeated as needed. Goby larvae and decapod larvae, for instance, are virtually all recovered alive in the first decant, few in the second, and rarely any in the third. Mollusc (Neritidae) larvae show abundantly in the decanted portion, but often a considerable number remain in the sediment. Caddis- and to a lesser extent mayfly larvae favour the sediment. The settled sediment can be easily subsampled using a suction tube and estimating the portion sampled (easily done on an area basis because the removed sediment leaves an empty area that is easy to estimate as a fraction of the total).

Once in the counting dish, the most reliable way of counting is to suction out individuals, one by one, for all OTUs (Operational Taxonomic Units) of principal interest. Other OTUs of peripheral interest to the study, such as mollusc larvae that often numbered in the hundreds or thousands, can be estimated, and the estimates can occasionally be checked against counts. If the estimates are terrible, the analyses will not likely show significance.

Counting pipette and anaesthetic

Key to live counting is a spring-loaded pipette (Fig. 2) made for the purpose. Several pipetting approaches were tried but the suction from rubber bulbs and standard lab devices is too slow, the effort makes the pipette hard to aim, and the plankton escape; and if too much is suctioned at once the operator cannot count the items. A quick-acting suctioning device that could decisively suction up a wary larva was made from a pasteur pipette fixed to a 10-ml syringe that in turn was modified to be spring-loaded. Pushing on the plunger compressed the spring and emptied the syringe, and releasing pressure allowed the syringe to suction up water. Releasing thumb pressure in controlled amounts is much more amenable to human control than applying it in controlled amounts. The plankton ejects easily, almost never did anything get stuck in the syringe, and the device stays clean for a long time. A quick rinse is easy by suctioning clean water, shaking, and expelling; but to allow use of a squirt bottle to occasionally rinse the syringe without having to remove the plunger, a 6 mm 'rinse hole' was punched in the syringe body just below where the plunger would stop when pulled all the way back.

Suctioning out goby larvae is much easier if the larvae are anaesthetised. I used 2-phenoxyethanol, from which I would make a stock dilution of about 0.1 to 1.0%, or a few drops of 2-phenoxyethanol in about 50 ml of water. It is important to make the stock dilution beforehand because 2-phenoxyethanol is an oily alcohol and takes a long time to mix. The stock dilution keeps, for practical purposes, indefinitely, and a couple of drops of this in the counting dish will anaesthetise goby larvae in a few seconds to a minute. Recovery occurs quickly after a water change. Identification much less reliable (and impossible if it requires reference to features that don't preserve) with preserved larvae, and anaesthetic is very helpful in keeping the live ones still. An emergency alternative anaesthetic might be ice water, but condensation on the bottom of the dish might interfere with the oblique/below illumination needed to see the larval features.

Goby larvae are best counted out in groups of any single type, which means one first needs to see the group that one will next count out; but the decapod larvae are very active and create problems by moving the goby larvae, interfering with the process of identification and counting. It is therefore best to remove decapod larvae first, but they do not succumb to 2-phenoxyethanol at the concentrations used. However a little practise seems sufficient to cope with them, and swirling the counting dish causes the inactive particles to move to the center, simplifying the process of suctioning out the decapod larvae from the edges.

Sampling protocol

Many cycles act together to influence the systems we study. It is common that people are advised to constrain sampling to, say a particular time of day or time of tide in order to eliminate variation from that source. But there are some major problems with that approach. First, the same time of day and tide do not recur often, and this reduces sampling opportunities. Second, how useful, really, is the sample set going to be if it cannot be related to the values (whether CPUE or temperature) outside a narrow range of times of day or other cycle? Third, taking the 'same time of day' approach can impose an artificial periodicity on the data and may limit the ability to resolve other cycles. The best approach is random sampling in time, but a close approximation is to use the world as your random number generator, and sample whenever manageable. (Logistical factors like vehicle availability, gear failures, time demands of other work, holidays, staff availability, etc., cannot be expected to correlate with the independent variable (number of larvae) being sampled, so there is no basis to declare a bias). Sampling times were therefore chosen haphazardly or opportunistically, within the conditions that permitted sampling.

The general difficulties of fieldwork at night hardly need explanation. A few remarks may help put into perspective any concern about the lower density of samples in the night and the inferences for night time. The analysis decomposes temporal pattern into a series of sinusoidal patterns; sinusoidality is the most parsimonious assumption for a cycle because a sinusoidal curve results from the intersection of a flat plane and a cylinder, so it is equivalent to the straight-line assumption for regression, and the regression can indeed be represented in such a form. Concern over a sparsely-sampled region of a sinusoidal curve is therefore exactly—no more, no less—the same as for a

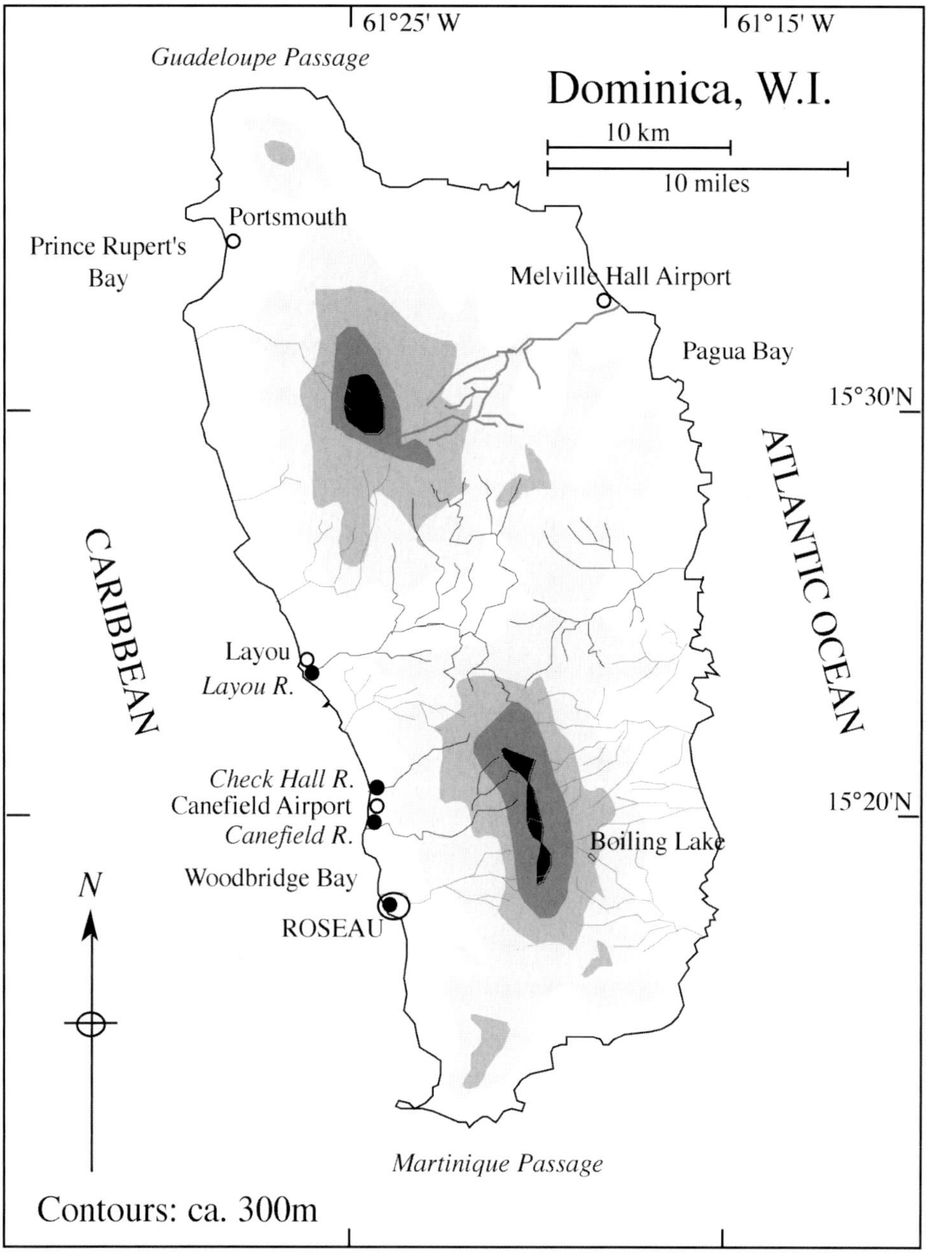

Figure 3. Sample sites for this study: Layou, Check Hall, Canefield, and Roseau Rivers. Only selected watersheds are shown. By volume, Layou and Roseau Rivers are similar and about twice either Check Hall or Canefield Rivers.

sparsely-sampled region of a straight-line regression. Residuals are the prime diagnostic test for this kind of issue. Concern over the lower representation of night-time samples would relate only to the diurnal cycle, first harmonic only. The peaks, it should be remembered, are the counterparts of the troughs, 180° out of phase. Peaks in pattern are not estimated individually or simply from peaks in observed abundance, instead it is a pattern that is estimated from all the data. The sine curve has parameters amplitude and phase, and nothing more. A reasonably dense sampling of half of a sinusoidal cycle is more than adequate to estimate the entire cycle. The sole issue is that of the assumption of sinusoidality, but as explained this is the simplest, most parsimonious cycle that can be proposed.

Sample locations
The work reported here was carried out in four rivers on the west coast Dominica, West Indies (Fig. 3). The *in situ* drift mortality study was carried out in lower reaches of Layou and Canefield Rivers.

Identification (species). Goby adults were identified according to Brockmann (1965) and Jordan & Evermann (1898). Brockmann (1965) supplies photographs as well as scale counts, leading to the identification of *Sicydium punctatum* Perugia, *S. antillarum* Ogilvie-Grant, and *Awaous taiasica* (Lichtenstein). *Sicydium punctatum* is often called *S. plumieri* (e.g., Erdman, 1986) and probably in error because the latter was described on the basis of a drawing, and there are no type specimens. Voucher specimens of *S. punctatum* recruits from Dominica are in the U.S. National Museum (catalogued as USNM 314002).

Identification (goby larval types). Larval type identification allowed analysing of individual types contained within the mix of "goby larvae". Features or characters used in larval typing are the location and pattern of each kind (black and yellow-green) of pigment, and the yolk size, colour, surface texture, and shape. These need to be seen with oblique lighting, so that larvae are seen against a dark field. Larval size is similar for most types (~1800 μm) but one type seems markedly larger (~2000 μm). Because the most important larval typing features do not preserve (at all), [1] all work had to be done with live samples; and [2] there had to be reliance on sketching and notes from live specimens. Few good photographs were taken, and the features rarely show as well in photographs as in life, but the future for this kind of work looks brighter, with advances in digital photography making it possible to confirm immediately that a good photograph has been taken. The best photographs are taken with lighting from the side against a dark field; transmitted light tends to hide differences. Good photographs can be taken under stereo/dissecting microscopes, or with compound microscopes and a depression slide.

OTUs—operational taxonomic units
Operational taxonomic units were employed as in Table 1, to accommodate identification to a level that was reasonable in terms of the central objective of the study, which was to determine features of sicydiine goby life-history that would be relevant to management and conservation.

Secondary Methods 1: **Temporal Cycles**
Periodic regression is described in previous works (Bliss, 1958; Batschelet, 1981; Bell *et al.*, 1995; Bell, 1997; Bell, 2004). The kind of situation with a periodic or circular x and a linear y is often termed "cylindrical" because the periodic x is transformed into the coordinates of the unit circle forming the base of the cylinder, and the linear y is visualised as marked as the height of the cylinder. If you like, imagine the dial of a 24-hour clock, with some values (e.g., air temperature) represented as columns of height y positioned on the rim of the clock according to the time of the observation. Visualising periodic regression is helped by remembering that a sine curve results, around the cylinder, if the cylinder is cut on a flat plane. That plane can be described by the coefficients of the sine and cosine components of a cycle.

Temporal variables (three periodic and one aperiodic) derived from the date observed with each sample are:

- DOY = day-of-year (0 to 364.999);
- TOD = time of day;
- LQ = phase in lunar cycle (with zero set at Last Quarter). For each lunar month, lunar phase is trued to the U.S. Naval Office (USNO) lunar phase tables because lunar months vary in length;

- NDOY = a sequential index of days past since the beginning of the study. The periodic variables were converted to a standard angular system (degrees or rads) so that sine and cosine could be taken for use in the regression analysis. The regression equation used for all taxa was:

Equation 1:

$$\ln(1+OTU/m^3) = B0 + B1NDOY + B2\sin`DOY + B3\cos`DOY + B4\sin`2DOY + B5\cos`2DOY + B6\sin`TOD + B7\cos`TOD + B8\sin`2TOD + B9\cos`2TOD + B10\sin`LQ + B11\cos`LQ + B12\sin`2LQ + B13\cos`2LQ + e$$

where the Bs are coefficients with numbered subscripts, sin` and cos` mean "proper" sine and cosine (i.e., taken after conversion of the natural periodic variable to a conventional angular system), and the numeral 2 (e.g., sin`2DOY) indicates a doubling of the index in order to obtain the second harmonic, and *e* is a normally-distributed error.

Second harmonics are commonly significant in climatic (etc.) data, and improve the fit. I prefer to have the same model for all groupings, and so I have kept the harmonics, whether significant or not. (And regarding dropping non-significant parts of a sine and cosine pair, and despite published bad advice to the contrary, dropping the non-significant part of a (sin, cos) pair is improper because where the peak is aligned with 0°, 90°, 180° or 270°, one will inevitably be non-significant; removing it illegitimately inflates the mean square because you used it to discover where the peak was. This all goes to the point that a cycle cannot be expressed in less than 2 dimensions.)

Residuals were used to help determine the suitability of the analysis and transformation used.

Secondary Methods 2: *In Situ* Mortality

Methods are as described in Bell (1994). In summary, the method is based on paired (one upstream and one downstream) quantitative stream drift samples with known distance and drift time between them, so that survival can be solved for each pair. The method estimates net disappearance rate from the plankton. That equates to mortality in taxa and stages that drift until they either die or reach the sea (or at least pass the lower sample station); that description fits drifting larvae of the anadromous taxa discussed here. Goby larvae in Dominica kept in aquaria were never seen to settle until death was imminent. The reach chosen should be one where reproduction can be assumed to be zero; but if that is violated, it only makes estimates of survival larger, therefore they are conservative. Production is indicated where S exceeds unity, and that again will be conservative to the extent of the actual mortality.

Results and Discussion

Larval types

Types can in principle be nominated on the basis of actually arising from a species (in which case the differences, however slight, are accepted), or on the basis of evident distinctiveness. The latter is used here because we do not have, from all known species, samples of their larvae matched to them.

Five credible types were defined from field samples: F, Y, W, PADBS, and PAF (Table 1, Fig. 4). These types are credible because they differ from each other by at least two characters. If these characters are taken separately for all the types, and randomly recombined, most combinations have no larvae to match them. Therefore the observed character combinationss cannot freflect independent random variation. Furthermore, amongst the many thousands of verified (captive spawning) *S. punctatum* larvae, all of type F. Field-collected nests from reaches inhabited by *S. punctatum* also showed virtually no within-nest variation, and the consistency of larvae from nests and captivity also tends to eliminate the possibility that larval types are determined by local conditions, or diet, etc. And, finally, the regressions show that types have differing temporal characteristics. This constancy suggests that the characters used (yolk colour, texture, size; pigment types, patterns and somatic distribution) are determined by the species, and therefore that the unique combinations of characters are

Table 1. List of Operational Taxonomic Units [OTUs]

Goby larval types

gl — all goby larvae

F — verified to correspond to *Sicydium punctatum*. Clear greenish smooth yolk, "fluor" (a bright yellow-green pigment that looks fluorescent, like fluorescein, although it isn't) in trunk (dorsal of midline). Subsumes possible but uncertain variation noted as subtypes **Fyg** (common) and **Frb** (rare),

Y — yellow spherical crusty yolk

W — clear yolk, no fluor

PADBS — postpelvic double brown spot. Rare.

PAF — post-anal fluor that is both dorsal and ventral to notochord (unlike F where fluor is centered on pelvic region and is only ventral to notochord). Rare (not found prior to 1997).

Other anadromous OTUs

shrABC ("shr" may be written "shrimp" for readability) or **SHRall** — candidates: all decapod larvae. Many species (Atyidae, Palaemonidae, and crabs) are in Dominica's rivers (Chace & Hobbs, 1969).

shrA or ShrimpA — Decapodan larva 2mm long, body depth through cephalothorax not more than 2x depth through anterior abdominal segments. This is the most abundant type.

shrB — decapodan larva > 2 mm, cephalothorax depth > 2 x depth of anterior abdominal segments. Second most abundant type but insufficient data for analysis.

shrC — like shrB except curled and with a cephalothoracic spike like a cypris larva, may be a later moult of B. Insufficient data for analysis.

moll — mollusc larvae, Neritidae; there seem to be at least two Dominican Neritids (Noblet & Damian, 1991), with possibly similar larvae, but one, possibly a smaller kind (lived ~7 yr in aquarium, stayed small but continuously produced larvae) liberates swimming larvae directly into the water and another species may lay eggs in capsules as is referred to elsewhere.

Nonanadromous or uncertain OTUs

mayfl — mayfly larvae

cadd — caddisfly larvae

cala — calanoid copepodites (diadromous? opportunistic?)

naup — nauplii

tick — acarines (likely ticks)

a sound way to differentiate and identify larvae. Time and growth will reduce the size of the yolk and there will be development of eyes and the jaw structure, but the key features distinguishing the five credible types remain clear during the short time spent in fresh water. We thus have ample reason to accept that the larval types are different species, even though we have not trued them all.

In addition to the five credible types based on two or more clear differences, one questionable sub-type distinction exists based on a single difference that is extremely subtle. Within Type F is a subtle difference between sub-types designated Fyg (most of F) and Frb. The difference is a brownish reddish tint to the otherwise yellow-green pigment, and a possibly slower reaction to the anaesthetic 2-phenoxyethanol. Frb is either a type, or a subtype, or a developmental error, or an artefact of lighting conditions, etc.; we cannot yet be sure which. The subtleness of the difference makes it very difficult to quickly and reliably distinguish Frb from Fyg, which is why the data are not good enough to analyse it separately.

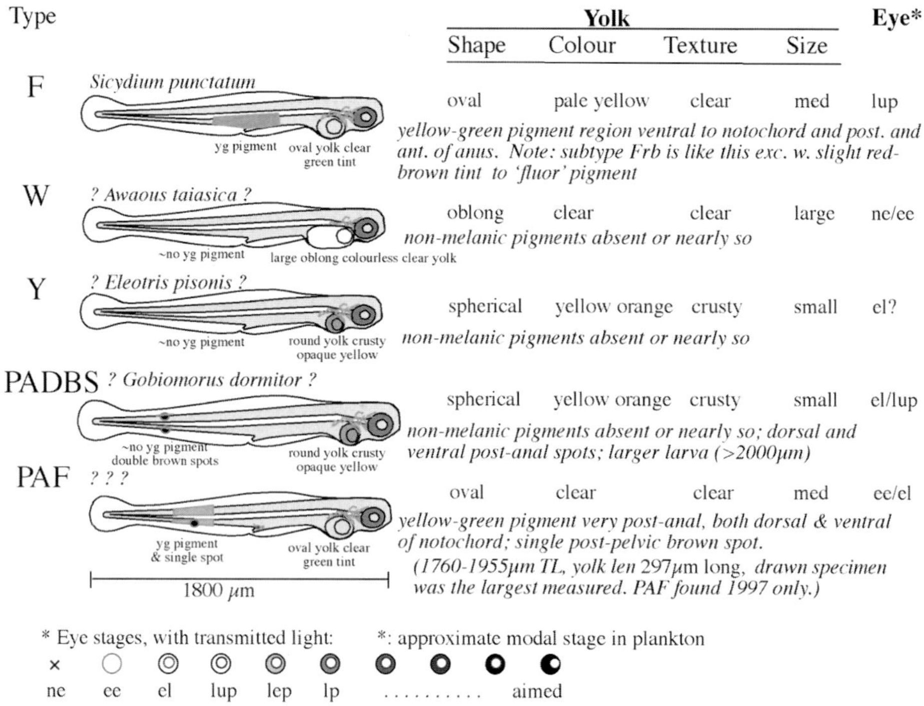

Figure 4. Characteristics of larval types recognised from Dominica, W.I. Type F corresponds to *Sicydium punctatum*. Circumstantial evidence prompts tentative linking of other types with species named between question marks. The yellow green pigment is sometimes referred to in notes as 'fluor' because its colour is like flourescein, although it does not fluoresce. PAF has a single brown spot, instead of the 2 in Ppdbs, and a very different pattern of 'fluor' pigment compared to F.

If the Frb sub-type were a good type, i.e. the sole match to a single species, the most plausible candidate would be a species in the same genus, which would be *Sicydium antillarum*; if not, the next most similar type seems to be PAF, but the low incidence of PAF (not seen prior to 1997 and even then uncommon) would be inconsistent with the presence of *Sicydium antillarum* at 1% to 5% of returning recruits (entering freshwater as postlarvae). Here I treat the Fyg/Frb distinction as insufficient to support a type, but the issue remains until the types are all accounted for by species.

The known numbers of credible larval types and of goby species match identically (unless subtype Frb is a good type).

The number of species recognised as adults may not be complete; it could be that a very rare larval type could represent a species that only occasionally occurs in Dominica, or that is rare enough to appear as larvae only once in a few hundred samples. Alternatively, there have been seen some adult morphs which look like *S. punctatum* but with a completely different, disorganized, colour pattern; they show a blotchy pattern of yellow/orange and brown/black, which interestingly are close to the colours of *Lentipes concolor*. These are very rare, and have been seen as postlarvae and as adults, in more than one river, and spanning decades; we do not know whether they are variants or hybrids. Species hybrids could generate unique larvae as rare as the hybrids. Or, if rare genetic colour morphs exist within a species, then it is possible that they might contribute larvae that show subtle variation from the normal species type.

Candidate species to match the types (see Fig. 4) are *Sicydium punctatum*, *S. antillarum*, *Eleotris pisonis*, *Awaous taiasica* [sensu Brockmann (1965) but disputed by Helen Larson (pers. comm.)], and *Gobiomorus dormitor* (very rare, only ~5 adults seen).

The present work was done 1989–1991 and 1997. Genetic matching of larvae and adults could match larvae with species, and Lindstrom (1999) has since done that nicely for several Hawaiian species. Captive spawnings of the other species present could also definitively resolve this issue. The credible types already permit a more interesting analysis of the goby larvae in the stream drift.

Temporal cycles of abundance

All but three taxonomic groupings show higher adjusted multiple R^2 when analysed as $\ln(1+N/m^3)$ as opposed to N/m^3, or concentration. Even those three exceptions were not markedly better as number/m^3, and therefore for consistency all taxonomic groupings are analysed as $\ln(1+N/m^3)$.

Many of us feel somewhat uneasy about transformations, and about interpreting the results, and would intuitively prefer to use un-transformed data, even at the cost of a slightly lower R^2 and reduced significance. However, there are hazards to not using the transformation, because certain aspects of the pattern may be exaggerated due to the distribution (i.e. non-normal distribution) of the data and consequently (and this rather than the distribution of data is what matters) the distribution of the residuals. Plankton data tend to be somewhat log-normally distributed, and these are no exception. Least-squares regression assumes normally-distributed residuals, and if data are not normal a few high values can distort a regression. For example, the mesor in a periodic regression establishes the central value to which all cyclic effects add variation (the net effect of a cycle over an entire period is zero). Mesors from a log regression are obviously in log scale, so, comparing log and untransformed regressions, the mesor of the regression of all goby larvae using $\ln(1+N/m^3)$ vs. the first and second harmonics of seasonal, lunar, and diurnal cycles yields a mesor of 3.903, which (exp(y)–1) is 48 larvae per cubic meter, whereas if we use the untransformed variable (N/m^3), the mesor is 298 larvae per cubic meter. But is 298 larvae the median, or modal number of larvae/m^3 found in the samples? No. The median is about 13, even the 75th percentile is still only 40, the arithmetic mean is 94, and the mode is whatever interval includes zero. A median value of 298 larvae/m^3 is simply inconsistent with the data. Therefore the 'untransformed' model does not fit, and we are compelled to use the transformation.

Some of the cycles in Figs. 5, 6, and 7 dip into negative values. (This happens, by the way, in the un-transformed regressions as well.) There is no such thing as a negative larva, so how do we interpret these? Negative values simply reflect the fact that periodic regression decomposes temporal trends into a number of specified cycles that are symmetrical, so, just as a straight-line function can dip into negative values if extended into a region of x where the expected values of y are zero (if actual values are not zero, this is either a result of a limitation of the model or the result of scatter around the expected value), a periodic regression is no different. We could ask the regression to replace those negative values with zeroes, but that is cumbersome and we can do it by eye just as well. Remember that, just as a straight line is the most parsimonious non-cyclic regression, a sine wave is the most parsimonious repeating pattern, because it is analogous to the intersection of a cylinder by a flat plane. Elimination those negative values would require defining a non-flat plane to intersect the cylinder, and the model would quickly gain extra terms (at least one per cycle, and probably more).

Periodic regression results are summarised in Tables 2 and 3. Residuals (Fig. 8) are acceptably close to normally distributed for all the regressions, with a few outliers that were not deleted from the analysis (other than their values, there was no justification to delete them). The model cycles indicated by the regressions are presented in Figs. 5, 6, and 7 (they are calculated from the mesor, the linear time term, and the coefficients for the first and second harmonics of the cycle being presented, and omitting all other cycles).

One has to be impressed by the degree to which variation can be explained by variables that we typically do not use. In the case of F, the larval type that is confirmed to be produced by *Sicydium punctatum*, the variation explained by time (periodic and linear terms) is 72%. Given the well-known variability of plankton data, this is little short of phenomenal.

All of the OTUs produce periodic regressions that are highly significant (2 at $p < 0.01$, the rest at $p < 0.001$). With random data, $p < 0.001$ would occur only in 1 out of 1,000 regressions. Many of the cycles represented in the regressions are also significant. Clearly, these regressions are not due to chance.

Table 2. **Summary of regressions:** abundance (as $\ln(1+N/m^3)$) of each OTU in terms of 3 cycles (seasonal, lunar, and diurnal) including their first and second harmonics, as well as a linear term, B(ndoy) to account for any long-term trend that might appear monotonic over this time scale. Each column represents one taxon and one regression. The table sections are: *Adjusted Multiple R-squared* which is a conservative estimate of the variation explained by the regression; *Probabilities*, for which the first row is the *p*-value for the overall regression and the next six rows refer to periodic components (under "*cycle*" and harmonic "***H***", 1st or 2nd) as noted; *Peak locations* give for each cycle the times of the peak ("peak") for each first harmonic, and the first peak ("peak1") for the second harmonic (the second peak of the second harmonic will be later by half a period of the main cycle); *Amplitudes*, which indicate the size of the effect of each cycle and harmonic; *Mesor* which is the mean value around which each cycle adds variation; and finally *B(ndoy)* which estimates the monotonic trend explained by a linear time term. Probabilities italicised are those that are significant at $p < 0.05$ or better; corresponding amplitudes are also italicised.

TABLE of p, amplitude, peaks

	OTU		anad. GL	anad. F	anad. Y	anad. W	anad. SHRall	anad. SHRA	anad. MOLL	non-CADD	non-MAYFL	non-CALA	non-NAUP	non-ACAR
Adj.Mult.Rsq:			0.382	0.722	0.403	0.53	0.432	0.288	0.375	0.37	0.223	0.142	0.295	0.147
Probabilities:			*<0.001*	*<0.001*	*<0.001*	*<0.001*	*<0.001*	*<0.001*	*<0.001*	*<0.001*	*<0.001*	*0.006*	*<0.001*	*0.004*
	cycle	***H***												
p	annual	**1**	*<0.001*	*<0.001*	0.588	0.069	*0.001*	*0.002*	*<0.001*	*0.036*	*0.006*	0.329	0.641	*0.015*
p		**2**	*0.002*	*0.021*	0.206	0.417	0.151	0.561	*0.016*	*0.003*	*0.032*	0.227	*0.005*	0.939
p	diurnal	**1**	*<0.001*	*<0.001*	*0.004*	*<0.001*	*<0.001*	*<0.001*	0.541	*0.028*	0.367	0.079	0.152	0.234
p		**2**	*0.006*	*0.018*	0.115	*0.011*	*<0.001*	*0.025*	0.314	*0.024*	0.502	*0.015*	*0.047*	0.179
p	lunar	**1**	*<0.001*	0.06	0.554	0.301	*0.001*	0.654	0.077	0.231	0.075	*0.041*	0.191	0.205
p		**2**	0.34	0.344	0.557	0.608	0.177	0.608	*0.005*	*0.007*	*0.023*	0.575	0.394	0.417
Peak locations														
peak	annual	**1**	jun06	feb24	jan17	aug30	jul09	jul10	jun18	feb18	apr07	sep09	sep04	may28
peak1	*365d*	**2**	may26	apr25	apr23	mar03	jun17	may22	jan25	may21	may31	jan10	feb13	mar08
peak	diurnal	**1**	0.90	1.0	0.82	0.78	0.91	0.94	0.04	0.84	0.79	0.12	0.08	0.69
peak1	*0-1*	**2**	0.40	0.44	0.24	0.32	0.43	0.49	0.29	0.37	0.36	0.14	0.09	0.27
peak	lunLQ	**1**	0.54	4.86	6.55	26.3	29.1	27.6	21.7	26.6	25.2	23.6	23.7	26.5
peak1	*29.5d*	**2**	9.4	14.2	1.7	11.3	3.6	6.1	8.2	11.1	8.0	13.7	5.5	8.9
Amplitudes														
	annual	**1**	*1.09*	*2.60*	0.80	1.01	*0.64*	*0.58*	*1.57*	*0.43*	*0.29*	0.29	0.15	*0.37*
		2	*0.60*	*1.41*	1.03	0.54	0.27	0.10	*0.84*	*0.47*	*0.26*	0.41	*0.95*	0.01
	diurnal	**1**	*2.04*	*1.72*	*0.85*	*0.67*	*2.23*	*1.59*	0.48	*0.48*	0.11	1.09	1.18	0.16
		2	*0.99*	*0.70*	0.51	*0.78*	*1.28*	*0.51*	0.45	*0.64*	0.13	*1.45*	*1.25*	0.21
	lunar	**1**	*0.62*	0.48	0.18	0.28	*0.46*	0.07	0.43	0.20	0.20	*0.56*	0.36	0.15
		2	0.2	0.17	0.19	0.08	0.18	0.08	*0.74*	*0.36*	*0.22*	0.17	0.31	0.10
Adj.Mult.Rsq:			0.38	0.72	0.40	0.53	0.43	0.29	0.38	0.37	0.22	0.14	0.30	0.15
Mesor			3.90	2.01	-0.75	0.58	2.21	1.59	3.64	1.30	0.53	2.11	4.43	0.60
B(ndoy)			-0	0	0	0	0	0	0.001	0.001	0	0	-0	0

The fact that these results are so significant may have implications for methodology. Plankton data are notorious for being very variable, so high R-squareds were unexpected (I had expected that 'good' results might have been R-squareds in the region of 0.2–0.3). In part the good fits found for several OTUs here may be an indication that some portion of the unexplainable variation we expect in plankton samples is not intrinsic. Instead, the notorious variability of plankton data may be a result of the normal pattern of plankton work: sample, preserve (with all the attendant loss of salience and identifiability in the sample), count later. Alternatively or as well, the periodic signals in data are perhaps, when ignored, large enough to swamp most other signals. Whereas the live-counting procedure results in data that are evidently very good, and periodic regression readily extracts that signal.

The results also have implications for how we approach something as simple as comparing two species' abundances at one site. If we sample, even with replicates, and find that species 2 is twice as abundant as species 1, that simple conclusion could mislead us because the abundance trends of different species can cross each other as the cycle progresses. For example, if we sampled around day 250, we would tend to find many more type W larvae than type F; but at day 100 the situation is very much reversed. We must accustom ourselves to think of abundance as not a number, but a pattern, just as was found for age-at-recruitment (Bell *et al.*, 1995). This also means we cannot compare abundances between sites without taking temporal pattern into account: even good replicated samples cannot suffice for comparison, unless either taken simultaneously amongst sites or corrected for a known pattern.

Responses of OTUs to annual, lunar and diurnal cycles

Operational taxonomic units (OTUs) are identified in Table 1, and goby larval types are further described in Fig. 4.

Curves (in Figs. 5, 6, and 7) incorporate both the first and second harmonics from the regressions, and can be perceived by inspecting the plots. For example, in Fig. 5: a curve showing a single peak (e.g., SHRA) indicate a dominance by the first harmonic, and curves showing two near-equal peaks (e.g., Naup) indicate dominance by the second harmonic, while curves showing two unequal peaks or a moderated peak (e.g., Moll) can result from an equal role of both.

Bear in mind that significance is harder to achieve with small amounts of data, as for the rarer OTUs, and the significance is not due to amplitude of the curve but the extent to which data are scattered about it. Also, although I usually disdain even talking about results that are not statistically significant, I may be slightly Bayesian and treat the curves/regressions as the best information available, and the presence of the exact information in the Tables should prevent me from misleading you. Note also that statistical significance I report for each cycle or harmonic is conservative: it is the lesser of the two p-values of the cycle's two components (sine and cosine), whereas the true probability for the cycle should be even smaller than that, but as yet I have no convenient way to combine them into a single, accurate, joint probability.

Discussion will follow Figs. 5, 6, 7, and Table 2. Residual plots are shown in Fig. 8. Responses to cycles vary, and indicate considerable diversity amongst OTUs—even amongst goby larvae.

Considering the anadromous taxa (fish, decapod shrimps, neritid molluscs), all but goby larval types Y and W (which hint at a response) show a significant response to the seasonal cycle, and all but neritid molluscs respond to the diurnal cycle. Regarding Y and W, W shows a marginal ($p = 0.07$) response on the first seasonal harmonic and the first peak of its (n.s.) second harmonic is close to; but Y, which shows neither harmonic as significant, produces a curve that is qualitatively similar to, but at a lower level than, Type F.

All goby larval types show a significant response to the primary diurnal cycle, but only type F responds significantly to the seasonal cycle and none to the lunar cycle. The lack of lunar response is interesting given that as postlarvae these species are such an example of lunar response: recruitment (of virtually all the anadromous taxa here) is timed to the 4th day following the last lunar quarter day. The aggregate of all goby larval types shows a response to the lunar cycle; but given that they individually decline to respond, this is more of an illustration of the hazard inherent in combining groups together.

Type F shows its strongest response to the annual and diurnal cycles (first harmonics of both); there is peak near day 90 of the year, a peak late at night, and very little variation on the lunar cycle.

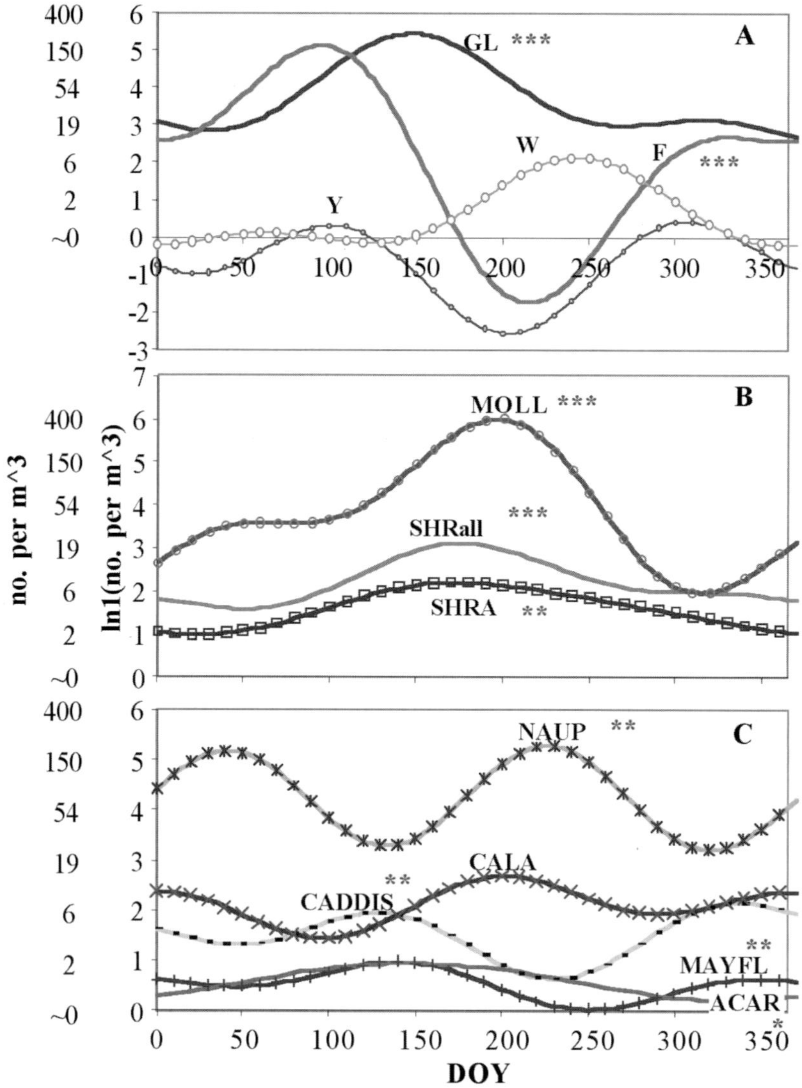

Figure 5. Seasonal cycles, based on regression using ln(1+N/m3). Each curve combines both first and second harmonics, and includes the mesor and the linear time term. A: anadromous gobies, by larval type; **B**: anadromous decapod crustaceans and neritid molluscs; **C**: nonanadromous taxa. Significance of each cycle (the more significant harmonic is given) indicated by *, **, and *** for $p<0.05$, 0.01, 0.001 respectively (applies to following figures also). Points are fitted values, i.e., x-values at which y is evaluated using the regression function.

Type Y and type W show a significant response only to the diurnal cycle, not to the lunar or seasonal (marginal for W) cycles.

Amongst goby larvae, types F and Y both have annual peaks near day 100, but all have or hint at a response on the second harmonic and therefore two peaks per year, the first being near Feb–Mar (Tables 2 and 3) and the second peak near day 300 (the second peak if calculated from the second harmonic alone, would always be 0.5 cycle later than the first, but that does not apply when the peaks result from two interacting cycles). Type W has a major peak nearly 150 days out of phase from the

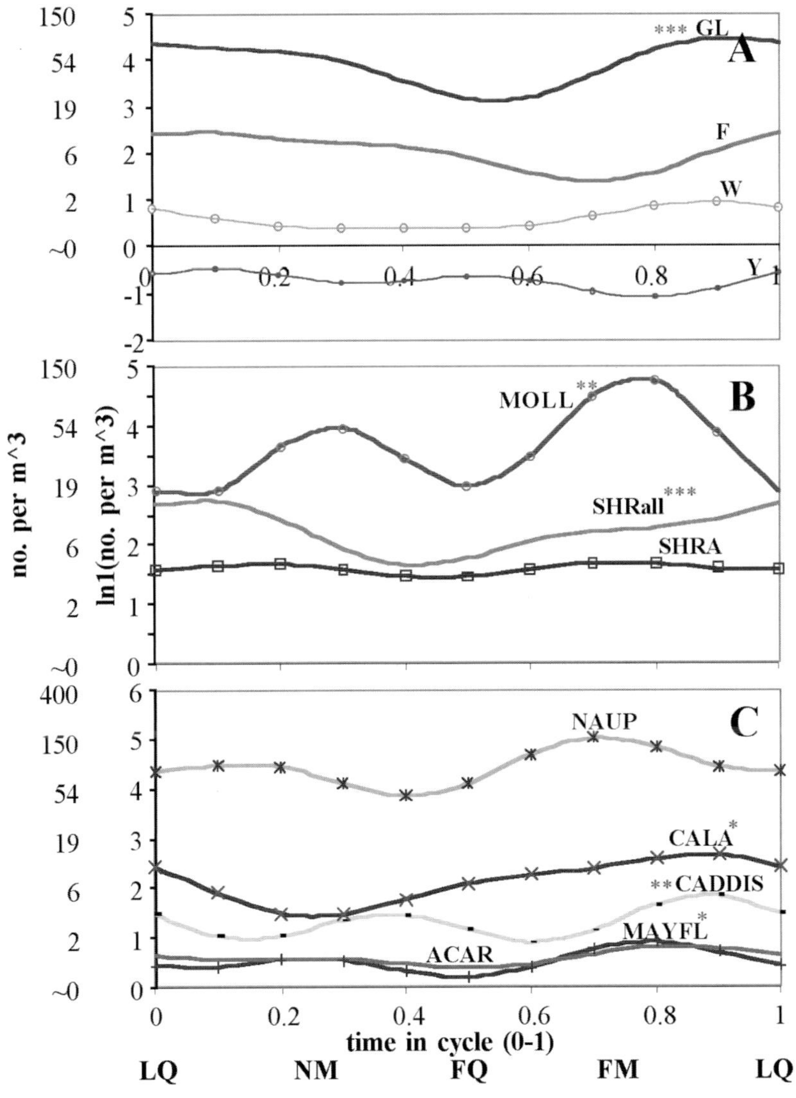

Figure 6. Lunar cycles, based on regression using ln(1+N/m3). Each curve combines both first and second harmonics. A: anadromous gobies, by larval type; **B**: anadromous decapod crustaceans and neritid molluscs; **C**: nonanadromous taxa.

day-100 peaks of F and Y, and a barely perceptible peak near day 70. These are descriptions a long way from being explained, however we can notice that although the second harmonic is not significant in all, it results in a first peak about the same time that water temperature (Fig. 9) shows its first peak. In regard to a possible temperature relationship, *Sicydium punctatum* from Dominica seem intolerant of temperatures as low as about 18 °C (pers. observ.); this was shown in an aquarium in Newfoundland that cooled during a power failure. At about 18 °C temperature they became torpid and unreactive, but when the temperature was brought back to 20 °C they resumed normal activity.

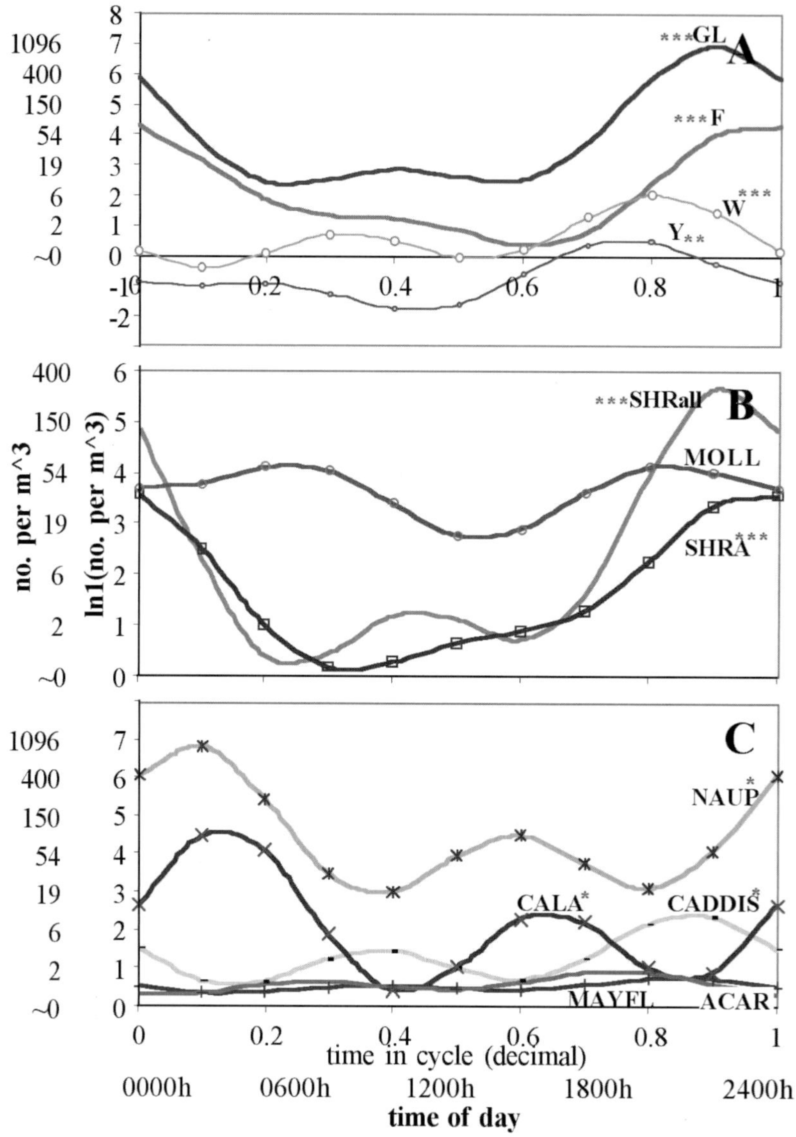

Figure 7. Diurnal cycles, based on regression using ln(1+N/m3). Each curve combines both first and second harmonics. **A**: anadromous gobies, by larval type; **B**: anadromous decapod crustaceans and neritid molluscs; **C**: nonanadromous taxa.

The temperatures I observed in Dominica during my study ranged from extremes of 20 °C to 30 °C, virtually exactly. Thus it is plausible that goby reproduction could respond to temperature.

In the non-goby OTUs I cannot see a similar response. Seasonally, calanoids and nauplii have similar patterns slightly out of phase, but they are nearly inverse to the patterns of caddis- and mayflies (similar to each other). Acarines show a weakly significant seasonal response on the first harmonic (only one peak).

Lunar cycles are not shown by individual goby OTUs; but are shown by mollusc (larvae of neritid snails) and decapod shrimps. Nauplii show a strong response to the second harmonic of year, with peaks near days 40 and 230 (Fig. 5), but only a small response to the first harmonic. A strong

Table 3. Summary and interpretation of periodic regressions for 13 operational taxonomic units (OTUs) or groupings. Data are from samples taken at lower reaches of the Layou, Check Hall, Canefield and Roseau rivers in Dominica, W.I. 1989–1991 and 1997. To fit the table, "sine" and "cosine" are sometimes abbreviated "s" and "c", and the meaning is the proper sine or cosine, i.e. after conversion to conventional angular units. B0, the intercept of common parlance, is called the mesor to reflect its properties in a periodic regression. Each regression has the form Y = B0 + B1*X1 + B2*X2 + ... B(n)*X(n), where Y is the ln(1+x) transform of counts/m^3 of each OTU, the Bs are the regression coefficients and the Xs represent the independent variables. The variables are named in the table header and their relation to a natural cycle is given. Peaks are calculated for each cycle, and in the case of the annual cycle are given as dates. For second harmonics, because they peak twice in their 'parent' cycles, the second peak is given in a second row. Adjusted Multiple R^2 (AMRsq) is a more conservative indicator of fit than R^2.

	B0	B1	B2	B3	B4	B5	B6	B7	B8	B9	B10	B11	B12	B13
CYCLE NAME:		na	**ANNUAL**		**SEMI-ANNUAL**		**DAILY**		**SEMI-DAILY**		**LUNAR**		**SEMI-LUNAR**	
Variables:	B0	NDOY	sDOY	cDOY	s2DOY	c2DOY	sTOD	cTOD	s2TOD	c2TOD	sLQ	cLQ	s2LQ	c2LQ
transform:	B0	NDOY	sin	cos	sin	cos	sin	cos	sin	cos	sin	cos	sin	cos
harmonic:			first		second		first		second		first		second	
period:			365	days	182.5	days	24	hours	12	hours	29.5	days	14.75	days
			for each OTU N, AMRsq and p are given; then Amplitude, Peak(s) follow under cycle headings above, then coefficients and p under variable headings above.											
ANADROMOUS: FISH														
OTU:		N	AMRsq	p										
GL		165	0.382	<0.001			**All goby larvae**							
amplitude (in units of Y)			1.0931		0.6028		2.0374		0.991		0.616		0.199	
Peak in cycle units			156.45	jun06	145.23	may26	21.545		9.6566		0.5347		9.368	
If 2nd harmonic, second peak:					327.73	nov24			21.657				24.12	
coeffs:	3.903	-0.001	0.474	-0.985	-0.578	0.171	-1.221	1.631	-0.933	0.334	0.07	0.612	-0.15	-0.132
p	<0.001	0.001	0.01	<0.001	0.002	0.364	<0.001	0.003	0.006	0.261	0.676	<0.001	0.34	0.429
OTU:		N	AMRsq	p										
F		62	0.722	<0.001			**Goby larvae of type F (*S. punctatum*)**							
amplitude (in units of Y)			2.605		1.4135		1.721		0.7027		0.4785		0.165	
Peak in cycle units			54.203	feb24	114.02	apr25	23.993		10.665		4.8512		14.15	
If 2nd harmonic, second peak:					296.52	oct24			22.665				28.90	
coeffs:	2.007	0	2.093	1.551	-0.998	-1.001	-0.003	1.721	-0.452	0.538	0.411	0.245	-0.042	0.16
p	0.009	0.005	<0.001	0.121	0.109	0.021	0.988	<0.001	0.162	0.018	0.06	0.289	0.804	0.344

Table 3 (continued) ...

	B0	B1	B2	B3	B4	B5	B6	B7	B8	B9	B10	B11	B12	B13
CYCLE NAME:		na	**ANNUAL**		**SEMI-ANNUAL**		**DAILY**		**SEMI-DAILY**		**LUNAR**		**SEMI-LUNAR**	
Variables:	B0	NDOY	sDOY	cDOY	s2DOY	c2DOY	sTOD	cTOD	s2TOD	c2TOD	sLQ	cLQ	s2LQ	c2LQ
transform:	B0	NDOY	sin	cos	sin	cos	sin	cos	sin	cos	sin	cos	sin	cos
harmonic:			first		second		first		second		first		second	
period:			365	days	182.5	days	24	hours	12	hours	29.5	days	14.75	days

for each OTU N, AMRsq and p are given; then Amplitude, Peak(s) follow under cycle headings above, then coefficients and p under variable headings above.

	B0	B1	B2	B3	B4	B5	B6	B7	B8	B9	B10	B11	B12	B13
ANADROMOUS: FISH														
OTU:		N	AMRsq	p										
Y		62	0.403	<0.001			**Goby larvae of type Y**							
amplitude (in units of Y)			0.7962		1.033		0.8552		0.5085		0.1828		0.185	
Peak in cycle units			17.33	jan17	112.61	apr23	19.581		5.6296		6.549		1.673	
If 2nd harmonic, second peak:					295.11	oct23			17.63				16.42	
coeffs:	-0.747	0	0.234	0.761	-0.693	-0.766	-0.783	0.344	0.098	-0.499	0.18	0.032	0.121	0.14
p	0.477	0.076	0.746	0.588	0.427	0.206	0.004	0.51	0.83	0.115	0.554	0.921	0.611	0.557
OTU:		N	AMRsq	p										
W		62	0.530	<0.001			**Goby larvae of type W**							
amplitude (in units of Y)			1.0129		0.5363		0.6648		0.7775		0.2822		0.08	
Peak in cycle units			241.96	aug30	61.517	mar03	18.623		7.5608		26.296		11.33	
If 2nd harmonic, second peak:					244.02	sep02			19.561				26.08	
coeffs:	0.583	0	-0.865	-0.527	0.458	-0.279	-0.656	0.108	-0.567	-0.532	-0.178	0.219	-0.079	0.009
p	0.391	0.223	0.069	0.562	0.417	0.473	<0.001	0.749	0.058	0.011	0.368	0.301	0.608	0.951
ANADROMOUS: NON-FISH														
OTU:		N	AMRsq	p										
ShrimpALL		162	0.432	<0.001			**Decapod shrimp larvae (all 3 recognised groupings: A,B,C)**							
amplitude (in units of Y)			0.6429		0.2701		2.2319		1.2751		0.4547		0.182	
Peak in cycle units			189.66	jul09	167.3	jun17	21.773		10.251		29.097		3.597	
If 2nd harmonic, second peak:					349.8	dec16			22.251				18.35	
coeffs:	2.213	0	-0.079	-0.638	-0.135	0.234	-1.229	1.863	-1.011	0.777	-0.039	0.453	0.182	0.007
p	<0.001	0.836	0.609	0.001	0.399	0.151	<0.001	<0.001	<0.001	0.003	0.784	0.001	0.177	0.964

Table 3 (continued) ...

	B0	B1	B2	B3	B4	B5	B6	B7	B8	B9	B10	B11	B12	B13
CYCLE NAME:		na	**ANNUAL**		**SEMI-ANNUAL**		**DAILY**		**SEMI-DAILY**		**LUNAR**		**SEMI-LUNAR**	
Variables:	B0	NDOY	sDOY	cDOY	s2DOY	c2DOY	sTOD	cTOD	s2TOD	c2TOD	sLQ	cLQ	s2LQ	c2LQ
transform:	B0	NDOY	sin	cos	sin	cos	sin	cos	sin	cos	sin	cos	sin	cos
harmonic:			first		second		first		second		first		second	
period:			365	days	182.5	days	24	hours	12	hours	29.5	days	14.75	days
			for each OTU N, AMRsq and p are given; then Amplitude, Peak(s) follow under cycle headings above, then coefficients and p under variable headings above.											
ANADROMOUS: NON-FISH														
OTU:		N	AMRsq	p										
ShrimpA		138	0.288	<0.001			**Decapod shrimp larvae type A only**							
amplitude (in units of Y)			0.5832		0.0952		1.5868		0.5125		0.0654		0.083	
Peak in cycle units			191	jul10	141.47	may22	22.456		11.776		27.58		6.097	
If 2nd harmonic, second peak:					323.97	nov20			23.776				20.85	
coeffs:	1.593	0	-0.085	-0.577	-0.094	0.015	-0.624	1.459	-0.06	0.509	-0.026	0.06	0.043	-0.071
p	<0.001	0.281	0.579	0.002	0.561	0.924	<0.001	0.001	0.823	0.025	0.843	0.654	0.743	0.608
OTU:		N	AMRsq	p										
MOLL		133	0.375	<0.001			**Neritid mollusc larvae (could be 2 indistinguishable spp.)**							
amplitude (in units of Y)			1.5668		0.8388		0.4786		0.4478		0.432		0.745	
Peak in cycle units			168.42	jun18	25.805	jan25	0.8777		6.8321		21.679		8.162	
If 2nd harmonic, second peak:					208.3	jul28			18.832				22.91	
coeffs:	3.641	0.001	0.376	-1.521	0.651	0.529	0.109	0.466	-0.189	-0.406	-0.43	-0.041	-0.245	-0.703
p	<0.001	0.048	0.155	<0.001	0.016	0.055	0.737	0.541	0.697	0.314	0.077	0.864	0.297	0.005
Arthopoda (non-anadromous)														
OTU:		N	AMRsq	p										
CADD		155	0.370	<0.001			**Caddisfly larvae**							
amplitude (in units of Y)			0.4264		0.4728		0.4837		0.6403		0.1962		0.356	
Peak in cycle units			48.902	feb18	140.01	may21	20.253		8.8776		26.56		11.12	
If 2nd harmonic, second peak:					322.51	nov19			20.878				25.87	
coeffs:	1.302	0.001	0.318	0.284	-0.47	0.051	-0.402	0.269	-0.639	-0.041	-0.115	0.159	-0.356	0.009
p	<0.001	<0.001	0.036	0.128	0.003	0.74	0.028	0.548	0.024	0.866	0.403	0.231	0.007	0.947

Table 3 (continued) ...

	B0	B1	B2	B3	B4	B5	B6	B7	B8	B9	B10	B11	B12	B13
CYCLE NAME:		na	**ANNUAL**		**SEMI-ANNUAL**		**DAILY**		**SEMI-DAILY**		**LUNAR**		**SEMI-LUNAR**	
Variables:	B0	NDOY	sDOY	cDOY	s2DOY	c2DOY	sTOD	cTOD	s2TOD	c2TOD	sLQ	cLQ	s2LQ	c2LQ
transform:	B0	NDOY	sin	cos	sin	cos	sin	cos	sin	cos	sin	cos	sin	cos
harmonic:			first		second		first		second		first		second	
period:			365	days	182.5	days	24	hours	12	hours	29.5	days	14.75	days
			for each OTU N, AMRsq and p are given; then Amplitude, Peak(s) follow under cycle headings above, then coefficients and p under variable headings above.											
Arthopoda (non-anadromous)														
OTU:		N	AMRsq	p										
MAYF		130	0.223	<0.001			**Mayfly larvae**							
amplitude (in units of Y)			0.2854		0.2554		0.114		0.1255		0.1946		0.215	
Peak in cycle units			96.959	apr07	150.95	may31	18.879		8.617		25.214		7.982	
If 2nd harmonic, second peak:					333.45	nov30			20.617				22.73	
coeffs:	0.525	0	0.284	-0.028	-0.226	0.119	-0.111	0.026	-0.123	-0.025	-0.154	0.119	-0.055	-0.208
p	0.012	0.067	0.006	0.809	0.032	0.246	0.367	0.925	0.502	0.861	0.075	0.184	0.52	0.023
OTU:		N	AMRsq	p										
ACAR118		0.147	0.004			**Acarines: ticks (or mites)**								
amplitude (in units of Y)			0.3664		0.012		0.1627		0.2093		0.151		0.101	
Peak in cycle units			147.81	may28	66.731	mar08	16.481		6.3859		26.541		8.923	
If 2nd harmonic, second peak:					249.23	sep07			18.386				23.67	
coeffs:	0.59	0	0.206	-0.303	0.009	-0.008	-0.15	-0.063	-0.042	-0.205	-0.089	0.122	-0.062	-0.08
p	0.006	0.002	0.042	0.015	0.939	0.944	0.234	0.825	0.826	0.179	0.33	0.205	0.51	0.417
Possible marine association														
OTU:		N	AMRsq	p										
CALA		115	0.142	0.006			**Calanoid copepods**							
amplitude (in units of Y)			0.2919		0.4114		1.0945		1.4546		0.5608		0.168	
Peak in cycle units			251.09	sep09	10.462	jan10	2.7704		3.3299		23.554		13.68	
If 2nd harmonic, second peak:					192.96	jul12			15.33				28.43	
coeffs:	2.105	0	-0.27	-0.111	0.145	0.385	0.726	0.819	1.433	-0.25	-0.535	0.168	-0.074	0.151
p	0.001	0.156	0.329	0.75	0.676	0.227	0.079	0.329	0.015	0.561	0.041	0.555	0.776	0.575
OTU:		N	AMRsq	p										
NAUP		140	0.295	<0.001			**Nauplii (could be of many families)**							
amplitude (in units of Y)			0.1529		0.9445		1.1789		1.2543		0.3618		0.309	
Peak in cycle units			246.99	sep04	43.501	feb13	1.8101		2.0716		23.685		5.451	
If 2nd harmonic, second peak:					226	aug15			14.072				20.20	
coeffs:	4.427	-0.001	-0.137	-0.068	0.942	0.069	0.538	1.049	1.109	0.586	-0.342	0.118	0.226	-0.211
p	<0.001	0.001	0.641	0.848	0.005	0.824	0.152	0.225	0.047	0.203	0.191	0.66	0.394	0.436

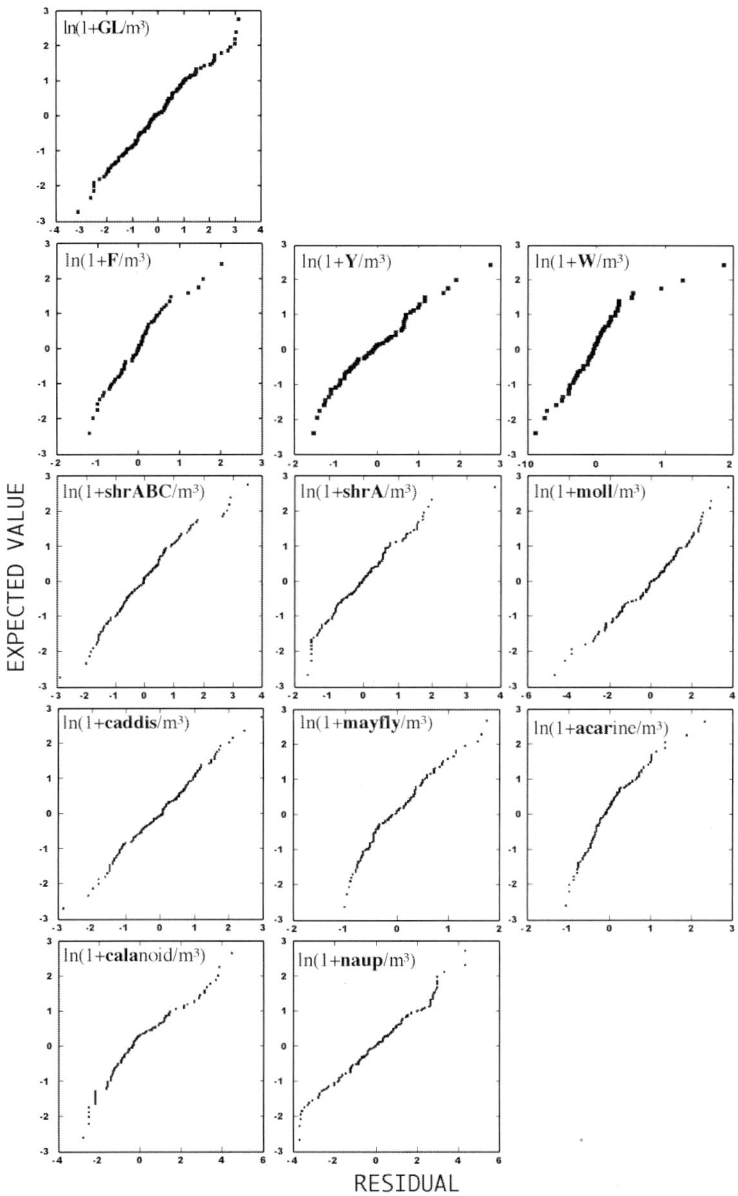

Figure 8. Residuals from regressions of ln(1+N/m3), according to Eq. 1 for each Operational Taxonomic Unit. The upper plot (GL) is an aggregate of all goby larvae including types having insufficient data for separate analysis; the second row are three goby larval types (F, Y, W); the third row are non-fish anadromous taxa (ShrimpABC subsumes three decapod larval types; ShrimpA is about 2mm long with cephalothorax not markedly larger than the set of abdominal segments, Shrimps B and C have insufficient data to analyse separately); the fourth row are insect larvae (caddis and mayfly) and acarines (which look more like ticks than mites); the fifth row is calanoid copepods and nauplii (which may be associated, or not).

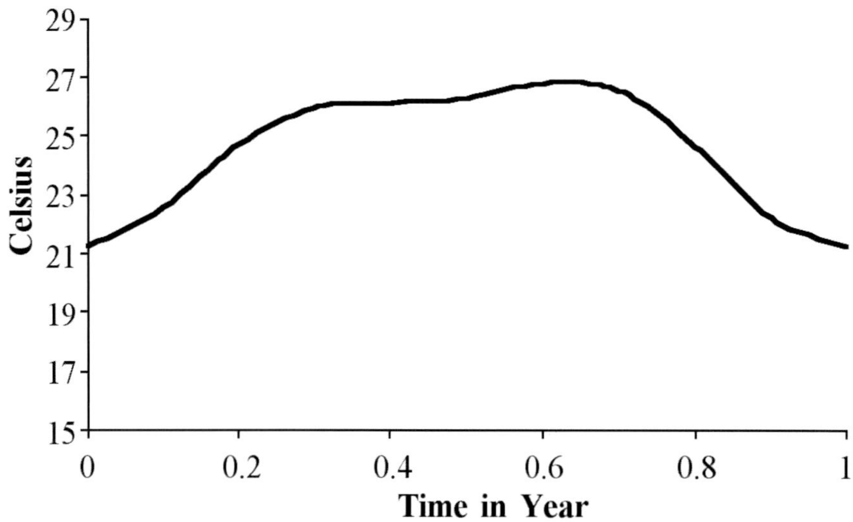

Figure 9. Seasonal water temperature pattern for same group of stations [on the lower Layou, Check Hall (CHR), Canefield (PHW), Roseau Rivers] as used for rheoplankton data in this paper.

semilunar response (Fig. 6) that hints at a response linked to tides; although in Dominica for a number of reasons the tides are very small, so the result raises the interesting question about what mechanism could be behind that association, and how many species of nauplii are subsumed here. The tidal cycle precesses against days, so it would be a mistake to think the strong response to time of day (Fig. 7) is related to tides.

In situ mortality

The mortalities reported by Bell (1994) for *S. punctatum* (identified as Type F in Tables and Figures) in Dominica (West Indies) exceed 50% per hour in the drift. Typical drift speeds are 0.3 m/s, or about 1 km/h, so the mortalities are on the order of 50% per kilometer as well. Goby larval types Y and W show even lower survival.

Conclusions

Stream drift study offers many advantages as a standard monitoring and investigative tool. The interesting general conclusion about cycles is that there is no standard response shown by all taxa. That the cycle is not stereotyped means they have to be characterised for each taxon. There are two main conservation-relevant reasons for investigating and analysing cycles. Firstly, cycles are much more likely to be present than absent. Any single sample, or group of samples that do not permit a cycle to be estimated, is by definition incapable of being reliably used to estimate total production over a cycle. Cycles therefore need to be characterised in order to make the best use of the sample data. Secondly, analysis of cycles can help identify anomalies. Anomalies are the indicators of the effect of unacknowledged factors, and can reveal short-term or spatially-limited conditions that affect production of larvae.

Stream drift study provides superior production data on in a much less invasive way than other methods and can be paired to develop *in-situ* mortality estimates. I raise also the possibility that by knowing the development rates of certain larval features (e.g. eye pigmentation, jaw development, reduction in oil/yolk) in fresh water, each larva can be given an approximate age (in hours); with sufficient data, mortality can be estimated from the descending limb of a histogram of abundances.

The high mortality rates found in Dominica (Bell, 1994, unpubl. data) are at present the only direct mortality estimates we have for drifting goby larvae anywhere. Whatever the mortality rate, it is cumulative, and, unless much lower than that found in Dominica, the strong implication is that fish at a distance from the sea are, egg for egg, at a substantial disadvantage. That raises the question of why large fish are found inland at all. Are they there because they were excluded from lower elevations? If there seem to be fewer fish at lower elevations, is it [a] because adults preferentially moved upstream and were not replaced, or [b] because they were lost (adults appear to be slow-growing and populations may take some time to recover) to the effects of anthropogenic disturbance? A drive to move ever further inland would need further explanation as an evolved behaviour if the consequence is exponentially reduced larval survival. The unavoidable conservation implication of mortality, because it is cumulative, is that, unless compensated for by greater egg production, the nesting habitats with the shortest drift times are the most important for the population.

Acknowledgements

I thank the fishes of Dominica that treated me well, the IDRC for the Young Canadian Researchers Award that funded the work in Dominica, Nigel Lawrence and the Fisheries Division of Dominica for their hospitality, J.A. Brown, and fellowships at Memorial University of Newfoundland. I also thank many friends and colleagues for support and challenge, amongst whom are: Pierre Pepin, W.G. Warren, Don Erdman, John D. Archbold, Martina Kelly, the Deorksen Institute, Horst Kaiser, Tony Booth, David Schneider, and the helpful comments of an anonymous reviewer.

Literature Cited

Allan, J.D. & E. Russek. 1985. The quantification of stream drift. *Canadian Journal of Fisheries & Aquatic Sciences* **42**: 210–215.

Atwood, T. 1791. *The history of the island of Dominica. Containing a description of its situation, extent, climate, mountains, rivers, natural productions, &c. &c. Together with an account of the civil government, trade, laws, customs, and manners, of the different inhabitants of that island. Its conquest by the French, and restoration to the British dominions*. J. Johnson, London. vii + 285 p.

Batschelet, E. 1981. *Circular statistics in biology*. Academic Press, London. xvi + 371 p.

Bell, K.N.I. 1994. Life cycle, early life history, fisheries, and recruitment dynamics of diadromous gobies of Dominica, W.I., emphasising *Sicydium punctatum* Perugia. Ph.D. Thesis, Memorial University of Newfoundland, St. John's. xviii + 275 p.

———. 1997. Complex recruitment dynamics with Doppler-like effects caused by shifts and cycles in age-at-recruitment. *Canadian Journal of Fisheries and Aquatic Sciences* **54**: 1668–1681.

———. 1999. An overview of goby-fry fisheries. *NAGA* **22**: 30–36.

———. 2004. *Introduction to circular variables & periodic regression in biology*. Razorbill Press, St. John's. ca. 50 p. (electronic book in PDF format).

———. & **J.A. Brown**. 1990. Aspects of the early life history of an anadromous goby in Dominica, W.I. Unpublished poster presented at conference: Ecological and Evolutionary Ethology of Fishes, Northern Arizona University, May 19–23, 1990.

———. & **J.A. Brown**. 1995. Active salinity choice and enhanced swimming endurance in 0 to 8-d-old larvae of diadromous gobies, with emphasis on *Sicydium punctatum* (Pisces), in Dominica, West Indies. *Marine Biology* **121**: 409–417.

———., **P. Pepin & J.A. Brown**. 1995. Seasonal, inverse cycling of length- and age-at-recruitment in the diadromous gobies *Sicydium punctatum* and *Sicydium antillarum* (Pisces) in Dominica, West Indies. *Canadian Journal of Fisheries and Aquatic Sciences* **52**: 1535–1545.

Blanco, G.J. 1956. Assay of the goby fry (ipon) fisheries of the Laoag River and its adjacent marine shores, Ilocos Norte Province. *Philippine Journal of Fisheries* **4**: 31–80.

Bliss, C.I. 1958. Periodic regression in biology and climatology. *Bulletin of the Connecticut Agricultural Experiment Station, New Haven* **615**: 1–55.

Brasher, A.M. 2003. Impacts of human disturbances on biotic communities in Hawaiian streams. *BioScience* **53**: 1052–1060.

Brockmann, F.W. 1965. *A taxonomic study of the Panamanian freshwater gobies of the genera Awaous and Sicydium.* M.S. Thesis, Florida State University, Tallahassee. v + 73 p.

Cambray, J.A. 1985. Early ontogeny of *Labeo capensis* (Pisces:Cyprinidae). *South African Journal of Zoology* **20**: 190–196.

Chace, F.A. & H.H.J. Hobbs. 1969. The freshwater and terrestrial decapod crustaceans of the West Indies with special reference to Dominica. *Bulletin of the United States National Museum* **292**, 258 p.

Copp, G.H. & B. Cellot. 1998. Drift of embryonic and larval fishes, especially *Lepomis gibbosus* (L.), in the upper Rhone River. *Journal of Freshwater Ecology* **4**: 419–424.

Ego, K. 1956. Life history of freshwater gobies. Project No. 4-4-R. Freshwater Game and Fisheries Management Research, Department of Lands and Natural Resources, Honolulu. 23 p.

Erdman, D.S. 1961. Notes on the biology of the gobiid fish *Sicydium plumieri* in Puerto Rico. *Bulletin Marine Sciences of the Gulf and Caribbean* **11**: 448–456.

———. 1986. The green stream goby, *Sicydium plumieri*, in Puerto Rico. *Tropical Fish Hobbyist* **1986** (February): 70–74.

Fitzsimons, J.M. & R.T. Nishimoto. 1990. Territories and site tenacity in males of the Hawaiian stream goby *Lentipes concolor* (Pisces: Gobiidae). *Ichthyological Exploration of Freshwaters* **1**: 185–189.

———., **R.M. Zink & R.T. Nishimoto.** 1990. Genetic variation in the Hawaiian stream goby, *Lentipes concolor*. *Biochemical Systematics and Ecology* **18**: 81–83.

Flecker, A.S., D.C. Taphorn, J.A. Lovell & B.P. Feifarek. 1991. Drift of characin larvae, *Bryconamericus deuterodonoides*, during the dry season from Andean piedmont streams. *Environmental Biology of Fishes* **31**: 197–202.

Herre, A.W. 1927. *Gobies of the Philippines and the China Sea.* Monographs of the Bureau of Science 23. Philippine Bureau of Science, Manila. 352 p.

———. 1958. Marine fishes in Philippine rivers and lakes. *Philippine Journal of Science* **87**: 65–88.

Iguchi, K. & N. Mizuno. 1990. Diel changes in larval drift among amphidromous gobies in Japan, especially *Rhinogobius brunneus*. *Journal of Fish Biology* **37**: 255–264.

———. **& N. Mizuno.** 1991. Mechanisms of embryonic drift in the amphidromous goby *Rhinogobius brunneus*. *Environmental Biology of Fishes* **31**: 295–300.

Jordan, D.S. & B.W. Evermann. 1898. *The fishes of north and middle America.* Bulletin of the United States Natural Museum. No. 47, Part III xxiv + p. 2183–2860. United States National Museum, Washington (reprinted 1963 for the Smithsonian Inst. by T.F.H. Publications Inc., Jersey City, New Jersey).

———. **& B. W. Evermann.** 1905. *The aquatic resources of the Hawaiian Islands.* Bulletin of the United States Fish Commission for 1903, Part I, *The shore fishes* Vol. XXIII. 392 p. United States Fish Commission, Washington.

Kinzie, R.A. 1993. Reproductive biology of an endemic, amphidromous goby *Lentipes concolor* in Hawaiian streams. *Environmental Biology of Fishes* **37**: 257–268.

Kohler, S.L. 1985. Identification of stream drift mechanisms: An experimental and observational approach. *Ecology* **66**: 1749–1761.

Lindstrom, D.P. 1999. Molecular species identification of newly hatched Hawaiian amphidromous gobioid larvae. *Marine Biotechnology* **1**: 167–174.

Manacop, P.R. 1953. The life history and habits of the goby, *Sicyopterus extraneus* Herre (Gobiidae) with an account of the goby-fry fishery of Cagayan River, Oriental Misamis [Province, Mindanao, Philippines]. *Philippine Journal of Fisheries* **2**: 1–60.

Montilla, J. 1931. The ipon fisheries of northern Luzon. *Philippine Journal of Science* **45**: 61–75.

Nishimoto, R.T. & J.M. Fitzsimons. 1986. Courtship, territoriality, and coloration in the endemic Hawaiian freshwater goby, *Lentipes concolor*, p. 811–817. *In*: Uyeno, T., R. Arai, T. Taniuchi & K. Matsuura (eds.), *Indo-Pacific fish biology.* Proceedings of the Second International Conference on Indo-Pacific fishes. Ichthyological Society of Japan, Tokyo.

Noblet, G.P. & R.T. Damian. 1991. On the status of *Biomphalaria* and *Thiara* snails and the threat of schistosomiasis and paragonimiasis on Dominica, West Indies. *Journal of Medical & Applied Malacology* **3**: 1–6.

Pavlov, D.S. 1994. The downstream migration of young fishes in rivers—mechanisms and distribution [English review]. *Folia Zoologica* **43**: 193–208.

Radtke, R.L., R.A. Kinzie & S.D. Folsom. 1988. Age at recruitment of Hawaiian freshwater gobies. *Environmental Biology of Fishes* **23**: 205–213.

Titcomb, M. 1977. *Native use of fish in Hawaii*. 2nd edition. University Press of Hawaii, Honolulu.

Biology of Hawaiian Streams and Estuaries. Edited by N.L. Evenhuis & J.M. Fitzsimons. *Bishop Museum Bulletin in Cultural and Environmental Studies* 3: 63–74 (2007).

Production, Marine Larval Retention or Dispersal, and Recruitment of Amphidromous Hawaiian Gobioids: Issues and Implications

CHERYL A. MURPHY[1]
Department of Oceanography and Coastal Sciences, Energy, Coast and Environment Building, Louisiana State University, Baton Rouge, Louisiana 70803 USA

JAMES H. COWAN, JR.
Department of Oceanography and Coastal Sciences, and Coastal Fisheries Institute, Energy, Coast and Environment Building, Louisiana State University, Baton Rouge, Louisiana, 70803 USA; email: jhcowan@lsu.edu

Abstract

Freshwater habitat alteration can have detrimental effects on amphidromous Hawaiian fishes. Although much information has been collected on adult and post-larval life-stages, there is little information collected on the egg, yolk-sac, and marine larval stages. The focus of this paper is to highlight what is known and what remains to be determined about larval production, mechanisms of retention or dispersal of marine larvae, and factors governing recruitment of post-larvae to freshwater streams. We highlight areas that need further investigation, and suggest how such information would affect management of freshwater habitats and amphidromous Hawaiian fishes.

Introduction

The Hawaiian Archipelago consists of a series of remote volcanic islands in the north Pacific formed relatively recently in geological time (~5.8 million years ago), and is characterized by unique flora and fauna (Carson & Clague, 1995; Funk & Wagner, 1995; McDowell, 2003). The islands are among the most isolated of the central Pacific (Scheltema *et al.*, 1996; McDowell, 2003), and this isolation is unmistakably demonstrated in the high degree of species endemism (Fitzsimons & Nishimoto, 1990; Hourigan & Reese, 1987). Of the greater than 500 species of Hawaiian marine fishes, a little less than one third are endemic to the Hawaiian Islands, and of the 5 extant species of freshwater fishes, 4 are found only in Hawai'i (Fitzsimons & Nishimoto, 1990; Hourigan & Reese, 1987; Randall, 1998). All these indigenous freshwater fish species are amphidromous and spend most of their lives in freshwater. Three gobies, *Lentipes concolor*, *Sicyopterus stimpsoni*, *Stenogobius hawaiiensis*, and an eleotrid *Eleotris sandwicensis* are endemic, whereas a fourth goby *Awaous guamensis* is panmictic in the western Pacific Ocean (Watson, 1992). Throughout this paper we will refer to all four species of gobies and the eleotrid collectively as *'o'opu*; the complete Hawaiian name for amphidromous gobioid fishes is *'o'opu wai* (Pukui, 1986).

The *'o'opu* have similar, but dramatic life-histories that are inextricably linked to the marine environment (Kinzie, 1988; Radtke & Kinzie, 1996). Adults spawn in streams and the eggs are tended by the male for 2-3 days until hatching; paternal care of eggs has been documented in all species except for *S. hawaiiensis* (Lindstrom, 1999). After hatching, the larvae are transported to the ocean where they reside in the plankton for several months (Radtke *et al.*, 1988). Post-larvae, or early juveniles, migrate back into freshwater *en masse* and may use freshets from flashfloods as olfactory cues (Nishimoto & Kuamo'o, 1997). How far the post-larva migrates up the stream is species-dependent. *Eleotris sandwicensis* and *S. hawaiiensis* remain near sea-level and are absent upstream of the first precipitous waterfall (Nishimoto & Kuamo'o, 1997). *Awaous guamensis, S. stimpsoni,* and *L. concolor* use the goby-typical fused pelvic fins as a type of suction cup, and are able to climb waterfall obstacles. *Awaous guamensis* and *S. stimpsoni* are limited to waterfalls smaller than 20 m and of ele-

1. Present address: Department of Ecology and Evolutionary Biology, University of Toronto, 25 Harbord St., Toronto, Ontario, Canada M5S 365; email: camurphy@zoo.utoronto.ca

vations less than 150 m; at elevations over 300 m, *L. concolor* is the only native fish found in the streams (Fitzsimons & Nishimoto, 1990). Incredibly, *L. concolor* has been sighted at elevations as high as 1 km, in a location above a waterfall with a sheer drop of 300 m (Englund & Filbert, 1997).

Amphidromy appears to be a common life-history strategy for gobioid species that colonize many of the islands in the Indo-Pacific and Caribbean regions (Keith, 2003). Some authors suggest that the amphidromous lifestyle is an indication that Hawaiian fishes first share ancestry with other amphidromous freshwater fishes in the Indo-Pacific, and these amphidromous freshwater fishes are derived from the marine environment (McDowell, 2003). Therefore, what is known about the biology of other species of amphidromous gobies should be transferable to the Hawaiian *'o'opu*, with consideration of species-specific differences and the degree of geographical isolation that is unique to the Hawaiian Islands.

The Hawaiian Islands were formed by volcanic activity and have areas of high elevation (volcano peaks) that divide the islands into two distinct climates. Fresh water is predominantly restricted to small, high-gradient streams where the flow of water is dependent on orographic rainfall associated with trade winds and seasonal storms (Nishimoto & Kuamo'o, 1997). As a result, perennial streams capable of supporting populations of *'o'opu* are located only on the windward sides of the islands (Fig. 1; Nishimoto & Kuamo'o, 1997), which are characterized by rain, lush vegetation, and many streams. The leeward sides are typically arid, are disproportionately populated by humans, and have been extensively developed for agriculture and tourism. Consequently, to support leeward side development, freshwater streams have undergone extensive modifications such as diversion, channelization, damming, and pollution (runoff). As a result, Hawaiian freshwater streams have declined in habitat quality and size (Brasher, 2003; Lindstrom, 1999).

Declines in suitable habitat is a problem that is much more pronounced on isolated islands like the Hawaiian Archipelago, because amphidromous species exhibit small population sizes even under pristine conditions. Mismanagement of the freshwater resource could potentially decimate populations of *'o'opu* (Brasher, 2003; Radtke *et al.*, 2001).

To maintain viable populations of amphidromous gobies, effective freshwater management strategies are critical. Paradoxically, effective management of freshwater resources requires additional information on the marine larval stage. Much of the juvenile and adult life of the amphidromous *'o'opu* has been studied (Keith, 2003), but there is little information available on the marine larval stage. Therefore, the focus of this paper is to review what is known about the marine larval stage, and to suggest possible mechanisms that may govern recruitment to fresh water. We will identify two main lines of scientific enquiry which we believe are critical to effective management of the *'o'opu* resource. One such line is to identify mechanisms of larval retention/dispersal around the islands of Hawai'i. The second line of enquiry explores production and recruitment processes and determines whether individual freshwater streams serve as sources or sinks for *'o'opu* (see McRae, this issue). The mechanisms of larval production, retention/dispersal, and recruitment to fresh waters undoubtedly are governed both by biological and physical processes. We will describe some likely mechanisms, and outline both the biological and physical processes involved. We will also highlight areas that need more information, either through carefully designed experiments, sampling regimes, or modeling.

Discussion

Larval Production

The reproductive biology of only two *'o'opu* has been described (*A. guamensis* and *L. concolor*; see review by Keith, 2003). In comparison to other gobiids, amphidromous gobies are unusual; females produce many small, rather than a few larger eggs (Ha & Kinzie, 1996; Miller, 1984). As such, amphidromous gobiids are more like marine pelagic fishes that are periodic life history strategists, than typical coral reef or freshwater gobies that exhibit equilibrium-like strategies (Winemiller & Rose, 1992). A periodic strategy infers that survival of larvae is unpredictable, and environmental variations can play a dominant role in survival of a cohort.

The fecundity of *A. guamensis* is high and ranges from 56,000 to 690,000 eggs per female, with

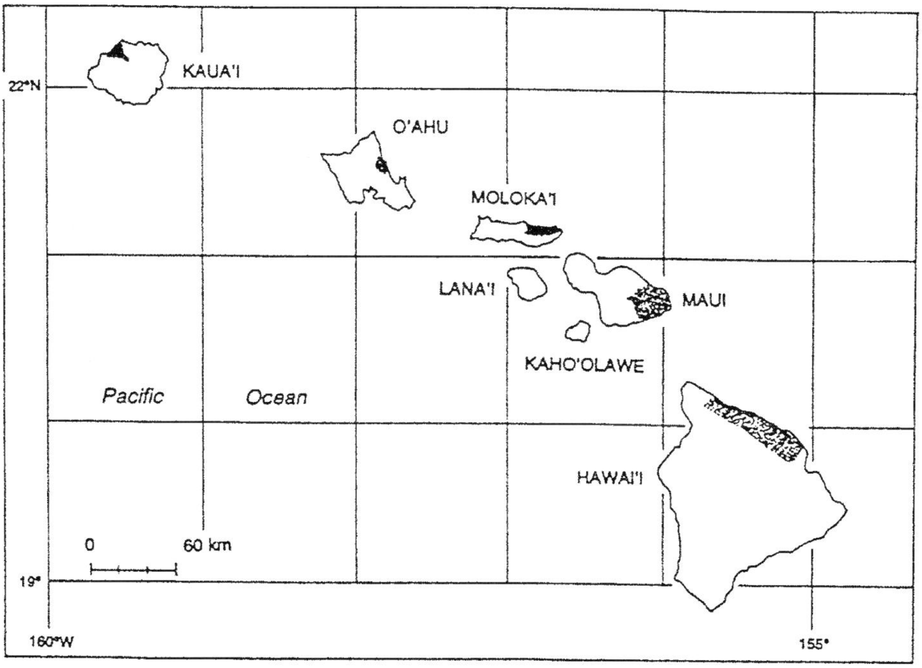

Figure 1. Main Hawaiian Islands. Stippled areas represent highest concentration of perennial streams (reprinted with permission from Nishimoto & Kuamo'o, 1997).

a very small average egg size of 0.3 mm (Ha & Kinzie, 1996). Iteroparous females spawn once between the months of August to December (Ha & Kinzie, 1996). Less fecund, batch-spawning females of *L. concolor* produce slightly larger eggs than *A. guamensis* (0.4–0.5 mm in diameter), may spawn up to 4 times a season, and produce up to 14,000 eggs each time (Kinzie, 1993). The spawning season for *L. concolor* ranges from October to June (Kinzie, 1993). There is very little information available on the reproductive biology of the other three species of *'o'opu*, but it is assumed that all five species have similar reproductive strategies (Lindstrom, 1999). Clearly, detailed reproductive studies on the remaining three species would demonstrate if such assumptions are valid. For example, because annual fecundity varies by up to 50-fold between *L. concolor* and *A. guamensis*, knowledge of whether the other *'o'opu* have fecundities that fall within this range, or are more or less fecund has implications for recruitment potential and dispersal (discussed below in Passive Transport).

Eggs hatch into yolk-sac larvae, a stage that appears to be sensitive to the amount of time spent in freshwater. After hatching, yolk-sac larvae passively drift downstream to the ocean and become part of the plankton. The duration of the yolk-sac stage for some species of *'o'opu* was reported as 4 days (Lindstrom, 1998), and in other amphidromous Japanese gobiids in the genus *Rhinogobius*, the yolk-sac of newly hatched larvae usually is consumed within 3–7 days (Hirashimi & Tachihara, 2000). Laboratory experiments on the yolk-sac stage suggest that as larvae mature, they become less able to tolerate freshwater (summarized in Keith, 2003). Therefore, the length of time spent in fresh water can have a direct impact on mortality rates of newly-hatched larvae and can potentially select for specific reproductive sites in streams that are at favorable distances from the ocean (Iguchi & Mizuno, 1999; Keith, 2003). We can also surmise that man-made insults to streams such as channelization, diversions, or constructions of dams (Brasher, 2003; Holmquist *et al.*, 1998), could alter the length of time that yolk-sac larvae spend in fresh water and thereby affect mortality rates and hence recruitment success (Houde, 1989).

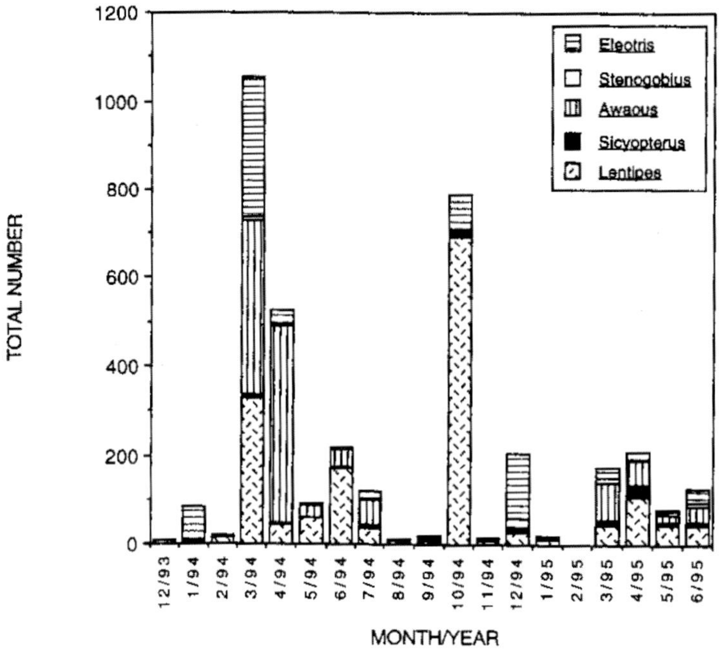

Figure 2. Monthly distribution of goby postlarvae by species trap-collected at Hakalau stream, Hawai'i, December 1993 to June 1995 (reprinted with permission from Nishimoto & Kuamo'o, 1997)

Larval Retention or Dispersal

The marine planktonic stage of the *'o'opu* is the least understood of all life-stages, and it is unknown if larvae are retained close to shore or are transported out to sea. We can think of many possible scenarios pertaining to the fate of the planktonic larvae, of which we discuss three. In the first and perhaps most likely situation, some of the planktonic larvae are not retained near shore and are passively transported out to sea, perhaps to other islands, but many remain in the 'vicinity' of their natal stream. When the *'o'opu* reach competency, they are cued to settle by some type of zeitgeber and initiate active migration toward and up their natal or nearby streams (Boehlert & Munday, 1988); migration to nearby streams ensures some gene flow. It also is possible that instead of actively swimming towards the coast, post-larvae can be transported onshore passively as has been demonstrated for other systems (e.g., Norcross & Shaw; 1984; Hare & Cowen, 1996). In the second scenario, *'o'opu* are transported far enough from the natal stream mouth to become entrained in ocean currents around the islands of Hawai'i. Once competency is reached, settlers locate non-natal, distant stream mouths, usually on other islands, and initiate immigration. Lastly, *'o'opu* are actively retained close to the shore, either through behavioral or physical processes, and when larvae become competent to settle, they easily locate their natal or nearby stream mouths for immigration. We will discuss the evidence that supports or refutes each scenario and briefly describe how each may lead to a different freshwater management scheme.

Before we discuss the three scenarios pertaining to larval retention and dispersal, we will first discuss the larval stage duration. Long larval stage durations are generally associated with organisms that are dispersed over long distances (Bradbury & Snelgrove, 2001). However, recent studies on marine reef fish by using a variety of techniques indicate that long distance dispersal does not occur with every species of fish with teleplanic larva (Taylor & Hellberg, 2003; Jones *et al.*, 1999; Swearer *et al.*, 1999). Caribbean reef gobiids do not disperse and remain close to natal areas, a finding determined by using mitochondrial cytochrome b DNA (Taylor & Hellberg, 2003). Fluorescent tags

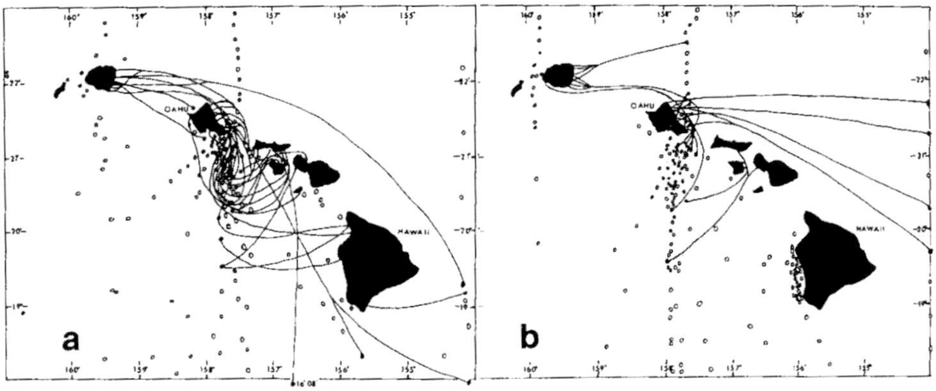

Figure 3. Drift-bottle deployments, recoveries and inferred trajectories. Open circles represent points of release but no recovery. Solid circles are sites of release with inferred trajectory to recovery location. **a:** Jan–May, **b:** Jun–Dec [from Barkley *et al.* (1964); reprinted with permission from Lobel (1989)].

applied to otoliths also showed that a larger than random sample of larvae of the damselfish *Pomacentrus amboinensis* settle back onto natal reefs (Jones *et al.*, 1999). Elemental composition analysis of the otoliths of settling larval Caribbean bluehead wrasse (*Thalassoma bifasciatum*) suggested that the larva remain close to shore (Swearer *et al.*, 1999). Therefore, the relationship between larval stage duration and dispersal is not clear, and we will discuss how long larval durations fit into each dispersal/retention scenario.

Larval Stage Duration

In general, the length of the larval stage for Hawaiian amphidromous gobies is longer than other fish with teleplanic larvae in the Pacific (Radtke *et al.*, 2001) and elsewhere (e.g., Dominican gobies, Bell *et al.*, 1995). The marine larval stage lasts 119–151 d for *S. hawaiiensis*, 150–169 d for *A. guamensis*, and 63–106 d for *L. concolor* (Radtke *et al.*, 1988; Radtke *et al.*, 2001), as determined via otolith analysis. Larval stage duration of *L. concolor*, however, appears similar to the Dominican gobies; all are sicydiine gobies and the marine larva stage duration may be a phylogenetic trait (Radtke *et al.*, 2001; Bell *et al.*, 1995), albeit that variations can occur in response to the influences of temperature, salinity, and prey density (Bradbury & Snelgrove, 2001),

Otolith studies of amphidromous gobies in Dominica, West Indies, demonstrate distinct seasonal cycles in age at recruitment and growth (Bell *et al.*, 1995). For two of the Hawaiian species, *A. guamensis* and *S. hawaiiensis,* otolith samples were taken from a small number of fish from one location, so determination of seasonal cycles in length of the larval duration was not possible (Radtke *et al.*, 1988). However, a much more extensive study considering geographic location, temperature, annual and seasonal cycles was done on *L. concolor* (Radtke *et al.*, 2001). Radtke *et al.* (2001) reported indications of seasonal changes in larval length; planktonic larval duration was shorter during warmer months. However, the size at settlement was also smaller during summer months and indicates that perhaps *L. concolor* do not use the marine larval stage to maximize size before recruitment. Interestingly, newly recruited *L. concolor* collected on the island of Maui consistently had shorter marine larval durations than those caught on the other islands, an observation that indicates some localized retention (Radtke *et al.*, 2001). Further research should determine whether other species of *'o'opu* show similar variability in marine larval duration. Such information could provide a means by which to distinguish biology, such as species differences, from the constraints imposed by the environmental conditions (i.e., ocean currents). If all species show similar patterns, we may infer that larval duration is regulated by the physical environment.

The hypothesis that *'o'opu* are retained close to natal streams may support the idea that the extended marine larval duration is not an adaptation for dispersal (Strathmann, 1985; Hohenlohoe, 2004). Rather, some authors suggest that the length of the marine larval stage is dependent upon

Table 1. Drift bottles released in a quadrangle around Hawaiian Archipelago [reprinted with permission from Scheltema *et al.* (1996); summarized from Barclay *et al.* (1964)].

Time of year	# of release points*	Total # bottles released	% bottles recovered
Jan–Feb	180	1005	2.7
Mar–May	59	925	12.5
Jun–Jul	182	1840	3.5
Sep–Dec	—	200	6.0

*From March through May returns were received from 52.5% of release points. During other parts of the year returns were only received from only 9-11% of release points.

available food; less food translates into a longer duration (Strathmann, 1986). However if larval fishes remain close to shore, an area typically rich in nutrients and food resources, larval fishes should grow faster and recruit at larger sizes (Swearer *et al.*, 1999). Since three species of *'o'opu* have long larval durations, it is not likely that food availability regulates larval stage duration if *'o'opu* are retained locally. However, because the length of the larval duration for only three species of *'o'opu* have been estimated, and only one (*L. concolor*) in sufficient detail, we can not rule out food availability as a regulator of marine larval duration. Interestingly, one species that has not been studied in great detail is *S. stimpsoni*, a species that has a much larger size at recruitment than any of the other 4 species (Nishimoto & Kuamo'o, 1997). Alternatively, other authors suggest that *'o'opu* need a long larval duration to complete development and make the necessary preparations to convert from salt water to fresh water (Radtke *et al.*, 2001). The observation that *L. concolor* has a similar marine larval duration as other sicydiine gobies located in the Caribbean indicates that the biological requirements of amphidromy, rather than physical oceanographic processes, necessitate long developmental periods (Radtke *et al.*, 2001; Bell *et al.*, 1995).

Passive Transport

Two aspects of *'o'opu* biology support passive transport scenarios: high fecundity and long larval duration. Passive transport scenarios assume that larval fishes act as inanimate particles that drift with prevailing currents. Larvae can be moved by wind-driven or other types of currents, but there exist no physical conditions, nor behavioral responses that act to concentrate and locally retain the larvae. As mentioned previously, *'o'opu* are unusually fecund for gobiid species; high fecundity increases the chance that at least a few individuals are capable of reaching suitable habitats for recruitment. *'O'opu* larvae are teleplanic, with larval stage durations of longer than two months which suggests a high dispersal potential (Radtke *et al.*, 2001; Bradbury & Snelgrove, 2001).

Hawaiian fishes, in general, have longer larval pelagic stage durations than other marine fish in the Pacific (Victor, 1986), which may be a response to the isolation of the Hawaiian Archipelago (Radtke *et al.*, 2001). There are advantages and disadvantages to having a long larval stage. Advantages include access to a food sources that do not overlap with adult conspecifics, higher levels of competency and growth prior to settlement, and the absence of benthic predators (Doherty *et al.*, 1983; Strathmann, 1990; Radtke *et al.*, 2001). However marine fish larvae can experience higher mortality rates attributable to starvation, predation by pelagic predators, and unpredictable environmental conditions (Jackson & Strathmann, 1981), which may increase cumulative mortality if the larval stage is extended (Cushing, 1975). Dispersal may also reduce local adaptation, and makes potential settlers less able to distinguish good sites from poor, sometimes at high costs (Strathmann *et al.*, 1981). Indeed, *'o'opu* will recruit to any freshwater source, regardless of habitat quality (Holmquist *et al.*, 1998). Nevertheless, the selection pressure for a long larval duration must be strong for it to persevere; the result is the potential to be transported great distances. Long larval durations may be one of the mechanisms that initially brought gobies to the isolated island chain (McDowell, 2003).

If all *'o'opu* larvae were transported great distances, however, Pacific stocks of *'o'opu* would be panmictic; endemism implies genetic isolation. Some larval transport via ocean currents between the Hawaiian Islands is a plausible scenario because there are no significant differences in the genetic composition of species on individual islands which indicates a sufficient mixing of fishes (gene flow) during the larval phase to preserve the unit stock (Zink *et al.*, 1996; Chubb *et al.*, 1998); a mixing of just 10% would ensure a genetically homogenous population (Zink *et al.*, 1996; Chubb *et al.*, 1998). Notably, of the *'o'opu* studied, the nonendemic species, *A. guamensis*, has the longest stage duration (Radtke *et al.*, 1988).

Retention by Ocean Currents

Long larval stage durations of *'o'opu* may be an adaptation to ensure some dispersal to maintain gene flow, but also to facilitate recruitment to the limited number of freshwater streams located in specific areas in the Hawaiian Island chain (Fig. 1). As such, larval duration may be coupled to physical oceanographic processes. Physical oceanographic processes such as prevailing currents, tides, boundary layers, and mesoscale eddies may function to retain larvae within the Hawaiian waters (Cowen *et al.*, 2000). *Awaous guamensis* spawns from August through to December (Ha & Kinzie, 1996) and has a larval period of 155 days or approximately 5 months (Radtke *et al.*, 1988); post-larval *A. guamensis* recruit from March to May, with most returning in March (Fig. 2), a pattern consistent with larval stage duration. Similarly, *L. concolor*, spawns October through May (Kinzie, 1993), and has a marine larval stage of approximately 3 months (Radtke *et al.*, 2001); *L. concolor* larval stage duration is consistent with the observed January to August recruitment pattern for the most part. However, a large pulse of recruitment occurs in October, a few months outside the hypothesized return period (Fig. 2). Reasons for this large pulse in unexplained recruits may be attributable to a lack of synchrony in space and time between studies of reproductive biology and recruitment.

A distinct seasonal pattern of currents, driven by trade winds, is observed around the Hawaiian Archipelago. Barkley *et al.* (1964) released 4000 drifters around the islands. Drifters that were released in proximity to an island tended to be recovered on the shore of nearby islands. The recovery of drifters showed distinct seasonal differences, with the highest recovery (12.5%) between the months of March and May (Table 1).

Barkley *et al.* (1964) also observed a distinct shift in ocean currents; the currents tended to flow to the northwest and parallel to Archipelago from January to May, and shift in June (with changes in the trade winds) to become more perpendicular to the Archipelago (Fig. 3; summarized by Lobel 1989). These currents, combined with other oceanographic data such as salinity and dynamic topography, demonstrate a closed gyral circulation around the Hawaiian Archipelago from March-April (summarized in Lobel, 1989). Barkley *et al.* (1964) also reported that most drifters returned to the coast between March and May (Table 1), which is coincident with the time when the majority of *'o'opu* post-larvae recruit to streams (Fig. 2). The second largest recruitment of post-larvae occurs between October and December (Fig. 2), which matches the second highest return of drifters (Table 1).

Additionally, mesoscale eddies, that form on occasion around the islands, contain high concentrations of planktonic larvae (Lobel & Robinson, 1988). While no larval *'o'opu* were reported from these eddies, these features could function in retaining larvae near the islands if *'o'opu* originating in the coastal zone become advected into the eddy field. Eddies entrain and retain drifting plankton and larvae and remain near the islands for sufficient duration for some larvae to complete the pelagic phase (Lobel & Robinson, 1988). Eddies, in addition to retaining larvae, also retain larval food items. As such, local hydrographic conditions may create a stable retention mechanism that can account for Hawaiian endemism, somewhat analogous to the genetically distinct Atlantic herring stocks maintained by stable hydrographic retention areas (Iles & Sinclair, 1982).

Consistent and convergent ocean current patterns can produce an effective barrier to genetic exchange (Hohenlohe, 2004). However, the currents around Hawai'i are highly variable in direction and speed and change from year to year (Lobel & Robinson, 1986; Qui *et al.*, 1997; Radtke *et al.*, 2001). Unpredictable ocean currents may explain the lack of genetic differentiation of *'o'opu* collected from different islands (Zink *et al.*, 1996; Chubb *et al.*, 1998; Radtke *et al.*, 2001).

The issue of retention is far from resolved, and indirect evidence suggests that strong localized retention is not widespread among the *'o'opu*. For example, marine larvae that are retained locally based on directed swimming are more prevalent in species that are non-pelagic spawners and produce larger and more highly developed larvae (Brogan, 1994). If possible, a comparison of the distribution of organisms that have non-motile planktonic phases, such as algae, to the distribution of *'o'opu* should indicate the degree to which *'o'opu* can regulate their dispersal (Bradbury & Snelgrove, 2001).

Behaviorally Mediated Localized Retention

It also is possible that localized retention of *'o'opu* larvae is behaviorally mediated (Boehlert & Munday, 1988; Atema *et al.*, 2002), a line of inquiry that deserves further study. Studies of some marine reef gobiid larvae suggest that larvae have the ability to actively maintain position from coastline, which thereby prevents dispersal (Leis, 1982). Indeed, the marine larvae of some species of amphidromous gobies from Dominica, West Indies, are able to maintain position at depth by choosing water masses based on salinity; actively choosing a water mass by migrating vertically can affect horizontal transportation (Boehlert & Munday, 1988; Bell & Brown, 1995). Additionally, reports that late stage marine larval fish are the strongest swimmers amongst the plankton and have the ability to swim faster than the background current, supports the hypothesis that *'o'opu* can actively remain near natal stream (Stobutzki & Bellwood, 1997)

The observed behavior of larval *'o'opu* in response to the physical environment also suggests they are retained close to the shore. In Hawai'i, the time between March and May is characterized by periods of heavy rainfall, and it has been hypothesized that post-larvae recruit in response to freshets that attract post-larvae to streams (Nishimoto & Kuamo'o, 1997). If attraction to freshwater is the dominant mechanism underlying migration toward shore and ultimate recruitment, we may infer that larval *'o'opu* are retained nearshore and are able to detect a sudden influx of fresh water.

From a management perspective, different transport and retention scenarios have significant implications. In the unlikely scenario that larval *'o'opu* are not retained by any mechanism and passively drift long distances, recruitment depends on the chance of post-larvae drifting close enough to respond to and reach an inland stream. In this case, management may need to protect spawning stock biomass in an effort to ensure the production of suitable numbers of pre-recruits. In the scenario that all *'o'opu* contribute to a single larval stock that is retained around the Hawaiian Islands, some freshwater habitat may be sacrificed for human use, while other streams are maintained as *'o'opu* sources (Diffendorfer, 1998). Some altered habitats (dammed, diverted) may act as *'o'opu* sinks, as amphidromous gobies will recruit to any freshwater source, regardless of habitat quality (Holmquist *et al.*, 1998). Finally, if larval *'o'opu* are retained much closer to natal streams, then the pattern of disrupted streams is critical. Localized retention may require management on a much finer scale (Palumbi, 1999). Prior studies on juvenile creek chubs demonstrated that the spatial distribution of source areas on a landscape, along with dispersal from those areas, can have dramatic impacts on fish populations (Schlosser, 1998). Many disrupted streams (larval sinks) adjacent to one another may cause an unrecoverable population decline.

Larval Recruitment

After spending several months as ichthyoplankton, post-larval *'o'opu* recruit *en masse* to perennial streams located on the windward sides of the Hawaiian Islands. There exist much circumstantial evidence and anecdotes on recruitment, but not much has been published. One study by Nishimoto & Kuamo'o (1997) documented recruitment to one stream on the island of Hawai'i. In this study, all five species of *'o'opu* recruited to the stream during third quarter lunar phase, generally after a flash flooding episode, but with distinct seasonality (Fig. 2). Recruits of *L. concolor* were most abundant, followed by *A. guamensis* and *E. sandwicensis*. In comparison, very few *S. stimpsoni* and *S. hawaiiensis* entered the stream (Nishimoto & Kuamo'o, 1997). Each species began upstream migration at different times of the day, both sicydiine gobiids, *L. concolor* and *S. stimpsoni,* recruited just after sunrise. *S. hawaiiensis* and *E. sandwicensis* predominantly began upstream migration as the tide rose just after dark whereas *A. guamensis* recruited throughout the day, but showed a distinct pulse just after dark at high tide (Nishimoto & Kuamo'o, 1997).

The recruitment pattern of *'o'opu* suggests that both physical and biological processes are involved (Nishimoto & Kuamo'o, 1997). Although the recruitment study was conducted on only one stream, and on only one island, there is enough evidence to suggest that physical processes such as diurnal cycles, moon phases, tides, flash floods, and season play a role in shaping patterns of recruitment of *'o'opu*. Variations in such physical processes have been shown to create patterns in recruitment for different species of wrasses (*Labridae*) from Barbados (Sponaugle & Cowen, 1997).

Indirect evidence also suggests that biological processes are involved in recruitment. Such evidence manifests in species differences, such as size at recruitment and time of recruitment. Also, post-larval *A. guamensis* school and migrate upstream together; such behavior suggests that predation may affect recruitment processes (Tate, 1997). Other biological processes that may affect recruitment, such as density dependence, remain to be determined.

The issue of recruitment is complicated by the lack of information on the mechanisms of marine larval retention/dispersal. If the marine larvae are dispersed far from natal streams, the recruitment processes are decoupled from local spawning which makes the population more or less open (Caley *et al.*, 1996). However, if marine larvae remain close to natal streams, the recruitment processes can be directly related to local spawning events, and the local population can be considered more or less closed. In reality, the populations probably do not fully resemble either endpoint, each of which requires a different population modeling approach to inform the management process. We can construct population models of *'o'opu* if we know whether the local population is open or closed, or somewhere in between. If the local population is closed, or semi-closed, stage-based matrix projections models can be constructed for each stream. To construct such models, we need estimates of demographic traits, such as fecundities, abundances, and growth and mortality rates for each lifestage (Caswell, 2001). If the local population is open, a number of different population modeling approaches have been developed for settling marine sessile invertebrates with long larval durations (e.g., Alexander & Roughgarden, 1996). Individual-based modeling to determine population dynamics is another modeling approach with higher predictive potential because transport, food-availability, behavior, and predator/prey interactions can be incorporated. However, individual-based modeling is a data-intensive approach, and there may not be enough information available at the present time (Hinckley *et al.*, 1996; Grimm, 1999; Werner *et al.* 2001).

Future Studies

The processes regulating production, marine larval retention or dispersal and recruitment are not well understood for amphidromous Hawaiian fishes. We can construct population models to assist management of *'o'opu*, but such models require critical demographic information such as timing and rates of reproduction for each species. Mortality rates and duration of the yolk-sac larva stage are also types of demographic information that are critical to model development. As discussed in detail in this paper, it is obvious that much information is needed on the marine larval phase of the amphidromous gobies, and we have mentioned a few examples where a little data would go a long way. For example, valuable information would include the length of the larval period for every species and whether growth and recruitment patterns vary seasonally. Because larval *'o'opu* can be identified through molecular markers (Lindstrom, 1999), quantification of abundance at location is possible. Therefore, deploying extensive ichthyoplankton surveys year round in areas close to the river mouths and amongst the cyclonic eddies and gyres off the coast may assist in determining the mechanisms behind dispersal or retention. Models can then be developed to predict the abundance of recruits from a source population by considering mesoscale features of ocean currents that interact with coastal habitat structure (e.g., Possingham & Roughgarden, 1990; Alexander & Roughgarden, 1996). Synthesis of all this information should provide the State of Hawai'i with an effective management tool and answer some fundamental questions in marine larval ecology.

Acknowledgments

The authors thank Hawaii's Department of Land and Natural Resources, Division of Aquatic Resources (DAR), Dr. Bob Nishimoto (DAR) and Dr. Mike Fitzsimons (LSU) for the invitation to participate in the Symposium on the Biology of Hawaiian Streams and Estuaries. We also thank Mark McRae (LSU) for early discussions that helped get this paper out of our heads and onto paper and Dr. Mike Fitzsimons and an anonymous reviewer for very helpful comments on the manuscript.

Literature Cited

Alexander, S.E. & J. Roughgarden. 1996. Larval transport and population dynamics of intertidal barnacles: a coupled benthic/oceanic model. *Ecological Monographs* **66**: 259–275.

Atema J., M.J. Kingsford & G. Gerlach. 2002. Larval reef fish could use odour for detection, retention and orientation to reefs. *Marine Ecology Progress Series* **241**: 151–160.

Barkley, R.A., B.M. Ito & R.P. Brown. 1964. Releases and recoveries of drift bottles and cards in the Central Pacific. *United States Fish & Wildlife Service Special Scientific Report* **492**, 31 p.

Bell, K.N.I. & J.A. Brown. 1995. Active salinity choice and enhanced swimming endurance in 0 to 8-d-old larvae of diadromous gobies, including *Sicydium punctatum* (Pisces), in Dominica, West Indies. *Marine Biology* **121**: 409–417.

———., **P. Pepin & J.A. Brown.** 1995. Seasonal, inverse cycling of length- and age-at recruitment in the diadromous gobies *Sicydium punctatum* and *Sicydium antillarum* in Dominica, West Indies. *Canadian Journal of Fisheries and Aquatic Sciences* **52**: 1535–1545.

Boehlert, G.W. & W. Munday. 1988. Roles of behavior and physical factors in larval and juvenile fish recruitment to estuarine nursery area. *American Fisheries Society Symposium* **3**: 51–67.

Bradbury, I.R. & P.V.R. Snelgrove. 2001. Contrasting larval transport in demersal fish and benthic invertebrates: the roles of behaviour and advective processes in determining spatial pattern. *Canadian Journal of Fisheries & Aquatic Sciences* **58**: 811–823.

Brasher, A.M.D. 2003. Impacts of human disturbances on biotic communities in Hawaiian streams. *Bioscience* **53**: 1052–1060.

Brogan, M.W. 1994. Distribution and retention of larval fishes near reefs in the Gulf of California. *Marine Ecology Progress Series* **115**: 1–13.

Caley, M.J., M.H. Carr, M.A. Hixon, T.P. Hughes, G.P. Jones & B.A. Menge. 1996. Recruitment and local dynamics of open marine populations. *Annual Review of Ecology & Systematics* **27**: 477–500.

Carson, H.L. & D.A. Clague. 1995. Geology and biogeography of the Hawaiian Islands, p. 14–29. *In*: W.L. Wagner & V.A. Funk (eds.), *Hawaiian biogeography: evolution on a hot spot archipelago.* Smithsonian Institute, Washington.

Caswell, H. 2001. *Matrix population models.* 2nd Ed. Sinauer Associates: Sunderlands, Massachusetts.

Cowen, R.K., K.M.M. Lwiza, S. Sponaugle, C.B. Paris & D.B. Olson. 2000. Connectivity of marine populations: open or closed? *Science* **287**: 857–859.

Chubb, A.L., R.M. Zink, & J.M. Fitzsimons. 1998. Patterns of mtDNA variation in Hawaiian freshwater fishes: the phylogeographic consequences of amphidromy. *Journal of Heredity* **89**: 8–16.

Cushing, D.H. 1975. *Marine Ecology and Fisheries.* Cambridge University Press, London. 278 p.

Doherty, P.J., D.M. Williams & P.F. Sale. 1983. The adaptive significance of larval dispersal in coral reef fishes. *Environmental Biology of Fishes* **12**: 81–90.

Diffendorfer, J.E. 1998. Testing models of source-sink dynamics and balanced dispersal. *Oikos* **81**: 417–433.

Englund R. & R. Filbert. 1997. Discovery of the native stream goby, *Lentipes concolor*, above Hawaii's highest waterfall, Hi'ilawe Falls. *Bishop Museum Occasional Paper* **49**: 62–64.

Fitzsimons, J.M. & R.T. Nishimoto. 1990. Territories and site tenacity in males of the Hawaiian stream goby *Lentipes concolor* (Pisces: Gobiidae). *Ichthyological Exploration of Freshwaters* **1**: 185–189.

Funk, V.A. & W.L. Wagner. 1995. Biogeographic patterns in the Hawaiian Islands, p. 379–419. *In*: W.L. Wagner and V.A. Funk (eds.), *Hawaiian biogeography: evolution on a hot spot archipelago.* Smithsonian Institute, Washington.

Grimm, V.1999. Ten years of individual-based modeling in ecology: what have we learned and what could we learn in the future? *Ecological Modeling* **115**: 129–148.

Ha, P.Y. & R.A. Kinzie III. 1996. Reproductive biology of *Awaous guamensis,* an amphidromous Hawaiian goby. *Environmental Biology of Fishes* **45**: 383–396.

Hare, J.A. & R.K. Cowen. 1996. Transport mechanisms of larval and pelagic juvenile bluefish (*Pomatomus saltatrix*) from South Atlantic Bight spawning grounds to the Middle Atlantic Bight nursery habitats. *Limnology & Oceanography* **41**: 1264–1280.

Hinckley, S., A.J. Hermann & B.A. Megrey. 1996. Development of a spatially-explicit, individual-based model of marine fish early life history. *Marine Ecolology Progress Series* **13**: 47–68.

Hohenlohe, P.A. 2004. Limits to gene flow in marine animals with planktonic larva: models of *Littorina* species around Point Conception, California. *Biological Journal of the Linnean Society* **82**: 169–187.

Holmquist, J.G., J.M. Schmidt-Gengenbach & B.B. Yoshioka. 1998. High dams and marine-freshwater linkages: effects on native and introduced fauna in the Caribbean. *Conservation Biology* **12**: 621–630.

Hirashimi, K. & K. Tachihara. 2000. Embryonic development and morphological changes in larvae and juveniles of two land-locked gobies, *Rhinogobius* spp. (Gobiidae), on Okinawa Island. *Japanese Journal of Ichthyology* **47**: 29–41.

Houde, E.D. 1989. Subtleties and episodes in the early life of fishes. *Journal of Fish Biology* **35** (Supplement A) 29–38.

Hourigan, T.F. & E.S. Reese. 1987. Mid-ocean isolation and the evolution of Hawaiian reef fishes. *Trends in Ecology & Evolution* **2**: 187–191.

Iguchi, K. & N. Mizuno. 1999. Early starvation limits survival in amphidromous fishes. *Journal of Fish Biology* **54**: 705–712.

Iles, T.D. & M. Sinclair. 1982. Atlantic herring: stock discreteness and abundance. *Science* **215**: 627–633.

Jackson, G.A. & R.R. Strathmann. 1981. Larval mortality from offshore mixing as link between precompetent and competent periods of development. *American Naturalist* **118**: 16–26.

Jones, G.P., M.J. Milicich, M.J. Emslie & C. Lunow. 1999. Self-recruitment in a coral reef fish population. *Nature* **402**: 802–804.

Keith, P. 2003. Biology and ecology of amphidromous Gobiidae of the Indo-Pacific and the Caribbean regions. *Journal of Fish Biology* **63**: 831–847.

Kinzie, R.A. III. 1988 Habitat utilization by Hawaiian stream fishes with reference to community structure in oceanic island streams. *Environmental Biology of Fishes* **22**: 179–192.

———., 1993. Reproductive biology of an endemic, amphidromous goby *Lentipes concolor* in Hawaiian streams. *Environmental Biology of Fishes* **37**: 257–268.

Leis, J.M. 1982. Nearshore distributional gradients of larval fish (15 taxa) and planktonic crustaceans (6 taxa) in Hawaii. *Marine Biology* **72**: 89–97.

Lindstrom, D.P. 1998. Reproduction, early development and larval transport dynamics of amphidromous Hawaiian gobioids. Unpublished Ph.D. Dissertation. University of Hawai'i.

———. 1999. Molecular species identification of newly hatched Hawaiian amphidromous gobioid larva. *Marine Biotechnology* **1**: 167–174.

Lobel, P.S. 1989. Ocean current variability and the spawning season of Hawaiian reef fishes. *Environmental Biology of Fishes* **24**: 161–171.

———. **& A.R. Robinson**. 1986. Transport an entrapment of fish larvae by ocean mesoscale eddies and currents in Hawaiian waters. *Deep Sea Research* **33**: 483–500.

Lobel, P.S. & A.R. Robinson. 1988. Larval fishes and zooplankton in a cyclonic eddy in Hawaiian waters. *Journal of Plankton Research* **10**: 1209–1223.

McDowell, R.M. 2003. Hawaiian biogeography and the islands' freshwater fish fauna. *Journal of Biogeography* **30**: 703–710.

Miller, P.J. 1984. The tokology of gobiid fishes, p. 119–153, *In*: G.W. Potts & R.W. Wootton (eds.). *Fish Reproduction: Strategies & Tactics*. Academic Press, New York.

Nishimoto, R.T. & D.G.K. Kuamo'o. 1997. Recruitment of goby postlarvae into Hakalau stream, Hawai'i, Island. *Micronesica* **30**: 41–49.

Norcross, B.L. & R.F. Shaw. 1984. Oceanic and estuarine transport of fish eggs and larvae: a review. *Transactions American Fisheries Society* **113**: 153–165.

Palumbi, S.R. 1999. The prodigal fish. *Nature* **402**: 733–734.

Possingham, H. P. & J. Roughgarden. 1990. Spatial population dynamics of a marine organism with a complex life cycle. *Ecology* **71**: 973–985.

Pukui, M.K. & S.H. Elbert. 1986. *Hawaiian dictionary*. University of Hawai'i Press, Honolulu, 572 p.

Qui, B., D.A. Koh, C. Lumpkin & P. Flament. 1997. Existence and formation of the North Hawaiian Ridge Current. *Journal of Physiological Oceanography* **27**: 431–444.

Radtke, R.L., R.A. Kinzie, III & S.D. Folsom. 1988. Age at recruitment of Hawaiian freshwater gobies. *Environmental Biology of Fishes* **23**: 205–213.

———. **& R.A. Kinzie III**. 1996. Evidence of a marine larval stage in endemic Hawaiian stream gobies from isolated high-elevation locations. *Transactions American Fisheries Society* **125**: 613–621.

———., **R.A. Kinzie III & D.J. Shafer**. 2001. Temporal and spatial variation in length of larval life and size at settlement of the Hawaiian amphidromous goby *Lentipes concolor*. *Journal of Fish Biology* **59**: 928–938.

Randall, J.E. 1998. Zoogeography of shore fishes of the Indo-Pacific region. *Zoological Studies* **37**: 227–268.

Scheltema, R.S., I.P. Williams & P.S. Lobel. 1996. Retention around and long-distance dispersal between oceanic islands by planktonic larvae of benthic gastropod Mollusca. *American Malacology Bulletin* **12**: 67–75.

Schlosser, I.J. 1998. Fish recruitment, dispersal, and trophic interactions in a heterogeneous lotic environment. *Oecologia* **113**: 260–268.

Sponaugle, S. & R.K. Cowen. 1997. Early life history traits and recruitment patterns of Caribbean wrasses (Labridae). *Ecological Monographs* **67**: 177–202.

Stobutzki, I.C. & D.R. Bellwood. 1997. Sustained swimming abilities of later pelagic stages of coral reef fishes. *Marine Ecology Progress Series* **149**: 35–41.

Strathmann, R.R. 1985. Feeding and nonfeeding larval development and life-history evolution in marine invertebrates. *Annual Review of Ecology & Systematics* **16**: 339–361.

———. 1986. What controls the type of larval development? *Bulletin of Marine Science* **39**: 616–622.

———. 1990. Why life histories evolve differently in the sea. *American Zoologist* **30**: 197–207.

———., **E.S. Branscomb & K. Vedder**. 1981. Fatal errors in set as a cost of dispersal and the influence of intertidal flora on set of barnacles. *Oecologia* **48**: 13–18.

Swearer, S.E., J.E. Caselle, D.W. Lea & R.R. Warner. 1999. Larval retention and recruitment in an island population of a coral-reef fish. *Nature* **402**: 799–802.

Tate, D.C. 1997. The role of behavioral interactions of immature Hawaiian stream fishes (Pisces: Gobioidea) in population dispersal and distribution. *Micronesica* **30**: 51–70.

Taylor, M.S. & M.E. Hellberg. 2003. Genetic evidence for local retention of pelagic larvae in Caribbean reef fish. *Science* **299**: 107–109.

Victor, B.C. 1986. Duration of the planktonic larval stage of one hundred species of Pacific and Atlantic wrasses (family Labridae). *Marine Biology* **90**: 317–326.

Watson, R.E. 1992. A review of the gobiid genus *Awaous* from insular streams of the Pacific plate. *Ichthyological Explorations of Freshwaters* **3**: 113–118.

Werner, F.E., J.A. Quinlan, R.G. Lough & D.R. Lynch. 2001. Spatially-explicit individual based modeling of marine populations: a review of the advances in the 1990s. *Sarsia* **86**: 411–421.

Winemiller, K.O. & K.A. Rose. 1992. Patterns of life-history diversification in North American fishes: implications for population regulation. *Canadian Journal of Fisheries and Aquatic Sciences* **49**: 2196–2218.

Zink, R.M, J.M. Fitzsimons, D.L. Dittman, D.R. Reynolds & R.T. Nishimoto. 1996. Evolutionary genetics of Hawaiian freshwater fish. *Copeia* **1996**: 330–335.

Biology of Hawaiian Streams and Estuaries. Edited by N.L. Evenhuis & J.M. Fitzsimons. *Bishop Museum Bulletin in Cultural and Environmental Studies* 3: 75–85 (2007).

Limitations of Early Seaward Migration Success in Amphidromous Fishes

KEI'ICHIRO IGUCHI
National Research Institute of Fisheries Science, Komaki, Ueda, Nagano 386-0031, Japan; keyichi@fra.affrc.go.jp

Abstract

Amphidromous fishes are abundant in insular streams and rivers which have a comparatively short length, whereas they are less common in continental rivers which are relatively larger in scale. This presents a challenging question about the adaptive role of amphidromy to freshwater habitats. This study overviews amphidromy, with a focus on the early survival of amphidromous fishes. Similarity in diel periodicities during downstream and vertical migrations of larval gobies suggests that, in running waters, seaward migration starts automatically as a result of phototaxis inherited from marine ancestors. Goby larvae feed on plankton, and during the downstream migration, loss of larvae from starvation prior to reaching the comparatively plankton rich brackish and marine zone can be regarded as a function of the distance they must be carried by water currents or the time spent without food. Smaller eggs may have an adaptive significance with respect to larvae reaching the sea more quickly, but such an evolutionary "countermeasure" is limited by the higher vulnerability of smaller embryos. Topographical features such as long, slow-flowing lower courses make typical continental rivers unsuitable as a habitat for amphidromous fish due to the high costs of dispersal. Disruptive selection, which is expected to favor both large- and small-sized eggs, may allow an evolutionary breakthrough by the abandonment of a migratory life cycle. In fact, some fishes which originated from amphidromous species have established effective landlocked populations within longer rivers. These fluvial species generally spawn large eggs, from which well-developed larvae emerge that are able to support themselves in running water.

Introduction

Fishes that inhabit freshwaters are conventionally classified according to their capacity to tolerate different salinities, and this can be useful in understanding their ecological and evolutionary status in local aquatic communities. Myers (1949) analyzed the categories of freshwater fishes and listed some definitions. Many fishes that are strictly confined to low salinity are defined as primary freshwater fish. Those that are strictly confined to freshwater, but are relatively salt tolerant, at least for a short period, are defined as secondary freshwater fish. However, some species are more adaptable and tolerant of a wide range of salinities. Those that regularly migrate between freshwater and seawater at a definite stage of their life cycle are defined as diadromous fishes. Marine species which move freely into freshwater are defined as peripheral fish. The dynamics of immigration and local extinctions are important factors in the ecology and evolution of species and populations dwelling on islands (MacArthur, 1972; Otto & Endler, 1989; Futuyma, 1998). Compared with freshwater fish faunas in mainland areas, those on islands can be characterized by a low occurrence of genuine primary and secondary freshwater species. Instead, euryhaline species fill this vacant niche in insular fish communities, especially in the middle and upper courses of rivers and streams where species of diadromous fishes are often dominant (Hayashi *et al.*, 1992; Shinomiya & Ike, 1992). Because suitable habitats on islands have less of an opportunity to be invaded by genuine freshwater fishes from mainland streams and rivers, insular freshwater fish faunas more plausibly depend upon euryhaline migratory fishes.

Diadromy, a term used to describe migrations of fishes between freshwater and the sea, includes three sub-categories: anadromy, catadromy, and amphidromy (Myers, 1949). Their definitions are as follows: anadromous fishes spend most of their lives in the sea and migrate to freshwater to breed, catadromous fishes spend most of their lives in freshwater and migrate to the sea to breed, and amphidromous fishes that are spawned in freshwater, migrate downstream to the sea and then go upstream at a juvenile stage for further growth and reproduction (McDowall, 1992, 1997a). Salmon and eels, representative of anadromous and catadromous fishes, respectively, can inhabit continental areas as well as insular areas, showing their ability to reproduce in large rivers. In contrast, habitats of amphidromous species are limited largely to islands with small streams such as those known from Japan, Hawai'i, Australia, and New Zealand (McDowall, 1997b). Amphidromy has developed among a wide variety of taxa involving several families such as Osmeridae, Galaxiidae, Aplochitonidae, Prototroctidae, Cottidae, and Gobiidae (McDowall, 1988). Like anadromous and catadromous fishes, amphidromous fishes have a potential to disperse through the ocean to some extent during their marine stages. However, establishment of populations of amphidromous fishes in large rivers running through continents is a rare occurrence. Such an asymmetry in the distribution of amphidromous fishes, biased toward insular habitats, is a challenging question requiring assessment.

Gobies comprise one of the largest families of marine fishes, have diverse life histories (Nelson, 1984), and often become predominant members of the freshwater fish faunas on oceanic islands in the tropics and subtropics [e.g. the Hawaiian Chain; Nishimoto & Kuamo'o (1991)]. It is readily comprehensible that amphidromous species are derived from marine ancestors. In fact, some amphidromous gobies known from the Japan Archipelago have coastal relatives neighboring with them near the stream mouth (Mukai, 2001; Senou *et al.*, 2004). Early migration of amphidromous gobies begins with passive drift (i.e. movement due to currents) just after hatching. The frequency of drifting larvae changes hourly within a day, and different species exhibit similar periodicities (Iguchi & Mizuno, 1990). Such diel drift patterns are principally explicable by larval phototaxis (Iguchi & Mizuno, 1991). Rather than being particular to amphidromous fishes, larval phototaxis is also prevalent among marine fishes and governs their diel vertical migrations (Woodhead & Woodhead, 1955). This leads to the idea that the mechanism which triggers downstream movement in amphidromous fishes did not evolve after invasion of freshwater habitats but is a pre-adaptive behavioral characteristic inherent in their marine ancestors. Drift and vertical migrations in larval fishes, therefore, are hypothesized to be homologous behaviors that evolved in different conditions, running water and lentic water. The aim of this study is to examine this hypothesis through the comparison of diel phenomena shown within a single species that inhabits both lotic and lentic habitats. Then, evolution and limitations of amphidromy are discussed on the basis of the life histories of gobies.

Materials and Methods

Study fish

Rhinogobius sp. CB, Shima-yoshinobori in Japanese, is a member of the *Rhinogobius* species complex formerly called *R. brunneus* Cross Band type (Kawanabe & Mizuno, 1989; Shimizu *et al.*, 1993; Aonuma *et al.*, 1998). This goby is common throughout Japan, inhabiting steep and moderately sloped reaches with stony riverbeds (Mizuno, 1960a). In marked contrast to continental rivers, watercourses on the Japan Archipelago are short in length but high in average water velocity, owing to the insular topography characterized by small coastal plains and mountains close to the coast. To evaluate watercourse landscapes in Japan, a reach type classification defined by the combination of the arrangement of rapids and pools (A or B) with the mode of flow from rapids to pools (a, b or c) is often adopted for convenience (Kani, 1944; Mizuno & Kawanabe, 1981). Aa-type reaches have two or more sets of a rapids and a pool in a single reach separated by clear falls, and usually occurs along steep courses in mountains. Bb-type reaches have a couple of turbulent rapids and pools in a single reach with no waterfall separating them, and usually appears along gentle courses in plains. Along the gentlest and lowest courses, Bc-type reaches which lack the distinguishable boundary of rapid and pool can be found. This reach type is rare in Japan, but constitutes the main part of continental rivers. Reach types, therefore, determine the capacity of watercourses to function as conveyors of larvae. Amphidromous gobies usually use Bb and Aa-type reaches as habitats for growth and breeding. However, landlocking of amphidromous fishes in lakes, reservoirs or ponds results in the establishment of new populations without the seaward migration (Mizuno, 1960a).

Rhinogobius species complex spawns on the undersurfaces of stones, and males, as in many other gobiid species, are the sole providers of parental care, maintaining nests and taking care of eggs until hatching (Breder & Rosen, 1966; Ito & Yanagisawa, 2003). Newly hatched larvae of *Rhinogobius* sp. CB are characterized by their pigmentation pattern, a yolk with oil globules somewhat larger than the eye in diameter and membranous fins without rays (Sakai & Yasuda, 1978). They are approximately 4 mm long in notochord length and have a specific gravity more than one (i.e. negatively buoyant in freshwater). They show a positive phototaxis to 500 lux ambient light but a negative response to light more than 5000 lux, even though their swimming ability is negligible (1.5 cm/sec maximum in still water) (Iguchi & Mizuno, 1991). Their diel drift pattern is variable along river courses, and corresponds to the environmental conditions present in the reach they occur in. In the case of the Mina River, which is typical of short rivers in Japan at just 12 km in length, downstream migration occurs throughout the day in upper courses with Aa-type reaches, whereas drifting larvae becomes nocturnal with a peak in number soon after sunset in the lower courses with Bb-type reaches (Iguchi & Mizuno, 1990). As is commonly the case in streams and rivers in Japan, the diel drift pattern of larval gobies corresponds to the reach type. Diel periodicity in larval drift is not specific to gobies but is known from other amphidromous fishes including ayu *Plecoglossus altivels* (Tago, 1999).

Field survey

Larvae of *Rhinogobius* sp. CB were collected from the Shimanto River, Mina River and Yoshifuji Pond in 1984. Sampling sites established in each location were different in topographical features. The Shimanto River is 190 km in length (the largest river on Shikoku Island, Japan) and has adult gobies distributed as far as 100 km upstream of the river mouth. The sampling site in the Shimanto River was located just upstream of the salt wedge, 9.1 km from the river mouth (32°59'N, 132°56'E). In the Mina River, which is 12 km in length, adult gobies occur as far as 5 km upstream of the river mouth; a sampling site was located in a tidal zone estuary in the river mouth (33°2'N, 133°3'E). The Yoshifuji Pond (33°50'N, 132°47'E) is a small reservoir (325,000 m^3 volume) constructed in 1957 for agricultural use in which gobies have been landlocked.

Fish collection was conducted two or three times at each locality from June to August corresponding to the breeding season of *Rhinogobius* sp. CB. In each sampling trial, netting was repeated at hourly intervals over a 24-hour period beginning at noon. Drifting larvae in the Shimanto River were captured using a drift net with a 30 x 30 cm mouth anchored for 10 min. Larvae in the surface layers of both the mouth of the Mina River and the Yoshifuji Pond were captured using a plankton net with a mouth of 25 cm in diameter by making ten 10m-castings in succession from the shore. Goby samples were fixed in 10% formalin immediately after being caught for later sorting and counting in the laboratory. Samples that contained a mix of the four sympatric *Rhinogobius* species were tentatively divided into two size classes; larvae that were approximately 4 mm notochord length were assorted into *Rhinogobius* sp. CB or DA, and those that were approximately 3 mm notochord length were assorted into *Rhinogobius* sp. CO or LD. Further identification of larvae relied on the pattern of pigmentation on the dorsal surface (M. Miwa & N. Mizuno, 1978, unpubl. data). Detection of their external characteristics required the aid of a microscope.

Results

At the lower course of the Shimanto River, the number of drifting larvae increased rapidly after sunset, peaked between 1900 and 2300 hours and then decreased, while few larvae also drifted during the daytime (Fig. 1A). Small numbers of larvae were observed to have already exhausted their yolk. Surface-netting in the tidal area of the Mina River revealed diel changes in the density of larval gobies with lower amount of remaining yolks. Larvae in the river mouth were nocturnal, appearing in the surface layer soon after sunset with a peak between 1900 and 2300 hours (Fig. 1B). In the Yoshifuji Pond, larval numbers in the surface layer increased immediately after sunset, peaking between 1800 and 2200 hours and then quickly decreasing, while small numbers also emerged in the daytime (Fig. 1C). Every sampling trial confirmed that diel periodicity was consistent across localities.

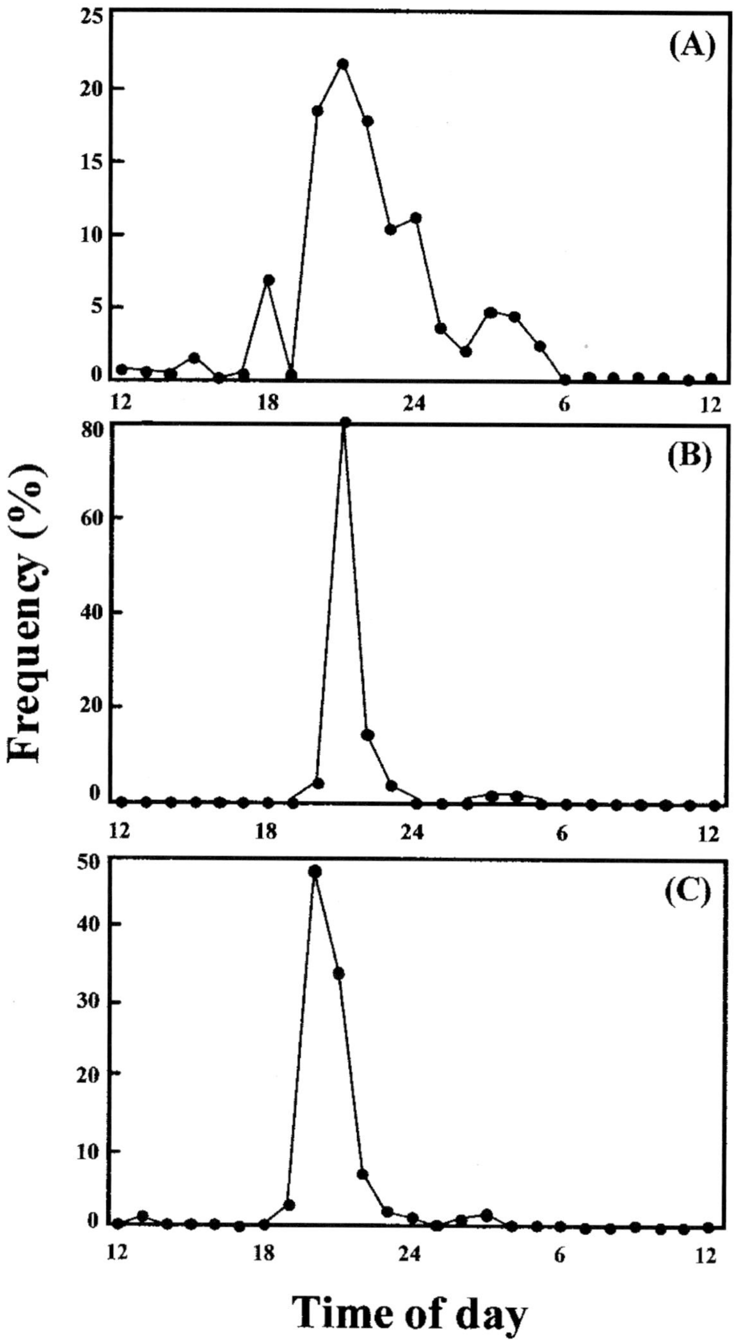

Figure 1. Diel changes in the frequency of larvae captured from (**A**) the Shimanto River on 11–12 July 1983 (N = 862), (**B**) the Mina River on 23–24 July 1983 (N = 200), and (**C**) the Yoshifuji Pond on 14–15 June 1983 (N = 652).

Discussion

Origin of drift migration

It is difficult for the appearance of larvae in the surface layer of water to be explained by horizontal movements from elsewhere because of the fact that larval *Rhinogobius* sp. CB have such poor swimming ability. Changes in the surface density of larvae in the river mouth of the Mina River as well as in the Yoshifuji Pond are therefore attributable to the vertical movement of larvae. Vertical migration is a widespread phenomenon among zooplankton including pelagic larvae of fishes (Motoda, 1972; Seliverstov, 1973; Takahashi, 1981; Fortier & Legget, 1983). In Lake Biwa, the largest lake in Japan, larvae of *Rhinogobius* sp. nocturnally ascend during their vertical migration (Nagoshi, 1982). Their lacustrine life has been estimated to have evolved during the diluvial epoch, 100,000 years ago (Takaya, 1963). On the other hand, for the population in the Yoshifuji Pond, it has only been, at most, 26 years since the transition from an amphidromous life to a landlocked life. Larvae in the mouth of the Mina River are still actively migrating to the sea. Thus, early vertical migration seems to be an instinctive action of goby free embryos regardless of what type of habitat they occur in. Similar to the larvae in the Shimanto River, those in the river mouth of the Mina River and in the Yoshifuji Pond share a pattern of diel periodicity, which indicates a common mechanism.

Diel drift patterns of larval gobies in comparatively small rivers are variable along courses, as was observed in the Mina River (Iguchi & Mizuno, 1991), and can be attributed to larval reactions to ambient light together with their vertical mobility in running water (Iguchi & Mizuno, 1991). Once leaving the nest, daytime larvae in upper courses can do nothing but be carried away by water currents, no matter how they may prefer to avoid strong illumination. The poor swimming ability of larvae is responsible for their all-day drifting through Aa-type reaches. Sizable pools formed in the lower courses of rivers allow larvae their directional movement, and negative phototaxis lets them halt drifting until twilight induces them to rise from the dark adjacent to the bottom. Phototaxis explains the marked increase in the number of drifting larvae in the Bb-type reaches that exist in relatively large rivers in Japan such as the Shimanto River. Likewise, the vertical position of larvae in lentic water changes in a similar pattern. Close to evening, positive phototaxis to weak illumination stimulates larvae to swim upward in the water column, and at dusk a high density of them occur in the surface layer. At night, larvae that stop moving sink due to their negative buoyancy and disappear from the surface layer. Periodical appearance of larvae at the surface layer observed at the river mouth of the Mina River and in the Yoshifuji Pond is, thus, explained chiefly by phototaxis. It can be concluded, then, that for amphidromous gobies, seaward drift migration and vertical migration can be considered homologous phenomena expressed under different conditions, running water and lentic water. A key factor in this homology is phototaxis, suggesting that early seaward migration in amphidromous gobies is driven by instinctive behaviors inherited from marine ancestors.

Early survival in rivers

Adhesive eggs, which need a hard substrate to attach to, are common among marine and freshwater gobies and are also known from many amphidromous fishes other than gobies [e.g., sculpins (genus *Cottus*), Goto (1990); ayu, Ishida & Ohoshima (1959)]. Such a phylogenetic constraint in spawning mode, which requires appropriate substrata on which to attach eggs, results in amphidromous fishes moving upstream until encountering reaches with stony riverbeds suitable for spawning grounds (Tamada, 2000). Herbivores found among amphidromous fishes include specified grazers such as the monk goby *Sicyopterus japonicus* and *ayu* (Fukui, 1979; Iguchi & Hino, 1996). Attached algae needs stony substrata on which to grow and moderate flow velocities to maintain their growth (Tanimizu *et al.*, 1981). These types of spawning and grazing habitats are less abundant in the lowest part of typical rivers in Japan because of the combination of muddy riverbeds and low flow velocities. Many species of amphidromous gobies have evolved pelvic fins fused to form a sucking disk that is helpful to climb up the steep waterfalls that frequently interrupt insular streams and rivers (Nishijima & Shikatani, 1994).

Low availability of food due to the lack of zooplankton in running water practically prohibits seaward drifting larvae from foraging even after the absorption of their yolk (Tsukamoto, 1991). The degree of starvation is, therefore, correlated with the distance from the hatching site to the sea or the

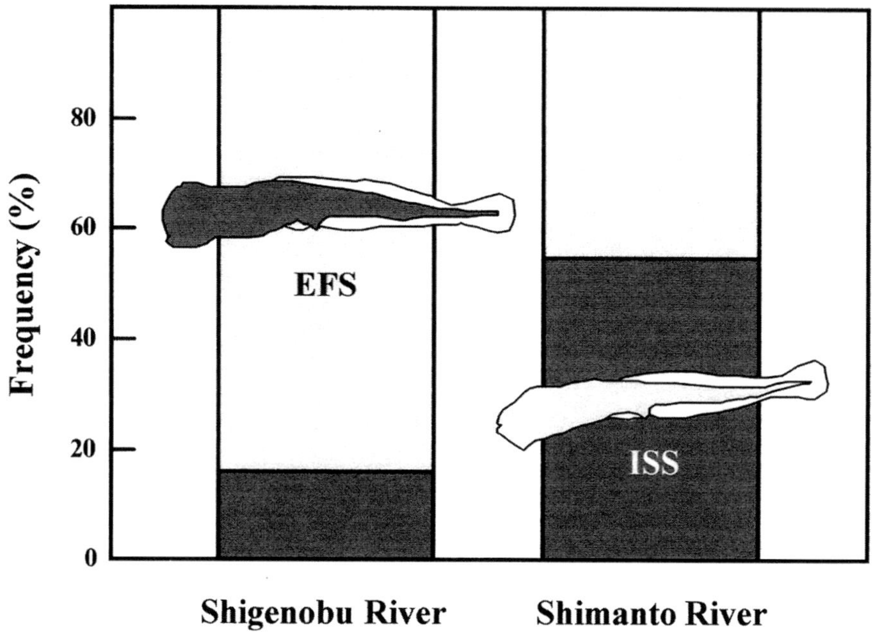

Figure 2. Frequency of larvae in endogenous feeding state (EFS) and irrecoverable starvation state (ISS) compared between samples (each N-100) from the headwater area in the Shigenobu River (33°46'N, 133°48'E) and from the beginning of the salt wedge of the Shimanto River (32°59'N, 132°56'E). Reanalysis was based data given in Iguchi & Mizuno (1999).

time spent in migration toward the sea. Based on the morphological transition of larvae from endogenous feeding state to irrecoverable starvation state, Iguchi & Mizuno (1999) hypothesized that early survival of amphidromous fishes varies according to the length of river. A test using *Rhinogobius* sp. CB individuals from different populations strongly supported the hypothesis that the longer the distance of migration without food, the higher the mortality rate for larvae (Fig. 2). Total survival of larvae tends to be overestimated, because dead individuals sink to the bottom before passing by a sampling site (see Moriyama *et al.*, 1998). Starved larvae are too weak to feed even if food ultimately becomes available (Blaxter & Ehrlich, 1974) and are vulnerable to potential predators that forage selectively on an easy catch (Paradis *et al.*, 1996). A convincing argument can be made, then, that rivers with long, slow-flowing lower courses are unsuitable habitats for amphidromous gobies. Even though adults are successful in spawning in upper courses, their larvae must travel a long way to the sea to the detriment of early survival. This may be a critical reason why amphidromous fishes are abundant in insular streams and rivers but hardly maintain populations in large continental rivers. Generally, insular streams and rivers are comparatively short with short, slow-flowing courses, which enables larvae to finish their travel to the sea quickly.

Pathway to another life

It is a natural assumption that young amphidromous gobies entering the mouth of a freshwater habitat would keep on going upstream until encountering a vacant niche for growth and reproduction. From an evolutionary standpoint, amphidromy will be favored, when invading individuals are able to enhance their fitness while balancing this with the costs of migration (Gross, 1987). Reproductive success associated with the early survival of larvae varies among parental individuals with the dis-

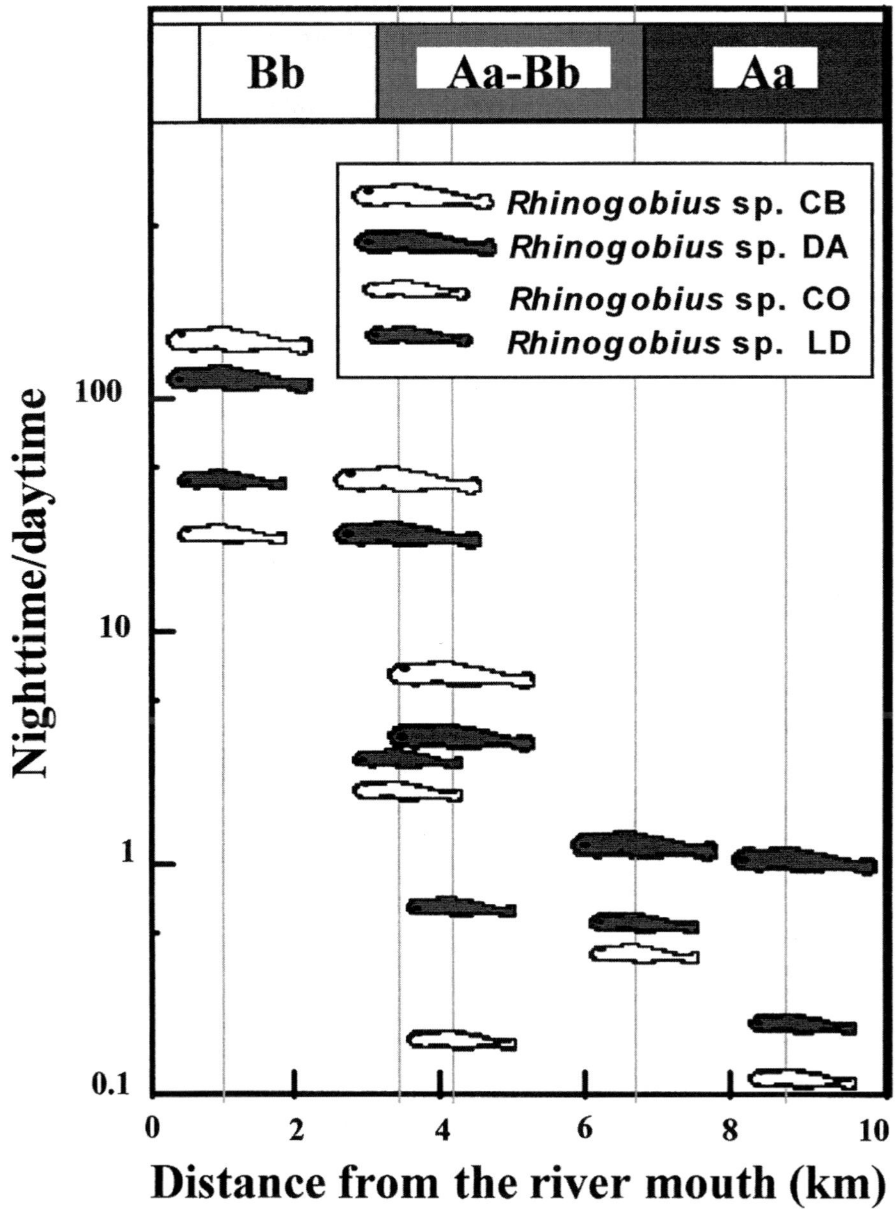

Figure 3. Night/day ratios of drifting larvae based on hourly collections along the the Mina River courses from June to July 1983. *Rhinogobius* sp. CB and DA are 4 mm in notochord length, while CO and LD are 3 mm. Natural photoperiod (L:D) was approximately 14:10. Upper horizontal bars indicate reach types. Reanalysis was conducted based on Iguchi (1985).

tance their offspring are required to drift. By replacing distance with time spent in-transit, it can be predicted that accelerated seaward migration will reduce early mortality from starvation. In one view, nocturnal activity of larvae wastes daytime hours available for further downstream movement. Meanwhile, daytime drift occurs inevitably according to the body size of larvae, a characteristic that primarily regulates swimming ability (Tsukamoto *et al.*, 1975). Downstream movement of smaller larvae during the daytime, even though involuntary, must shorten the total time spent migrating to sea. Some *Rhinogobius* species sympatric within a river have different egg sizes that correspond to differences in their distributional ranges, and species with small eggs prefer habitats in upper tributaries (Mizuno *et al.*, 1979; Tamada 2000). Egg size has an influence on the occurrence of daytime drift (Iguchi, 1985); larvae from small eggs gain more distance during the day than those from large eggs, although overall trends become nocturnal as they move downstream (Fig. 3). Egg size within a species is variable between populations from different rivers, suggestively involved in the distributional difference in spawning site (Tamada, 2005). Having smaller eggs that hatch into smaller larvae which are more likely to drift during the day results in higher dispersal efficiency and appears to be advantageous to populations living in the upper courses of streams and rivers.

The intraspecific variation in egg size indicates that larval body size can be an adaptive trait selected for in the process of invading. The trade-off between size and number of eggs leads to an increase in fecundity with decreasing egg size (Smith & Fretwell, 1974), which may be beneficial to some extent. However, the advantage of small eggs would attain a plateau rapidly because of the minimum size to exist. Monk gobies, famous for being the smallest larvae among vertebrates, often ascend to the headwaters of insular streams and rivers. It is, nevertheless, rare that populations that inhabit the upper courses of continental rivers are able to successfully reproduce (Dôtu & Mito, 1955). A complete turnabout in adaptive strategy, required to provide countermeasures against early mortality, is essential for populations of amphidromous fishes to successfully expand their distributions into habitats in longer rivers. This evolutionary turnabout is the abandonment of amphidromy. In terms of egg size evolution, sympatric speciation is subject to disruptive selection rather than directional selection (Kishi, 1978). It is estimated that large eggs from which larvae hatch at a well-developed stage and that are able to support themselves in running water allows for the evolution of a non-migratory freshwater life. Actually, fluvial species that have larger eggs than their amphidromous congeneric relatives have been identified from gobies; *Rhinogobius flumineus* (Mizuno, 1960b), *Rhinogobius* sp. BB, and *Rhinogobius* sp. YB (Hirashima & Tachihara, 2000), as well as from sculpins; *Cottus nozawae*, and *Cottus pollux* large-egg type (Goto, 1990; Goto & Andoh, 1990). In conclusion, the shift from an amphidromous to a fluvial life is highly likely to allow fish with large eggs to establish populations in longer rivers running through continents (e.g. *Rhinogobius* species from the mid Mekong River, Chen *et al.*, 1999).

Acknowledgments

Thanks to K. Fukuyama, M. Ito, A. Iwata, K. Omori, T. Ohtake, H. Yuge, and the members of Ecology Research Laboratory, Ehime University for their help in the field investigation and discussion while I was a student. I also thank N. Mizuno for his support during the study and his hearty criticism.

Literature Cited

Aonuma, Y., A. Iwata, T. Asahida, H. Ida & T. Kobayashi. 1998. Genetic variation of *Rhinogobius* fishes (Pisces: Gobiidae) around the East China Sea with note on its zoogeography. *DNA Polymorphism* **6**: 116–122.

Blaxter, J.H.S. & K.F. Ehrlich. 1974. Changes in behaviour during starvation of herring and plaice larvae, p. 575–588. *In*: J.H.S. Blaxter (ed.), *Early life history of fish*. Springer-Verlag, Heidelberg.

Breder, C.M. & D.E. Rosen. 1966. *Modes of reproduction in fishes*. Natural History Press, New York. 941 p.

Chen, I.S., M. Kottelat & P.J. Miller. 1999. Freshwater gobies of the genus *Rhinogobius* from the Mekong Basin in Thailand and Laos, with description of three new species. *Zoological Studies* **38**: 19–32.

Dôtu, Y. & S. Mito. 1955. Life history of a gobioid fish, *Sicydium japonicum* Tanaka. *Scientific Bulletin of Faculty of Agriculture, Kyushu University* **15**: 213–221.

Fortier, L. & W.C. Legget. 1983. Vertical migrations and transport of larval fish in a partially mixed estuary. *Canadian Journal of Fisheries and Aquatic Science* **40**: 1543–1555.

Fukui, S. 1979. On the rock-climbing behavior of the goby, *Sicyopterus japonicus*. *Japanese Journal of Ichthyology* **26**: 84–88.

Futuyma, D.J. 1998. *Evolutionary biology*. 3rd ed. Sinauer Associates, Sunderland. 751 p.

Goto, A. 1990. Alternative life-history styles of Japanese freshwater sculpines revisited. *Environmental Biology of Fishes* **28**: 101–112.

———. **& T. Andoh**. 1990. Genetic divergence between the sibling species of river-sculpin, *Cottus amblystomopsis* and *C. nozawae*, with special reference to speciation. *Environmental Biology of Fishes* **28**: 257–266.

Gross, M.R. 1987. Evolution of diadromous fishes. *American Fisheries Society Symposium* **1**: 14– 25.

Hayashi, M., T. Itoh, H. Hayashi, K. Hagiwara & K. Kimura. 1992. Inland-water fish fauna and zoogeographical features in Amami Island, southern Japan. *Scientific Report of Yokosuka City Museum* **40**: 45–63.

Hirashima, K. & K. Tachihara. 2000. Embryonic development and morphological changes in larvae and juveniles of two land-locked gobies, *Rhinogobius* spp. (Gobiidae), on Okinawa Island. *Japanese Journal of Ichthyology* **47**: 29–41.

Iguchi, K. 1985. Larval drift of amphidromous gobies, especially *Rhinogobius brunneus*. Unpublished Master of Science Thesis, Ehime University, Matsuyama, Japan.

———. **& T. Hino**. 1996. Effect of competitor abundance on feeding territoriality in a grazing fish, the ayu *Plecoglossus altivelis*. *Ecological Research* **11**: 165–173.

———. **& N. Mizuno**. 1990. Diel changes of larval drift among amphidromous gobies in Japan, especially *Rhinogobius brunneus*. *Journal of Fish Biology* **37**: 255–264.

———. **& N. Mizuno**. 1991. Mechanisms of embryonic drift in the amphidromous goby, *Rhinogobius brunneus*. *Environmental Biology of Fishes* **31**: 295–300.

———. **& N. Mizuno**. 1999. Early starvation limits survival in amphidromous fishes. *Journal of Fish Biology* **54**: 705–712.

Ishida, R. & Y. Ohoshima. 1959. On the adhesive power of the eggs of ayu, *Plecoglossus altivelis* T. & S. *Bulletin of the Japanese Society of Scientific Fisheries* **24**: 696–899.

Ito, S. & Y. Yanagisawa. 2003. Mate choice and mating pattern in a stream goby of the genus *Rhinogobius*. *Environmental Biology of Fishes* **66**: 67–73.

Kani, T. 1944. Ecology of torrent-inhabiting insects, p. 171–317. *In*: H. Furukawa (ed.), *Fauna flora in Japan, Insects 1*. Kenkyu-sha, Tokyo.

Kawanabe, H. & N. Mizuno. 1989. *Freshwater fishes of Japan*. Yamatokeikoku-sha, Tokyo. 719 p.

Kishi, Y. 1978. Evolution of egg size. *Biological Science* **30**: 48–56.

MacArthur, R.H. 1972. *Geographical ecology*. Harper & Row Publishers, New York. 269 p.

McDowall, R.M. 1988. *Diadromy in fishes: migration between freshwater and marine environments*. Croom Helm, London. 308 p.

———. 1992. Diadromy: origins and definitions of terminology. *Copeia* **1992**: 248–251.

———.1997a. The evolution of diadromy in fishes (revised) and its place in phylogenetic analysis. *Reviews in Fish Biology and Fisheries* **7**: 443–462.

———. 1997b. Is there such a thing as amphidromy? *Micronesica* **30**: 3–40.

Mizuno, N. 1960a. Study on fresh water goby, *Rhinogobius brunneus* Gill, with a proposition on the relationship between landlocking and speciation of some fresh water gobies in Japan. *Memoirs of the College of Science* (B) **27**: 97–115.

———. 1960b. Description of a new freshwater goby from Japan. *Memoirs of the College of Science* (B) **27**: 117–119.

———. & **H. Kawanabe**. 1981. A topographical classification of streams, with an introduction widely used in Japan. I. Reach type, stream zone and stream type. *Verhandlungen der Internationalen Vereinigung für Theoretische und Angewandte Limnologie* **21**: 913.

———., **S. Uehara & M. Maki**. 1979. Studies on a fresh water fish, *Rhinogobius brunneus* (Gobiidae) IV. Habitat segregation among sympatric of 4 colour types. *Japanese Journal of Ecology* **29**: 137–147.

Moriyama, A., Y. Yanagisawa, N. Mizuno & K. Omori. 1998. Starvation of drifting goby free embryos due to retention in the upper stream. *Environmental Biology of Fishes* **52**: 321–329.

Motoda, S. 1972. Zooplankton ecology with particular reference to the behaviour of diurnal vertical migration. *Journal of Oceanography Society of Japan*. **28**: 278–292.

Mukai, T. 2001. Hybridization and introgression in the speciation process of fishes. *Japanese Journal of Ichthyology* **48**: 1–18.

Myers, G.S. 1949. Usage of amphidromous, catadromous and allied terms for migratory fishes. *Copeia* **1949**: 89–97.

Nagoshi, M. 1982. Diel vertical migration of zooplankters and fish larvae in Lake Biwa. *Bulletin of the Faculty of Fisheries, Mie University* **9**: 1–10.

Nelson, J.S. 1984. *Fishes of the world*. 2nd ed. John Wiley & Sons, New York. 523 p.

Nishijima, S. & N. Shikatani. 1994. Streams on islands. *In*: S. Ikehara & S. Shokita (eds.), *Crystal streams on Ryukyu Islands*. Okinawa-syuppan, Okinawa.

Nishimoto, R.T. & D.G.K. Kuamo'o. 1991. The occurrence and distribution of the native goby (*Lentipes concolor*) in Hawaiian Island streams with notes on the distribution of other native fish species. *In*: W. Devick (ed.), *Proceedings of the 1990 Symposium on Freshwater Stream Biology and Fisheries Management*, Hawaii Department of Land and Natural Resources, Honolulu.

Otto, D. & J.E. Endler. 1989. *Speciation and its consequences*. Sinauer Associates, Sunderland. 679 p.

Paradis, A.R., P. Pepin & J.A. Brown. 1996. Vulnerability of fish eggs and larvae to predation: Review of the influence of the relative size of prey and predator. *Canadian Journal of Fisheries and Aquatic Science* **53**: 1226–1235.

Sakai, H. & F. Yasuda. 1978. Development of eggs and larvae of the freshwater goby, *Rhinogobius brunneus*. *Japanese Journal of Ichthyology* **25**: 92–100.

Seliverstov, A.S. 1973. Vertical migrations of larvae of the Atlanto-Scandian herring (*Clupea harengus* L.), p-. 253–262. *In*: J.H.S. Blaxter (ed.), *Early life history of fishes*, Springer-Verlag, Heidelberg.

Senou, H., K. Yano, T. Suzuki & K. Shibukawa. 2004. *A photographic guide to the gobioid fishes of Japan*. Heibonsya, Tokyo.

Shimizu, T., N. Taniguchi & N. Mizuno. 1993. An electrophoretic study of genetic difference of a Japanese goby, *Rhinogobius brunneus*. *Japanese Journal of Ichthyology* **39**: 329–343.

Shinomiya, A. & T. Ike. 1992. Fish fauna of rivers in Amamioshima Island. *Memoirs of Faculty of Fisheries in Kagoshima University* **41**: 77–86.

Smith, C.C. & S.D. Fretwell. 1974. The optimal balance between size and number of offspring. *American Naturalist* **108**: 499–506.

Tago, Y. 1999. Estimation of abundance of ayu larvae during seaward drifting at the Shou River. *Nippon Suisan Gakkaishi* **65**: 718–727.

Takahashi, S. 1981. Vertical distribution and diel migration of isaza, *Chaenogobius isaza*, Pisces in Lake Biwa. *Zoological Magazine* **90**: 145–151.

Takaya, Y. 1963. Stratigraphy of Paleo-Biwa group and the paleogeography of Lake Biwa with special reference to the origin of the endemic species in Lake Biwa. *Memoirs of the College of Science, University of Kyoto* (B) **30**: 81–119.

Tamada, K. 2000. Distributions of the spawning grounds of four amphidromous gobies (*Rhinogobius* spp.) in the Aizu River, Wakayama Prefecture, Japan. *Japanese Journal of Ichthyology* **47**: 55–59.

Tamada, K. 2005. Clutch and egg size in *Rhinogobius* sp. DA inhabiting small rivers. *Japanese Journal of Ichthyology* **52**(1): 17–20.

Tanimizu, K., T. Miura & M. Higashi. 1981. Effect of water movement on the photosynthetic rate of algal community attached to reed stem. *Verhandlungen der Internationalen Vereinigang für Theoretische und Angewandte Limnologie* **21**: 584–589.

Tsukamoto, K. 1991. Age and growth of ayu larvae *Plecoglossus altivelis* collected in the Nagara, Kiso and Tone rivers during the downstream migration. *Nippon Suisan Gakkaishi* **57**: 2013–2022.

Tsukamoto, K., T. Kajihara & M. Nishiwaki. 1975. Swimming ability of fish. *Bulletin of the Japanese Society of Scientific Fisheries* **41**: 16.

Woodhead P.M.J. & D. Woodhead. 1955. Reactions of herring larvae to light: a mechanism of vertical migration. *Nature* **176**: 349–350.

Biology of Hawaiian Streams and Estuaries. Edited by N.L. Evenhuis & J.M. Fitzsimons. *Bishop Museum Bulletin in Cultural and Environmental Studies* 3: 87–98 (2007).

The Potential for Source-Sink Population Dynamics in Hawaii's Amphidromous Fishes

MARK G. MCRAE
Museum of Natural Science, Louisiana State University, Baton Rouge, LA 70803 U.S.A.; e-mail: mmcrae@paws.lsu.edu

Abstract

To ensure successful reproduction, Hawaii's amphidromous gobioid fishes (*'o'opu*) must have their offspring successfully migrate from instream hatching sites to oceanic larval habitats. The differential ability of the five species of *'o'opu* to climb waterfalls influences how far inland each species can invade in individual streams and determines their instream distributions. Given the short amount of time that *'o'opu* free-embryos can persist in fresh water, a possible negative correlation exists between the distance that an individual lives from the stream-mouth and the likelihood that the offspring that individual produces will successfully migrate to the ocean. Recruitment of post-larvae originally spawned in streams that function as source habitats can sustain local populations of *'o'opu* that live in streams that function as sink habitats. For this reason, management techniques that utilize population density or biodiversity as indicators of habitat quality are not applicable for the management of *'o'opu* populations. A clearer understanding of the role that the pre-marine phase is playing in the population dynamics of *'o'opu* will allow for more responsible freshwater resource management in Hawai'i.

Introduction

The fishes, crustaceans, and mollusks that make up Hawaii's indigenous stream-dwelling macrofauna are believed to be derived from marine ancestors and to have retained a pelagic marine larval phase (Radtke & Kinzie, 1996); all nine species have amphidromous life cycles. All of the five species of fishes are in the suborder Gobioidei and are referred to collectively as *'o'opu*. *Lentipes concolor*, *Sicyopterus stimpsoni*, *Stenogobius hawaiiensis*, and *Awaous guamensis* are in the family Gobiidae, while *Eleotris sandwicensis* is in the family Eleotridae. Amphidromy is a form of diadromy, or a life cycle that is split between freshwater and marine environments (McDowall, 1992). Specifically, juveniles and adults live in stream or river habitats, where demersal spawning takes place. Newly hatched free-embryos drift downstream towards their oceanic feeding sites. Once in the ocean, the larvae become a part of the pelagic zooplankton community until they mature into post-larvae (Radtke *et al.*, 1988). Post-larvae must then migrate back into fresh water where they metamorphose into juveniles (Nishimoto & Kuamo'o, 1997). Due to this dependence on marine environments, none of the indigenous species of fishes and macroinvertebrates that live in Hawaii's streams and rivers can be considered primary or secondary freshwater organisms (McDowall, 1997; Balon & Bruton, 1994).

Indeed, these species' life-history characteristics are much more similar to those for species which inhabit nearshore marine environments such as coral reefs and intertidal zones (Leis, 1991; Pfister, 1999). Similar to patch reefs and discrete tide pool habitats, each stream represents a habitat patch (Hanski, 1991). The groups of individuals inhabiting each stream make up local populations which are portions of the larger genetically and ecologically connected meta-population (Hanski, 1991; Chubb *et al.*, 1998). It is the marine larval phase which connects these local populations.

Stochastic phenomena in marine environments can lead to extremely high mortality rates for ichthyoplankton through mechanisms such as starvation, predation, or advection away from suitable larval or adult habitats (Doherty, 1991; Houde, 1989). Stochastic events have strong effects on observed patterns in recruitment, and ultimately, adult population and community structure (Doherty & Fowler, 1994; Pfister, 1999). Organisms with amphidromous life cycles are similarly subject to high mortality rates that take place during their marine larval stage (Radtke & Kinzie, 1996). Unlike nearshore marine species, however, amphidromous species have an additional life-history stage during which they may experience high mortality rates (Bell & Brown, 1995; Iguchi & Mizuno, 1999). The time period which begins at hatching and extends until the free-embryo passes out of the mouth of the stream into the ocean represents a critical step in this complex life cycle, during which biological, hydrologic, hydrographic, geomorphometric and anthropogenic stream features may affect migratory success and survival (Iguchi & Mizuno, 1990; Bell & Brown, 1995; Lindstrom, 1998; Moriyama *et al.*, 1998; Way *et al.*, 1998; Iguchi & Mizuno, 1999; March *et al.*, 2003). The life stage during which free-embryos must migrate to the ocean can be thought of as a "pre-marine" stage (Font, 1996), and will be referred to as such throughout the remainder of this paper.

The purpose of this review is to explore the existing literature on pre-marine stage amphidromous organisms, with a focus on the amphidromous fishes of the Hawaiian Islands. A description of the range of stream habitats that must be traversed by seaward migrating larvae will be followed by a description of the instream distribution patterns of the adult fishes. A review of what is currently known about the behavior, physiology, and ecology of pre-marine amphidromous fishes will then be summarized to gain insight into the potential implications that the unique aspects of this life history stage have on the population ecology and conservation of these species.

Lotic Habitats in the Hawaiian Islands

Polhemus *et al.* (1992) outlined a classification for continuous perennial stream reaches on islands in the tropical Pacific in which three main reach-type divisions were described based on gradient, physico-chemical characteristics, and faunal composition. Headwater reaches were described as being high in elevation (> 800 meters) with high gradient (> 30%), and substratum comprised mainly of bedrock. Midreach areas are generally between 50 and 800 meters in elevations, between 5% and 30% gradient, with substrata predominantly boulders, rocks and gravel. Terminal reaches were defined as "watercourses below sharp gradient that bars upstream migration of itinerant marine fishes" (Polhemus *et al.*, 1992). Elevations in terminal reaches are usually <50m and gradients <5%, with substratum consisting mainly of cobble, gravel, sand, and sediment. Mizuno & Kawanabe (1981) outlined a topographically based classification system for stream reaches in Japan which defines reach types by two topographic criteria. The first criterion is the number of riffle/pool sequences in stream segments between meander bends. Type "A" reach types have more than one riffle/pool sequence between bends, while Type "B" reaches have only one. The second criterion relates to the gradient and turbulence of riffles or cascades connecting the pools. Type "a" are waterfalls, Type "b" are turbulent, high-gradient riffles or cascades, and Type "c" are smooth, low-gradient riffles. Thus, a steep reach having two or more plunge pools separated by waterfalls in each segment between meander bends is defined as Type Aa (Fig.1), whereas a reach of moderate gradient with only one pool and turbulent riffle between bends is defined as Type Bb. Mizuno & Kawanabe's (1981) finer-scaled classification system will be used, along with the broader-scale classification of Polhemus *et al.* (1992) to describe the range of stream and reach types that amphidromous fishes inhabit in the Hawaiian Islands.

Two main stream types exist in the Hawaiian Islands, and they represent the extremes in a spectrum of stream morphologies which ranges from low discharge, high gradient streams to relatively high discharge, long streams with terminal reaches that can form estuarine habitats near the stream mouth (Nishimoto & Kuamo'o, 1991). On the geologically younger islands such as Hawai'i and Maui, short, steep, straight streams are the most common stream type (Nishimoto & Kuamo'o, 1991), with Type Aa and Ab reaches extending from the mouth of the stream up to the headwaters. In these relatively young streams, waterfalls, cascades, and plunge pools are abundant (Fig. 1), with the stream often entering the ocean as a terminal waterfall – the stream falls off the edge of a cliff

Figure 1. Midreach of Umauma Stream, Island of Hawai'i, demonstrating an example of "Aa" topography.

either into the ocean or onto a narrow rocky beach (Fig. 2). On the relatively older islands such as O'ahu and Kaua'i, larger stream systems are more common than they are on the younger islands. Long, low gradient terminal reaches exist upstream of sinuous mixohaline estuaries (Nishimoto & Kuamo'o 1991), above which can exist long midreach areas dominated by Type Bc and Type Bb reaches which grade into type Ab reaches farther inland. Type Aa reaches are encountered in these "terminal estuary streams" in upper midreach and headwater areas which can be several kilometers inland from the mouth of the stream (Fig. 3).

Instream Distributions of Hawaiian Amphidromous Gobies

The distributions of juvenile and adult amphidromous fishes in Hawaiian streams are strongly influenced by the ability of each species' postlarvae to surmount obstacles to upstream migration (Nishimoto & Kuamo'o, 1997). Three of the five species have the ability to climb the vertical rock

Figure 2. Terminal waterfall at the mouth of Manoloa Stream on the Island of Hawai'i.

faces associated with waterfalls when they are post-larvae and juveniles by using pelvic fins which are fused to form a suction disk. *Eleotris sandwicensis* lacks fused pelvic fins while *S. hawaiiensis* has fused pelvic fins, but lacks the associated musculature necessary for climbing waterfalls (Nishimoto & Kuamo'o, 1997). Both species, therefore, are precluded from dispersing upstream of any precipitous waterfall and are restricted in their distributions to terminal reaches and terminal estuaries. *Awaous guamensis* and *S. stimpsoni* can climb and disperse beyond moderately high waterfalls (less than ~20 m) and are found in Type Bc, Bb, and Aa midreaches up to an elevation of approximately 150 m (Fitzsimons & Nishimoto, 1990). Vertical waterfall height does not appear to be a limiting factor for *L. concolor* as they migrate upstream. *Lentipes concolor* has been sighted at elevations higher than 1 km (Fitzsimons & Nishimoto, 1990) and in stream reaches that were located above a waterfall with a more than 300 m vertical height (Englund & Filbert, 1997; personal observation).

Because the instream distributions of the five amphidromous Hawaiian gobioids are largely determined by their waterfall-climbing ability, it follows that the distribution patterns will vary with the type of stream into which they recruit (Nishimoto & Kuamo'o, 1991). In large streams with terminal estuaries, it is common to find juveniles and adults of all five species, with *E. sandwicensis* and *S. hawaiiensis* occurring only in the terminal estuary, terminal reach, and lower midreach areas. *Awaous guamensis* and *S. stimpsoni* are common in the upper regions of the terminal reach and in midreach areas, with *L. concolor* farther upstream in the upper midreaches (Fig. 4). In these terminal estuary streams, *L. concolor* appear to migrate inland until they encounter their preferred Type Aa habitats (Kinzie, 1988) and are therefore often located farther inland than any of the other species

Figure 3. Example of a terminal estuary stream. Wailoa Stream, Waipi'o Valley, Island of Hawai'i. Photo by D.G.K. Kuamo'o.

(Nishimoto & Kuamo'o, 1991; Kinzie, 1988; Kido *et al.*, 2002). Similar waterfall-delineated in-stream distribution patterns have been described for assemblages of amphidromous gobies in Micronesia (Parham, 1995; Nelson *et al.*, 1997).

In terminal-waterfall streams, the two non-climbing species, *E. sandwicencis* and *S. hawaiiensis* are either absent or restricted to the short stretch of stream running across the narrow rocky beaches that are often at the base of coastal cliffs (Nishimoto & Kuamo'o 1991; Fig. 2). The number of species that occur upstream of terminal waterfalls can range from one to three and depends on the height of the waterfall. In streams with the highest terminal falls, *L. concolor* can be the only

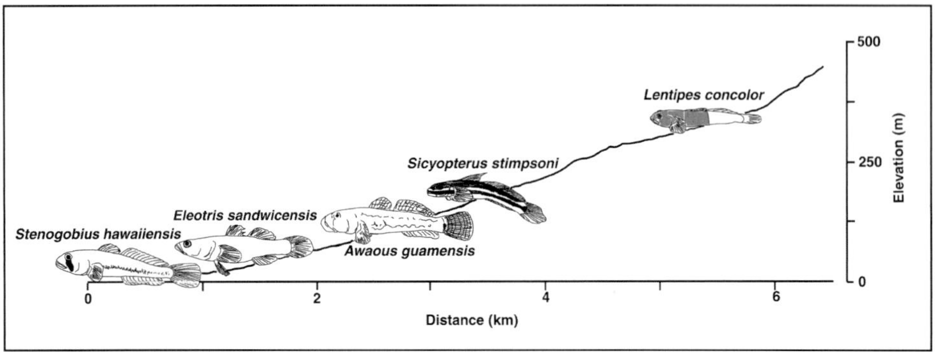

Figure 4. Elevational profile of a terminal-estuary stream on the Big Island of Hawai'i (Hakalau Stream). Adapted with permission from Nishimoto and Kuamo'o (1991).

species present, and they often occur in high densities a relatively short linear distance from the mouth of the stream, presumably because they encounter preferred Type Aa stream reaches as soon as they arrive at the top of the falls (Fig. 5).

When one considers the wide range of distances and reach types that seaward migrating pre-marine amphidromous fishes must travel on their way to oceanic feeding sites, it becomes clear that there is a high potential for this pre-marine stage to play a critical role in the population dynamics of these species. This is especially true for populations of *L. concolor*, which can be found a few meters away from the ocean in small, terminal-waterfall streams, or several kilometers away from the ocean in the upper midreaches and headwaters of larger terminal-estuary streams (Nishimoto & Kuamo'o, 1991). The following section synthesizes the research that has been done on the behavior and survival of pre-marine amphidromous fishes that relates to potential survival limitations.

Pre-Marine Amphidromous Fishes

Amphidromous fishes emerge from demersally spawned eggs in a highly under-developed state and are referred to as free-embryos (Balon, 1990; Balon & Bruton, 1994). During the pre-marine stage, free-embryos are unable to feed and survive on nutrition from a yolk sac until they arrive at their oceanic feeding sites (Lindstrom, 1998; Bell & Brown, 1995). *Awaous guamensis*, *L. concolor*, and the Caribbean species *Sicydium punctatum* were all observed to die after being held in fresh water for only four days (Ego, 1956; Lindstrom, 1998; Bell & Brown, 1995). Tomihama (1972) reported that *S. stimpsoni* free-embryos died after five days in fresh water. *Rhinogobius brunneus* (Japan) free-embryos lived for seven days in fresh water (Iguchi and Mizuno, 1999), but were observed to reach an irrecoverable starvation state (ISS) after approximately six days. If it is assumed that the Hawaiian species similarly reach ISS at some point before death in fresh water, it is clear that pre-marine *'o'opu* have a discrete "window of opportunity" to move from their freshwater hatching sites to their oceanic feeding sites.

Temporal patterns in seaward migration of pre-marine amphidromous fishes have been observed at both the lunar and daily scale. Lindstrom (1998) reported that spawning in captive populations of *L. concolor* peaked during the new moon, which corresponds to observed new-moon peaks in the seaward migration of *L. concolor* free-embryos captured during stationary drift-net sampling (McRae, unpublished data). Lindstrom (1998) also reported that hatching of eggs from captive *L. concolor* took place after sunset. Post-sunset hatching may explain post-sunset peaks in the daily drift patterns observed for that species in streams on the Island of Hawai'i (McRae, unpublished data). New moon and post-sunset peaks in migration possibly evolved as ways to avoid capture by visually oriented planktivores that live in lower stream reaches and estuaries. No significant seasonal pattern in seaward drift has been observed in *'o'opu* free-embryos (Way *et al.*, 1998; Lindstrom, 1998).

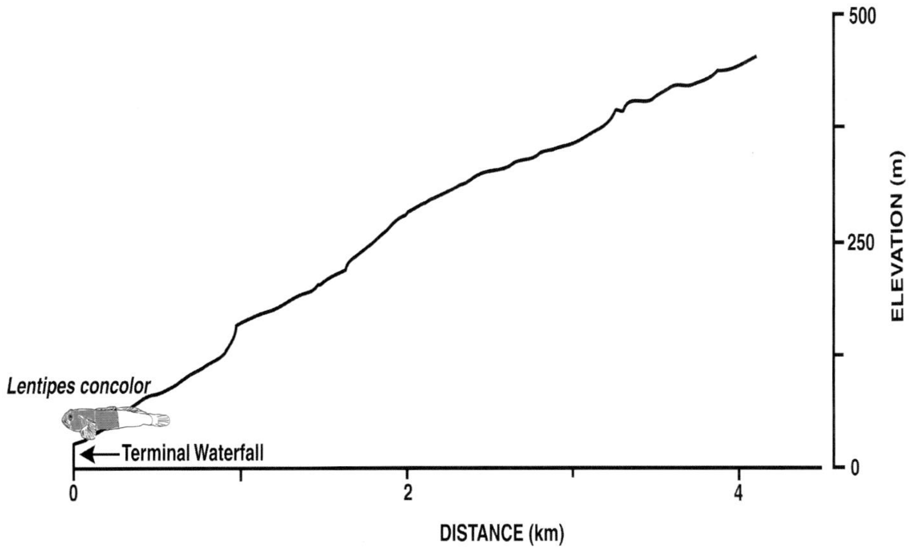

Figure 5. Elevational profile of a terminal-waterfall stream (Manoloa Stream) on the Big Island of Hawai'i. Adapted with permission from Nishimoto & Kuamo'o (1991).

It has been hypothesized that pre-marine amphidromous fishes can experience high mortality rates during their downstream journey towards the ocean. *Sicydium punctatum* free-embryos captured in streams on the island of Dominica, W.I. were generally <24 hours old (Bell & Brown, 1995). Bell & Brown (1995) also estimated that mortality rates for newly hatched *S. punctatum* were as high as 50% hour^{-1}. Moriyama *et al.* (1998) concluded that in normal or low river flow conditions, most *R. brunneus* free-embyos hatched at high gradient upstream reaches likely starved to death before reaching the sea.

Adults of the amphidromous species *A. guamensis* in Hawai'i (Ego, 1956; Harrison *et al.*, 1991; Kido & Heacock, 1991) and *Plecoglossus altivelis* in Japan (Iguchi *et al.*, 1998) migrate to spawning sites in lowland stream reaches near to stream mouths presumably as a strategy that reduces the distance and time required for free-embryos to reach the ocean. However, for species such as *L. concolor*, which often inhabit stream reaches that are upstream of major waterfalls, such downstream migrations have not been observed. For these species, the distance that free-embryos are required to migrate downstream roughly estimates total drifting time (Iguchi & Mizuno, 1999) and as a result likely has a strong effect on the overall fitness of individuals in local populations. The following section describes features of Hawaiian streams that may lengthen total drifting time and contribute to increased risk of mortality for pre-marine amphidromous gobies.

Instream Obstacles to Seaward Migration

Stream features that exist between hatching sites and the ocean can potentially increase the amount of time that *'o'opu* free-embryos spend in-transit on their way to their oceanic feeding sites and consequently can decrease their chances for survival. Low water velocity habitats in midreach, terminal reach, and terminal estuaries can retain the passively migrating free-embryos in freshwater habitats. Terminal reaches and estuaries also often have high densities of introduced freshwater and itinerant marine zooplanktivorous fishes (McRae, 2001) that likely prey upon drifting *'o'opu*.

Low water velocity habitats, such as plunge pools, side pools, and backwater eddies are abundant in the Aa and Ab reaches that are common in the headwaters and upper midreaches of Hawaiian streams. On the geologically young islands of the Hawaiian Archipelago, waterfalls are abundant throughout the midreaches of perennial streams. Their associated plunge pools can be very deep and wide in spite of the relatively low discharge of these watersheds (Fig. 1). Each of these plunge pools, during all but the most intense freshets, represents a naturally semi-impounded habitat, inside which water velocities can be practically imperceptible. These midreach low-water-velocity habitats are likely significant obstacles to pre-marine gobies as they migrate to sea (Moriyama *et al.*, 1998). Plunge-pools can also contain large populations of introduced poeciliid fishes (e.g., guppies, *Poecilia reticulata* and swordtails, *Xiphophorus helleri*) that may prey upon the free-embryonic gobies that are entrained in these low-water-velocity habitats.

Similarly, long terminal stream reaches and terminal estuaries not only increase the amount of time that pre-marine *'o'opu* spend in transit to the ocean, but they also are often home to large populations of native and introduced zooplanktivorous freshwater and marine fishes (McRae, 2001). The endemic āholehole, *Kuhlia xenura*, feed actively on food items drifting downstream and are common in terminal stream reaches in Hawai'i. Their adaptation for nocturnal activity patterns potentially allow them to feed during the post-sunset peak in seaward migration of pre-marine gobies. Introduced mosquitofish, *Gambusia affinis*, are well known larvivores and zooplanktivores (Arthington, 1991; Komak & Crossland, 2000) that likely feed very heavily on any free-embryonic amphidromous organisms that end up in the slack-water habitats they prefer (McRae, 2001). Shortfin mollies, *Poecilia mexicana*, swordtails, and tilapia, *Oreochromis mossambicus* are common introduced fishes in Hawaii's terminal streams and estuaries (Devick, 1991; McRae, 2001) that are known to be at least opportunistically zooplanktivorous.

Conclusions

Population Ecology

The term "metapopulation" has been defined as a system of local populations connected by dispersing individuals (Hanski, 1991). The local populations of amphidromous fishes that live in individual streams throughout the Hawaiian Islands likely represent portions of a larger, interconnected metapopulation (Kinzie, 1993; Font, 1996). Linked closely with this concept is the idea that habitat-specific demographic rates of local populations can have significant impacts on the growth and regulation of the overall metapopulation (Pulliam, 1988). "Sink" habitats are habitat patches (e.g., individual streams) where within-habitat reproduction is insufficient to balance local mortality. Local populations in sink habitats can nonetheless persist through the input of continual immigration from more-productive "source" habitats. Source habitats are, in general, net exporters of individuals, while sinks are net importers of individuals (Pulliam, 1988).

Large streams with terminal estuaries may be playing a critical role as source habitats for metapopulations of non-climbing *'o'opu* (*S. hawaiiensis* and *E. sandwicensis*) throughout the Hawaiian Islands. The low gradient, low water velocity habitats that are preferred by nonclimbing *'o'opu* are abundant in the long, sinuous terminal reaches of these streams, and these habitats are located close to stream mouths, where free-embryos can have easy access to marine larval habitats. The large, high discharge systems produce large freshwater plumes that extend into the nearshore marine environment and likely act as strong olfactory navigational cues for post-larvae of all five species to recruit into these streams (Nishimoto & Kuamo'o, 1997). Individuals of the three climbing *'o'opu* (*A. guamensis*, *S. stimpsoni*, and *L. concolor*) disperse upstream until they encounter the large amounts of suitable adult habitat that exists in the inland mid and headwater reaches. These relatively large stream systems may represent ecological sinks for the three species of climbing goby that recruit into these habitat patches because a large proportion of their offspring may starve to death and/or be eaten before reaching the ocean. This may be especially true for *L. concolor* because of their tendency to occur farther inland in streams of this type than any other species of *'o'opu* (Fig. 4). It is possible that the turbulent flow characteristics of these inland, high gradient reaches, together with the length, low-gradient, and high predator densities of terminal reaches and estuaries may retain *L. concolor* free-embryos at instream sites longer than their three to four day migratory "window-of-opportunity" (Lindstrom, 1998). Large inland populations of *L. concolor* and *S. stimpsoni* (Nishimoto & Kuamo'o, 1991; Kido *et al.*, 2002) are possibly sustained by recruitment of post-larvae originally spawned in smaller streams by relatively small, near-shore adult populations.

Conversely, the smaller terminal waterfall streams are probably important ecological sources for the three waterfall climbing species. This is especially likely for *L. concolor* which can be found in very high population densities a very short distance from the ocean in these streams (Fig. 5). Thus, the majority of the free-embryos that are produced in terminal waterfall streams are likely successful in migrating downstream to their oceanic larval habitats. These small, high gradient systems, however, are almost assuredly ecological dead ends, or sinks, for individuals of the two non-climbing amphidromous fishes that may recruit into them.

There are many streams throughout the Hawaiian Islands that are intermediate in length, gradient, and reach-type diversity and are not categorized adequately by either the terminal-waterfall or terminal-estuary stream descriptions. The distribution of *L. concolor* and/or *S. stimpsoni* in these watersheds can often start a few to a few hundred meters above the stream mouth and extend for several kilometers inland. If individual fitness is defined by the contribution made to future generations (Smith, 1995), then there could be a negative correlation between fitness and distance from the stream mouth for individuals of the waterfall-climbing species of *'o'opu* in Hawaiian streams.

Conservation

Most fishes that live in continental streams and rivers are primary freshwater fishes that complete their entire life cycles within individual streams. As such, the fitness or health of individual populations in continental streams can be correlated with population size. Similarly, in these "closed" systems, it can be argued that population size is an indicator of habitat quality. These relationships between population size and population fitness or habitat quality have led to the development of management techniques that allow a comparison of the relative health (or biological integrity) of individual habitats (Karr & Chu, 1999). Specific management plans can then be developed for individual habitats based on these assessments.

The unique aspects of the life-history strategies and population ecology of the indigenous fishes that live in Hawaii's streams and rivers necessitate unique approaches in the management of these species. The profound differences in the biological characteristics and ecological requirements that exist between adult fishes and free-embryos result in no direct correlation between adult local population density and the ability of the individuals in that local population to successfully reproduce. A similar lack of a relationship exists between local population density and habitat quality. If a habitat is defined as an area in which a local population can successfully live and reproduce, then for Hawaiian amphidromous fishes, population size or density alone does not necessarily correspond to habitat quality. Habitat quality in individual streams must be assessed with regard to whether the habitat patch is a net exporter (source) or net importer (sink) of individuals. Such definitions of habitat quality depend on many variables including distance of reproducing individuals from the ocean, overall stream gradient between hatching sites and the ocean, water velocity-depth relationships between hatching sites and the ocean, overall densities and distributions of predators downstream from hatching sites, *as well as* the local population density of adult *'o'opu* that occupy that stream. When such variables are acknowledged as being critical measures of habitat quality in a stream, it becomes clear that densities of adult fishes in lower-quality, sink habitats may actually be greater at times than in the high-quality source habitats (Van Horne, 1983; Pulliam, 1988). If amphidromous fishes in Hawai'i exist in open populations, then the scale at which habitat quality assessments should be made increases beyond the level of individual streams (Caley *et al.*, 1996).

Patterns in the instream distributions of adult amphidromous fishes in Hawaiian streams are species-specific (Nishimoto & Kuamo'o, 1991). Stream characteristics that define high-quality source habitats for one species (e.g. terminal waterfall streams for *L. concolor*) can define low-quality sink habitats for another species (e.g., terminal waterfall streams for *E. sandwicensis*). The overall biodiversity of native amphidromous fishes in a single habitat patch (stream) is not, therefore, necessarily correlated to habitat quality. A stream that contains the entire compliment of Hawaii's amphidromous fishes may only be a source for those species that live nearest to the mouth of the stream. For those species that live relatively far inland in these streams, recruitment into the habitat could far exceed successful seaward migration. Conversely, a stream could only have one species of fish present, and yet such a stream could "export" far more individuals of that one species than it "imports".

Most anthropogenic perturbations in streams take place in lowland, nearshore reaches that lie between hatching sites and the ocean. Such habitat alterations could negatively impact the species of *'o'opu* that inhabit the affected reaches (e.g., *S. hawaiiensis* and *E. sandwicensis*), as well as species that may live farther upstream (e.g., *L. concolor* and *S. stimpsoni*). Channelization or water extraction can alter stream morphologies and flow characteristics which could decrease the chances of successful seaward migration of *'o'opu* free-embryos. The high densities of introduced larvivorous fishes that are commonly found in lowland terminal reaches could also result in significant increases in the mortality of pre-marine *'o'opu*. Human disturbances in lowland areas can also eliminate near-ocean, high quality adult habitats for *L. concolor* and *S. stimpsoni* while allowing upstream, inland populations to persist in undisturbed adult habitats in the same watershed. The high population densities of these species that exist in the inland reaches of such streams could actually be utilizing the lowest quality habitats present in that watershed because they might only rarely, if ever, have their offspring successfully migrate to sea.

While a handful of landmark studies have focused on the pre-marine stage in the life cycle of Hawaii's amphidromous fishes (Ego, 1956; Lindstrom, 1998) detailed descriptions of the biology, behavior, and ecology of the pre-marine stages of all *'o'opu* species are needed. What are the spatial and temporal patterns in seaward migration for all five species? Are some streams or stream reaches functioning as source habitats and others as sink habitats? Are introduced predators consuming significant numbers of larval *'o'opu*? What is the appropriate scale at which to assess habitat quality for amphidromous fishes in Hawai'i? These, and related questions need to be addressed so that resource managers in Hawai'i will be equipped to make responsible decisions regarding stream protection and restoration.

Acknowledgements

Discussions with L.K. Benson, C.T. Chong, D.G.K. Kuamo'o, C.A. Murphy, and R.T. Nishimoto provided the inspiration and encouragement that made writing this review possible. The manuscript was improved by comments from L.K. Benson, J.M. Fitzsimons, and C.A. Murphy.

Literature Cited

Arthington, A.H. 1991. Ecological and genetic impacts of introduced and translocated fresh-water fishes in Australia. *Canadian Journal of Fisheries & Aquatic Science* **48**: 33–43.

Balon, E.K. 1990. Epigenesis of an epigeneticist: the development of some alternative concepts on the early ontogeny and evolution of fishes. *Guelph Ichthyological Review* **1**: 1–42.

———. **& M.N. Bruton**. 1994. Fishes of the Tatinga River, Comoros, with comments on freshwater amphidromy in the goby *Sicyopterus lagocephalus*. *Ichthyological Exploration of Freshwaters* **5**: 25–40.

Bell, K.N.I. & J.A. Brown. 1995. Active salinity choice and enhanced swimming endurance in 0 to 8-d-old larvae of diadromous gobies, including *Sicydium punctatum* (Pisces), in Dominica, West Indies. *Marine Biology* **121**: 409–417.

Caley, M.J., M.H. Carr, M.A. Hixon, T.P. Hughes, G.P. Jones & B.A. Menge. 1996. Recruitment and local dynamics of open marine populations. *Annual Review of Ecology & Systematics* **27**: 477–500.

Chubb, A.L., R.M. Zink & J.M. Fitzsimons. 1998. Patterns of mtDNA variation in Hawaiian freshwater fishes: the phylogeographic consequences of amphidromy. *Journal of Heredity* **89**: 8–16.

Devick, W.S. 1991. Patterns of introductions of aquatic organisms into Hawaiian fresh waters, p. 189–213. *In*: Devick, W. (ed.), *New directions in research, management and conservation of Hawaiian freshwater stream ecosystems: Proceedings of the 1990 symposium on stream biolo-*

gy and fisheries management, State of Hawai'i Department of Land and Natural Resources, Division of Aquatic Resources, Honolulu.

Doherty, P.J. 1991. Spatial and temporal patterns in recruitment, p. 261–293. *In*: Sale, P.F. (ed.), *Ecology of fishes on coral reefs*. Academic Press, San Diego.

———. **& T. Fowler**. 1994. An empirical test of recruitment limitation in a coral reef fish. *Science* **263**: 935–939.

Ego, K. 1956. Life history of fresh water gobies. Project no. F-4-R. Fresh Water Game Fish Management Research, Department of Land and Natural Resources, Territory of Hawai'i, Honolulu. 24 p.

Englund, R. & R. Filbert. 1997. Discovery of the native stream goby, *Lentipes concolor*, above Hawaii's highest waterfall, Hi'ilawe Falls. *Bishop Museum Occasional Papers* **49**: 62–64.

Fitzsimons, J.M. & R.T. Nishimoto. 1990. Territories and site tenacity in males of the Hawaiian stream goby *Lentipes concolor* (Pisces: Gobiidae). *Ichthyological Exploration of Freshwaters* **1**: 185–189.

Font, W.F. 1996. Post symposium reflections, p. 299–304. *In*: Devick, W.S. (ed.), *Will stream restoration benefit freshwater, estuarine, and marine fisheries? Proceedings of the October 1994 Hawai'i stream restoration symposium*, State of Hawaii Department of Land and Natural Resources, Division of Aquatic Resources, Honolulu.

Hanski, I. 1991. Metapopulation dynamics: brief history and conceptual domain. *Biological Journal of the Linnean Society* **42**: 3–16.

Harrison, J.T., R.A. Kinzie III, G.C. Smith, P.Y. Tamaru, D. Heacock, L. Honigan, A. Newman, J. D. Parrish & M. Kido. 1991. Baseline studies of the Hanalei River estuary and other north shore rivers, Kauai. Final Report prepared for the State of Hawaii Board of Land and Natural Resources by agreement with the University of Hawaii Environmental Center, Contract No. 25853, Honolulu. 165 p.

Houde, E.D. 1989. Subtleties and episodes in the early life of fishes. *Journal of Fish Biology* **35**: 29 –38.

Iguchi, K. & N. Mizuno. 1990. Diel changes of larval drift among amphidromous gobies in Japan, especially *Rhinogobius brunneus*. *Journal of Fish Biology* **37**: 255–264.

———. **& N. Mizuno**. 1999. Early starvation limits survival in amphidromous fishes. *Journal of Fish Biology* **54**: 705–712.

———., **F. Ito, M. Yamaguchi & N. Matsubara**. 1998. Spawning downstream migration of ayu in the Chikuma River. *Bulletin of National Research Institute of Fisheries Science* **11**: 75–84.

Karr, J.R. & E.W. Chu. 1999. *Restoring life in running waters: better biological monitoring*. Island Press, Washington, District of Columbus. 206 p.

Kido, M.H. & D.E. Heacock. 1991. The spawning ecology of *'o'opu nakea* (*Awaous stamineus*) in Wainiha River and other selected north shore Kauai rivers, p. 142–157. *In*: Devick, W. (ed.), *New directions in research, management and conservation of Hawaiian freshwater stream ecosystems: Proceedings of the 1990 symposium on stream biology and fisheries management*, State of Hawaii Department of Land and Natural Resources, Division of Aquatic Resources, Honolulu. 318 p.

———., **K.Y. Kaneshiro & C. Wichman**. 2002. Persistent patterns of community structure for native Hawaiian stream animals. Final Report to the Hawaii Division of Aquatic Resources.

Kinzie, R.A. III. 1988. Habitat utilization by Hawaiian stream fishes with reference to community structure in oceanic island streams. *Environmental Biology of Fishes* **22**: 179–192.

———. 1993. Reproductive biology of an endemic, amphidromous goby *Lentipes concolor* in Hawaiian streams. *Environmental Biology of Fishes* **37**: 257–268.

Komak, S. & M.R. Crossland. 2000. An assessment of the introduced mosquitofish (*Gambusia affinis holbrooki*) as a predator of eggs, hatchlings and tadpoles of native and non-native anurans. *Wildlife Research* **27**: 185–189.

Leis, J.M. 1991. The pelagic stage of reef fishes: the larval biology of coral reef fishes. *In*: Sale, P.F.

(ed.), *Ecology of fishes on coral reefs*. Academic Press, San Diego.

Lindstrom, D.P. 1998. Reproduction, early development, and larval transport dynamics of amphidromous Hawaiian gobioids. Unpublished Ph.D. dissertation. University of Hawai'i, Honolulu.

March, J.G., J.P. Benstead, C.M. Pringle & F.N. Scatena. 2003. Damming tropical island streams: problems, solutions, and alternatives. *BioScience* **53**: 1069–1078.

McDowall, R.M. 1992. Diadromy: origins and definitions of terminology. *Copeia* **1992**: 248–251.

———. 1997. Is there such a thing as amphidromy? *Micronesica* **30**: 3–14.

McRae, M.G. 2001. Microhabitat use in an assemblage of native and introduced stream fishes in Wailoa Stream, Island of Hawai'i. Unpublished Master's Thesis, Louisiana State University, Baton Rouge.

Mizuno, N. & H. Kawanabe. 1981. A topographical classification of streams, with an introduction of the system widely used in Japan. I. Reach type, stream zone and stream type. *Verhandlungen der Internationalen Vereinigung fur Limnologie* **21**: 913.

Moriyama, A, Y. Yanagisawa, N. Mizuno & K. Omori. 1998. Starvation of drifting goby larvae due to retention of free embryos in upstream reaches. *Environmental Biology of Fishes* **52**: 321–329.

Nelson, S.G., J.E. Parham, R.B. Tibbatts, F.A. Camacho, T. Leberer & B.D. Smith. 1997. Distributions and microhabitats of the amphidromous gobies in streams of Micronesia. *Micronesica* **30**: 83–91.

Nishimoto, R.T. & D.G.K. Kuamo'o. 1991. The occurrence and distribution of the native goby (*Lentipes concolor*) in Hawai'i Island streams with notes on the distribution of other native fish species, p. 77–95. *In*: W. Devick (ed.), *New directions in research, management and conservation of Hawaiian freshwater stream ecosystems: Proceedings of the 1990 symposium on stream biology and fisheries management*. State of Hawaii Department of Land and Natural Resources, Division of Aquatic Resources, Honolulu.

———. **& D.G.K. Kuamo'o.** 1997. Recruitment of goby postlarvae into Hakalau stream, Hawai'i Island. *Micronesica* **30**: 41–49.

Parham, J.E. 1995. Habitat use by an assemblage of tropical oceanic island streamfish. Unpublished Master's Thesis, University of Guam, Mangilao.

Pfister, C.A. 1996. The role and importance of recruitment variability to a guild of tide pool fishes. *Ecology* **77**: 1928–1941.

Polhemus, D.A., J. Maciolek & J. Ford. 1992. An ecosystem classification of inland waters for the tropical Pacific islands. *Micronesica* **25**: 155–173.

Pulliam, R.H. 1988. Sources, sinks, and population regulation. *American Naturalist* **132**: 652–661.

Radtke, R.L. & R.A. Kinzie III. 1996. Evidence of a marine larval stage in endemic Hawaiian stream gobies from isolated high-elevation locations. *Transactions of the American Fisheries Society* **125**: 613–621.

———., **R.A. Kinzie III & S.D. Folsom.** 1988. Age at recruitment of Hawaiian freshwater gobies. *Environmental Biology of Fishes* **23**: 205–213.

Smith, R.L. 1995. *Ecology and field biology*. HarperCollins College Publishers, New York. 740 p.

Tomihama, M.T. 1972. The biology of Sicydium stimpsoni, freshwater goby endemic to Hawai'i. Unpublished Honors Thesis. University of Hawai'i.

Van Horne, B. 1983. Density as a misleading indicator of habitat quality. *Journal of Wildlife Management* **47**: 893–901.

Way, C.M., A.J. Burky, J.M. Harding, S. Hau & W.K.L.C. Puleloa. 1998. Reproductive biology of the endemic goby *Lentipes concolor*, from Makamaka'ole Stream, Maui and Waikolu Stream, Moloka'i. *Environmental Biology of Fishes* **51**: 53–65.

Biology of Hawaiian Streams and Estuaries. Edited by N.L. Evenhuis & J.M. Fitzsimons. *Bishop Museum Bulletin in Cultural and Environmental Studies* 3: 99–111 (2007).

Applications of Stable Isotope Analysis to Tracing Nutrient Sources to Hawaiian Gobioid Fishes and Other Stream Organisms

KEITH A. HOBSON
Canadian Wildlife Service, 115 Perimeter Road, Saskatoon, Saskatchewan, Canada, S7N 0X4; email: keith.hobson@ec.gc.ca

R. JAN F. SMITH†
Department of Biology, University of Saskatchewan, Saskatoon, Saskatchewan, Canada, S7N 5E2

PETER SORENSEN
Department of Fisheries, Wildlife and Conservation Biology, University of Minnesota, 1980 Folwell Ave., St. Paul, Minnesota 55108-6124, USA

Abstract

The measurement of naturally occurring stable isotopes of carbon ($^{13}C/^{12}C$, measured as $\delta^{13}C$) and nitrogen ($^{15}N/^{14}N$, $\delta^{15}N$) can be used to distinguish sources of nutrients to freshwater stream fishes and other organisms in cases where such sources are isotopically distinct. We used $\delta^{13}C$ measurements of fish muscle tissue to quantify relative contributions of allochthonous (leaf litter) and autochthonous (algae) nutrients to adult gobioid fishes at Hakalau Stream, Hawaii, 1996–2002. We also used $\delta^{15}N$ measurements to delineate trophic interactions among stream organisms. We detected a strong reliance on benthic algae vs. leaf litter. The Hawaiian gobioid fishes are amphidromous and have a marine larval phase. Newly arriving larval recruits into Hakalau stream were examined isotopically to investigate if this approach could be used to quantify their relative dependence on marine and freshwater nutrient contributions. On the basis of a three-source, two isotope mixing model, the stable isotope ratios of these recruits strongly resembled those of the freshwater algal-based food web and suggest that larvae spend sufficient time in the nutrient-rich plume to register ingestion of stream-borne nutrients. However, further studies are now needed to contrast our stable isotope values of larval gobioids with known completely marine pelagic larvae.

Introduction

Fundamental to an understanding of the ecology of freshwater Hawaiian streams is the establishment of the relative importance of various sources of nutrients to higher trophic-level organisms. Within food webs supporting fishes, it is also important to delineate trophic interactions among organisms in order to understand community composition, competition, and carrying capacity. Previously, these questions were difficult to answer due to limitations inherent in conventional approaches to dietary analyses of most taxa. Recently developed isotopic tracing techniques provide the means of identifying recent nutrient sources for consumers (Lajtha & Michener, 1994; Kelly, 2000), and permit inferences about the spatial and temporal distribution of organisms which move between isotopically distinct habitats (Hobson, 1999). In both terrestrial as well as marine and freshwater systems, stable-carbon isotope ratios ($^{13}C/^{12}C$) differ according to a variety of biogeochemical processes. However, in general, marine food webs tend to be more enriched in ^{13}C compared to freshwater and terrestrial C-3 food webs (Chisholm *et al.*, 1982; Hobson & Sealy, 1991; Hobson *et al.*, 2000). This phenomenon is related to the fact that heavier (i.e., those containing a ^{13}C vs. a ^{12}C atom) atmospheric CO_2 molecules tend to be dissolved into marine waters as bicarbonate ions and leave atmospheric sources of CO_2 more depleted. Further, within marine systems, inshore or benthically linked food webs tend to be more enriched in ^{13}C than pelagic food webs (Hobson & Welch, 1992; France, 1995). Within freshwater or marine systems, algal-based food webs also tend to be enriched in ^{13}C compared to phytoplankton-based food webs. Stable-nitrogen isotope ratios ($^{15}N/^{14}N$) in marine food

Figure 1. Overview of the mouth of our study site, Hakalau Stream, Hawai'i. Photograph courtesy of Dr. Anne Hansen (University of Colorado).

webs also tend to be more enriched than in terrestrial or freshwater food webs (Michener & Schell, 1994; Hobson *et al.*, 1999). However, ^{15}N in consumer tissues shows a stepwise increase with trophic level, and $\delta^{15}N$ measurements have been used successfully to delineate trophic relationships among organisms, especially within marine systems (Hobson & Welch, 1992; Michener & Schell, 1994). The use of at least a dual-isotope approach in food web studies can be an effective tool in clarifying spatial and temporal aspects of the feeding history of organisms (Lajtha & Michener, 1994).

We examined stable isotope patterns in 5 species of native Hawaiian gobioid amphidromous fish and their supporting food web. These included *Lentipes concolor*, *Sicyopterus stimpsoni*, *Stenogobius hawaiiensis*, and *Awaous guamensis*, members of the Family Gobiidae and a fifth species from the Family Eleotridae, *Eleotris sandwicensis*. Previously, Sorensen and Hobson (in press) presented results of isotopic analyses of a similar dataset aimed at examining specifically the question of dietary histories of feeding in new larval recruits. Here, we were interested more in identifying sources of nutrients to freshwater adults with particular emphasis on evaluating the role of (allochthonous) leaf litter and (autochthonous) epiphytic algae. In addition, we examined newly arriving larval recruits from marine habitats to see if there was evidence of use of freshwater-derived nutrients from the river plume prior to upstream migration using newly developed isotope mixing models. Although few Hawaiian streams have well-developed estuaries, observations suggest that fresh water might exert a considerable influence on the coastal environment because large amounts of colored organic matter can be seen hundreds of meters from river mouths much of the year. The contrast can be dramatic because the ocean is nutrient-poor and clear. Finally, we surveyed a number of representative inshore and offshore marine species to examine the range of isotopic variation potentially influencing amphidromous species during their marine phase.

Methods

Our study was conducted at Hakalau Stream (Island of Hawai‘i, 19°53'55.5"N, 155°07'32.0"W; photograph Fig. 1; map in Schoenfuss *et al.*, 1997). Hakalau Stream is relatively typical of many Hawaiian streams. It is of moderate size, and, while it has been impacted by agriculture, there has been little development and it continues to attract large numbers of all native amphidromous species. Accordingly, it has also been the subject of much study (e.g., Tate, 1997; Nishimoto & Kuamo‘o, 1997; Radtke *et al.*, 2001). Stream flow in Hakalau Stream is highly variable, and while it is not gauged, gauged flows in nearby Honoli‘i Stream (19°46'00"N, 155°09'16"W) which is similar in size suggest it likely fluctuates between approximately 50 m^3 sec^{-1} in early spring to about 1% of this value in the summer. We sampled and analyzed juvenile and adult gobies and possible nutritional sources for them within Hakalau Stream between1996–2002 for one month during peak (February –April; Nishimoto & Kuamo‘o, 1997) migration each year. On several years sampling was further restricted because of flooding and drought, and in 2000 and 2001 no samples could be obtained. For three years, we also examined nutrient particulate organic matter (POM) at inshore and offshore marine locations.

For one month each spring, amphidromous migratory post-larvae were individually collected by researchers equipped with snorkels and small hand nets at four locations within Hakalau Stream and identified by using the key of Tate *et al.* (1992). Sampling locations included the freshwater interface where Hakalau Stream empties directly into the Pacific Ocean, approximately 100 m upstream from the freshwater/marine interface, and a riffle located above the bridge where adults of two species (*A. guamensis* and *S. stimpsoni*) were found in abundance (adults are not often found below this location). Another site was located approximately 2 km above several terminal waterfalls (300 m above sea level) where only adult *L. concolor* were found (19°52'38.3"N, 155°09'15.1"W). Each year we attempted to collect at least 5 adults of each species at the upstream sites as well at least 10 new recruits from both the river mouth and bridge sites. New recruits were frozen within 8 h of capture, while adults were euthanized and a 1 g sample of caudal muscle removed and frozen.

To assess possible nutrient sources for juvenile and adult gobies, we sampled in both coastal, offshore marine, and freshwater environments. Because the relative roles of allochthonous and autochthonous production has not been assessed in Hawaiian freshwater streams which tend to contain large amounts of leaf litter and other detritus as well as freshwater algae (which covers the rocks), we sampled both. Reports that adult gobies consume a combination of algae, stream invertebrates, and detritus (Kido, 1996, 1997; Way & Burky, 1991) further emphasized that both detritus and algae might be important sources of nutrients. Both allochthonous (leaf litter and associated detritus floating in the water and accumulating on the bottom) and autochthonous material (epiphytic algae growing on rocks) were sampled in the lower regions of Hakalau by handnet in 1999 and 2002. We obtained several samples of POM from nearby coastal regions by trawling a 0.5-m plankton mesh net (333- mesh) from a small boat at a distance of 0.5–1.0 km offshore within 5 m of the surface in 1999, 2000, and 2002. In all cases, samples were taken a few km to the southeast and seemingly out of the immediate influence of freshwater (salinities were in excess of 35‰). Another set of oceanic POM samples was taken by an ocean-going research ship using a 0.5 m 333- plank - ton net and towing for three 30-min intervals in the top 10 m approximately 10 km offshore of Hakalau Stream (approximately 19°10'N, 156°45'W) in 2002. All POM samples contained zooplankton the larger of which were removed under a microscope. In addition to these samples, some reference organisms (fishes and molluscs) were also collected from both Hakalau Stream and offshore fisheries.

For analysis, samples were freeze-dried, powdered, and lipids removed by using a 2:1 chloroform: methanol solution. For POM, we treated samples with 1N HCl to remove carbonates. Leaf litter and stream algae were processed by sorting and cleaning in distilled H_2O, dried at 60 °C, and powdered in an analytical mill. For final analysis, residual solvents were removed from processed samples by drying in an oven (60 °C), and then 1 mg of each powdered homogenized tissue was loaded into a tin cup and combusted at 1200 °C in a Europa ANCA-GST elemental analyzer (Europa Scientific Ltd., Crewe, UK). We analyzed the resultant gas with a Europa 20:20 continuous-flow isotope ratio mass spectrometer (CFIRMS; Europa Scientific Ltd., Crewe, UK) with every 5 unknowns separated by 2 laboratory standards. The isotopic composition of tissues is reported in δ notation as parts per thousand according to·

$$\delta X = [(R_{sample}/R_{standard}) - 1] \times 1{,}000$$

where X is ^{13}C, or ^{15}N, and R is the corresponding ratio $^{13}C/^{12}C$ or $^{15}N/^{14}N$. The standard for carbon was the PeeDee Belemnite and for nitrogen atmospheric air (AIR). Replicate measurements of internal laboratory standards indicate measurement errors of 0.1‰ for carbon and 0.3‰ for nitrogen isotope measurements.

Results

Stream-resident adult fish and primary production

With the possible exception of 1998, adult *A. guamensis* and *L. concolor* fishes had very similar isotope ratios (Fig. 2). Nevertheless, slight shifts in isotopic values were apparent between years for all three species, and an overall ANOVA measured a significant effect of species ($F_{3,56}$ = 10.68, p<0.001) and year ($F_{3,56}$ = 3.18, p = 0.033) on adult goby $\delta^{13}C$ values. There was no interaction between species and year ($F_{3,56}$ = 2.10, p = 0.113). *Sicyopterus stimpsoni* was slightly more enriched in ^{13}C compared with the other 3 species which did not differ among themselves (Tukey's, p>0.05). Accounting for species differences, the years 1997 and 1998 were more enriched in ^{13}C than the years 1999 and 2002 (p<0.05), each of which represented a homogenous subset.

We found a small but significant effect of species ($F_{3,56}$ = 97.48, p<0.001) and year ($F_{3,56}$ = 21.56, p<0.001) on adult goby $\delta^{15}N$ values (Table 1). No interaction between species and year was detected for adult goby $\delta^{15}N$ values ($F_{3,56}$ = 0.86, p = 0.435). All species formed homogenous subsets in the following order of lowest to highest mean $\delta^{15}N$ values: *S. stimpsoni*, *A. guamensis*, *L. concolor*, and *E. sandwicensis*. Accounting for species differences, the years 1998 and 2002 formed one homogenous subset for $\delta^{15}N$ values, and the years 1997 and 1999 two other homogenous subsets, respectively.

Stable isotope values of epiphytic algae did not differ among years. However, leaf litter differed slightly but significantly in $\delta^{15}N$ values between the two years of sampling, 1999 and 2002 ($t = 3.85$, $p = 0.004$). This was also the case for $\delta^{13}C$ values ($t = 2.21$, $p = 0.03$). Leaf litter was significantly less enriched in ^{13}C than algae ($t>8.1$, $p<0.001$, for all comparisons) (Table 2).

For marine POM, we found no effect of year on either $\delta^{15}N$ ($t = 1.49$, $p = 0.21$) or $\delta^{13}C$ ($t = 1.03$, $p = 0.41$) values of inshore POM (Table 2). Offshore POM had somewhat lower $\delta^{15}N$ values compared with the combined (2002 and 1999) inshore values ($t = 4.08$, $p = 0.03$). No difference was found for inshore and offshore marine POM $\delta^{13}C$ values ($t = 0.36$, $p = 0.73$).

New recruits

We found a high level of similarity between the isotope values of all gobioid fish, with all mean annual values falling between 6.1 and 7.1‰ for $\delta^{15}N$, and -19.0 and -17.3‰ for $\delta^{13}C$ values (Table 3). There was a significant effect of year ($F_{4,237} = 16.21$, $p<0.001$) and species ($F_{2,237} = 6.37$, $p = 0.002$) on the $\delta^{13}C$ values of arriving gobies but no interaction between year and species ($F_{7,237} = 1.9$, $p = 0.07$). Overall, *A. guamensis* were slightly more enriched in ^{13}C than either *L. concolor* or *S. stimpsoni* ($p<0.05$). Accounting for species, 1998, 1999, and 2002 formed a single homogenous subset ($p<0.05$) whereas 1996 and 1997 formed a more enriched homogenous subset.

For $\delta^{15}N$ values, we determined a significant effect of year ($F_{4,237} = 10.15$, $p<0.001$) and species ($F_{2,237} = 8.72$, $p<0.001$) with a significant interaction between these two factors ($F_{7,237} = 5.60$, $p<0.001$). The year 2002 formed one homogeneous subset with the lowest $\delta^{15}N$ values; 1996, 1997, and 1998 formed another subset with intermediate $\delta^{15}N$ values and 1999, the most enriched $\delta^{15}N$ values (Table 3). All three species formed homogeneous subsets with *S. stimpsoni* being the most depleted, *L. concolor* intermediate, and *A. guamensis* the most enriched in mean $\delta^{15}N$ values.

POM for offshore and inshore environments did not vary nor did values for epiphytic freshwater algae. Accordingly, we decided to combine values for all samples for the years for which we had the most complete data sets and which also showed the least inherent variation: 1999–2002. Given that the stable isotope values of new recruits did not closely coincide with any specific nutrient source, we felt that the most parsimonious approach was to consider an isotopic mixing model to evaluate the roles of all possible nutrient sources in an unbiased manner. Once again, we employed data only from those years for which we had data for all three dietary endpoints, 1999 and 2002. Mean isotopic endpoints and elemental concentration values for the concentration-dependent 3-source, two-isotope model of Phillips & Koch (2001) were derived from values shown in Table 4. We assumed trophic fractionation factors of 1‰ for carbon and 3.4‰ for nitrogen diet-tissue stable isotopes (Michener & Schell, 1994). This model allowed us to estimate the percent contributions from each of these sources to larval goby biomass for the year(s) of interest. Models were created by using both inshore and offshore POM for each species. Although some variation was apparent in the results, the models all suggested that newly arriving gobies derived 80–90% of their nutrients from a combination of freshwater algae and marine POM with leaf litter having a minor role (Table 4). The calculated contributions of freshwater algae and POM were similar. For example, our model for *A. guamensis* and offshore POM suggested this species might be acquiring nearly 50% of its nutrients from marine sources, about 40% from freshwater algae, and approximately 10% from leaf litter (Table 4, Fig. 3). Models using inshore POM tended to suggest a greater role for freshwater algae than offshore models.

Marine species

Reference species from the marine pelagic zone had $\delta^{13}C$ values between -15.8 and -16.9‰ and $\delta^{15}N$ values ranging from 7.1 to 12.5‰ (Table 5). As expected, marine benthic species were more enriched in their $\delta^{13}C$ values, ranging from -16.7 to -12.9‰ (Table 5). Juvenile flag-tails, that live in the lower reaches of the stream had $\delta^{13}C$ values close to other stream animals, while larger individuals that live farther out from the stream were closer to marine pelagic species in their $\delta^{13}C$ values (Table 1). $\delta^{15}N$ values approximately matched expected trophic levels. For example, in the marine pelagic grouping the large marlin, *Makaira nigricans,* had a higher $\delta^{15}N$ value than smaller scombrids such as the mackerel scad (Table 5). Among the marine benthic species the predatory hard-headed hawk fish had higher $\delta^{15}N$ ratios than the more herbivorous surgeon fishes.

Table 1. **Summary of stable isotope values (mean ± SD) measured for adult resident fishes caught in Hakalau Stream, Hawai'i 1996–2002.**

Species	Common name	Date	n	n‰^{15}N (‰)	‰^{13}C (‰)
Instream Residents					
Lentipes concolor	*'O'opu alamo'o*	1997	5	7.9 ±0.6	-18.9 ±1.4
		1998	8	7.8±0.33	-17.7±0.6
		1999	5	7.3±0.43	-19.5±1.0
		2002	4	6.4±0.6	-19.3±0.7
Sicyopterus stimpsoni	*'O'opu nōpili*	1997	6	5.5±0.24	-16.5±0.42
		1998	8	5.3±0.27	-17.1±1.02
		1999	1	5.5	-14.6
Awaous guamensis	*'O'opu nākea*	1998	3	6.9	-17.5±1.3
		1999	3	7.0±0.3	-18.3±1.0
		2002	5	5.7±0.4	-18.6±0.9
Eleotris sandwicensis	*'O'opu akupa*	1998	5	7.0±0.08	-18.4±0.68
		1999	1	7.5	-19.6

Discussion

Initially, we had expected that freshwater fishes resident to Hakalau Stream would possess much more depleted δ^{15}N and δ^{13}C values than marine organisms (Chisholm *et al.*, 1982), and so provide a convenient means of evaluating the relative contribution of marine-derived nutrients to the diet of larval amphidromous fishes. Instead, we found a significant algal contribution to the freshwater food web of this model stream which resulted in stream resident fishes more enriched in ^{13}C than expected from a C-3 temperate continental freshwater food web. This effect reduced from expected the carbon isotopic difference between those nutrients available to arriving larval gobies at the stream compared with marine POM. Nonetheless, we conservatively examined those arriving gobies for the years for which we had contemporary data on all of the three assumed sources of primary production available from the stream and from the marine environment. That analysis provided evidence that recruiting post-larval amphidromous gobiid fishes receive significant quantities of nutrients from the freshwater stream prior to entering the stream *per se.* Their stable isotope ratios were remarkably similar to that of stream resident adults which in turn were notably close to those of the large quantities of freshwater algae found in these streams. If true, then nutrient plumes exuding from freshwater streams in oceanic islands such as Hawai'i likely represent an important feeding and staging area for the many amphidromous fishes that populate these systems. This finding raises the intriguing possibility that Hawaiian freshwater ecosystems serve as a critical source of nutrients for larval amphidromous fishes while at sea, and that these systems have a high reliance on algae production.

Our use of a dual-isotope approach and the large number and variety of samples we collected over a five year period gives us confidence that our isotopic values and interpretations are not influenced by any short-term fluctuations in stream or inshore food web isotopic signatures. We are also confident that recruits we sampled had just arrived from the ocean and reflected larval values for several reasons. All species of new recruits had similar isotope ratios whether collected right at the front or 100 m upstream and were found to contain some diatoms of estuarine origin. Notably, the stable isotope ratios of these recruiting fishes did not appear to change during their initial short movement inland. Our data also suggest that this movement is very rapid, although perhaps not quite as rapid as suggested by Tate (1997).

Table 2. Stable isotope values (mean±SD) and elemental concentrations of dietary endpoints used in the two-isotope, three-input mixing model used in this study.

Common name	[C]	[N]	Site	Year	n	^{15}N (‰)	$\delta^{13}C$ (‰)
1. Marine Production							
POM	16%	1.7%	inshore	2002	3	5.5±0.3	-20.0±0.2
POM	16%	1.7%	offshore	2002	3	3.5±0.9	-19.5±0.7
POM*	16%	1.7%	inshore	1999	3	5.0±0.2	-19.3±0.2
2. Stream Production							
Leaf Litter	8%	1.7%	Hakalau	1999	3	1.8±0.8	-26.7±0.4
	8%	1.7%	Hakalaù	2002	6	-2.0±1.3	-28.7±1.8
Epiphytic algae	57.3%	4.5%	Hakalau	1999	3	2.5±0.7	-16.6±0.3
			Hakalau	2002	1	2.4	-16.8

* derived from microzooplankton sample assuming 3.4‰ and 1‰ isotopic fractionation from diet for nitrogen and carbon, respectively.

Table 3. Summary of stable isotope values (mean±SD) measured for arriving migrant gobies at Hakalau Stream, 1996-2002. New arrivals represent fish collected at the interface and at the bridge site.

Species	Common name	Date	n	$\delta^{15}N$ (‰)	$\delta^{13}C$ (‰)
Lentipes concolor	*ʻoʻopu alamoʻo*	1996	8	6.9±0.41	-17.9±0.25
		1997	5	6.5±0.22	-18.0±0.1
		1998	17	6.4±0.98	-18.2±0.93
		1999	28	6.9±0.5	-18.9±0.8
		2002	4	6.4±0.5	-18.6±0.3
Sicyopterus stimpsoni	*ʻoʻopu nōpili*	1996	23	6.1±0.4	-18.3±0.27
		1997	7	6.5±0.48	-18.3±0.07
		1998	27	5.8±0.52	-18.5±0.56
		1999	17	6.8±0.8	-19.0±0.7
		2002	2	6.2±0.5	-18.8±0.3
Awaous guamensis	*ʻoʻopu nākea*	1996	19	6.5±0.28	17.6±0.46
		1997	15	6.5±0.24	-17.3±0.5
		1998	17	6.7±0.23	-17.3±0.3
		1999	40	7.1±0.6	-18.9±0.5
		2002	9	6.0±0.5	-18.5±0.4
Eleotris sandwicensis	*ʻoʻopu ʻakupa*	1999	1	7.1	-23.5

Unfortunately, little comparative oceanic stable isotope data exists for primary production and zooplankton about the Hawaiian Archipelago (E.A. Laws, B. Popp, pers. comm.). In particular, we recognize that a rigorous test of our hypothesis is not possible without comparative isotopic data from purely marine fish larvae of similar ecology. Laws *et al.* (1999) reported $\delta^{15}N$ values for filtered particulate matter at the most marine sites off Honolulu to be 3.5‰, similar to our mean value of 3.0‰. However, stable-nitrogen isotope values of invertebrates and POM of the order of 8‰ have also been reported previously and are expected in nitrate-poor tropical waters (Wada & Hattori, 1976; Gould *et al.*, 1997). The carbon isotopic values we measured for marine POM were in gener-

Table 4. Results of mixing model estimates of percent biomass contribution to diets of returning larval gobies at Hakalau Stream, Hawai'i.

Species	n	Year	% marine	% stream leaf litter	% stream algae
Inshore Model					
A. guamensis	40	1999	50.3	16.3	33.4
S. stimpsoni	17		38.9	20.2	40.9
L. concolor	28		42.7	18.4	38.8
A. guamensis	9	2002	20.8	16.8	62.4
S. stimpsoni	2		27.2	16.7	56.1
L. concolor	4		31.9	14.3	53.8
Offshore Model					
A. guamensis	9	2002	46.3	11.8	41.9
S. stimpsoni	2		60.1	9.9	30.0
L. concolor	4		70.7	6.6	22.7

al agreement with those found previously for similar latitudes of the Pacific Ocean (Rau *et al.*, 1982, 1989; France, 1995). Parnell (2001) measured larval bivalves (*Spondylus tenebrosus*) off Honolulu and found they averaged about 3.6 ‰ for $\delta^{15}N$ and -20.2‰ for $\delta^{13}C$ values. These larvae were thus substantially depleted in both isotopes compared to our "arriving" larvae. Finally, the stable isotope values we measured from several species of marine benthic and pelagic fishes (data not shown) bore no resemblance to those of recruiting gobies, although such fishes are not likely to be ecological equivalents to the larval gobioids we examined. Together, the most parsimonious interpretation of our results is that larval amphidromous gobioid fishes derive nutrients from some of the same terrestrial sources as the stream resident fishes. If true, the mouth of Hakalau Stream, and likely that of many other tropic island streams, function as important nursery regions for the amphidromous fish that spend most of their lives within them.

Examining the stable isotope ratios of stream residents, we find clear evidence both for an important role of algae in Hawaiian freshwater ecosystems and dietary specialization by adults. The stable isotope ratios of adults fell between that of detritus and algae. The size of these adults suggested they had been in freshwater many (3–5) years (Ego, 1956). Although all recruits arrived in fresh water with similar isotope values, some divergence in these values was observed with age and suggested niche specialization similar to that indicated by Kido (1997). In particular, *S. stimpsoni* became relatively enriched in ^{13}C and depleted in ^{15}N. This agrees with observations that while all three gobiid species have overlapping omnivorous food habits which include algae, various invertebrates, and detritus (Ego, 1956; Lau, 1973; Kido, 1996, 1997), *S. stimpsoni* specializes in algae (Kido, 1997). Somewhat surprising was our finding of relatively higher $\delta^{15}N$ values in *L. concolor*, which suggest that it is the most carnivorous of these species, a possibility that was also raised in a cursory study by Lau (1973). Our finding that adults of these species as a whole rely heavily upon an algal-based food web throughout the Hawaiian Islands makes good sense given the high volume of algal material found in these systems; these results lend biochemical support to the gut analysis work of Kido (1996, 1997) from a stream in the island of Kaua'i.

This study provides biochemical evidence that amphidromous gobioid fishes aggregate at river mouths some time prior to their entry where they feed actively and acquire considerable body mass. Whether this means that many larval fish spend their entire marine existence within freshwater nutrient plumes, or rather they stage at river mouths after a period of drifting offshore is as yet unclear. A study of $\delta^{15}N$ changes in a landlocked Japanese goby suggests that half-life elemental turnover of this isotope in migrants is likely on the order of a month and associated with growth (Maruyama *et al.*, 2001); it seems likely that Hawaiian gobies whose stable isotope values appear to be at least 50% fresh water in origin, may spend at least a month in front of river mouths before entering. The similarity of the stable isotope values of all three species of recruits suggests that they likely consume

Table 5. Stable carbon and nitrogen isotope values of muscle tissue from various marine and marine-associated species from Hawai'i.

Species	Common name	n	$\delta^{13}C \pm SD$	$\delta^{15}N \pm SD$
Marine Pelagic:				
Etilis carbunculus	Ruby-colored snapper	1	-16.9	12.0
Thunnus albacares	Yellowfin tuna	3	-16.8	10.0
Katsuwonus pelamis	Skipjack tuna	1	-16.6	9.8
Aphareus rutilans	Ironjaw snapper	1	-16.4	9.5
Coryphaena hippurus	Mahimahi	2	-16.6	9.0
Acanthocybium solandri	Wahoo	1	-16.8	8.7
Trachiurops crumenopthalmus	Big-eyed scad	10	16.7±0.12	8.4±0.2
Decapterus pinnulatus	Mackerel scad	1	-16.8	7.5
Stenoteuthis oualaniensis	Deepwater squid	1	-16.4	7.1
Makaira nigricans	Blue marlin	1	-15.8	12.5
Marine Benthic:				
Cirrhitus pinnulatus	Hard-headed hawk fish	5	-12.9±0.6	10.5±0.2
Acanthurus sandvicensis	Sandwich Island surgeon	5	-13.3±1.7	8.6±0.4
Acanthurus achilles	Achilles surgeon fish	5	-14.6±0.4	7.8±0.6
Acanthurus dussumieri	Dussumier's surgeon fish	5	-16.7±2.4	8.6±0.8
Parupeneus multifasciatus	Red & black banded goat fish	5	-16.7±0.2	8.7±0.3
Polydactylus sexfilis	Pacific thread-fin	1	-16.4	9.2
Stream Mouth:				
Kuhlia xenura	Hawaiian flag-tail (30 mm)	9	-17.5±0.1	4.7±0.3
Polydactylus sexfilis	Pacific thread-fin	1	-18.3	6.4

very similar food in similar locations, with *A. guamensis* (whose $\delta^{13}C$ and$\delta^{15}N$ values were both slightly higher) perhaps being located slightly more inshore than the other species. In any case, that these larval fish do not distribute themselves widely as commonly imagined prior to stream entry, makes some ecological sense given the enormity of the Pacific Ocean and the fact that it is nutrient poor, while freshwater streams are nutrient-rich.

Our suggestion that larval gobies spend some time in river mouths, does not conflict with findings of previous studies. Thus, while Radtke *et al.* (1988, 1996) employed stable-isotopic analyses of the inorganic fraction of Hawaiian goby otoliths to describe an early, several-month long, larval phase in salt water, their data do not preclude that this marine stage might concentrate in diluted river nutrient plumes. This emphasises key differences in interpretations based on isotopic analyses of otoliths which give information on the nature of the elemental environment in which the fish spent time and the isotopic analyses of metabolically active fish tissues which give information on the nutritional sources contributing to fish biomass. Notably, Radtke *et al.* (1996) note surprisingly little variance in the age of 'larval settlement', an observation suggesting to us that many new recruits may not be coming from great distances. Similarly, while both Fitzsimons *et al.* (1990) and Zink *et al.* (1996) note a lack of genetic evidence for specific populations of gobies between islands, their results could be explained by as few as one recruit per generation drifting between streams (Zink, pers. comm.).

Conclusions

We have demonstrated evidence to suspect that the larvae of amphidromous Hawaiian fish, and likely those from other tropical island archipelagos, spend the majority of their early lives in river plumes and subsist on material associated with freshwater food webs. Further, we suggest that these food webs may derive much of their energy from algal primary production. Lastly, it seems reason-

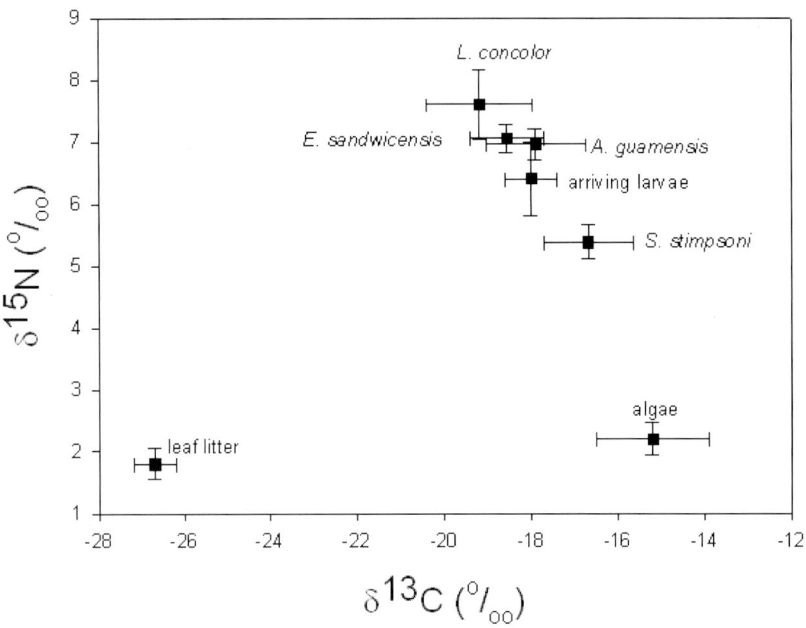

Figure 2. Relative isotopic positions of stream resident adult gobies and returning migrant larval gobies (all species combined) relative to Hakalau Stream primary production sources (1999–2002). This figure suggests a stronger algal (autochthonous) than leaf litter (allochthonous) input to the stream food web. For simplicity, only leaf litter values for 1999 shown. See Tables 1and 2 for individual isotope values.

able to hypothesize that the many species of amphidromous invertebrates likely pursue similar life history strategies which rely upon freshwater stream productivity while at sea, yet still allowing for some dispersal. If true, these hypotheses have important implications for both the ecology and management of these unique tropical ecosystems, many of which are now threatened by development and water removal, and previously were suspected to derive their energy from detritus and to have no role as nursery habitat. Instead, our data now suggest that each freshwater stream and its associated nutrient plume within these systems should be considered a coherent ecological unit that supports indigenous amphidromous organisms both as larvae and adults. Direct tests of these ideas by evaluating the food webs of these systems with a combination of laboratory feeding and rearing experiments together with the measurement of larvae of purely marine species off Hawai'i are now warranted.

The inherent problems involved with resolving marine and algal-based freshwater food webs due to overlap in stable isotope signatures suggests that future isotope studies should make use of other stable isotope measurements in addition to those of elements used here. In particular, we recommend the use of deuterium (δD) and sulphur ($\delta^{34}S$) isotope measurements in order to obtain tracers that may be independent of the nature of the freshwater algal vs. marine POM constraint.

Acknowledgements

We thank Jean Smith and Darrell Kuamo'o for their invaluable assistance collecting specimens. Both Mike Fitzsimons and Bob Nishimoto encouraged and supported our study from the beginning. This research also could not have been conducted without the continued support for research on Hawaiian stream organisms from Bill Devick and the Division of Aquatic Resources of the Hawai'i

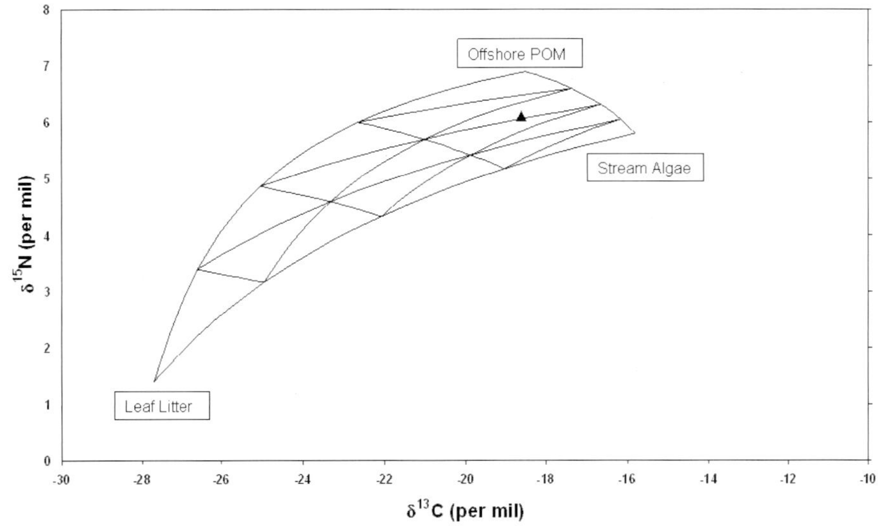

Figure 3. Example of the three-source, two-isotope mixing model used to estimate mean contribution of marine POM, instream algae and leaf litter to diets of returning larval gobies at Hakalau Stream. This figure shows the model parameters for 2002 using the offshore POM endpoint and larval returning *A. guamensis*. Note that the endpoint values represent the isotopic value of a hypothetical larva consuming 100% of this diet.

Department of Land and Natural Resources. Patricia Healy assisted with sample preparation, and Garth Parry provided isotopic analyses at the Department of Soil Science, University of Saskatchewan. Fishes were donated for sampling by Bruce Kaya, Merlin Pratt, Lance and Elaine Nishiura, Skippy Hau, and Wade Ishikawa. Offshore marine plankton were kindly collected by students and staff aboard the SSV Robert C Seamans operated by the Sea Education Association of Woods Hole, Massachusetts. This research was funded by a grant from the Division of Aquatic Resources, Department of Land and Natural Resources, State of Hawaii and the Research Corporation of the University of Hawaii (Dingell-Johnson Wallop-Breaux Sport Fish Restoration Project F-14-R-18), the University of Saskatchewan, and the Canadian Wildlife Service. We thank M. Fitzsimons and an anonymous reviewer for helpful comments on a previous version of this manuscript.

Literature Cited

Chisholm, B.S., D.E. Nelson & H.P. Schwarcz. 1982. Stable-carbon isotope ratios as a measure of marine versus terrestrial protein in ancient diets. *Science* **216**: 1131–1132.

Ego, K. 1956. Life history of freshwater gobies. Final report for Project no. 4-4R Freshwater Fish and Game Management Research, State of Hawai‘i, Honolulu, p. 97–114

Fitzsimons, J.M., R.M. Zink & R.T. Nishimoto. 1990. Genetic variation in the Hawaiian stream goby, *Lentipes concolor*. *Biochemical & Systematic Ecology* **18**: 81–83.

France, R.L. 1995. Carbon-13 enrichment in benthic compared to planktonic algae: food web implications. *Marine Ecology Progress Series* **124**: 307–312.

Gould, P., P. Ostrom & W. Walker. 1997. Food of flesh-footed shearwaters *Puffinus carneipes* associated with high-seas driftnets in the central North Pacific Ocean. *Emu* **97**: 168–173.

Hobson, K.A. 1999. Tracing origins and migration of wildlife using stable isotopes: A review. *Oecologia* **120**: 314–326.

———. & **S.G. Sealy**. 1991. Marine protein contributions to the diets of Northern Saw-whet Owls on the Queen Charlotte Islands, British Columbia: a stable isotope approach. *Auk* **108**: 437–440.

———., **M. Drever & G. Kaiser**. 1999. Norway rats as predators of burrow-nesting seabirds: insights from stable isotope analysis. *Journal of Wildlife Management* **63**: 14–25.

———, **J. Sirois & M.L. Gloutney**. 2000. Tracing nutrient allocations to reproduction using stable-isotopes: a preliminary investigation using the colonial waterbirds of Great Slave Lake. *Auk* **117**: 760–774.

———. & **H.E. Welch**. 1992. Determination of trophic relationships within a high Arctic marine food web using stable-isotope analysis. *Marine Ecology Progress Series* **84**: 9–18.

Kelly, J.F. 2000. Stable isotopes of carbon and nitrogen in the study of avian and mammalian trophic ecology. *Canadian Journal of Zoology* **78**: 1–27.

Kido, M.H. 1996. Diet and food selection in the endemic Hawaiian amphidromous goby, *Sicyopterus stimpsoni* (Pisces: Gobiidae). *Environmental Biology of Fishes* **45**: 199–209.

———. 1997. Food relationships between coexisting native Hawaiian stream fishes. *Environmental Biology of Fishes* **49**: 481–494.

Lajtha, K. & R.H. Michener (eds.) 1994. *Stable isotopes in ecology and environmental science*. Blackwell Scientific, Oxford. 316 p.

Lau, E.T.K. 1973. Dimorphism and speciation in the Hawaiian freshwater goby genus *Lentipes*. Unpublished M.Sc. Thesis. University of Hawaii, Honolulu. 83 p.

Laws, E.A., D. Ziemann & D. Schulman. 1999. Coastal water quality in Hawaii: the importance of buffer zones and dilution. *Marine Environmental Research* **48**: 1–21.

Maruyama, A., Y. Yamada, B. Rusuwa & M. Yuma. 2001. Change in stable nitrogen isotope ratio in the muscle tissue of a migratory goby, *Rhinogobius sp.*, in a natural setting. *Canadian Journal of Fisheries and Aquatic Science* **58**: 2125–2128.

Michener, R.H. & D.M. Schell. 1994. Stable isotope ratios as tracers in marine aquatic food webs, p. 138-157. *In*: Lajtha, K. & R.H. Michener (eds.), *Stable Isotopes in Ecology & Environmental Science*. Blackwell Scientific, Oxford.

Nishimoto, R.T. & D.G.K. Kuamo'o. 1997. Recruitment of goby postlarvae into Hakalau Stream, Hawai'i Island. *Micronesica* **30**: 41–49.

Parnell, P. 2001. The distribution of estuarine and oceanic water masses on the southern shore of O'ahu, Hawai'i: Ecological and coastal management implications, and novel methodology. *Limnology & Oceanography* **46**: 1468–1485.

Phillips, D.L. & P.L. Koch. 2002. Incorporating concentration dependence in stable isotope mixing models. *Oecologia* **130**: 114–125.

Radtke, R.L., R.A. Kinzie III & S.D. Folsom. 1988. Age at recruitment of Hawaiian freshwater gobies. *Environmental Biology of Fishes* **23**: 205–213.

———., **R.A. Kinzie III & D.J. Shafer**. 2001. Temporal and spatial variation in length of larval life and size at settlement of the Hawaiian amphidromous goby *Lentipes concolor*. *Journal of Fish Biology* **59**: 928–938.

———., **W. Showers, E. Moksness & P. Lenz**. 1996. Environmental information stored in otoliths: insights from stable isotopes. *Marine Biology* **127**: 161–170.

Rau, G.H., R.E. Sweeney & I.R. Kaplan. 1982. Plankton ^{13}C:^{12}C ratio changes with latitude: differences between northern and southern oceans. *Deep Sea Research* **29**: 1035–1039.

———., **T. Takahashi & D.J. Des Marais**. 1989. Latitudinal variations in plankton $\delta^{13}C$: implications for CO_2 and productivity in past oceans. *Nature* **341**: 516–518.

Schoenfuss, H.L., T.A. Blanchard & D.G.K. Kuamo'o. 1997. Metamorphosis in the cranium of postlarval *Sicyopterus stimpsoni*, an endemic Hawaiian stream goby. *Micronesica* **30**: 93–104.

Sorensen, P. & K.A. Hobson. 2005. Stable isotope analysis of amphidromous Hawaiian gobies suggest their larvae spend a substantial amount of time in freshwater river plumes. *Environmental Biology of Fishes* **74**: 31–42.

Tate, D.C. 1997. The role of behavioral interactions of immature Hawaiian stream fishes (Pisces:

Gobiodei) in populations dispersal and distribution. *Micronesica* **30**: 51–70.

———., **J.M. Fitzsimons & R.P. Cody**. 1992. Hawaiian freshwater fishes (Osteichthyes, Gobioidei) a field key to the species of larvae and postlarvae during recruitment into fresh waters. *Occasional Papers of the Museum of Natural Science* **65**, 10 p.

Wada, E. & A. Hattori. 1976. Natural abundance of ^{15}N in particulate organic matter in the North Pacific Ocean. *Geochimica et Cosmochimica Acta* **40**: 249–251.

Way, C.M. & A.J. Burky. 1991. A preliminary survey of macroinvertebrates and a preliminary assessment of the diet of the endemic Hawaiian goby (*'o'opu alamo'o, Lentipes concolor*), p. 158–164. *In*: Devick, W. (ed.), *Proceedings of the 1990 Symposium on Freshwater Stream Biology and Fisheries Management.* Hawaii Department of Land and Natural Resources, Honolulu.

Zink, R.M., J.M. Fitzsimons, D.L. Dittman, D.R. Reynolds & R.T. Nishimoto. 1996. Evolutionary genetics of Hawaiian freshwater fish. *Copeia* **1996**: 330–335.

Biology of Hawaiian Streams and Estuaries. Edited by N.L. Evenhuis & J.M. Fitzsimons. *Bishop Museum Bulletin in Cultural and Environmental Studies* 3: 113–123 (2007).

Morphometric and Genetic Confirmation of Two Species of *Kuhlia* (Osteichthyes: Kuhliidae) in Hawai'i

LORI K. BENSON MCRAE
Department of Biology, University of Tampa, 401 W. Kennedy Blvd., Tampa, Florida 33606-1490, USA; email: lbenson@ut.edu

Abstract

Flagtails, members of the Genus *Kuhlia*, are Indo-Pacific fishes found in marine and freshwater habitats. Known locally as *āholehole*, they are important food fishes in the Hawaiian Islands. Local fishermen have noted the presence of two morphotypes occurring sympatrically in Hawai'i, although at the beginning of this study only one species, *Kuhlia sandvicensis*, was identified in the scientific literature. During the course of this study, Randall & Randall (2001) published a revision of this genus, which included a description, based on meristic evidence, of the "big-eyed" morphotype as *K. xenura*. A small-eyed species with dark reticulations on the dorsal surface of the head and nape of living fish retains the name *K. sandvicensis*. For this study, genetic and morphometric comparisons confirm the presence of two species of *Kuhlia* in Hawai'i. Also, *K. sandvicensis*, rather than *K. marginata*, is confirmed herein as the species present on Johnston Island. A classification equation was developed and can be used to identify preserved fish from past studies and to aid fishery managers in identifying fish whose reticulations have faded or whose eyes appear to be intermediate in size. The variables that best discriminate between the two species are body depth, eye diameter, and interorbital distance. Genetic data concur with morphology in the recognition of two species of Hawaiian kuhliids; DNA sequences reveal significant distinctions between the two types of *Kuhlia*. In addition to the morphological and DNA analysis, the ecology of juveniles of both Hawaiian *Kuhlia* was investigated. Preliminary analysis of data indicates overlap of juveniles in certain tidepool habitats, but *K. xenura* appears to be the only species utilizing the lower reaches of freshwater streams as nursery habitat. Due to the former recognition of Hawaiian *Kuhlia* as one species, management strategies currently in place are possibly more relevant for one species than the other. Thus, conservation plans for both *āholehole* should be reconsidered in light of these and Randall and Randall's findings that these two "types" are separate species with genetically distinct populations and different nursery habitat preferences.

Introduction

The monogeneric family Kuhliidae contains approximately 10 species of fishes found in subtropical and tropical fresh, estuarine and marine waters of the Pacific (Randal & Randall, 2001). *Kuhlia* are important fishes in Hawai'i, both for their popularity as sportfish and because they are culturally important species that were once used by Hawaiians in traditional ceremonies. Until recently, *K. sandvicensis* was the only species listed in the scientific literature for the Hawaiian Islands, and it was believed to be endemic to Hawai'i. However, local fishermen have long recognized two types of *Kuhlia* or "*āholehole*" from differences in eye size. During underwater observations for doctoral research on the behavioral ecology of Hawaiian kuhliids, I noted other varying external features among small and large-eyed animals. The small-eyed fish have wide black reticulations on the head that extend posteriorly as two black lines along either side of the dorsal fin, a white patch on the posterior section of the soft dorsal fin, a more brightly marked black and white tail, and are less deep-bodied than the larger eyed animals. The small-eyed types are silver dorsally and whitish below, and the big-eyed types are often pale olive along the dorsum. The big-eyed fish also have reticulations on the head, but these are much fainter and very narrow (Benson & Fitzsimons, 2002) (Fig. 1), and their eyes are iridescent red along their upper edge. Near the completion of this study, a revision of the family Kuhliidae by Randall & Randall (2001) included a description of the big-eyed morphotype as a second species of *Kuhlia* in

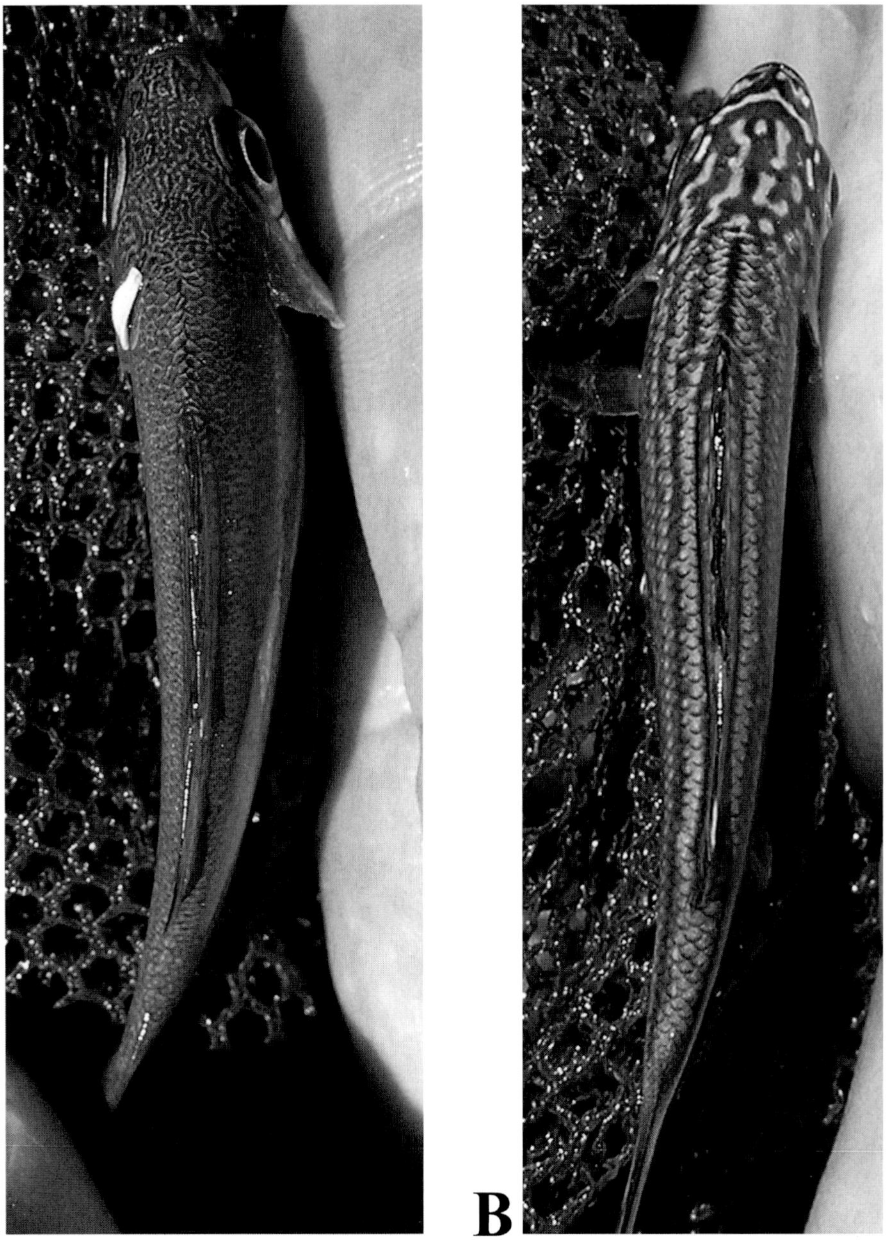

A B

Figure 1. Dorsal view of head reticulations for big-eyed *Kuhlia xenura* (**A**) and small-eyed *Kuhlia sandvicensis* (**B**) in Hawai'i.

Hawai'i. The available name for these fish is *Kuhlia xenura*, and the small-eyed type retains the name *K. sandvicensis*, based primarily on a lateral line scale count in the original description (Steindachner, 1876). The authors noted the eye size difference between the two species and included meristic data for various morphological characters. Because these characters overlap somewhat in the two species, additional morphometric features and DNA sequence variation are employed here to confirm the validity of recognizing two species in Hawaiian waters.

Materials and Methods

Collection methods and disposition of specimens

Fishes were taken by a variety of methods (seines, castnets, and hook and line) from sites on the islands of O'ahu and Hawai'i from 1997 to 2000. For specific collection sites, see Benson's doctoral dissertation (2002). The U.S. Fish and Wildlife Service provided specimens collected in March 2000 from Johnston Island, 724 km WSW of Hawai'i. Johnston is the nearest island group to Hawai'i, and prior to this study the fishes there were listed in the scientific literature as *K. marginata*. After collection, all fishes were frozen immediately for later morphometric and DNA analysis. At the conclusion of the study, specimens were fixed in 10% formalin, placed in 70% ethanol as the final preservative, and catalogued into the Ichthyology Collection at the Louisiana State University Museum of Natural Science (LSUMZ 12307-12332, 12334-12341).

Morphometric analysis

Specimens were sexed (whenever possible) and measured using either hand-held electronic digital calipers (Mitutoyo Plasti-Cal) or a measuring board. Measurements included standard length (*L*), fork length (*F*), snout length (*S*), head length (*H*), body depth (*D*), eye diameter (*E*), and interorbital distance (*I*). These variables were chosen because they are standard classification measurements in ichthyology (e.g., Hubbs & Lagler, 1958), and they are the characters that appear most different in these two morphotypes. A small amount of characters was preferred so that field identification and discriminant equation use would be as easy as possible for fishermen and biologists.

Measurement data were examined in SigmaPlot version 4.01 (SPSS, Inc. 1997. SigmaPlot 4.01. SPSS, Inc., Chicago, Illinois.) to test for allometric influences, which cause differing patterns of growth between juveniles and adults (Thorpe & Leamy, 1983). The character chosen to detect allometry was the ratio of body depth to standard length, as this character seems to change the most with increasing size for these fishes. Ratios were plotted against standard length, and a linear regression was conducted. The steep slope due to changes in depth to standard length ratio began to flatten out at lengths of 40 mm, so only specimens 40 mm standard length or greater were used in the multivariate analysis (Burbrink, 2000). While some fish used in these analyses are technically juvenile fish and could not be sexed, their body shape and proportions are consistent with those of adults. In a study of *Kuhlia* on Johnston Island, Gosline (1955) also included only fish greater than 40 mm in his meristic research, further supporting this size cutoff. The final sample size for *K. xenura* was 158; *K. sandvicensis* were harder to collect and less abundant, and the final sample size for this group was 71. Twenty-nine fish from Johnston Island were included in this study.

Data were log transformed prior to further analysis to stabilize variances in specimens of different sizes. A linear regression was then performed to plot depth and standard length against fork length, and the remaining variables against head length. The residuals for these five variables were used in subsequent multivariate analyses to eliminate the effect of multicollinearity and to better express the fishes' shape without the effects of body size (as per Freund & Wilson, 1997). Multivariate statistical analyses were conducted in SYSTAT version 8.0 (SPSS, Inc. 1998. Systat 8.0 Statistics. SPSS, Inc., Chicago, Illinois) and included a MANOVA, by using the Wilks' lambda statistic, to test for a difference between group means. In addition, a principal components analysis was conducted on the five remaining log residual variables to examine if the specimens readily fall into separate groups. A t-test on these factor scores was employed to determine the statistical significance of their morphological differences. A discriminant function analysis was employed to calculate the probability of correctly classifying each fish by type. In addition, a jackknifed classification matrix was provided; this resulted from a DFA which used functions computed from all data except the case being classified. Analyses not including all fish were also conducted, and the percentage that the unclassified fish were grouped correctly was noted. Automatic backward and forward stepwise analyses were used to determine which morphometric variables best discriminate between the two groups. The discriminant function analysis was conducted once again and used only the variables selected by the stepwise analyses. Percent correctness data from this output are also provided. Finally, a discriminant analysis was performed on the raw measurement data. From this output, clas-

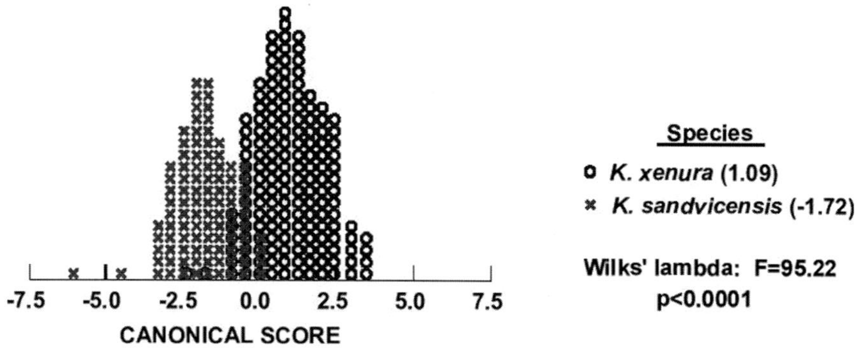

Figure 2. Discriminant function analysis of five morphometric variables used to classify two *Kuhlia* spp. from Hawai'i.

sification functions were used to derive a linear discriminant function equation for future use in classifying Hawaiian *Kuhlia* of unknown species. This equation consists of weighted coefficients that are multiplied by the morphological variables measured for each fish. The procedure results in a canonical score or value (*C*) that can be compared with a cutoff criterion differentiating the species.

DNA sequencing and analysis

DNA (from tail muscle) was isolated from frozen or DMSO preserved muscle tissue via Proteinase K digestion and phenol-chloroform extraction. The polymerase chain reaction (PCR) was used to amplify the entire cytochrome *b* gene (1140 base pairs) in two fragments that overlapped partially by using one primer for the light strand (L-14724) and one for the heavy strand (either cyt b-L or cyt b-HKu) (Table 1). Amplifications were carried out in 50 µl reactions which included 1 or 3 µl purified genomic DNA as template, 5 or 6 µl of MgCl2 Solution (25mM), 1 µl of 10mM dNTP's (Perkin Elmer), 0.25 µl AmpliTaq DNA Polymerase or AmpliTaq Gold Polymerase (5U/µl), and 2.5 µl each of the primers (all 10µM in concentration). The PCR was carried out in either a GeneAmp PCR System 2400 oil-free thermal cycler (Perkin Elmer) or in a PT C-200 Peltier Thermal Cycler (MJ Research). Methods were modified from Palumbi (1996) and included 30 cycles of amplification with primer annealing at 50 °C or 54 °C for 45 seconds. Five µl of the PCR reaction mixture was stained with ethidium bromide and loaded into a one percent agarose gel; electrophoresis was performed in order to visualize and verify the presence of the desired amplified product.

For specimens that exhibited successful amplifications, the remaining PCR product was then loaded into a one percent agarose gel (stained with ethidium bromide) for electrophoresis. The visualized bands were excised out of the gel and purified by using the BIO 101 GeneClean® Kit (BIO 101, Carlsbad, CA). Several primers were designed and used for the subsequent cycle sequencing reactions (Table 1). These reactions were performed on the PCR products by using the ABI PRISMBigDye® kit (PE Applied Biosystems, Foster City, CA). Cyt b-L and L-14724 served as

Table 1. Primers Used in PCR Amplification and Cycle Sequencing.

Primer	Nucleotide Sequence (5′to 3′)	Source
Cyt b-L	TGG RAC TGA GCT ACT AGT GTC	Reed *et al.*, 2002
L-14724	TGA CTT GAA RAA CCA YCG TTG	Palumbi *et al.*, 1991
L431	GAG GAC AAA TRT CYT TCT GAG G	Reed *et al.*, 2002
L431-Ku	GAG GAC AAA TRT CAT TTT GAG G	designed by author
H520	TGA GAG TGG CGT TGT CTA CT	designed by author
Cyt b-HKu	GAG CTA CTA GTG CAS CTT CAT T	designed by author

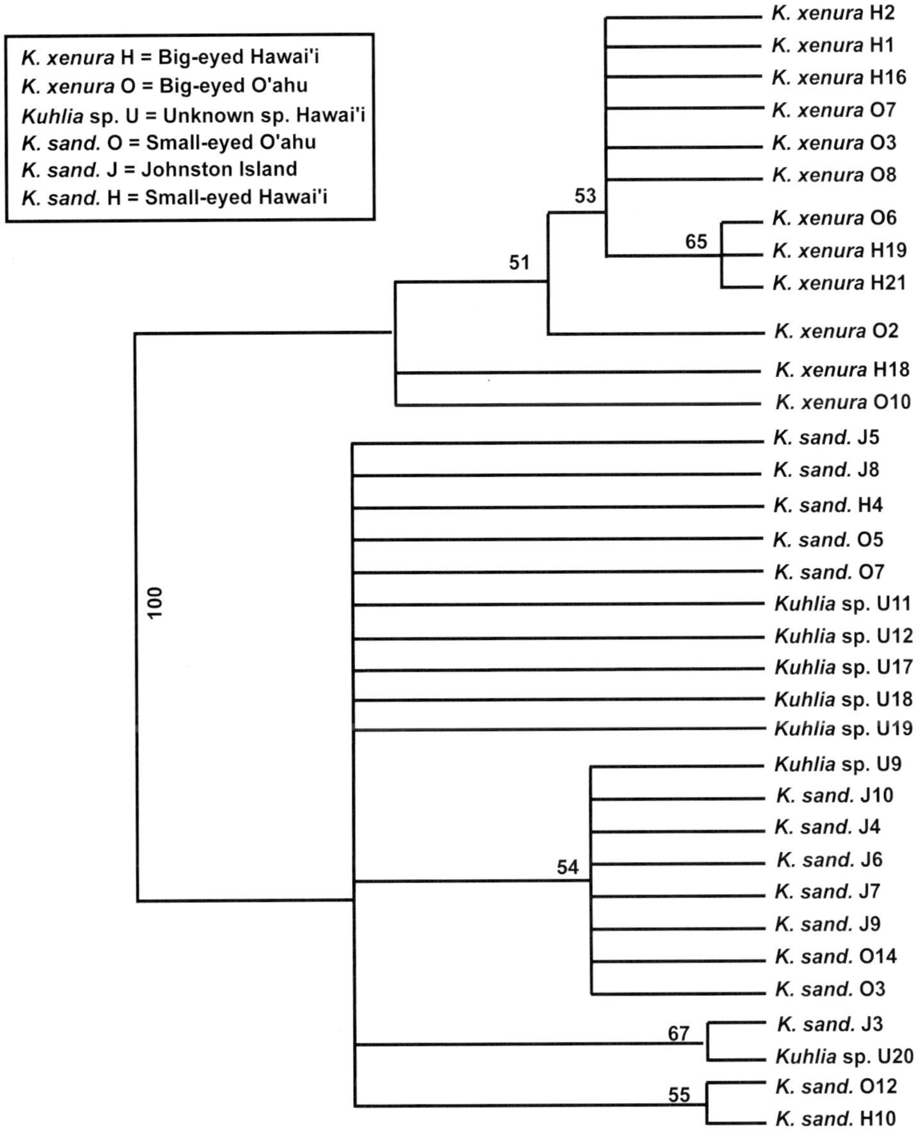

Figure 3. Maximum likelihood tree (manually rooted at midpoint) with bootstrap values (TVM+G model). Numbers at nodes indicate percent bootstrap support (of 100 replicates).

external primers for the heavy and light strands respectively; another general fish primer, L-431, was used to sequence the light strand internally. Initial sequences were examined and further primers were designed and renamed L431-Ku and Cyt b-HKu. A specific internal primer for the heavy strand was designed as well (H520). Subsequent sequencing used the L-14724 primer along with the three primers designed by the author. Cycle sequencing was conducted in 10 µl reaction volumes, which included 2 µl or 3 µl PCR product, 3.2 µl of 1µM primer, and either 2 µl (light strand) or 3 µl (heavy strand) of Big Dye reaction premix. The reaction was carried out in one of three machines, depend-

Table 2. Classification matrix and jackknifed classification matrix for complete discriminant function analysis with Johnston Island fish coded separately (A) and with Johnston Island fish grouped with *K. sandvicensis* from Hawai'i (B). Numbers under the species/location names represent fish classified as each type by the DFA analysis.

A

Type	*K. xenura*	*K. sandvicensis*	Johnston	%Correct
K. xenura	141	2	15	89
K. sandvicensis	3	52	16	73
Johnston	0	5	24	83
Total	144	59	55	84

B

	Number of fish classified as:		
Type	*K. xenura*	*K. sandvicensis*	%Correct
K. xenura	144	14	91
K. sandvicensis	6	94	94
Total	150	108	92

ing on availability: a GeneAmp PCR System 2400 oil-free thermal cycler (Perkin Elmer Applied Biosystems, Norwalk, Connecticut), a PT C-200 Peltier Thermal Cycler (MJ Research), or a HYBAID Omn-E Thermal Cycler. The following sequencing protocol (modified from Hillis *et al.*, 1996) was used: 10 seconds at 96 °C, 5 seconds at either 48 °C or 50 °C, 4 minutes at 60 °C, and storage at 4 °C or frozen until cleanup. Sequencing products were precipitated by using a sodium acetate/ethanol cleanup protocol, as per the manufacturer's directions. Purified cycle-sequence reaction products were sequenced with an ABI model 377-XL Automated Sequencer (Perkin Elmer Applied Biosystems).

Sequences were visualized, and heavy and light strand fragments were aligned for each specimen by using the program Sequencher 3.1 (Gene Codes Corporation, Ann Arbor, Michigan). After cleanup, 34 useable sequences were analyzed for this study: six big-eyed *K. xenura* from the Big Island of Hawai'i, six big-eyed *K. xenura* from O'ahu, two small-eyed *K. sandvicensis* from the Big Island of Hawai'i, five small-eyed *K. sandvicensis* from O'ahu, eight Johnston Island *Kuhlia* (previously thought to be *K. marginata*), and seven fish of intermediate eye size that were classified as unknown *Kuhlia*. Seven of these 34 sequences had less than 1140 base pairs, and these ranged in length from 1116 to 1136 base pairs. A BLAST search (Altschul et al., 1990) in GenBank was conducted, and the most similar sequence in the database was from *Zingel streber* (Song et al., 1998), a percomorph fish in the same suborder (Percoidei) as the kuhliids. This sequence was downloaded and used to help align the study specimens. Sequences representing each study group were deposited in GenBank (see Acknowledgments). Analyses of nucleotide composition and percent informative sites were conducted with the program MEGA version 2.1 (Kumar, S., T. Koichiro, I. Jakobsen, & M. Nei. 2001. MEGA2: Molecular Evolutionary Genetics Analysis software. Arizona State University, Tempe, Arizona).

Uncorrected pairwise sequence divergences between all taxa were calculated in PAUP* 4.0 (Swofford, D.L. 1998. PAUP*. Phylogenetic Analysis Using Parsimony [* and other methods]. Version 4. Sinauer Associates, Sunderland, Massachusetts). Based on sequence divergence, a neighbor joining tree (Saiton & Nei, 1987) was built using *Zingel streber* as an outgroup to root the tree. Because the outgroup was highly divergent from the study group, it was excluded from further analyses in this study. The sequencing dataset (minus the outgroup taxon) was also run in ModelTest v. 3.06 (Posada & Crandall, 1998) to determine the optimal DNA substitution model for a maximum

Table 3. Range and average of percent sequence divergences (uncorrected p x 100) among and within all operational taxonomic units (OTUs) involved in this study. Unless otherwise indicated, *K. sandvicensis* refers to only Hawaiian fish.

Pair of OTUs	Range in % Sequence Divergences	Average % Sequence Divergence
Interspecific Variation		
Kuhlia xenura vs. *K. sandvicensis*	7.982–9.219	8.523
Johnston Island vs. *K. sandvicensis*	0.000–0.360	0.173
K. xenura vs. Johnston Island	7.982–9.045	8.454
K. xenura vs. Johnston and *K. sandvicensis*	7.982–9.219	8.488
Comparisons involving fish of unknown type		
Unknown *Kuhlia* sp. vs. *K. sandvicensis*	0.000–0.618	0.256
Unknown *Kuhlia* sp. vs. *K. xenura*	7.982–9.158	8.489
Unknown *Kuhlia* sp. vs. Johnston Island	0.000–0.527	0.231
Johnston Island vs. Unknown and *K. sandvicensis*	0.000–0.618	0.202
K. xenura vs. Unknown and *K. sandvicensis*	7.982–9.219	8.506
K. xenura vs. three remaining types	7.982–9.219	8.489
Intraspecific Variation		
K. sandvicensis	0.088–0.361	0.213
K. xenura	0.179–1.405	0.705
Johnston Island	0.000–0.263	0.113
Unknown *Kuhlia* sp.	0.000–0.617	0.314
K. sandvicensis, Johnston Island, and Unknown fish	0.000–0.617	0.506
Comparisons involving outgroup		
K. xenura vs. Outgroup *(Zingel streber)*	18.519–19.491	19.155
K. sandvicensis vs. Outgroup *(Zingel streber)*	18.881–19.386	19.208
Johnston Island vs. Outgroup *(Zingel streber)*	19.123–19.386	19.342
Unknown *Kuhlia* sp. vs. Outgroup *(Zingel streber)*	19.211–19.496	19.350
Unknown and *K. sandvicensis* vs. Outgroup	18.881–19.496	19.289
Unknown/*K. sandvicensis*//Johnston vs. Outgroup	18.881–19.496	19.308

likelihood analysis. The model chosen was "TVM+G" and the parameters specified were: nucleotide frequencies A= 0.2566, C = 0.3058, G = 0.1540, T = 0.2836, and a gamma shape parameter (·) of 0.0702. Transition and transversion rates were specified as follows: A-C = 0.19, A-G = 3.09, A-T = 0.06, C-G = 0.26, C-T = 3.09, and G-T = 1. Support for the internal nodes in the maximum likelihood tree was assessed by a bootstrap analysis using 100 replicates. Trees were visualized and manually rooted at the midpoint by using the program TREEVIEW (Page, 1996).

Results

Morphometric analysis

Using combinations of the aforementioned morphological variables, the DFA procedure was able to group the three types (*K. xenura*, *K. sandvicensis*, and Johnston Island) separately with 84% correctness. There was some overlap between the Johnston Island fish previously identified as *K. marginata* and the *K. sandvicensis* specimens. Sixteen of the 71 *K. sandvicensis* specimens were mistakenly classified as Johnston Island fish; overall percent classification correctness was 73%. Johnston Island and *K. xenura* specimens were correctly classified a higher percentage of the time (Table 2A). Results for the jackknifed classification matrix were identical (Table 2A). Because the Johnston Island fish appear very similar to *K. sandvicensis* individuals, and because genetic evidence indicated that they were the same species, the analysis was repeated with the Johnston Island fish coded as *K. sandvicensis*. With all five variables as part of the MANOVA, the Wilks' lambda statistic revealed a significant difference in group means for the two types of fish (F = 95.2187, P<0.001). The DFA grouped the two species of *Kuhlia* with 92% correctness (*xenura* = 91% and *sandvicensis* = 94%). Group means and overlap are shown in Figure 2. Fourteen of 158 *K. xenura* and six of 100 *K. sandvicensis* were misclassified (Table 2B). Again, results for the jackknifed classification matrix were identical (Table 2B). When the same analysis was performed with 30 randomly selected individuals coded with no species designation, the DFA grouped the fishes that were defined a priori with 93% correctness. Twenty-eight of the 30 fishes (93%) that were not coded by species were classified correctly as either *K. xenura* or *K. sandvicensis*.

Automatic backward stepwise discriminant analysis resulted in three variables that could classify the two species with 94% correctness overall and with 93% correctness for the jackknifed classification procedure. The final analysis included eye diameter, interorbital distance, and body depth as variables with discriminatory value; these three variables had the three highest F-to-remove values (all greater than 44.9). The other two variables, standard length and snout length, had low F-to-remove values (below 1.0) and were therefore removed from the DFA. Automatic forward stepwise discriminant analysis added depth to the analysis first, as it had the highest F-to-enter value. Interorbital distance and eye diameter were then added; again the final output did not include standard length or snout length. The number of misclassifications and the percent correctness were identical to that of the automatic backward stepwise procedure. Classification functions provided with the DFA output provide a classification function coefficient for each of the variables. These results indicate that *K. sandvicensis*, the small-eyed type of āholehole, has overall smaller eyes and body depths relative to their length, and they possess a larger distance between their eyes along the dorsal surface than does *K. xenura*.

From the second analysis involving raw measurement data, classification functions were used to derive a discriminant equation that can classify fish of unknown type. The equation for the canonical value, *C*, is:

$$C = -2.887 + (0.679D) - (0.166L) - (2.406I) + (3.079E) + (0.638S) - (0.73H).$$

Values of $C < 0$ were classified as *K. sandvicensis*, whereas fish with values of $C > 0$ were classified as *K. xenura*. When this equation was applied to fish of known species, 237 of 258 specimens (approximately 92%) were classified correctly.

Principal components analysis resulted in three axes that accounted for 94.25% of the variation in the morphometric dataset. Based on factor loading scores, principal component one represented the three head measurement variables, which were strongly and positively correlated to one another. Analysis of the second and third component axes was not straightforward. The second axes indicated that standard length (a strong negative loading score) and body depth (a positive loading score) were negatively correlated. Meanwhile, the third principle component shows a strong positive loading for depth and a weaker positive loading for standard length. A paired t-test was conducted on each of the first three factors to determine if any morphological differentiation, by species, existed. Probability values were, for factors one, two, and three, 0.059, 0.000, and 0.000. With axes two and three statistically significant, and axis one nearly so, differentiation by species is supported by these morphological data.

DNA sequencing and analysis

Among 1140 sites, 127 base pairs were variable (11.1%) and 106 were parsimony informative (9.3%). This corresponds to 16 of 380 variable amino acids (4.2%), five of which were parsimony informative (1.3%). Base substitutions were most frequent at the third position of the codons (109 variable sites), while substitution rates at the first and second positions were nearly identical (8 and 10 respectively). All sequences began with the start codon ATG, and no stop codons were present in any of the *Kuhlia* sequences examined for this study. For the *Kuhlia* sequences examined here, nucleotide compositional bias existed, especially at the second and third positions in each codon.

Overall, the average percent nucleotide composition was 28.7 for T, 31.0 for C, 25.2 for A, and 15.1 for G. This low percentage of guanine in cytochrome *b* sequences is reported for at least one other group of teleosts in the genus *Scomberomorus* (Spanish Mackerels) (Banford *et al.*, 1999).

Twenty-five distinct haplotypes were noted from the 34 ingroup taxa. All 12 specimens of *K. xenura* had unique haplotypes. One haplotype was shared by five of the eight Johnston Island fish; this haplotype was also shared by a fish from O'ahu that was identified as *K. sandvicensis*. In addition, a second Johnston Island haplotype was shared by a *K. sandvicensis* from O'ahu and two fish of unknown morphotype from Hawai'i. The remaining two Johnston Island fish each had unique haplotypes. Pairwise comparisons between *K. sandvicensis* and *K. xenura* revealed raw percent sequence divergences between 7.98% and 9.22%, with the average percent sequence divergence at 8.52% (Table 3). Comparisons between *K. xenura* and the Johnston Island *Kuhlia* yielded similar percent sequence divergences, with the mean at 8.45%. However, comparisons between Hawaiian *K. sandvicensis* and the Johnston Island *Kuhlia* yielded very different results, with an average percent sequence divergence of only 0.17%. Several Johnston Island fish had haplotypes identical to those for some of the *K. sandvicensis* specimens, and the range in sequence divergences between these two groups was from zero percent to 0.36%. This is similar to the range and mean for comparisons within the *K. sandvicensis* fish or when comparing the Johnston Island fish to one another. When *K. sandvicensis* and Johnston Island fish were grouped together and compared to *K. xenura*, the average percent sequence divergence was 8.49%. For all fish coded as "unknown type," their sequences were always highly divergent from the *K. xenura* specimens (mean of 8.49%), and they were either identical to or at the most only 0.62% different from the *K. sandvicensis* and Johnston Island fish. Finally, all *Kuhlia* examined were roughly 19% divergent (uncorrected) from the outgroup *Z. streber*. A maximum likelihood analysis provided a tree with two monophyletic groups, a *K. xenura* clade and a clade containing both Hawaiian *K. sandvicensis* and the fish from Johnston Island. All eight of the unknown specimens were grouped within the *K. sandvicensis*/Johnston Island clade. The maximum likelihood analysis had a bootstrap value of 100% supporting the two aforementioned clades (Fig. 3).

Discussion

Discriminant function analysis proved to be an effective procedure for distinguishing and classifying species. With Johnston Island fish coded as *K. sandvicensis*, the analysis grouped the two *Kuhlia* species with 94% correctness based strictly upon morphometric data. While DFA maximizes differences between species and then classifies unknowns to these groupings, it does not answer the question of whether these groupings should be designated to begin with (Beuttell & Losos, 1999). In this study, genetic analysis supports our groupings, as does principal components analysis, which resulted in statistically significant factor score differences between the two species.

As for body shape differences and characteristics, both stepwise discriminant procedures indicated that eye diameter, depth, and interorbital distance were important characters for discriminating the two species. Overall, the eyes and body depths of *K. sandvicensis* are smaller than in *K. xenura* specimens of the same length. Conversely, the smaller-eyed *K. sandvicensis* have a larger interorbital distance than *K. xenura* specimens of the same size. One aspect of future research could be to add more specimens, of a larger size range, from Johnston Island. Interestingly, *K. xenura* specimens were more likely to be misclassified as Johnston Island fish than as *K. sandvicensis*. A further morphological comparison of the two populations of *K. sandvicensis* might indicate shape differences, although DNA sequence evidence, at least for *cyt b*, indicates that some level of gene flow is occurring between Hawai'i and Johnston.

The DNA sequencing study indicated high percent sequence divergences between the two proposed *Kuhlia* species in Hawai'i. Uncorrected *cyt b* divergences, which averaged 8.52%, suggest species level differences. According to Johns & Avise (1998), 90% of sister species pairs show at least 2% sequence divergence in their *cyt b* genes. Furthermore, the Johnston Island specimens (previously identified as *K. marginata* in Gosline, 1955 and Randall *et al.*, 1985) are as closely related to the *K. sandvicensis* specimens as they are to each other. In addition, they are as divergent from the *K. xenura* specimens as are the *K. sandvicensis* individuals (mean of 8.45%). Maximum likelihood analysis revealed a tree with two well-supported clades (bootstrap values of 100), which correspond to the two species. In addition, the Johnston Island individuals were included within the clade containing *K. sandvicensis*, whereas the *K. xenura* specimens were reciprocally monophyletic. These results corroborate Randall & Randall's (2001) assertion that Johnston Island fish, formerly classified as *K. marginata*, a widespread fish throughout much of Oceania, are actually *K. sandvicensis*. Obtaining *K. sandvicensis* specimens for DNA analysis from other island groups besides Johnston and Hawai'i, and determining definitively their distribution in the process, would be prudent.

Morphological and DNA sequence data provide strong evidence that there are two species present in Hawai'i, with one of them being the same species that is present in Johnston Island. These data support Randall & Randall's 2001 conclusion that the "big-eyed" morphotype is correctly renamed *K. xenura*, and the small-eyed morphotype and the Johnston Island fish are *K. sandvicensis*. Even though morphological differences between the two species are subtle, it is possible, in some cases, to identify fish based solely on superficial appearance. However, correct identification of specimens is occasionally difficult, as features like reticulations and stripes often fade when the fish are frozen or preserved (Randall & Randall, 2001; pers. observ.). Use of a classification equation, which is less time consuming and costly than DNA sequence analysis, is advantageous. Because traditional meristic counts in the two forms of *Kuhlia* overlap, this morphometric classification equation will be helpful in telling apart fish whose stripes or reticulations may have faded after death or whose eye sizes are intermediate. In addition, it may be possible, by using this equation, to analyze voucher specimens from past *Kuhlia* studies, where investigators did not designate the "type" on which they were working.

Young *K. xenura* are ubiquitous in the lower reaches of Hawaiian streams, but this species' use of fresh water is facultative (Benson & Fitzsimons, 2002). Conversely, *K. sandvicensis* has not been observed in freshwater streams; members of this species likely encounter reduced salinities only in tide pools where there is freshwater input from subsurface runoff. Now that two species of *Kuhlia* have been identified in Hawai'i and because there appear to be discrete habitat differences for them (Benson & McRae, unpubl. data), management strategies currently in place must be examined to assure that they adequately protect both species. Previous studies upon which management decisions have been based should be reconsidered. The single study on reproduction in Hawaiian *Kuhlia* by Tester & Takata (Tester, A.L. & M. Takata. 1953. A contribution to the biology of the *āholehole,* a potential baitfish. Industrial Research Advisory Council Grant no. 29, 1953. Hawaii Marine Laboratory, 54 pp.), for example, is problematic because it is not known which of the two species of *Kuhlia* was used in the study. Their status as popular food fishes, coupled with the evidence that *K. xenura* appears to be endemic to the Hawaiian Islands, makes proper identification, monitoring, and management practices essential for their conservation.

Acknowledgments

Our appreciation is extended to W. S. Devick, Division of Aquatic Resources, State of Hawaii, for generous support of this research via a grant administered by the Research Corporation of the University of Hawai'i (U.S. Fish and Wildlife Service Sport Fish Restoration Project F-14-R-18). Collection of specimens was approved by Hawai'i DAR. Specimen collection was done in accordance with the Animal Use Protocol (98-105) approved by the LSU Institutional Animal Care and Use Committee. For additional assistance, I acknowledge M. Fitzsimons, M. McRae, M. Bebler, F. Burbrink, M. Hafner, K. Han., W. Ishikawa, D. Kuamo'o, R. Moyle, R. Nishimoto, L. Nishiura, J. Parham, D. Reed, F. Sheldon, A. Styring, and C. Wilson. GenBank Accession Numbers are AY359409-AY359442.

Literature Cited

Altschul, S.F., W. Gish, W. Miller, E.W. Myers, & D.J. Lipman. 1990. Basic local alignment search tool. *Journal of Molecular Biology* **215**: 403–410.

Banford, H.M., E. Bermingham, B.B. Collette & S.S. McCafferty. 1999. Phylogenetic systematics of the *Scomberomorus regalis* (Teleostei: Scombridae) species group: molecules, morphology and biogeography of spanish mackerels. *Copeia* **1999**: 596–613.

Benson, L.K. 2002. Aspects of the behavioral ecology, life history, genetics, and morphology of the Hawaiian kuhliid fishes. Unpublished doctoral dissertation in Zoology, Louisiana State University, Baton Rouge, Louisiana.

———. **& J.M. Fitzsimons**. 2002. Life history of the Hawaiian fish *Kuhlia sandvicensis* as inferred from daily growth rings of otoliths. *Environmental Biology of Fishes* **65**: 131–137.

Beuttell, K. & J.B. Losos. 1999. Ecological morphology of Caribbean anoles. *Herpetological Monographs* **13**: 1–28.

Burbrink, F.T. 2000. Systematics of the Polymorphic North American Rat Snake (*Elaphe obsolete*). Unpublished doctoral dissertation in Zoology, Louisiana State University, Baton Rouge, Louisiana.

Freund, R.J. & W.J. Wilson. 1997. *Statistical Methods*. Academic Press, Inc., San Diego. 684 pp.

Gosline, W.A. 1955. The Inshore Fish Fauna of Johnston Island, a Central Pacific Atoll. *Pacific Science* **9**: 442–480.

Hillis, D.M., B.K. Mable, A. Larson, S.K. Davis & E.A. Zimmer. 1996. Nucleic Acids IV: sequencing and cloning, p. 321–381 *In*: Hillis, D.M., C. Moritz, and B.K. Mable (eds.), *Molecular Systematics*. 2nd ed. Sinauer Associates, Inc., Sunderland, Massachusetts.

Hubbs, C.L. & K.F. Lagler. 1958. *Fishes of the Great Lakes Region*. University of Michigan Press, Ann Arbor. 213 pp.

Johns, G.C. & J.C. Avise. 1998. A comparative summary of genetic distances in the vertebrates from the mitochondrial cytochrome *b* gene. *Molecular Biology & Evolution* **15**: 1481–1490.

Page, R.D.M. 1996. TreeView: An application to display phylogenetic trees on personal computers. *Comparative and Applied Biological Sciences* **12**: 357–358.

Palumbi, S.R.. 1996. Nucleic Acids II: The Polymerase Chain Reaction, p. 205–247. *In*: Hillis, D.M., C. Moritz & B.K. Mable (eds.), *Molecular Systematics*. 2nd ed. Sinauer Associates, Inc., Sunderland, Massachusetts.

———., **A. Martin, S. Romano, W.O. McMillan, L. Stice & G. Grabowski**. 1991. *The simple fools guide to PCR*. Department of Zoology and Kewalo Marine Laboratory, University of Hawai'i, Honolulu, Hawai'i.

Posada, D. & K.A. Crandall. 1998. MODELTEST: Testing the model of DNA substitution. *Bioinformatics* **14**: 817–818.

Randall, J.E., P.S. Lobel, & E.H. Chave. 1985. Annotated checklist of the fishes of Johnston Island. *Pacific Science* **39**: 24–80.

———. **& H.A. Randall**. 2001. Review of the fishes of the genus *Kuhlia* (Perciformes: Kuhliidae) of the Central Pacific. *Pacific Science* **55**: 227–256.

Reed, D.L., M. deGravelle & K. Carpenter. 2002. Molecular systematics of the jacks (Perciformes: Carangidae) based on mitochondrial cytochrome *b* sequences using parsimony, likelihood, and Bayesian approaches. *Molecular Phylogenetics and Evolution* **23**: 513–524.

Saiton, N. & M. Nei. 1987. The neighbor-joining method: a new method for reconstruction of phylogenetic trees. *Molecular Biology & Evolution* **4**: 406–425.

Song, C.B., T.J.Near & L.M Page. 1998. Phylogenetic relations among percid fishes as inferred from mitochondrial cytochrome b DNA sequence data. *Molecular Phylogenetics and Evolution* **10**(3): 343–353.

Steindachner, F. 1876. *Moronopsis argenteus* var. *sandvicensis*. *Sitzungsbericht der K. Akademie der Wissenschaften in Wien* **74**: 205.

Thorpe, R.S. & L. Leamy. 1983. Morphometric studies in inbred and hybrid house mice (*Mus* sp.): multivariate analysis of size and shape. *Journal of Zoology, London* **199**: 421–432.

Biology of Hawaiian Streams and Estuaries. Edited by N.L. Evenhuis & J.M. Fitzsimons. *Bishop Museum Bulletin in Cultural and Environmental Studies* 3: 125–141 (2007).

The Importance of Functional Morphology for Fishery Conservation and Management: Applications to Hawaiian Amphidromous Fishes

Heiko L. Schoenfuss
Department of Biological Sciences, St. Cloud State University, St. Cloud, Minnesota 56301, USA; email: hschoenfuss@stcloudstate.edu

Richard W. Blob
Department of Biological Sciences, Clemson University, Clemson, South Carolina 29634, USA; email: rblob@clemson.edu

Abstract

The functional capabilities of species can have a strong effect on their ability to maintain sustainable populations in anthropogenically altered environments. This is especially true among species comprising freshwater ichthyofaunas on small oceanic islands, where fishes frequently traverse boundaries between different ecosystems. Many aspects of functional performance in fishes relate directly to morphological features and limitations. In this review, we use the ichthyofauna of the Hawaiian Islands as an example to demonstrate how the habits, habitat, and in-stream distribution of five amphidromous fish species relate to the functional morphology of locomotion and feeding. Furthermore, we evaluate how performance limitations related to functional morphology might affect each species under changing environmental conditions. Based on this analysis we make four key recommendations. (1) Hawaiian stream management plans need to be tailored to island-specific conditions. (2) Estuaries need to be studied as a critical habitat because of the interface between fishes and human populations in these areas. (3) New recruits of *Sicyopterus stimpsoni* might serve as a key indicator species to monitor long-term health of Hawaiian estuaries. (4) The broad use of the Index of Biological Integrity (IBI) to assess Hawaiian stream health needs to be critically evaluated.

Introduction

Successful conservation and management of ichthyofaunas relies on the integration of ecological, economic, political, and sociocultural information as plans of action are developed (Krueger & Decker, 1993; Wolter, 2001; Arlinghaus *et al.*, 2002). The ecological context for management and conservation decisions includes both abiotic and biotic components. Measurement and analysis of many of these factors, such as flow rates, sediment loads, nutrient state, and fish density, can be performed by natural resource managers based on field data or laboratory chemical techniques. However, an additional, critical biotic component of the ecological context for ichthyological conservation and management that can require distinctly different analytical methods is the functional morphology of fish species. Studies of functional morphology, the functional analysis of structural features (i.e., anatomy) of animals (Schwenk, 2000), can provide tremendous insight into how fishes perform the behaviors that allow them to survive, and how the performance of a species might be impeded when environmental parameters are altered. Morphology can be one of the most critical factors affecting the performance of an individual, which in turn can determine its survival and fitness (Arnold, 1983, 2003; Irschick, 2003). Because constraints on function related to species morphology have the potential to result in dramatic performance failure of a species, especially after anthropogenic alterations to the environment (Holland, 1986; Wolter, 2001; Wolter & Arlinghaus, 2003), it is critical to consider the functional morphology of fish species as an important biotic factor during the process of making conservation and management decisions.

The performance of a wide range of behaviors can be considered in studies of functional morphology, but two that are generally considered among the most critical for fishes are locomotion and feeding (Webb, 1984; Wainwright *et al.*, 2002; Walker, 2000; Fulton *et al.*, 2001; Domenici, 2003; Ferry-Graham & Wainwright, 2003; Westneat, 2004). Locomotion is critical for fishes because they must move to obtain the food they need as well as other resources, such as mates and shelter; feeding is critical because of the need to obtain nutrients to survive. A lack of knowledge about the functional morphology of either of these behaviors could impede management efforts and lead to erroneous conclusions about the quality of a stream habitat when fish diversity is used as part of an Index of Biological Integrity (IBI) (Novotny *et al.*, 2005). Such gaps in knowledge are a particular concern for ichthyological systems that span multiple habitats. Many fish species inhabit a range of different habitats, either seasonally or during ontogeny (Azevedo & Simas, 2000; Keith, 2003), and such species could be at risk from anthropogenic habitat changes in any of the environments they inhabit or traverse. The multi-habitat amphidromous fishes of the Hawaiian Islands provide an excellent example of how knowledge of fish functional morphology can provide informed decisions in stream management. This paper will provide a critical review of the application of functional morphological data to the conservation and management of Hawaiian stream fishes.

Only five indigenous species of fishes live as adults in Hawaiian streams (McDowall, 2003, 2004). This low species diversity is likely a result of the geographic isolation of these islands and their recent volcanic origin (far away from continental shelves that possess diverse and abundant fish communities). All five species belong to the mostly oceanic Gobiidae or Eleotridae. All undergo an extended oceanic larval phase (3–6 months) before undertaking amphidromic migrations into adult stream habitats (Radtke *et al.*, 1988; Radtke & Kinzie, 1991), and three of the five species climb waterfalls during instream migrations to adult habitats and spawning grounds (Nishimoto & Fitzsimons, 1999; Schoenfuss & Blob, 2003). Despite the intensive physiological and energetic demands of vertical climbing over long distances, fishes with similar lifestyles and locomotory abilities have been described from other Pacific islands such as Guam (Fitzsimons *et al.*, 2002), Pohnpei and Palau (Nelson *et al.*, 1997), Reunion (Voegtle *et al.*, 2002), and Japan (Fukui, 1979). Moreover, Hawaiian streams are similar in many respects to streams on other volcanic islands. These similarities include abiotic factors, such as small watersheds with brief retention times that are fed by periodic rain-storms with subsequent cyclic patterns of extremely high and low velocity flows (Juvik & Juvik, 1998; Fitzsimons *et al.*, 2002). Common biotic features of streams on volcanic islands include limited species diversity when compared to continental streams of comparable latitude, low nutrient loads, and a fish fauna comprised largely of migratory species that spend part of their life cycle in the oceanic or near-shore environment (Nelson *et al.*, 1997; Fitzsimons *et al.*, 2002). As a result, applications of functional morphology in the conservation and management of Hawaiian taxa might serve as an example for fisheries managers on many oceanic islands. Thus, in addition to highlighting the importance of functional morphological data for the management and conservation of Hawaiian stream fishes, a broader goal of this review is to provide a model for the application of functional morphology as a tool in stream fish management on other volcanic island systems.

In this review, we will examine functional morphology across life history stages for all five native Hawaiian freshwater fishes by synthesizing published information, as well as new data, to identify critical aspects of functional performance for each species. We will then describe how knowledge of functional morphology might influence the decision making process in managing Hawaii's ichthyofauna. Finally, we will outline future research needs with respect to the Hawaiian freshwater and estuarine ichthyofauna.

Functional morphology and natural history of Hawaiian stream fishes: Background for species management

General Observations

All five species of native Hawaiian stream fishes belong to the superfamily Gobioidea (Pezold, 1993) and fall within two distinct, broadly recognized groups: the family Eleotridae, which contains *Eleotris sandwicensis*, and the subfamily Sicydiinae (Pezold, 1993) or *Sicydium* group (Birdsong *et*

Table 1. **Comparison of life history traits of five amphidromous Hawaiian gobies.**

	Eleotris sandwicensis	*Stenogobius hawaiiensis*	*Awaous guamensis*	*Sicyopterus stimpsoni*	*Lentipes concolor*
Oceanic phase	3–5 months	3–5 months	3–5 months	6 months	3-5 months
Larval diet	planktivore	planktivore	planktivore	planktivore	planktivore
Recruit size*	13.5 mm	14.1 mm	16 mm	23.6 mm	14.5 mm
Adult diet	carnivore	detritus feeder	omnivore	herbivore	algivore & insectivore
Adult size*	130 mm	40–50 mm	140 mm	75 mm	67 mm
Climbing ability	none	none	weak	good	excellent
Station holding	weak	weak	larvae: good adults: weak	good	excellent
Adult habitat	estuary	estuary	low/mid-stream	mid-stream	upper-stream

*Mean standard length.

Table 2. **Climbing speed (cm/sec; mean ± standard error) for larvae of *Awaous guamensis* and *Lentipes concolor* on three substrates of increasing coarseness (modified after Blob *et al.*, 2006).**

Substrate	*Awaous guamensis*		*Lentipes concolor*	
	n	climbing speed	n	climbing speed
Smooth	12	0.068 ± 0.0041	8	0.59 ± 0.024
Fine grained	12	0.21 ± 0.028	17	0.82 ± 0.084
Coarse grained	12	0.71 ± 0.22	10	1.2 ± 0.27

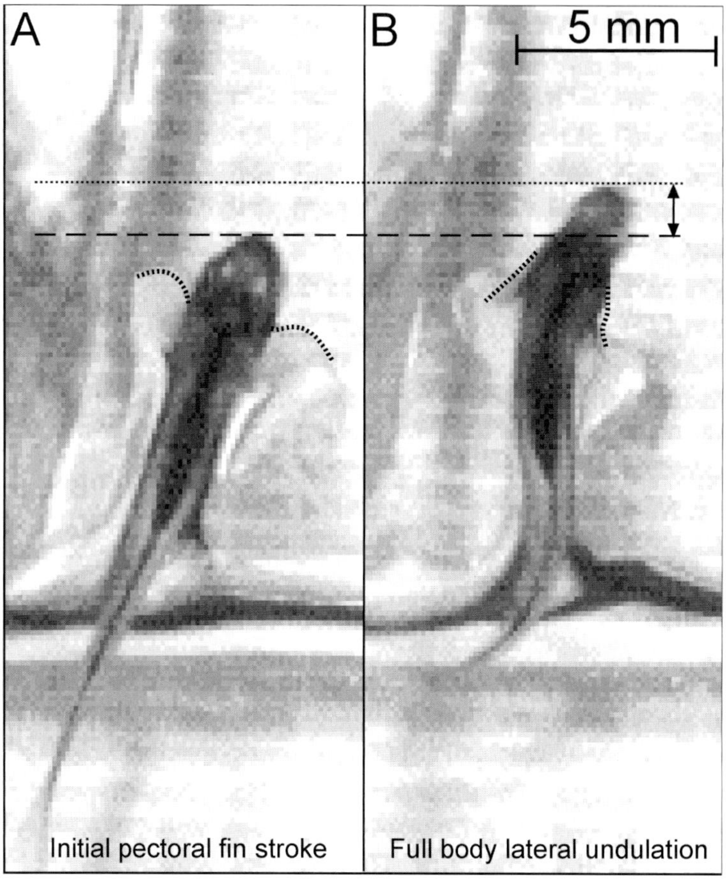

Figure 1. "Powerburst" climbing style common to juvenile *Awaous guamensis* and *Lentipes concolor* as recorded on a nearly vertical sheet of Plexiglas. **A:** Fish attached to the substrate via the pelvic suction disc. Pectoral fins (accentuated by dotted line) are outstretched in advance of a rapid adduction that begins the locomotory bout. **B:** The fish advances through rapid axial undulations of the entire body after rapid pectoral fin adduction. Arrow indicates the total advancement a fish makes during one climbing cycle. Adapted from Schoenfuss & Blob (2003).

al., 1988) which contains the remaining four species (*Stenogobius hawaiiensis*, *Awaous guamensis, Sicyopterus stimpsoni,* and *Lentipes concolor*). The close relationship of these four species has been confirmed in several recent studies using both molecular (Thacker, 2002) and morphological analyses of phylogeny (Parenti & Thomas, 1998). All 5 species share an amphidromous life cycle that includes a larval oceanic period of approximately 6 months for *S. stimpsoni* and approximately 3–5 months for the remaining four species (Table 1) (Radtke *et al.*, 1988; Radtke & Kinzie, 1991). All five species are likely pelagic during their oceanic phase, and the terminal position of the mouth in the oceanic larvae of these species indicates their reliance on a planktonic diet. A pelagic larval phase is also consistent with the ontogeny of related oceanic gobioid species (Privitera, 2002; Keith, 2003). Although several studies have suggested that larvae from a single source can be distributed among multiple streams (Schoenfuss *et al.*, 2004) and even multiple islands (Zink, 1991; Zink *et al.*, 1996; Chubb *et al.*, 1998), most larvae likely remain close to the stream from which they were washed out shortly after hatching (Keith, 2003; Taylor & Hellberg, 2003). For at least one Hawaiian species, *S. stimpsoni*, the gut contents of larvae entering stream mouths indicate a near-shore food source dur-

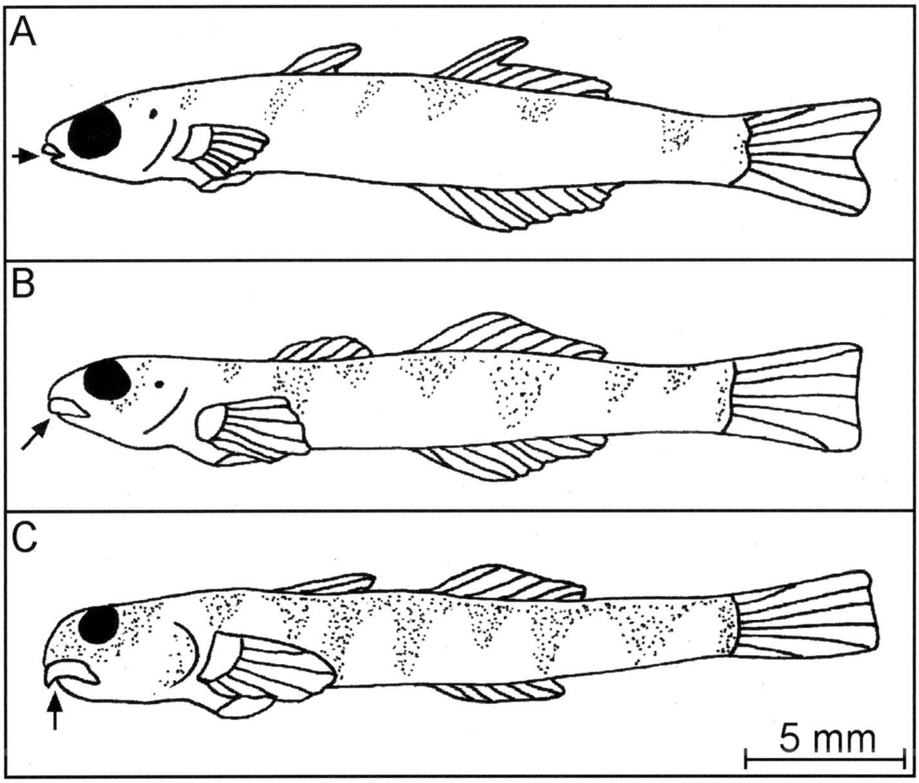

Figure 2. Anatomical changes in larval *Sicyopterus stimpsoni* entering the freshwater system (arrow in each drawing indicates mouth position). **A:** Incoming recruit captured at the stream mouth; **B:** Larvae in fresh water for 16 h; **C:** Juvenile *Sicyopterus stimpsoni* 36 hours after entering freshwater and now able to climb small vertical barriers.

ing the days prior to their arrival in the stream (Julius *et al.*, 2005). There is little evidence from the available literature, or personal observations, that the larval oceanic development of the five Hawaiian species differs from that of most other near-shore oceanic gobies (Azevedo & Simas, 2000; Munday *et al.*, 2002; Privitera, 2002; Keith, 2003). However, the in-stream development of these five species is quite distinctive among gobioid fishes.

***Eleotris sandwicensis* (*'o'opu 'akupa*)**

The only piscivore among native Hawaiian amphidromous fishes, *E. sandwicensis* lack fused pelvic fins and are unable to climb vertical obstacles in Hawaiian freshwater streams (Nishimoto & Kuamo'o, 1997). The poor station holding ability of these fish (Fitzsimons *et al.*, 1997) and their tendency to hide on pebbled portions of the stream bottom as an ambush predator also limit their range to stream estuaries. *Eleotris sandwicensis* appear to be poor continuous swimmers, with larvae relying on tides and wave action to be moved into the stream (Tate, 1997), and adults largely limited to short bursts of swimming to capture prey. Gut contents of *E. sandwicensis* include a mixture of animal and plant material (Kido, 1996b), but much of the plant material might enter the digestive tract as "by-catch" with the captured larvae of other Hawaiian amphidromous fishes. Although this species grows to a substantial size that is only rivaled by *A. guamensis* in the Hawaiian freshwater system, its station holding ability does not seem to change throughout its life and is the lowest among all native Hawaiian stream fishes (Table 1) (Fitzsimons *et al.*, 1997).

***Stenogobius hawaiiensis* (*'o'opu naniha*)**

Although *S. hawaiiensis* share an amphidromous life cycle with the three species of Hawaiian climbing gobies, a weak pelvic suction disc precludes them from climbing and effectively limits their range to the estuaries (Nishimoto & Kuamo'o, 1997). In addition, the limited station holding ability of *S. hawaiiensis* (Table 1) (Fitzsimons *et al.*, 1997) likely prevents movement into swifter upstream currents even in those Hawaiian streams not obstructed by waterfalls. These fish use their fused row of teeth in the upper jaw and conical teeth in the lower jaw to acquire small quantities of sand and mud from the stream bottom (see review: Kido, 1996b). Digestible items then are filtered out of ingested sediment by the well-developed gill-arch sieve (Kido, 1996b). The anatomy of this species appears to approximate the generalized gobioid "bauplan" described by Birdsong (1975), but its biology has generally received less study than that of other Hawaiian stream fishes.

***Awaous guamensis* (*'o'opu nākea*)**

Awaous guamensis is the only nonendemic Hawaiian freshwater fish and has been found across Micronesia (Fitzsimons *et al.*, 2002). In contrast to *E. sandwicensis* and *S. hawaiiensis*, *A. guamensis* is able to climb obstacles during its return migration from the ocean into freshwater habitats (Nishimoto & Fitzsimons, 1999; Schoenfuss & Blob, 2003). However, in comparison to other climbing species of Hawaiian gobies (*S. stimpsoni* and *L. concolor*), *A. guamensis* has the poorest station holding ability (Fitzsimons *et al.*, 1997) and its upstream range usually falls short of that of either of the other climbing species (Fitzsimons & Nishimoto, 1991). The larval development of *A. guamensis* mirrors that of *L. concolor*, and migrating larvae of the former are usually only slightly larger than those of the latter (Table 1) (Nishimoto & Kuamo'o, 1997).

To climb obstacles in streams, juvenile *A. guamensis* employ a technique known as "powerburst" climbing (a strategy that is also used by larval *L. concolor*) (Schoenfuss & Blob, 2003). While attached to a vertical substrate, *Awaous guamensis* begin bouts of powerburst climbing with a single, rapid adduction of the outstretched pectoral fins (Fig. 1A). However, most of the climbing bout is powered by undulating the entire long axis of the body to propel the fish against the falling water of a waterfall (Fig. 1B). Because this locomotory style is quite rapid, it is likely powered mainly by fast-twitch (white) muscle fibers. This climbing style may be energetically expensive, but it allows immediate mobility of the larvae out of the predator-rich estuaries (Schoenfuss & Blob, 2003). However, *A. guamensis* consistently performed below the levels of the other powerburst climber, *L. concolor*, when climbing performance was measured over approximately 20 body lengths on an artificial waterfall (Table 2) (Blob *et al.*, 2006).

Among the five species of native Hawaiian stream fishes, *A. guamensis* undergo the most significant change in body shape during juvenile growth (*S. stimpsoni* undergo a substantial metamorphosis, but this occurs at the end of the larval period and is not associated with an overall change in body shape). During growth, *A. guamensis* become significantly larger than either of the other two species of climbing Hawaiian gobies, and their shape becomes considerably more plump and robust. As a result, fluid dynamic drag is likely to increase substantially (Vogel, 1994) and probably contributes to the dramatic decrease in the ability of *A. guamensis* to hold station as they grow larger than approximately 50 mm (Fitzsimons *et al.*, 1997). Both juveniles and adults of *A. guamensis* remain in lower stream reaches and adults do not appear to climb (Blob *et al.*, in press). Adults and juveniles both feed primarily on filamentous algae harvested together with its epiphytes through a biting action of the jaws (Steinmann, 1996; Schoenfuss, pers. observ.).

***Sicyopterus stimpsoni* (*'o'opu nōpili*)**

Sicyopterus stimpsoni is unique among Hawaiian amphidromous fishes in that it undergoes a dramatic metamorphic transformation prior to gaining the ability to climb and reach its adult midstream habitats. The early development and larval anatomy of *S. stimpsoni* mirror that of the other Hawaiian stream fishes and, indeed, most gobioid fishes (Fig. 2A). However, oceanic larval development lasts up to 180 days for *S. stimpsoni* (Radtke *et al.*, 1988; Radtke & Kinzie, 1991), substantially longer than for the other species of Hawaiian amphidromous fishes. As a consequence, recruits

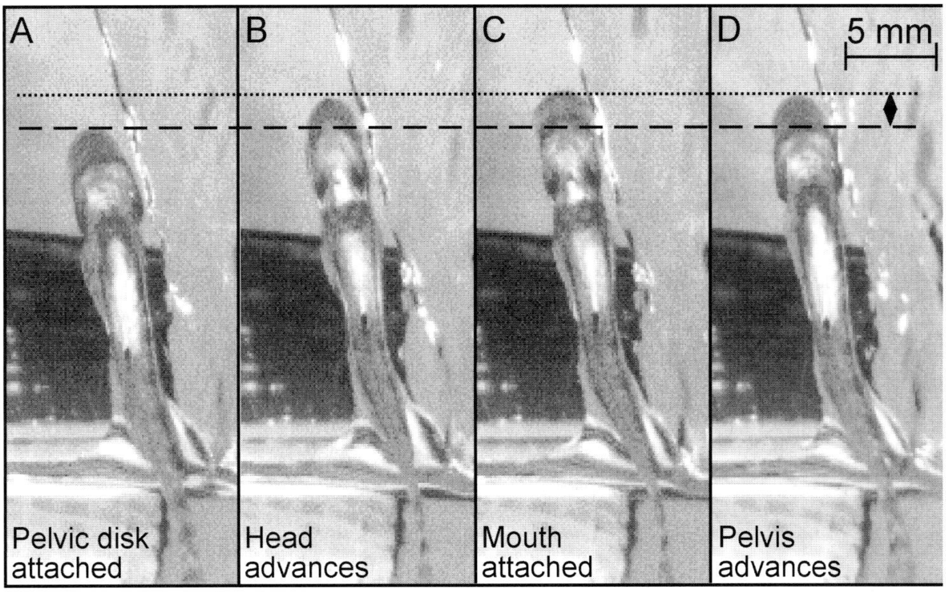

Figure 3. "Inching-up" climbing style of *Sicyopterus stimpsoni* as recorded on a nearly vertical sheet of Plexiglas. **A**: fish attached to the substrate via the pelvic suction disc; **B**: the head of the fish advances and the mouth makes contact with the surface; **C**: the mouth of the fish seals against the substrate to anchor the animal; **D**: the pelvic and post-pelvic structures advance. Diamond in Figure 3D indicates the total advancement a fish makes during one climbing cycle. Adapted from Schoenfuss & Blob (2003).

of *S. stimpsoni* returning into the stream system (Fig. 2B) are about 1.5 times the length and 2–3 times the weight of the other four species (Nishimoto & Kuamo'o, 1997; Schoenfuss & Blob, 2003); moreover, *S. stimpsoni* recruits are able to actively swim through the stream mouth even with its high currents (Tate, 1997). Once in fresh water, larvae of *S. stimpsoni* are initially confined to the estuaries as their larval anatomy precludes climbing. While confined to the estuary, *S. stimpsoni* larvae are exposed to significant predatory pressure from estuarine species like *E. sandwicensis* as well as marine fish species that frequently invade the estuary (e.g., *Kuhlia xenura*). However, within 36–48 h of entering the freshwater environment, *S. stimpsoni* larvae undergo a dramatic metamorphosis during which the mouth shifts from its larval, terminal position to its adult sub-ventral position as the upper lip greatly increases in size (Schoenfuss *et al.*, 1997). After this metamorphosis (Fig. 2C), *S. stimpsoni* juveniles are able to climb waterfalls up to 100 m tall and will quickly move upstream into their adult habitats. This shift and re-structuring of the cranium of *S. stimpsoni* comprises one of the most dramatic rapid metamorphic events among all vertebrates. However, the metamorphosis is also an extremely expensive process energetically: *S. stimpsoni* larvae lose approximately 15% body weight during its 36–48-h progression (Schoenfuss *et al.*, 1997). The demands of the metamorphic transformation, as well as the energy-depleted state of the animal following the transformation, make *S. stimpsoni* particularly vulnerable to adverse environmental effects.

Waterfall climbing in *S. stimpsoni* differs markedly in style from *L. concolor* and *A. guamensis*. The changes in mouth structure during metamorphosis allow *S. stimpsoni* juveniles to use their mouth as a second suction disc during climbing. *Sicyopterus stimpsoni* climb by using an "inching-up" climbing technique where they alternately attach themselves to the rock surface with the mouth or their pelvic suction disc (Fig. 3A–D) (Schoenfuss & Blob, 2003). By using this climbing style, fish remain constantly attached to the surface of the waterfall and avoid the need for energetically costly bursts of swimming against falling water. In addition, the slow speed of motion during inch-

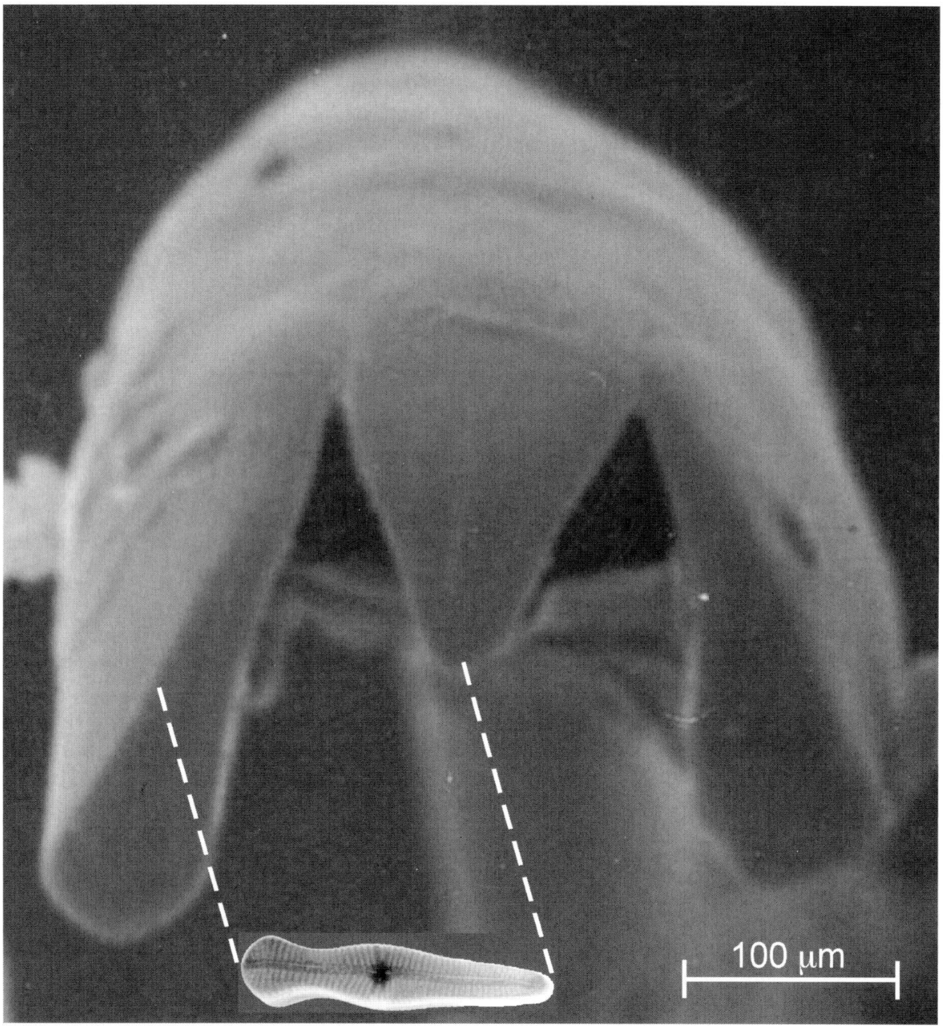

Figure 4. Scanning electron micrograph of a tri-cuspid tooth of *Sicyopterus stimpsoni* and of the diatom *Gomphonema truncatum* commonly found in the digestive tract of *Sicyopterus stimpsoni*. The diatom is illustrated at the same magnification as the fish tooth and positioned to demonstrate the similarity in diatom width and distance between the cusps of the tooth shown.

ing climbing suggests that it may be possible for *S. stimpsoni* to power climb by primarily using slow oxidative (red) muscle fibers and aerobic metabolism, preventing fatigue (Schoenfuss & Blob, 2003). The evolution of this distinctive juvenile climbing style in *S. stimpsoni* [and potentially other species of *Sicyopterus*, e.g., *S. japonicus*—Fukui (1979)] may have been correlated with the increased energetic cost associated with moving against falling water in juveniles with a greater body mass than that of other species of climbing gobies (e.g., Hawaiian *A. guamensis* and *L. concolor*). Because post-metamorphosed *S. stimpsoni* have already greatly depleted energy reserves prior to climbing (Schoenfuss *et al.*, 1997), aerobic "inching-up" might be the only energetically feasible mode of climbing that juvenile *S. stimpsoni* can use to overcome in-stream obstacles like waterfalls (Schoenfuss & Blob, 2003).

The metamorphosis of *S. stimpsoni* also signals a change in diet for this species. The sub-ventral mouth position of *S. stimpsoni* allows feeding on short-stalk diatoms. With the streamlined body of *S. stimpsoni* oriented parallel to flow, it is possible for these fish to feed even in high velocity currents. Diatoms are scraped off rock surfaces by cyclically raking rows of tricuspid premaxillary (upper jaw) teeth over the substrate (Schoenfuss, 1997). The two lateral cusps of the upper-jaw teeth in *S. stimpsoni* are slightly longer than the median cusp (Kido, 1996b) and all three cusps form an anterior semi-circle around the hollow crown of each tooth (Fig. 4). The distance between the central and each lateral cusp closely matches the size of the diatoms that comprise the main dietary resource for *S. stimpsoni* [Fig. 4; see also Julius *et al.* (2005)]. These teeth are continuously replaced from several rows of teeth located superior to the active row (T. Maie, pers. comm.). The unique feeding apparatus of *S. stimpsoni* allows this species to monopolize short-stalk diatoms as a high quality food source that is unavailable to other Hawaiian freshwater fishes in which jaw morphology seems to impede close contact between the teeth and the surfaces of feeding rocks. The unusual jaw structure of *S. stimpsoni* may also be related to an unusual behavior in this species: the guarding and "gardening" of feeding rocks by adult fishes (Fitzsimons *et al.*, 2003). The repeated grazing of feeding rocks as well as the location of feeding rocks in the highest stream currents will delay the growth of longer-stalked and filamentous algae that can block access to the low stalked diatoms that comprise most of the diet of *S. stimpsoni* (Julius *et al.*, 2005). These longer-stalked and filamentous algae grow over a 4- to 6-week period following high flow events on most rock surfaces in Hawaiian streams (Julius *et al.*, 2005). It is noteworthy that peak in-stream migration of oceanic larvae into Hawaiian freshwater streams occurs after storms (Nishimoto & Kuamo'o, 1997) that result in increased stream flow and a removal of filamentous algae from potential feeding surfaces (Julius *et al.*, in review). These events may expand the available surfaces for growth of short-stalked diatoms to non-guarded small rocks and pebbles in lower current areas of the stream, as well as provide areas for migrating juveniles to feed.

Once in its adult lower- to mid-stream habitats, *S. stimpsoni* maintain their exclusively herbivorous diet, retain a highly convoluted gut, and continuously replace worn teeth in the upper jaw (Kido, 1996b). Adult *S. stimpsoni* will readily climb to avoid perceived danger (Schoenfuss, pers. obs.). Adult *S. stimpsoni* climb by using the inching mechanism employed by juveniles, though studies of high speed video footage suggest that adults also incorporate pectoral fin adduction into each climbing cycle to increase propulsive thrust (Blob *et al.*, in press.).

***Lentipes concolor* ('o'opu 'alamo'o)**

Lentipes concolor inhabit the highest reaches of Hawaiian freshwater streams (Englund & Filbert, 1997; Nishimoto & Fitzsimons, 1999) and frequently are the sole vertebrate inhabitants of these stream reaches. This species also exhibits the greatest station-holding ability among Hawaiian freshwater fishes (Table 1) (Fitzsimons *et al.*, 1997). Similar to the other two species of climbing Hawaiian amphidromous gobies, the life cycle of *L. concolor* is marked by amphidromous migrations before reaching maturity. *Lentipes concolor* return into streams after an oceanic larval period that is likely similar in duration (Radtke *et al.*, 1988) to that of the closely related *A. guamensis* (Parenti & Thomas, 1998). Significantly smaller than *S. stimpsoni*, migrating larvae of *L. concolor* rely on wave action and high tides to be swept into the estuaries of Hawaiian streams (Tate, 1997). Larvae traverse estuaries quickly and immediately climb if they encounter waterfalls or other vertical in-stream barriers. Like *A. guamensis*, *L. concolor* rely on the "powerburst" style of climbing to scale in-stream barriers (Schoenfuss & Blob, 2003). Little is known about the up-stream migration of *L. concolor* except that juveniles of different size classes are sometimes encountered climbing waterfalls along the way to known adult habitats (Yamamoto & Tagawa, 2000). Size distribution of *L. concolor* within adult habitats ranges from newly arrived fish weighing little more than 1g to adults large enough to suggest residence for several years (Fitzsimons & Nishimoto, 1990).

Lentipes concolor retain a torpedo-shaped body throughout life and will climb if threatened or displaced. However, *L. concolor*, which are powerburst climbers as juveniles, appear to dramatically alter their locomotory mechanics to climb as adults. High-speed footage indicates that climbing adult *L. concolor* greatly reduce axial undulation compared to juveniles, propelling themselves with pectoral fin adduction during each climbing cycle and producing climbing kinematics that strongly resemble those of adult *S. stimpsoni* (Blob *et al.*, in press.; Yamamoto & Tagawa, 2000).

Unlike *S. stimpsoni*, adult *L. concolor* exhibit significant dietary breadth. A large portion of the gut contents of *L. concolor* consists of insects, and the teeth in the upper jaw of *L. concolor* are conical (Kido, 1996b), similar to those of insectivorous fish species. However, other components of gut contents in *L. concolor* consist of a variety of items that are picked out of the water column as they float by in the stream. To catch floating food, *L. concolor* rises from a resting location on the bottom of a pool and quickly engulfs the food item. Rapid capture of floating food appears to be facilitated by the lever mechanics of the jaw skeleton in this species. A preliminary sample of measurements of jaw mechanical advantage (ratio of the lever arm for jaw adductor A2 to the distance from the jaw joint to the jaw tip: Westneat, 2003) indicates that the jaws of *L. concolor* have the lowest mechanical advantage of any Hawaiian goby and, thus, have the best advantage for rapid (as opposed to forceful) closing (Maie *et al.*, in prep.).

Discussion: How Functional Morphology Can Inform Efforts to Manage and Conserve Hawaiian Streams

As noted in the Introduction, functional morphological analyses can provide valuable information for species conservation and management efforts. Our discussion in reference to Hawaiian stream fishes will focus on the functional performance of two major categories of behavior, locomotion and feeding.

Locomotion

Many studies in fishery research have examined the swimming performance of fishes as it relates to species management (e.g., under varying flow regimes or differing exposure to environmental contaminants: see Hammer, 1995; Wolter & Arlinghaus, 2003 and references therein). Very little is known about the swimming performance of Hawaiian stream fishes (or other closely related species: e.g., Todd, 1975; Bell & Brown, 1995), and this aspect of functional morphology represents an important area for future investigation. However, the importance of climbing locomotion in the life history of some Hawaiian stream fishes (as well as several stream fishes on other oceanic islands) is well known and has received recent study (Schoenfuss & Blob, 2003; Blob *et al.*, 2006). The significance of climbing as an aspect of locomotory functional morphology introduces a variety of factors that must be considered for effective management and conservation of these species.

Differences in locomotory performance among fish species, particularly in their ability to scale vertical barriers, can strongly affect the diversity of fishes found throughout a stream. Vertical barriers, both natural and man-made, are found in many Hawaiian freshwater streams; because of differences in climbing ability among species, such barriers limit the upstream diversity of fishes in a predictable fashion. Modest barriers of only a few feet are sufficient to preclude non-climbing species (*E. sandwicensis* and *S. hawaiiensis*) from stream reaches upstream from a barrier. If a barrier (a waterfall or man-made structure) exceeds several meters in height, it typically will also exclude *A. guamensis* from upstream reaches for two reasons. First, juvenile *A. guamensis* use energetically demanding powerburst climbing (Schoenfuss & Blob, 2003) but climb more slowly than the other powerburst climbing species, *L. concolor* (Table 2) (Blob *et al.*, 2006). This may limit the height of barriers that this species can scale and, thereby, limit its penetration into upstream habitats. Second, *A. guamensis* have the lowest ability to hold station among Hawaiian climbing gobies (Fitzsimons *et al.*, 1997) and do not appear capable of climbing vertical barriers as adults (Blob *et al.*, in press). Thus, in contrast to *S. stimpsoni* and *L. concolor*, which retain the ability to climb as adults (Blob *et al.*, in press), adult *A. guamensis* appear to be the climbing species most likely to be swept downstream and the least able to re-penetrate habitats above barriers. Because of the functional limitations that barriers place on the distribution of fish species in Hawaiian streams, stream surveys need to consider the potential influence of barriers when evaluating ichthyofaunal diversity. The increased use of Indices of Biological Integrity (IBI) in stream quality assessment, for example, might erroneously reduce the value of a stream where physical barriers exclude fish species as described above.

Differences in locomotory functional morphology among Hawaiian amphidromous species also can be correlated with differences in the ability of these species to climb barriers with specific physical characteristics. Our work has shown that the roughness of a barrier surface affects the climbing performance of juvenile powerburst climbers, but not the performance of the juvenile inching climber *S. stimpsoni* (Table 2) (Blob *et al.*, 2006). Over the range of surface textures that we tested (smooth to coarse sand, up to an average grain size of 1206 m), increases in surface roughness did not affect climbing speed in juvenile *S. stimpsoni*, but significantly improved the climbing speed of powerburst climbing species. We believe that this pattern of performance occurs because rougher surfaces allow juvenile powerburst climbers to gain better purchase on the substrate with their pectoral fins and, thus, prevent slippage (and loss of propulsive force) during the initial rapid fin adduction at the onset of climbing bouts (Blob *et al.*, 2006). In contrast, juvenile inching climbers do not experience such effects because they do not use their pectoral fins during climbing (Schoenfuss & Blob, 2003). Surface textures of obstacles could be selected to facilitate or improve the likelihood of climbing success among powerburst climbing juveniles. Decisions about construction materials for in-stream structures should be made with an awareness of such potential long-term effects on the locomotory performance and recruitment of stream fishes.

Available data on the functional morphology of locomotion in juvenile Hawaiian stream fishes suggest several possible effects of in-stream barriers on these species, but barriers also have the potential to affect adult fishes in Hawaiian streams. Although adult locomotion has received less study than that of juveniles, our work suggests that in-stream barriers might become insurmountable as fish grow in size (Maie *et al.*, in press). For example, *A. guamensis* appear to lose the ability to climb as adults (Blob *et al.*, in press), potentially because their great increase in mass and shift to a plumper body shape during growth leads to an increase in fluid dynamic drag (Vogel, 1994) that powerburst climbing mechanisms cannot overcome. This possibility is further suggested by observations indicating that neither Hawaiian fish species that retains climbing ability into adulthood appears to use juvenile powerburst climbing mechanisms (Blob *et al.*, in press). Adult *S. stimpsoni* climb using the inching mechanism employed by juveniles, though adults may also incorporate pectoral fin adduction into each climbing cycle and add propulsive thrust (Blob *et al.*, in press). In addition, our data indicate that *L. concolor*, which is a powerburst climber as a juvenile, dramatically alters climbing as an adult by reducing axial undulation in comparison to juveniles, adding pectoral fin adduction during each climbing cycle, and producing climbing kinematics that resemble those of adult *S. stimpsoni* (Blob *et al.*, in press). The incorporation of pectoral fin adduction into adult climbing by both *S. stimpsoni* and *L. concolor* indicates that the surface roughness of stream barriers could affect climbing success in adults of both of these species, much as it does in juvenile powerburst climbers that use the pectoral fins to initiate climbing bouts (Blob *et al.*, 2006).

The functional morphology of locomotion in Hawaiian stream fishes is relevant not only to how these species respond to stream structures that might present barriers to climbing, but also to how they are likely to respond to changes in stream flow patterns. Hawaiian streams exhibit considerable natural variation in flow: sudden, violent rains and flash floods can quickly and dramatically increase flow rates, volume, and turbidity (Fitzsimons & Nishimoto, 1995). However, flow patterns can also be altered anthropogenically, as stream water diversion for agriculture or development can drastically reduce flow volumes. The three climbing species of Hawaiian gobies show excellent ability to hold station against high velocity flows of 70 cm/sec or more (Fitzsimons *et al.*, 1997). Such flows remove parasite-bearing exotic species, which are often introduced into streams but, without pelvic sucking discs, cannot withstand intense stream flow (Fitzsimons *et al.*, 1997). As a result, allowance for periodic high velocity flows appears to be a critical issue for stream management in order to remove exotic sources of parasites and promote the health of native Hawaiian stream fishes. In addition, high flow regimes and frequent flash floods limit the instream range of the piscivores *E. sandwicensis* and *Kuhlia xenura* in streams without vertical obstacles. As a result, patterns of stream flow and frequency of flash floods should affect amphidromous fishes differently on islands with frequent vertical barriers (i.e., Hawai'i) than on islands with larger river systems (i.e., Kaua'i).

Feeding

The dramatic performance of several Hawaiian stream fishes makes the importance of locomotory functional morphology readily apparent for conservation efforts. However, recent data have shown that the functional morphology of feeding also can exert considerable influence on the in-stream distribution of Hawaiian stream fishes and, thus, should be considered as management and conservation strategies are developed for these species.

One of the most direct ways in which the significance of feeding functional morphology is evident in Hawaiian stream fishes involves predation on the four species of larval gobies. The estuaries of Hawaiian freshwater streams are frequented by several piscivorous predators, including the marine *āholehole K. xenura* and the amphidromous eleotrid *E. sandwicensis*. Data from Tate (1997) suggest that *E. sandwicensis* in particular are highly effective predators that have the potential to exert substantial selective pressure on migrating larvae. On the Island of Hawai'i, where estuaries are short before waterfalls exclude predaceous *E. sandwicensis* from upstream habitats, *A. guamensis* and *L. concolor* might not experience substantial selection due to predation as these powerburst climbing species typically climb out of the estuary very quickly (Tate, 1997). On the other hand, larval *S. stimpsoni* are unable to escape the estuary until after metamorphosis, when morphological changes make it possible for the juveniles to climb (Schoenfuss, 1997). *Sicyopterus stimpsoni* are likely exposed to higher predation pressure on Hawai'i than powerburst climbers despite short stream estuaries. However, these interactions may differ on islands in which streams have different physical characteristics. For example, estuaries of streams on Kaua'i are much longer than those on Hawai'i, and they allow predatory *E. sandwicensis* to penetrate further upstream (Schoenfuss & Blob, unpubl. data) where the larvae of the three climbing species are likely exposed to more extensive predation pressure. Predatory effects of *E. sandwicensis* on *S. hawaiiensis* are unknown on both Hawai'i and Kaua'i, though the fact that these species are found in close proximity in streams at all life stages suggests that predation by *E. sandwicensis* does not affect the sustainability of *S. hawaiiensis* populations. Nonetheless, the interactions of Hawaiian stream fishes as predators and prey illustrate two important points. First, they demonstrate the potential for interactions between the functional morphology of feeding (predation) and the functional morphology of locomotion (escape) to affect selection on populations. Second, they demonstrate that differences in the physical features of streams (e.g., long or short estuaries) can influence the importance of particular aspects of functional morphology across populations. This suggests that efforts at stream management and conservation will require a degree of specific tailoring to local conditions.

Feeding habits may also contribute to structuring interactions among age classes of Hawaiian stream fishes. For example, although juvenile and adult *A. guamensis* appear to eat similar material (Julius *et al.*, in prep.), high stream flows may reduce competition for food between juveniles and adults because station holding ability decreases as *A. guamensis* increases in size (Fitzsimons *et al.*, 1997). Feeding performance may also be important in determining habitat use among different size classes of *S. stimpsoni*. The feeding habits of *S. stimpsoni* are distinctive due to its ventral mouth position and its tendency to scrape its upper jaw over rocks in order to graze on short-stalked algae and diatoms. The growth of filamentous algae can impede access to shorter stalked food items, particularly for juvenile fishes (Julius *et al.*, 2005). However, high stream flows delay the growth of long algal strands, and periodic flash floods can "reset the clock" of algal growth by cleaning rocks of filamentous taxa (Julius *et al.*, 2005). Adult *S. stimpsoni* appear able to withstand longer periods of low flows as they actively remove filamentous algae from their feeding rocks (a behavior termed "gardening": Fitzsimons *et al.*, 2003). Adults defend large feeding rocks against incursions from other individuals, including juvenile *S. stimpsoni* (Fitzsimons *et al.*, 2003), which are forced to feed off smaller rocks and pebbles. Moreover, juvenile *S. stimpsoni* do not appear able to remove filamentous algae from feeding rocks (Fitzsimons *et al.*, 2003). Thus, to successfully migrate upstream, juvenile *S. stimpsoni* may depend substantially on frequent (3- to 5-week intervals) flash floods to clean filamentous algae from small feeding rocks and pebbles (Julius *et al.*, 2005). Water management efforts should take this cycle into account, as failure of this cycle could contribute to poor recruitment of migrating juveniles into adult stream reaches and reduce the diversity of fish age classes in streams.

Differences in the functional morphology of feeding between species of Hawaiian gobies may also play a strong role in determining the stream habitats that these species exploit. For example, there appears to be dietary overlap between *A. guamensis* and *L. concolor* (Kido, 1996b; Julius, this issue). However, *A. guamensis* feeds in slower flowing water near the streambed, whereas *L. concolor* rises in the water column to pick food items out of the water (Schoenfuss, pers. observ.). Previously noted differences in jaw lever mechanics between these species may contribute to the differences in their feeding habits, as the jaws of *L. concolor* appear better suited than those of *A. guamensis* for the rapid movements required to snatch floating prey out of fast flowing water (T. Maie, in prep.). In addition, feeding requirements and abilities may sometimes exclude particular species of Hawaiian gobies from populating otherwise high quality freshwater habitats. A dramatic example of such exclusion was examined by Schoenfuss *et al.* (2004) in a unique freshwater habitat on the southeastern aspect of the Island of Hawai'i. Here, a fracture in the lava shield contains a spring-fed freshwater environment that is host only to *S. stimpsoni* despite excellent water quality. *Eleotris sandwicensis* and *S. hawaiiensis* may be unable to subsist in this habitat due to a lack of gravel and sand in the stream-bed that could provide food (for *S. hawaiiensis*) or cover for predatory strikes (for *E. sandwicensis*). Long-stalked algae are lacking in this habitat because of its restricted light regime (tall volcanic rocks flank the freshwater stream), precluding *A. guamensis,* which feeds heavily on these taxa (Kido, 1996b; Julius, this issue). Finally, insects and other floating food items are largely absent in this habitat and preclude *L. concolor*. Thus, despite excellent water quality, feeding restrictions appear to limit colonization of this locality to *S. stimpsoni*, which grazes on abundant low stalked algae and thrives at a population density considerably greater than that in other Hawaiian streams. Although the low diversity of fish species in this locality might suggest that it represents low quality habitat, this example illustrates how knowledge of functional morphology (in this case for feeding) provides alternative explanations for patterns of presence or absence of species in stream systems.

Conclusions and Recommendations

The above discussion demonstrates that several aspects of functional morphology affect the distribution and sustainability of all five species of Hawaiian freshwater fishes. These factors are related in principle to locomotory limitations and dietary restrictions among Hawaiian freshwater fishes. Based on the published literature and our observations, we recognize several key findings. (1) Stream flow rates and frequency of maximum flow events are crucial to sustaining a healthy Hawaiian ichthyofauna. (2) The marine/freshwater interface and the estuarine environment are of critical importance to all five species of Hawaiian freshwater fishes. (3) Limitations in fish diversity within Hawaiian freshwater streams might be a result of locomotory or dietary constraints that are not attributable to environmental degradation.

These key findings lead us to recommend several avenues of future research to ensure the health not only of the Hawaiian freshwater ichthyofauna, but also for other organisms in the nearshore environment. (1) The Hawaiian Islands vary dramatically in geological morphology, leading to distinct, island-specific flow regimes and cycles of maximum flow events. Stream management plans need to be tailored to these local conditions. One freshwater management plan will not be applicable to all islands, and future research should carefully examine how environmental differences between the islands relate to population distribution in fish species. (2) Estuaries are a critical habitat for Hawaiian freshwater fishes and sustain several marine species as well. Estuaries are also a common interface between Hawaiian streams and the local human population. These interactions need to be studied in greater detail to develop management plans that maximize public access while minimizing impact for the native ichthyofauna. (3) *Sicyopterus stimpsoni* might serve as a key indicator species to monitor the long-term health of Hawaiian estuaries, as its migrants spend the greatest amount of time in this habitat among species that traverse it. In addition, the metamorphosis of this species results in an especially energy depleted and vulnerable life stage in the estuarine habitat, making it the most vulnerable of the species that pass through it. Finally, this species has been studied extensively in the past 15 years, and future efforts could be focused on its interactions within the estuarine ecosystem. (4) The broad use of the Index of Biological Integrity (IBI) to assess Hawaiian stream health needs to be critically evaluated and possibly amended to reflect the uniqueness of the Hawaiian fish fauna.

Acknowledgments

We thank the organizers of this conference for inviting our participation, and the State of Hawaii for funding our previous research and this conference. We also thank the Stearns Manufacturing Company for additional support of our research along with faculty research grants from St. Clous State University and Clemson University. For advice and logistical support during fieldwork we thank R. Nishimoto, M. Fitzsimons, D. Kuamo'o, B. Kaya, L. Nishimura, T. Sakihara, T. Shimoda, and T. Shindo. We also thank M. Julius and R. Rai (St. Cloud State University) and T. Maie, C. Mims, and M. Wright (Clemson University) for their contributions to data collection and analysis. The manuscript was improved through helpful suggestions by two anonymous reviewers.

Literature Cited

Arlinghaus, R., T. Mehner & I.G. Cowx. 2002. Reconciling traditional inland fisheries management and sustainability in industrialized countries, with emphasis on Europe. *Fish & Fisheries* **3**: 261–316.

Arnold, S.J. 1983. Morphology, performance, and fitness. *American Zoologist* **23**: 347–361.

———. 2003. Performance surfaces and adaptive landscapes. *Integrative & Comparative Biology* **43**: 367–375.

Azevedo, J.M.N. & A.M.V. Simas. 2000. Age and growth, reproduction and diet of a sublittoral population of the rock goby *Gobius paganellus* (Teleostei, Gobiidae). *Hydrobiologia* **440**: 129–135.

Bell, K.N.I. & J.A. Brown. 1995. Active salinity choice and enhanced swimming endurance in 0 to 8-d-old larvae of diadromous gobies, including *Sicydium punctatum* (Pisces), in Dominica, West Indies. *Marine Biology* **121**: 409–417.

Birdsong, R.S. 1975. The osteology of *Microgobius signatus* Poey (Pisces: Gobiidae), with comments on other gobiid fishes. *Bulletin of the Florida State Museum Biological Sciences* **19**: 136–186.

———., **E.O. Murdy & F.L. Pezold.** 1988. A study of the vertebral column and median fin osteology in gobioid fishes with comments on gobioid relationships. *Bulletin of Marine Science* **42**: 174–214.

Blob, R.W., R. Rai, M.L. Julius & H.L. Schoenfuss. 2006. Functional diversity in extreme environments: effects of locomotor style and substrate texture on the waterfall climbing performance of Hawaiian gobiid fishes. *Journal of Zoology* **268**: 315–324.

———., **K.M. Wright, M. Becker, T. Maie, T.J. Iverson, M.L. Julius & H.L. Schoenfuss.** In press. Ontogenetic change in novel functions: waterfall climbing in adult Hawaiian gobiid fishes. *Journal of Zoology*

Chubb, A.L., R.M. Zink & J.M. Fitzsimons. 1998. Patterns of mtDNA variation in Hawaiian freshwater fishes: the phylogeographic consequences of amphidromy. *Journal of Heredity* **89**: 8–16.

Domenici, P. 2003. Habitat, body design, and the swimming performance of fish, p. 137–160. *In*: Bells, V.L., J.-P. Gasc & A. Casinos (eds.), *Vertebrate biomechanics and evolution*. BIOS Scientific Publishers Ltd, Oxford.

Englund, R. & R. Filbert. 1997. Discovery of the native stream goby, *Lentipes concolor*, above Hawaii's highest waterfall, Hi'ilawe Falls. *Bishop Museum Occasional Papers* **49**: 62–64.

Ferry-Graham, L.A. & P.C. Wainwright. 2003. Evaluating suction feeding performance in fishes: implications for evolutionary diversification, p. 101–115. *In*: Bells, V.L., J.-P. Gasc, & A. Casinos (eds.), *Vertebrate biomechanics and evolution*. BIOS Scientific Publishers Ltd., Oxford.

Fitzsimons, J.M., M.G. McRae, H.L. Schoenfuss & R.T. Nishimoto. 2003. Gardening behavior in amphidromous Hawaiian fish *Sicyopterus stimpsoni* (Osteicthyes: Gobiidae). *Ichthyological Exploration of Freshwaters* **14**: 185–191.

———. **& R.T. Nishimoto.** 1990. Territories and site tenacity in males of the Hawaiian stream goby *Lentipes concolor* (Pisces: Gobiidae). *Ichthyological Exploration of Freshwaters* **1**: 185–189.

———. **& R.T. Nishimoto.** 1991. Behavior of gobioid fishes from Hawaiian fresh waters, p. 106-124. In: Devick, W.S. (ed.), *New directions in research, management, and conservation of Hawaiian freshwater stream ecosystems*. State of Hawaii, Honolulu.

———. **& R.T. Nishimoto.** 1995. Use of fish behavior in assessing the effects of Hurricane Iniki on the Hawaiian island of Kaua'i. *Environmental Biology of Fishes* **43**: 39–50.

———., **J.E. Parham & R.T. Nishimoto.** 2002. Similarities in behavioral ecology among amphidromous and catadromous fishes on the oceanic islands of Hawai'i and Guam. *Environmental Biology of Fishes* **65**: 123–129.

———., **H.L. Schoenfuss & T.C. Schoenfuss.** 1997. Significance of unimpeded flows in limiting the transmission of parasites from exotics to Hawaiian stream fishes. *Micronesica* **30**: 117–125.

Fukui, S. 1979. On the rock-climbing behavior of the goby, *Sicyopterus japonicus*. *Japanese Journal of Ichthyology* **26**: 84–88.

Fulton, C.J., D.R. Bellwood & P.C. Wainwright. 2001. The relationship between swimming ability and habitat use in wrasses (Labridae). *Marine Biology* **139**: 25–33.

Hammer, C. 1995. Fatigue and exercise tests with fish. *Comparative Biochemistry & Physiology* **112A**: 1–20.

Holland, L.E. 1986. Effects of barge traffic on distribution and survival of ichthyoplankton and small fishes in the Upper Mississippi River. *Transactions of the American Fisheries Society* **115**: 162–165.

Irschick, D.J. 2003. Measuring performance in nature: implications for studies of fitness within populations. *Integrative and Comparative Biology* **43**: 396–407.

Julius, M.L., R.W. Blob & H.L. Schoenfuss. 2005. The survival of *Sicyopterus stimpsoni*, an endemic amphidromous Hawaiian gobiid fish, relies on hydrological cycles of streams: evidence from changes in the algal composition of diet through growth stages. *Aquatic Ecology* **39**: 473–484.

Juvik, S.P. & J.O. Juvik. 1998. *Atlas of Hawai'i*. 3rd ed. University of Hawai'i Press, Honolulu, 333 p.

Keith, P. 2003. Biology and ecology of amphidromous Gobiidae of the Indo-Pacific and Caribbean regions. *Journal of Fish Biology* **63**: 831–847.

Kido, M.H. 1996a. Diet and food selection in the endemic Hawaiian amphidromous goby, *Sicyopterus stimpsoni* (Pisces: Gobiidae). *Environmental Biology of Fishes* **45**: 199–209.

———. 1996b. Morphological variation in feeding traits of native Hawaiian stream fishes. *Pacific Science* **50**: 184–193.

Krueger, C.C. & D.J. Decker. 1993. The process of fisheries management, p. 33–54. *In*: C.C. Kohler & W.A. Hubert (eds.), *Inland Fisheries Management in North America*. American Fisheries Society, Bethesda, Maryland.

Maie, T., H.L. Schoenfuss & R.W. Blob. In press. Ontogenetic scaling of body proportions in waterfall-climbing gobiid fishes from Hawai'i and Dominca: implications for locomotor function. *Copeia*.

McDowall, R.M. 2003. Hawaiian biogeography and the islands' freshwater fish fauna. *Journal of Biogeography* **30**: 703–710.

———. 2004. Ancestry and amphidromy in island freshwater fish faunas. *Fish & Fisheries* **5**: 75–85.

Munday, P.L., S.J. Pierce, G.P. Jones & H.K. Larson. 2002. Habitat use, social organization and reproductive biology of the seawhip goby, *Bryaninops yongei*. *Marine & Freshwater Research* **53**: 769–775.

Nelson, S.G., J.E. Parham, R.B. Tibbatts, F.A. Camacho, T. Leberer & B.D. Smith. 1997. Distribution and microhabitats of the amphidromous gobies in streams of Micronesia. *Micronesica* **30**: 83–91.

Nishimoto, R.T. & J.M. Fitzsimons. 1999. Behavioral determinants of the instream distribution of native Hawaiian stream fishes, p. 813–818. *In*: Seret, B. & J.-Y. Sire (eds.), *Proceedings of the 5th Indo-Pacific Fish Conference*. Noumea.

———. **& D.K.G. Kuamo'o**. 1997. Recruitment of goby postlarvae into Hakalau stream, Hawai'i Island. *Micronesica* **30**: 41-49.

Novotny, V., A. Bartosova, N. O'Reilly & T. Ehlinger. 2005. Unlocking the relationship of biotic integrity of impaired waters to anthropogenic stresses. *Water Research* **39**: 184–198.

Parenti, L.R. & K.R. Thomas. 1998. Pharyngeal jaw morphology and homology in sicydiine gobies (Teleostei: Gobiidae) and allies. *Journal of Morphology* **237**: 257–274.

Pezold, F. 1993. Evidence for a monophyletic Gobiinae. *Copeia* **1993**: 634–643.

Privitera, L.A. 2002. Reproductive biology of the coral-reef goby, *Asterropteryx semipunctata*, in Kaneohe Bay, Hawaii. *Environmental Biology of Fishes* **65**: 289–310.

Radtke, R.L. & R.A. Kinzie III. 1991. Hawaiian amphidromous gobies: Perspectives on recruitment processes and life history events, p. 125–141. *In*: Devick, W.S. (ed.), *New directions in research, management, and conservation of Hawaiian freshwater stream ecosystems*. State of Hawaii, Honolulu.

———., **R.A. Kinzie & S.D. Folsom**. 1988. Age at recruitment of Hawaiian freshwater gobies. *Environmental Biology of Fishes* **23**: 205–213.

Schoenfuss, H.L. 1997. Metamorphosis of the Hawaiian stream goby *Sicyopterus stimpsoni*: a structural, functional, and behavioral analysis. Ph.D. Dissertation. Louisiana State University. 145 p.

———. **& R.W. Blob**. 2003. Kinematics of waterfall climbing in Hawaiian freshwater fishes (Gobiidae): vertical propulsion at the aquatic-terrestrial interface. *Journal of Zoology* **261**: 191–205.

———., **M.L. Julius & R.W. Blob**. 2004. Colonization of a recent, volcanically formed freshwater habitat: an example of primary succession. *Ichthyological Exploration of Freshwaters* **15**: 83–90.

———., **T.A. Blanchard & D.K.G. Kuamo'o**. 1997. Metamorphosis in the cranium of postlarval *Sicyopterus stimpsoni*, an endemic Hawaiian stream goby. *Micronesica* **30**: 93–104.

Schwenk, K. 2000. *Feeding: form, function, and evolution in tetrapod vertebrates*. Academic Press, San Diego.

Steinmann, A. 1996. Effects of grazing on freshwater benthic algae, p. 341–373. *In*: Stevenson, R.J., M.L. Bothwell & R.L. Lowe (eds.), *Algal ecology: freshwater benthic ecosystems*. Academic Press, San Diego.

Tate, D.C. 1997. The role of behavioral interactions of immature Hawaiian stream fishes (Pisces: Gobiodei) in population dispersal and distribution. *Micronesica* **30**: 51–70.

Taylor, M.S. & M.E. Hellberg. 2003. Genetic evidence for local retention of pelagic larvae in a Caribbean reef fish. *Science* **299**: 107–109.

Thacker, C.E. 2002. Molecular phylogeny of the gobioid fishes (Teleostei: Perciformes: Gobioidei. *Molecular Phylogenetics & Evolution* **26**: 354–368.

Todd, E.S. 1975. Vertical movements and development of the prolarvae of the eleotrid fish, *Dormitator latifrons*. *Copeia* **1975**: 564–568.

Voegtle, B., M. Larinier & P. Bosc. 2002. Etude sur les capacités de franchissement des cabot bouche-rondes (*Sicyopterus lagocephalus*, Pallas, 1770) en vue de la conception dispositifs adaptés aux prises d'eau du transfert Salazie (île de la Réunion). *Bulletin Français de la Pêche et de la Pisciculture* **364**: 109–120.

Vogel, S. 1994. *Life in moving fluids: the physical biology of flow*. Princeton University Press, Princeton, New Jersey. 467 p.

Wainwright, P.C., D.R. Bellwood & M.W. Westneat. 2002. Ecomorphology of locomotion in labrid fishes. *Environmental Biology of Fishes* **65**: 47–62.

Walker, J.A. 2000. Does a rigid body limit maneuverability? *Journal of Experimental Biology* **203**: 3391–3396.

Webb, P.W. 1984. Body form, locomotion, and foraging in aquatic vertebrates. *American Zoologist* **24**: 107–120.

Westneat, M.W. 2003. A biomechanical model for analysis of muscle force, power output, and lower jaw motion in fishes. *Journal of Theoretical Biology* **223**: 269–281.

———. 2004. Evolution of levers and linkages in the feeding mechanisms of fishes. *Integrative & Comparative Biology* **44**: 378–389.

Wolter, C. 2001. Conservation of fish species diversity in navigable waterways. *Landscape & Urban Planning* **53**: 135–144.

———. **& R. Arlinghaus.** 2003. Navigation impacts on freshwater fish assemblages: the ecological relevance of swimming performance. *Reviews in Fish Biology & Fisheries* **13**: 63–89.

Yamamoto, M.N. & A.W. Tagawa. 2000. *Hawai'i's Native and Exotic Freshwater Animals*. Mutual Publishing, Honolulu, p. 200.

Zink, R.M. 1991. Genetic variation within and between populations of *Lentipes concolor* from Hawai'i and Kaua'i, p. 96–105. *In*: Devick, W.S. (ed.), *New directions in research, management, and conservation of Hawaiian freshwater stream ecosystems*. State of Hawaii, Honolulu.

———., **J.M. Fitzsimons, D.L. Dittman, D.R. Reynolds & R.T. Nishimoto.** 1996. Evolutionary genetics of Hawaiian freshwater fishes. *Copeia* **1996**: 330–335.

Biology of Hawaiian Streams and Estuaries. Edited by N.L. Evenhuis & J.M. Fitzsimons. *Bishop Museum Bulletin in Cultural and Environmental Studies* 3: 143–156 (2007).

Mugilids in the Muliwai: a Tale of Two Mullets

Robert T. Nishimoto, Troy E. Shimoda & Lance K. Nishiura
Hawaii Division of Aquatic Resources, 75 Aupuni Street, Hilo, Hawai'i 96720 USA;
emails: robert.t.nishimoto@hawaii.gov, troy.e.shimoda@hawaii.gov, lance.k.nishiura@hawaii.gov

Abstract

The *'ama'ama* or striped mullet, *Mugil cephalus*, is circumglobal in tropical and warm temperate seas. This species is common in estuaries, inland lagoons, and rivers. There are only two species of native mullet in Hawai'i. The *uouoa* or sharpnose mullet, *Neomyxus leuciscus*, is a small species and common along the rocky coastline. The striped mullet is an estuarine inhabitant as larvae and juveniles; the adults are found in deeper waters, usually where there is significant freshwater outflow, as stream surface flow or by groundwater intrusion. The *'ama'ama* is a significant species in traditional Hawai'i. This fish was coveted by royalty, and there are numerous words in the Hawaiian language describing the life stages and migration pattern. The species was once cultured in Hawaiian fish ponds but now is more significant as a recreational fishery species. Because of the general decline in coastal fish species, the Hawai'i Division of Aquatic Resources partnered with the Oceanic Institute to develop a prototype marine stock augmentation program for the *'ama'ama* fishery. A centralized hatchery spawned adults, reared the larvae, tagged young fingerlings with coded wire tags, and released them into juvenile habitats, such as Wailoa River estuary and Waiākea Pond in Hilo. The release of 20,000 hatchery fingerlings/year had no negative effect on the native mullet stocks, but made a significant contribution to the fishery. The size, location, and season at release of hatchery mullet fingerlings were critical to the success of the effort. The alien mullet, *Valamugil engeli* (*kanda*), accidentally released on O'ahu from 1955–1958, are reported to occur in great numbers in many estuaries in the main Hawaiian Islands. *Kanda* are very common in Hilo Bay, Wailoa, River and Waiākea Pond, a significant juvenile fish habitat for coastal marine species. The effort to protect the decline of the native mullet by classical fishery management rules has inadvertently provided greater protection for the alien mullet. The impact of the alien on the native mullet is now being studied, and, concomitantly, novel approaches are being discussed to control this invasion.

Introduction

Mullets comprise one of the more speciose and characteristic fish families (Mugilidae) in tropical and warm temperate coastal waters, estuaries, and lagoons. This family is made up of 80 species belonging to 17 genera (FishBase 2000). These schooling fish are highly euryhaline or diadromous (McDowall, 1988), iliophagous, and are generally found in mostly protected, soft-bottom habitats. Their marine phase is short and consists mostly of spawning individuals and the early larval stages. Mullets, especially *Mugil cephalus*, are an important food fish worldwide and have been the subject of numerous fishery studies and research on the aquaculture of the species. The striped mullet is the best-studied species, but its life cycle is poorly understood (McDowall, 1988).

Hawai'i, with its depauperate fish fauna, has only two indigenous species of mullets; the sharpnose mullet, *Neomyxus leuciscus,* and the circumglobal *M. cephalus*. The sharpnose mullet, locally known as *uouoa*, rarely exceeds 30 cm, occurs in small schools close to shore, and is rarely, if ever, observed in estuaries. The striped mullet or *'ama'ama*, is a much larger fish, reported to reach 90 cm; however, the average size is more like 50 cm (Randall, 1996). The striped mullet is known to penetrate low gradient streams, such as certain streams on Kaua'i, but is more common in protected habitats, especially in estuaries (*muliwai*). There are three types of estuaries in Hawai'i (Juvik & Juvik, 1998). The two most common are at stream mouths or where surface flow is absent but with significant groundwater discharge, such as Hilo Harbor. The other type is the more classical large embayment, such as Pearl Harbor on O'ahu.

Figure 1. Makoa and the Mullet, print by Dietrich Varez. Note live mullet in hand being carried from Waiākea Pond in Hilo to the young King Kamehameha in Kawaihae, south Kohala. Used here with permission from the artist via Volcano Art Center, Volcano, Hawai'i.

In Hawai'i the *'ama'ama* reach sexual maturity at about 28 cm, at about 3 years old, and migrates offshore during the winter months to spawn in the ocean. The pre-juveniles, averaging 20 mm standard length (SL), appear at intertidal estuarine habitats 30–45 days after hatching at sea (Major, 1978). The recruiting fingerlings use turbidity gradients as an orientation cue (Cyrus & Blaber, 1987) along with tidal transport as a mechanism to move into juvenile habitats. Pre-juveniles, averaging 17–35 mm SL, are very common in the shallow intertidal habitats in the spring but disappear by the end of June. Major (1978) reported that pre-juveniles inhabit shallow areas and tolerate highly fluctuating salinity and water temperatures as a pre-adaptation to avoid piscine predators. The fingerlings metamorphose into juveniles at 50 mm SL, abandon the extreme conditions in the shallows, and move into deeper waters. Concomitantly, there is a lengthening of the intestine and morphological changes to the teeth and lips, a feeding adaptation to benthic macrophagous omnivory (Blaber, 1987).

The *'ama'ama* has a long history in Hawai'i. This species was revered by Hawaiian royalty as a food fish harvested from fishponds and later as a valuable commercial product in the early 1900s. In Hilo, there is a specialized hook-and-line recreational fishery targeting this species, and recently it was the subject of a very successful stock enhancement project to test the efficacy of marine stock augmentation by releasing hatchery-raised fingerlings.

In this report we will (1) provide, a brief background history of *'ama'ama*, in traditional Hawai'i and as an important species in the fisheries; (2) summarize ongoing research on the use of stock enhancement in managing the recreational mullet fishery; and (3) document the invasion of *kanda*, an alien mullet introduced to the Hawaiian Islands from the South Pacific in 1955.

1. Background

'Ama'ama in traditional Hawai'i

The mullet was prized as a food fish for royalty. Most were collected from coastal fishponds, constructed by placing a stonewall enclosing a small bay at a stream mouth. Recruiting fingerlings were

Table 1. Hawaiian language terminology describing the different size classes (A) and migratory behavior (B) of the native striped mullet, *Mugil cephalus*, on O'ahu.

A

Hawaiian Name	Size Class
Pua 'ama, Pua po'ola	Finger size, new recruits
Kahaha, Pahaha	Hand length, juvenile stage
'Ama'ama	20 cm, estuary resident
'Anae	30+ cm, reproductive adult

B

Behavioral Group	Migration Route	Behavioral Activity
'Anae-holo	'Ewa to Lā'ie	Spawning migration, full-bodied
'Anaepali	Lā'ie to 'Ewa	Return migration, skinny

passively collected by lifting the sluice gates during the rising tide and corralling the fish in the pond. The gate was lowered to keep fish from leaving but still allowed tidal circulation. During the reign of King Kamehameha I, it was common for some Hawaiian Chiefs to select the swiftest runner to collect the *'ama'ama* from their favorite fishpond so the fish would still be alive when they returned (Wyban, 1992). As a tribute to the swiftness of these runners, the present day Volcano Art Center's annual Kīlauea Volcano Runs trademark depicts the legendary runner *Makoa* (Fig. 1), who carried the *'ama'ama* from the fish ponds in Hilo, over the saddle road between the Mauna Kea and Mauna Loa volcanoes, and delivered the still wriggling fish to the young King in Kawaihae, north Kona (Desha, 2000). In 1939 King Kamehameha III introduced a code of law where every commoner had access to the fish resources. The exceptions were a few species that were solely reserved for royalty (Jordan & Evermann, 1902). The *'ama'ama* of Hulehia (Hulēia?), Anehola (Anahola?), and Hanalei, for example, were taboo to the general population and shows that the mullet was a highly prized fish during the Hawaiian Kingdom period.

The Hawaiian language recognizes the different size classes of the *'ama'ama* (Pukui & Ebert, 1986) (Table 1A) but most intriguing is recognizing the traditional migratory route between Ewa to Lā'ie, O'ahu (Table 1B). The terms describe the traditional spawning migration and their return (Wyban, 1992).

In the late 1800s, many coastal fishponds were not tended and fell into disrepair as the population migrated to the city or other crops, such as rice and taro, became more profitable. In 1900, there were only 99 documented fishponds, and Chinese immigrants operated most. A census at the Honolulu fish market in 1900 reported that 35.6% of the fishes sold were the *'ama'ama*, however, there was no differentiation between mullet taken from fishponds or the coastal seas (Cobb, 1905). Mullet were the most expensive fish at the market and sold for 25 cents/lb. The number of fishponds used to cultivate *'ama'ama* and other estuarine species continued declining into the next century. In 2000, there were only two fishponds in production in Hawai'i. These fishponds collectively sold less than 1,000 lbs of *'ama'ama* in 2003 (Hawaii Division of Aquatic Resources, 2004).

'Ama'ama in the recreational fishery

Fishing for mullet by using local pole-and-line technique is a dying art if the popularity of this food fish is considered. This type of fishing, once easily recognized by the numerous, small wooden platforms, called stilt chairs, dotting the tidal flats (Hosaka, 1944) in Kāne'ohe Bay and Ala Wai Canal, is gone. These wooden chairs were not set randomly; instead, the locations were carefully selected and placed close to the daily migratory path of the mullet (Rizzuto, 1985). Bread was used for chumming and as bait. These platforms are now prohibited because of environmental regulations, and more than likely, there is not much interest for perpetuating this fishing technique. Small skiffs now

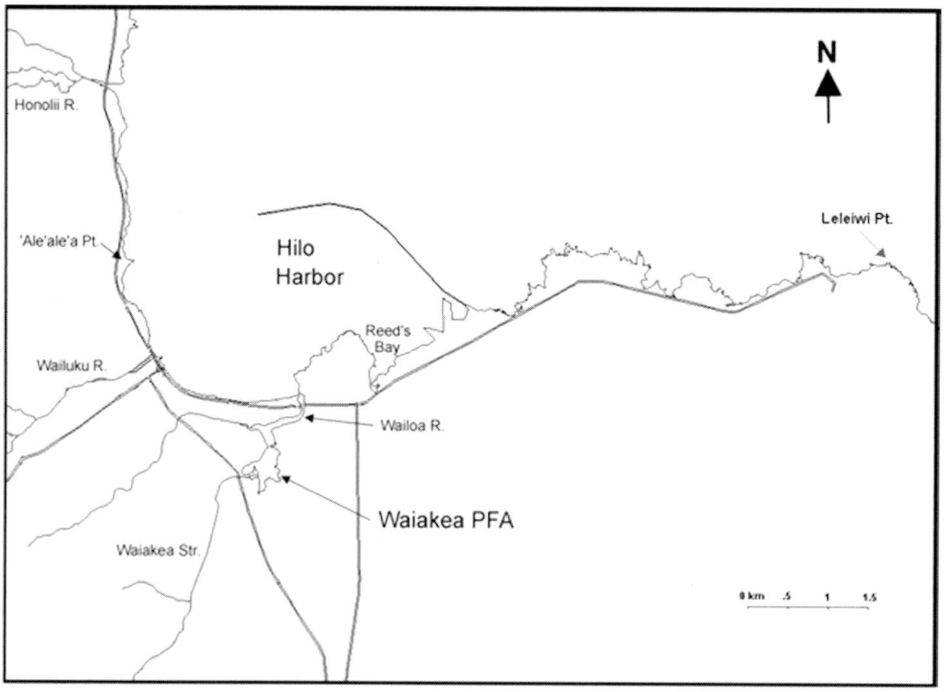

Figure 2. Map of Hilo Harbor, Wailoa River estuary, and Waiākea Pond.

replace such platforms. Hilo Harbor, especially the Waiākea Public Fishing Area (PFA), is one of the last strongholds of this type of mullet fishing. Fishers use a system of a delicately balanced bobber and tandem hooks baited with algae. Microscopic examination revealed that the bait is a wad of algae consisting almost exclusively of the chained diatom, *Melosira tropicalis* (Julius *et al.,* 2002). The catching of *'ama'ama* in Hilo is the only fishery in the world where diatoms are used as bait.

2. Stock Enhancement in the Management of the Hilo Recreational Mullet Fishery

History

Hawai'i coastal fishery stocks have seriously declined mostly due to anthropogenic impacts (Shomura, 1987). Observations by Okamoto (1994, pers. comm.), a seasoned fisher and Hawaii Fish & Game fishery biologist, noted that the average size of mullet caught in the 1940s averaged 3–4 lbs. He noticed a dramatic decline in average size over time and stated that mullet stocks are overfished and the brood stocks severely depleted. He hypothesized that the losses of shallow water nursery habitat and competition from the alien *kanda*, *Valamugil engeli*, have contributed to the decline.

The Hawai'i Fisheries Plan of 1990–1995 (Department of Land & Natural Resources, 1990) stated that the Division of Aquatic Resources (DAR) aims to restore to former abundance species whose numbers have become depleted, at least in part, because of loss or degradation of natural spawning and nursery areas. In 1990, DAR and the Oceanic Institute (OI) partnered to develop a collaborative project to help restore the declining coastal stocks by using marine stock enhancement technology. Initially, there was a series of public workshops to identify potential species that were candidates for stock enhancement. The two species that received the highest overall scores were the Pacific threadfin or *moi*, *Polydactylus sexfilis*, and the *'ama'ama, M. cephalus* (Leber, 1994). The latter species was selected because OI already had the technology to aquaculture the *'ama'ama* and there was a well-established recreational mullet fishery in Hilo, Hawai'i Island and a commercial net fishery in Kāne'ohe, O'ahu Island.

Methods and study site

'Ama'ama fingerlings were cultured at a central hatchery on O'ahu and shipped to the State Fisheries Research Station in Hilo for grow-out. Fingerlings of various sizes were batch tagged with internal Coded Wire Tags (CWT). Tagged fishes were kept for several days to allow recuperation from tagging stress. A total of 268,228 CWT mullet fingerlings were released at various locations in Hilo Bay from August 1990 to September 2000, except for 1996 when none were released. Hatchery release impact was assessed by creeling the recreational fisheries (starting in 1991) and by conducting bimonthly cast-net sampling (starting in 1990) at fixed stations in Waiākea Pond, Wailoa River, and Reeds Bay, all located within Hilo Harbor (Fig. 2). A 5-ft and 8-ft diameter cast nets with 3/8 inch stretch mesh, were used in combination to sample mullet fingerlings for the presence of CWT. A total of 15 throws were made at each station. The number of mullets was tallied by species, and the presence of CWT was detected with a magnetic tag detection unit. Fish with tags were collected and frozen. The CWT tags were excised, deciphered, and size, date, and release site information were recorded in a database. All native mullets were checked for tags and immediately returned to the wild if there were no tags. Creel census was conducted by interviewing mullet fishers in Hilo Harbor, but mostly in the Waiākea Public Fishing Area (PFA) where the fishery is concentrated.

Results and Discussion

The results were significant: 1. The prototype marine stock enhancement experiment demonstrated that even small-scale releases can have a significant impact on wild stock abundance (Leber *et al.*, 1995a); 2. The number of mullet entering the fishery was significant and was achieved annually; 3.The release of 20,000 fingerlings per year did not displace wild fish from the estuary (Leber *et al.*, 1995b); 4. The Wailoa River estuary, especially the boat launching ramp, was found to be an excellent release site; and 5. The most successful size at release was 70 mm total length (TL) and the optimum release period was during spring (Leber, 1995a; Leber *et al.*, 1997).

The number of CWT mullet in the fishers creel ranged from a low of 3.9% in 2003 to as high as 61.1% during 1999 (Fig. 3). The last batch of hatchery raised and tagged fish were released in 2000. The overall average increase on the recreational mullet fishery after 9 years of releasing hatchery-raised *'ama'ama* was 21.7%. Tagged mullet resided between 3 and 4 years in the Wailoa River estuary (Fig. 4). Most tagged mullet were caught during the second year after release and averaged 31 cm TL. The absence of 3–4 year (after tag and release) mullet from the Waiākea PFA fishery suggests that these fishes moved out of the estuary and presumably underwent an offshore spawning migration. There is no indication that these fish returned to the estuary; however, we have not conducted sampling or creel surveys outside of Hilo Harbor or along the coastal areas. Adult *M. cephalus* are often observed in smaller groups schooling along the coastal areas, and the larger individuals, around 55 cm, have been observed singly or in pairs, in deeper waters.

Hatchery mullet displayed a strong site fidelity to the Wailoa estuary. Most mullet tagged and released in the estuary resided there until they reached maturity. Small batches of CWT fingerlings released south (Leleiwi Point) and north (Honoli'i Stream) of Hilo Harbor and a small neighboring estuary, Reeds Bay, returned and resided in Waiākea Pond PFA until undertaking the spawning migration.

The Hilo mullet project verified the potential of stock enhancement as an effective tool to replenish diminishing stocks. However, stock enhancement should always be used in conjunction with rather than instead of classical fishery management options (Blankenship & Leber, 1995), such as bag limit and area closure, since the goal of stock enhancement is to augment and not replace wild stocks.

Based on the results of this project, several management measures were implemented to further DAR's mission of replenishing and conserving native fish stocks:

A. Minimum catch size was increased from 7 to 11 inches (fork length, FL) (= 12.75 in TL) to allow mullet to 'escape' the fishery and emigrate offshore to spawn.

B. Closed season was extended for one more month, from December to March, because larger mullet caught in late February and March still had mature gonads.

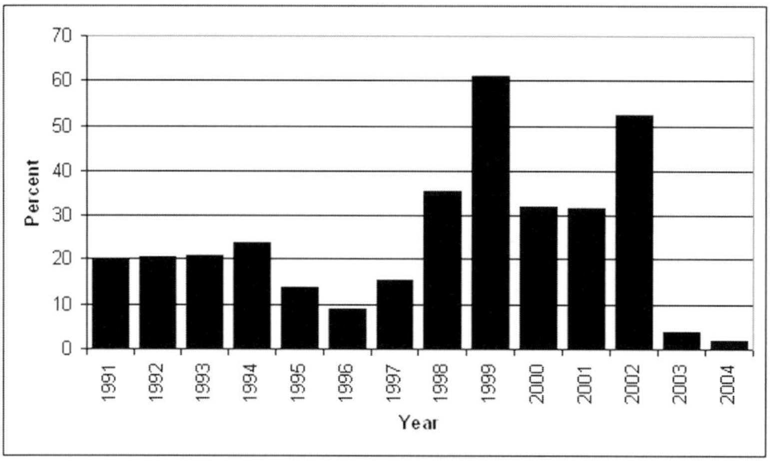

Figure 3. Percent tagged mullet in Hilo recreational fishery from 1991 to 2004 (n= 2,511).

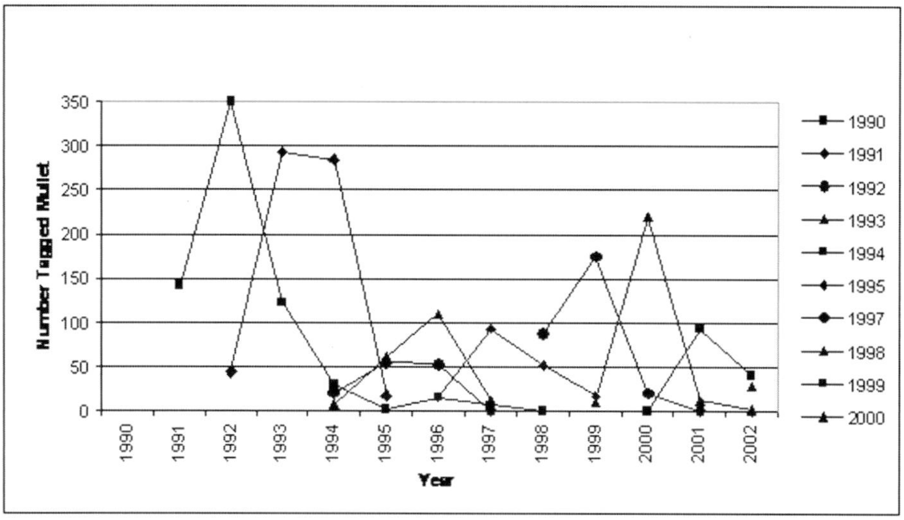

Figure 4. Residency time of tagged native mullet in Wailoa River estuary and Wai/a/*kea Pond from 1991 to 2002. Last release of tagged fish was in 2000.

C. The Waiākea PFA was extended seaward to include the Wailoa River estuary. Fishing gear was restricted to pole-and-line and the use of cast net and spearing was disallowed. The use of small mesh net to collect mullet fingerlings to stock private fishponds was not permitted. Wailoa River estuary was especially significant as a nursery habitat for newly recruited mullet fingerlings.

D. Bag limit was lowered to 10 fishes per day in Hilo Bay, Wailoa River estuary and Waiākea PFA.

E. Plans were developed to re-establish shoreline vegetation shoreline vegetation around the perimeter of Wailoa River and Waiākea has been thoroughly removed to accommodate recreational park users for aesthetic reasons. We plan to reestablish shoreline vegetation to provide cover for juvenile fishes and crustaceans. Newly metamorphosed mullet seek shallow estuarine intertidal after recruiting into the estuary. Vegetative cover is essential to lessen the impact of avian and piscine predators (Majors, 1978).

3. Invasion of the Alien Mullet *V. engeli* into the Habitat of Native *M. cephalus*

Introduction

Hawai'i has experienced waves of alien introductions starting in the early 1900's. The first several waves were considered purposeful introduction, mostly as recreational game fish, for commercial harvesting, or to provide live bait for the tuna fishery. More recently, alien introductions are being viewed with great concerns and recent introductions now have been accidental releases, escapees, or animals inadvertently transported between islands or even introduced from other continents, such as Asia.

The alien *kanda*, locally as Summer or Marquesan mullet, was first introduced to O'ahu between 1955–1958 (Randall & Kanayama, 1972, Hawaii Fish & Game fish introduction list). This fish was unknowingly mixed with several shipments totaling 143,800 Marquesan sardines *Sardinella marquesensis* (Maciolek, 1984), which were released at several sites on O'ahu as supplemental bait for the live-bait skipjack tuna fishery (Murphy, 1960). The fishers were not enthusiastic about using Marquesan sardines because they were not as effective for attracting skipjack tuna as the native anchovy or *nehu*, *Encrasicholina purpurea*. The native anchovy is not very common now in Hilo Bay and was presumably out competed by the introduced sardine *S. marquesensis* (Okamoto, 1994 pers. comm.) or affected adversely by anthropogenic activity, such as sedimentation or the introduction of upland pollutants (U.S. Army Engineer District, 1980). The Marquesan sardine has been replaced by the introduced goldspot sardine, *Herklotsichthys quadrimaculatus*.

The population of the *kanda* exploded and invaded the native mullet habitat in many estuary and bays in the Hawaiian Islands. This alien mullet was reported to outnumber the native three to one on Kaua'i from 1975–1978 (Maciolek, 1984) and made up nearly half of the catch in an O'ahu estuary in 1981 (Maciolek & Timbol, 1981). In 2000, about a third of the mugilids sampled (n = 2806) on the south shore beaches of O'ahu (Kapahulu Groin to the Diamond Head Lighthouse) were kanda (Iwai, 2004). Only 4 individuals of the native striped mullet were collected in this 14 month sampling program.

Methods and study site

CWT sample- Bimonthly cast-net sampling was started in 1990 at fixed stations in Waiākea Pond, Wailoa River and Reeds Bay to assess hatchery release impacts by noting the frequency of mullet with CWT. Sympatric species, especially the alien *kanda*, were counted and recorded. A 5 ft and an 8 ft diameter, 3/8 inch mesh cast nets were used in combination to sample juvenile and fingerling mullets. A total of fifteen throws were made at each station.

Alien mullet- Because of the increasing dominance of the alien mullet in the nursery habitat of the recreational mullet fishery, we started a monthly sampling program in May 2001 to gather baseline ecological and life history information the *kanda* mullet. We used a similar sampling protocol as in our standard sampling project except samples were collected monthly with only the 8-ft net. Three substations were established in each site in the estuary/pond (Wailoa River estuary and Waiākea PFA) and open bay (Hilo Bayfront).

Results and Discussion

CWT Sample

Long-term sampling data available from 1990 to 2005 clearly show decline of striped mullet in association with an apparent concurrent appearance of *kanda* around 1997 (Fig. 5). It seems that the disappearance of the native mullet is correlated with the proliferation of the alien mullet. However, both

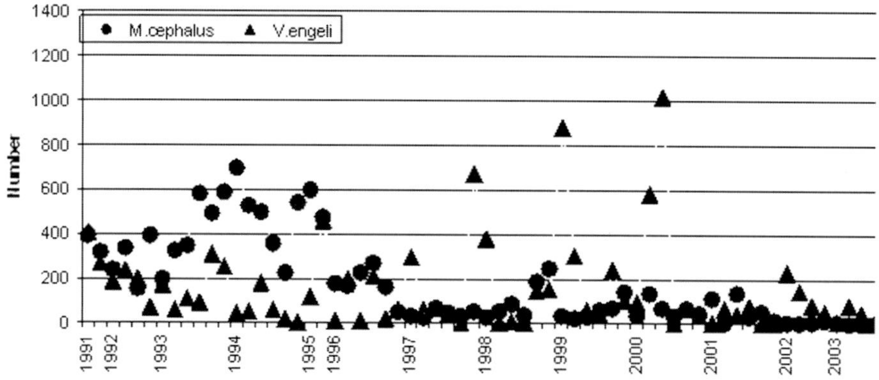

Figure 5. Total number of striped mullet and *kanda* in Wailoa River and Waiākea Pond from 1990 to 2005.

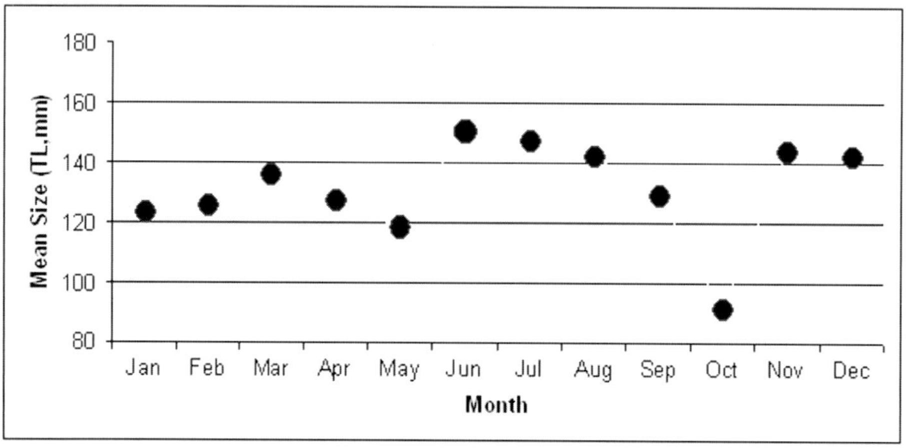

Figure 6. Combined average monthly mean size of *kanda* sampled in Wailoa River from 2001 to 2004. Note small mean size in October (n = 11,812).

mullets were apparently rare in 1997 and 2001. The pattern of *kanda* abundance does not seem to simply be the alien dominating the native. The alternative explanations can be: (1) the *kanda* is filling the niche opened by the decline of the striped mullet, either by years of poor recruitment or possibly due to years of over fishing; (2) the striped mullet is simply being out competed by the alien *kanda*; or (3) there is no competitive interaction between the species of mullet and the trends are coincidental.

Alien mullet, Wailoa estuary

Kanda were most abundant around the area where the Wailoa River Flood Canal empties into Waiākea Pond. Their size ranged from 28 mm to 250 mm TL, however, the average size was around 136 mm TL. There was a dramatic drop in average size during the months of October from 2001 to 2004 (Fig. 6). The drop in average size and an increase in numbers caught suggest the arrival of a batch of new recruits into the Wailoa River estuary. Like the native striped mullet in the estuary, the schooling behavior of *kanda* is very obvious. They graze on the bottom, jump out of the water when disturbed, and travel in large, well-defined schools. It is very difficult to discern the two species without some practice. Preliminary results suggest that *kanda* reach sexual maturity at around 140 mm TL, unlike the native mullet, which matures at a much larger size.

Table 2. Number and frequency of fish species sampled in Hilo Bayfront from December 2002 to January 2005. Mullet species emphasized in this study are in **boldface**.

Family	Species	Number	Percent
Mugilidae	***Valamugil engeli***	2947	68.5
Clupeidae	*Herklotsichthys quadrimaculatus*	306	7.1
Kuhlidae	*Kuhlia xenura*	298	6.9
Carangidae	*Caranx ignobilis*	199	4.6
Mugilidae	***Mugil cephalus***	143	3.3
Polynemidae	*Polydactylus sexfilis*	128	3.0
Carangidae	*Selar crumenopthalmus*	87	2.0
Carangidae	*Caranx sexfasciatus*	84	2.0
Mullidae	*Mulloides flavolineatus*	35	0.8
Carangidae	*Caranx melampygus*	20	0.5
Atherinidae	*Atherinomorus insularum*	18	0.4
Scombridae	*Scomberoides laysan*	11	0.3
Portunidae	*Portunus sanguinolentus*	6	0.1
Albulidae	*Albula vulpes*	4	0.1
Mullidae	*Upeneus arge*	3	0.1
Belonidae	*Platybelone argalus*	3	0.1
Mullidae	*Parupeneus porphyreus*	2	0.1
Palaemonidae	*Palaemon pacificus stimpsoni*	2	0.1
Hemiramphidae	*Hemiramphus depaupratus*	2	0.1
Acanthuridae	*Achanthurus triostegus*	1	0.0
Lutjanidae	*Lutjanus fulvus*	1	0.0
Poecilidae	*Poecillia sphenops*	1	0.0
Pomacentridae	*Abudefduf abdominalis*	1	0.0

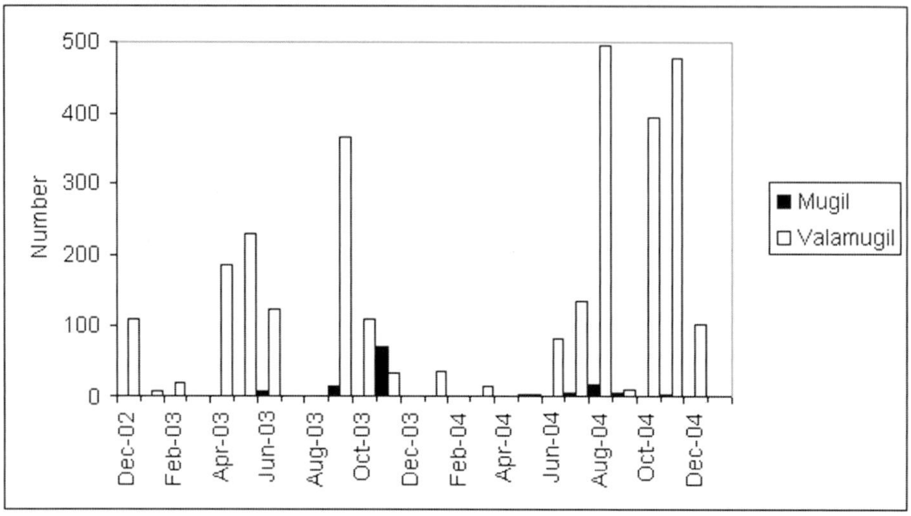

Figure 7. Number of striped mullet and *kanda* in Hilo Bayfront from December 2002 to December 2004.

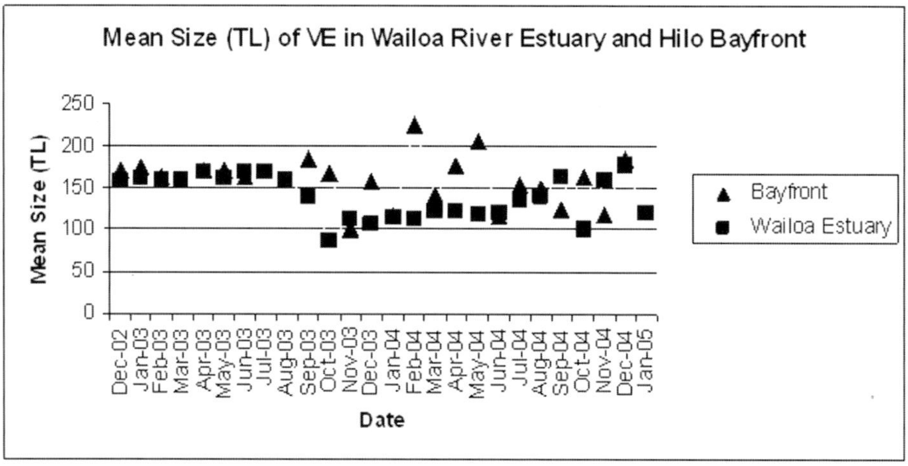

Figure 8. Mean size of *kanda* from Wailoa River Estuary and Hilo Bayfront from December 2002 to January 2005 (n = 3096).

Alien mullet, Hilo Bay

Mostly coastal marine species were collected in samples from the sandy bottom habitat along the Bayfront beach within the harbor. There was a total of 21 species from 15 families (Table 2). The most common family was the Carangidae, mostly piscivores, represented by 4 species. The goatfish family, Mullidae, had 3 species, and there were 2 species of mullet family. These fishes were mostly juveniles which represented species most targeted as adults by recreational fishers in the Hawaiian Islands.

The most numerically dominant species was the alien *kanda*, representing 68.6% of the total sample (n = 4294). Size ranged from 35 to 230 mm TL, but the overall average was 148.4 mm TL. A distant second was the goldspot sardine, *H. quadrimaculatus*, introduced to the Hawaiian Islands from the Marshall Islands in 1972. The *āholehole*, *Kuhlia xenura*, was the most common native

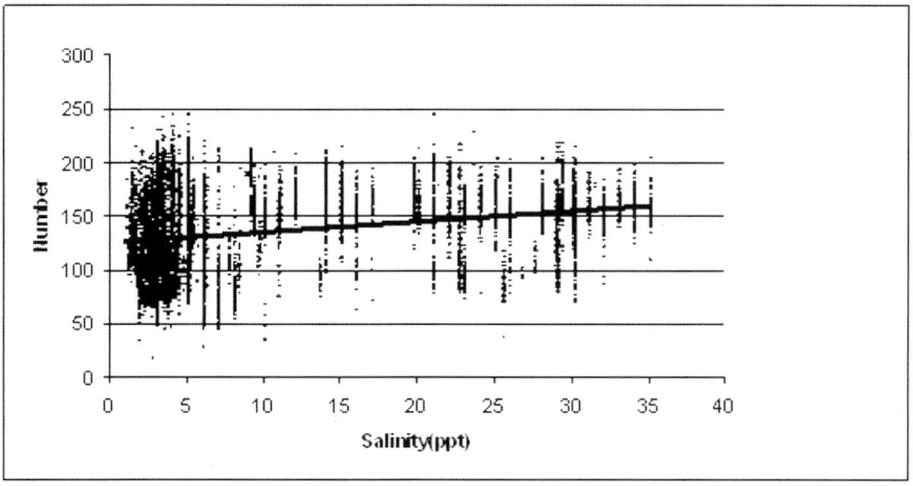

Figure 9. Correlation of number and salinity in the *kanda* in Wailoa River Estuary and Hilo Bayfront (n=16,097).

species at this site, and represented 6.9% of the total sample catch. The native striped mullet made up only 3.3% of the sample.

The number of *kanda* along the Hilo Bayfront far surpassed the number of native mullet (Fig. 7). The results are not surprising since the *'ama'ama* juveniles seem to prefer the more protected habitat in the estuary and pond. Generally, the smaller *kanda* are more common in the estuary and pond while the larger individuals prefer the marine conditions found in the harbor (Fig. 8). There is a significant correlation between size and salinity for *V. engeli (*Fig. 9), however, salinity may not dictate their preferred habitat. We believe that the alien *kanda*, like the native striped mullet, is euryhaline, and marine conditions may not be obligatory.

Mugilids in the Muliwai – lessons for management and restoration of Hawaii's coastal fisheries
The native Hawaiian culture, language, and legends are replete with knowledge of native plants and animals and their habitat. Much of this information probably was gathered by people who were closely bound to the land and the natural resources. Much of this knowledge appears in legends and proverbs. Hawaiian words depicting the different size classes of the mullet, from fingerlings to juveniles, and terms describing the traditional spawning migration route is proof of a broad knowledge of the life history of this prized species. Wai'anae and 'Anaeho'omalu, are places where the larger mullet probably congregated. Traditional fishponds were located at sites which more than likely attracted great number of fingerlings since the areas provided a protective habitat and optimal growing conditions. Even the placement of 'stilt chairs' on the reef flat marks the daily migratory route of the mullet and other species. These pieces of life history information, although disjointed, should be viewed as potential sites for restoration work or life history studies. For example, areas of traditional migratory routes could be identified and protected from overfishing. Bays with Hawaiian fishponds may be a potential nursery habitat that may need further protection. Gear restrictions at sites with potential brood stock populations should be considered. The wealth of information in the native culture is a valuable resource and should not be ignored because of its antiquity.

The project in Hilo Stock Enhancement of Marine Fish in Hawai'i provided the data essential to implement new regulations for the mullet fisheries in Hilo. Release and recapture of tagged fish identified Wailoa River and Waiākea Pond as a significant nursery habitat. Increasing the minimum size allow young mullet time to reach reproductive size and "escape" the fishery. Extending the closed season and decreasing bag limit was added insurance to assure the continued contribution of mature fish to the wild brood stock population. From the start, fishing community participation was critical to the success of project.

Long term monitoring for tagged mullet highlighted the increasing dominance of the alien *kanda* in the nursery habitat for many of the major coastal fish species. The adults of many of these species are targeted by recreational fishers. The inadvertent release of a single alien fish species on one island demonstrates how quickly aliens can disperse because *kanda* now can be found statewide. Kanda are a smaller species than the native mullet. Thus, increasing the minimum net mesh size benefited the native but, unfortunately, the alien mullets as well

Denuding the nursery habitat of riparian vegetation provided greater opportunities for recreational fishing (especially the highly prized game fish, such as the piscivorous carangids) and appeased recreational park users who favored a more unobstructed and "pleasant" view. The loss of protective habitat among thickly vegetated banks of the river and pond likely made newly recruited mullet fingerlings more vulnerable to piscivorous fishes and bird and may have negatively impacted native mullet stocks in the Hilo mullet fishery. Reestablishing the vegetative belt of the shallow intertidal will be the beginning of restoring the nursery habitat of the native mullet.

The *muliwai*, or stream mouth/estuary, is an essential nursery habitat, especially for the native mullet. Unfortunately, it also is a preferred habitat for the alien *kanda*. Coastal areas in bays, such as Hilo Bay, provide habitat for many coastal marine fish species. Estuary as nursery habitat has long been ignored in Hawai'i. Most management efforts have focused on adult stages. The estuary connects two major habitats, the ocean and the forest, and inattention to the management and restoration of this critical bridge may contribute to the collapse of the watershed.

The establishment of alien fishes in Hawai'i has been repeatedly reported in the literature. Our focus now needs to move from documentation to management and ultimately, restoration.

Fishery management and restoration can best be accomplished if the effort is steeped in the marriage of native knowledge and scientific research, and they are more effectively implemented by empowering the fishing community.

Acknowledgments

We thank Mr. William S. Devick, retired DAR Administrator, for his support and constant reminder about the need to manage aquatic resources from a watershed perspective. Often, his behind-the-scene actions spoke louder then his words. Thanks also to Dr. Ken Leber, (then) Oceanic Institute and (now) Mote Marine Laboratory, for his pioneering contribution of using the responsible approach to stock enhancement in Hawaii's coastal fisheries. This study was supported by funding from the United States Fish and Wildlife Service Sport Fish Restoration Program, Study on Coastal Fisheries Stock Enhancement, F-12-D-28.

Literature Cited

Blaber, S.J.M. 1987. Factors affecting recruitment and survival of mugilids in estuaries and coastal waters of southeastern Africa. *In*: J.M. Dadswell *et al.* (eds.), Common strategies of anadromous and catadromous fishes. *American Fisheries Society Symposium* **1**: 507–518.

Blankenship, H.L. & K.M. Leber. 1995. A responsible approach to marine stock enhancement. *In:* H. L. Schramm Jr. & R.G. Piper (eds.), Use and effects of cultured fishes in aquatic ecosystems. *American Fisheries Society Symposium* **15**: 165–175.

Cobb, J.N. 1905. The commercial fisheries of the Hawaiian Islands. *In*: Jordan, D.S. & B.W. Evermann. The aquatic resources of the Hawaiian Islands. *Bulletin United States Commercial*

Fisheries **23**: 577–765.

Cyrus, D.P. & S.J.M. Blaber. 1987. The influence of turbidity on juvenile marine fishes in estuaries. Part 2, laboratory studies. *Journal of Experimental Marine Biology and Ecology* **109**: 71–91.

Department of Land & Natural Resources. 1990. *The Hawaii Fisheries Plan, 1990–1995*. 54 p.

Desha, S.L. 2000. *Kamehameha and his warrior Kekuhaupio*. Kamehameha School Press. Honolulu, Hawaii. 559 p.

Fishbase 2000. World Fish Center (ICLARM).

Hawaii Division of Aquatic Resources. 2004. *Commercial marine mullet landings by method*. Inhouse report.

Hosaka, E.Y. 1944. *Sport fishing in Hawaii*. Bond's, Honolulu. 198 p.

Iwai, T.Y., Jr. 2004. *Preliminary investigation into the enhancement of Pacific Threadfin Moi* (Polydactylus sexfilis) *population within the Waikiki-Diamond Head Marine Reserve, Oahu, Hawaii*. Manuscript. 43 p.

Jordan, D.S. & B.W. Evermann. 1902. The fishes and fisheries of the Hawaiian Islands. *United States Fish Commission Report* **1901**: 353–380.

Julius, M.L., R. Nishimoto, B. Kaya, L Nishiura & T. Shimoda. 2002. *Using diatoms to catch fish*. Poster presentation at the World Aquaculture Society Conference. Madrid, Spain.

Juvik, S.P. & J.O. Juvik. 1998. *Atlas of Hawai'i*. Third edition. University of Hawai'i Press, Honolulu. 333 p.

Leber, K.M. 1994. *Prioritizing marine fishes for stock enhancement in Hawaii*. Oceanic Institute, Makapu'u, Hawai'i. 36 p.

———. 1995. Significance of fish size-at-release on enhancement of striped mullet fisheries in Hawaii. *Journal of the World Aquaculture Society* **26**: 143–153.

———., **D.A. Sterritt, R.N. Cantrell & R.T. Nishimoto**. 1995a. Contribution of hatchery-released striped mullet, *Mugil cephalus*, to the recreational fishery in Hilo Bay, Hawaii. *In:* K. Lowe (ed.), Proceedings of the first biennial symposium for the main Hawaiian Islands marine resource investigation. *Technical Report* **95-01**, *Department of Land and Natural Resources, Division of Aquatic Resources*.

———., **N.P. Brennan & S.M. Arce**. 1995b. Marine enhancement with striped mullet: are hatchery releases replenishing or displacing wild stocks? *In:* H.L. Schramm, Jr. & R.G. Piper (eds.), Uses and effects of cultured fishes in aquatic ecosystems. *American Fisheries Society Symposium* **15**: 376–387.

———., **H.L. Blankenship, S.M. Arce, & N.P. Brennan**. 1997. Influence of release season on size-dependent survival of cultured striped mullet, *Mugil cephalus*, in a Hawaiian estuary. *Fishery Bulletin* **95**: 267–279.

Maciolek, J.A. 1984. Exotic fishes in Hawaii and other islands of Oceania, p. 131–161. *In*: W.R. Courtenay, Jr. & J.R. Stouffer (eds.), *Distribution, biology, & management of exotic fishes*. John Hopkins University Press, Baltimore, Maryland.

Maciolek, J.A. & A.S. Timbol. 1981. Environmental features and macrofauna of Kahana Estuary, Oahu. *Bulletin of Marine Science* **31**: 712–722.

Major, P.F. 1978. Aspects of estuarine intertidal ecology of juvenile striped mullet, *Mugil cephalus*, in Hawaii. *Fishery Bulletin* **76**: 299–314.

McDowall, R.M. 1988. *Diadromy in fishes: migrations between freshwater and marine environments*. Timber Press, Portland, Oregon. 308 p.

Murphy, G.I. 1960. Introduction of the Marquesan sardine, *Harengula vittata* (Cuvier & Valenciennes) into Hawaiian waters. *Pacific Science* **14**: 186–187.

Pukui, M.K. & S.H. Ebert. 1986. *Hawaiian dictionary*. University of Hawaii Press, Honolulu, Hawaii. 572 p.

Randall, J.E. 1996. *Shore fishes of Hawaii*. University of Hawaii Press, Honolulu. 216 p.

———. **& R.K. Kanayama**. 1972. Hawaiian fish immigrants. *Sea Frontiers* **18**: 144–153.

Rizzuto, J. 1985. Fishing lines: Fishing for mullet. *Hawaii Fishing News* **1985**(*July*): 20–21.

Shomura, **R**.**S**. 1987. Hawaii's marine fisheries resources: Yesterday (1900) and today (1986). *National Marine Fisheries Service, Southwest Fisheries Center, Honolulu Laboratory. Administrative Report* **H-87-21**.

United States Army Engineer District, Honolulu. 1980. *Hilo area comprehensive study, geological, biological, and water quality investigations of Hilo Bay*. United States Army Corps of Engineers, October 1980. 248 p.

Wyban, **C**.**A**. 1992. *Tide and currents; fishponds of Hawaii*. University of Hawaii Press, Honolulu, Hawaii. 192 p.

Biology of Hawaiian Streams and Estuaries. Edited by N.L. Evenhuis & J.M. Fitzsimons. *Bishop Museum Bulletin in Cultural and Environmental Studies* 3: 157–169 (2007).

Parasites of Hawaiian Stream Fishes: Sources and Impacts

WILLIAM F. FONT
Department of Biological Sciences, Southeastern Louisiana University, Hammond, Louisiana 70402, USA; email: wffont@selu.edu

Abstract

Introduced freshwater fishes impact native Hawaiian stream fishes in two important ways. In addition to direct negative effects associated factors such as predation, competition, and interference, indirect effects may occur when exotic fishes transfer their parasites to native hosts. Six species of helminths that have been introduced with alien live-bearing fishes, including guppies, green swordtails, shortfin mollies, and mosquitofish which now parasitize the five species of gobioids that occur naturally in Hawaiian streams. Some of these exotic parasites form large populations and produce heavy infections in native fishes that can result in disease. Sources, host specificity, distribution, and life cycles of these parasites were studied to assess their potential for pathogenicity and to aid in the formulation of comprehensive conservation and management plans for native stream species in Hawai'i.

Introduction

Vitousek *et al.* (1997) regarded introduced species to be second only to habitat destruction as a threat to biodiversity. Although he was referring to the global distribution of alien species, his experience with the negative impacts of introductions was gained through his extensive research in Hawai'i. Much research has been conducted on species introduced either accidentally or deliberately into terrestrial ecosystems within the archipelago by humans. Maciolek (1984) and Devick (1991) addressed the problem of introduced species in Hawaiian streams. In addition to providing a list of introduced aquatic animals, they presented a chronology of introductions, details pertaining to the sources of each species, and reasons for which exotics were brought to the archipelago. Negative impacts of exotic fishes on native species have been documented globally for many species. Among the most well studied species, the mosquitofish *Gambusia affinis* has been introduced worldwide, principally for mosquito control, and has resulted in the reduction or elimination of populations of several native fish species (Rupp, 1996). Adverse factors that affect native fishes include predation, competition for food or other resources, and disruption of normal reproduction. Significantly, Devick (1991) raised the specter of an additional threat posed by exotic fishes, i.e. the possibility that they had introduced parasites or diseases that could affect native Hawaiian stream fishes. He was responsible for insuring that parasitology research was included in the ongoing comprehensive research program of the Hawaii Division of Aquatic Resources designed for the conservation and management of native stream fishes. That parasitology research program initiated in 1993 resulted in the first report of parasites of native Hawaiian freshwater gobioids and the association of some of these parasites with live-bearing fishes introduced into Hawaiian streams (Font & Tate, 1994).

Subsequent papers have provided the identity of both native and exotic parasites of Hawaiian stream fishes. Geographic distribution of these parasites in Hawaiian streams has been determined at different scales, both within streams and among streams throughout the archipelago (Font, 1997, 1998, 2003). Host specificity and seasonal dynamics of parasite populations was reported by Vincent & Font (2003a, b). The impact of parasites of near-shore coral reef fishes on stream fishes was studied by Rigby & Font (1997).

The objective of this paper is to integrate new findings regarding parasites in Hawaiian streams with information gained from prior studies in order to add to our understanding of the biology of these parasites and to appreciate the role that these parasites play in the conservation of Hawaii's native stream fishes.

Materials and Methods

Techniques for collecting fishes included using small shrimp seines (*'ōpae* net) or baited hand lines while snorkeling, dip nets, traps, and spears. Most fish were transported alive in aerated buckets and maintained in aquariums until examined for parasites, usually within 3 days of collection. Other fishes were preserved immediately in 10% formalin for subsequent examination. Standard parasitological techniques for necropsy and specimen preparation were employed (Pritchard & Kruse, 1982). Voucher specimens have been deposited in the United States National Museum Parasite Collection, Beltsville, Maryland.

Ecological terms are used in accordance with recommendations of Bush *et al.* (1997). Prevalence is the percentage of fish infected with a parasite. Abundance is the average number of helminths per fish.

Results and Discussion

Table 1 lists the 5 native gobioid stream fishes and the 4 species of introduced poeciliid fishes that were examined for helminth parasites. The 14 species of helminths parasitizing native Hawaiian gobioid stream fishes include both species that are native to the archipelago and species that have been introduced in association with anthropogenic activities. Two sources of native parasites have been discovered: Hawaiian marine fishes and native Hawaiian piscivorous birds. Non-native freshwater fishes, principally livebearers in the family Poeciliidae, have been implicated as sources of helminths that now parasitize native stream gobioids. Table 2 provides a summary of which helminths occurring in Hawaiian streams are native and which are exotic, as well as the sources of each, the manner in which stream fish become infected, life cycle stage that develops in the fish, and whether the fish serves as definitive (= final) or intermediate host for the parasite.

Native parasites

Although the number of species of native parasites exceeds the number of species of exotic parasites, the population sizes, distribution, and adverse impacts of exotics are actually greater than that of natives. With few exceptions, native helminths are rare in stream gobioids; most have a prevalence of less than 1% and a mean abundance of less than 1 worm per fish. Because the pathogenicity of a helminth parasite is often correlated with intensity (i.e., the number of worms per infected fish) and its biomass, the small number of native helminths infecting individual gobioids is unlikely to produce disease in their hosts (Font, 1997). Furthermore, the distribution of native parasites differs from that of exotic parasites. Although the distribution of native parasites spans the entire archipelago from Kaua'i to Hawai'i, Font (1997) reported that, in general, they were found in far fewer streams than exotic parasites. The distribution of native hosts accounts, in large measure, for the distribution of native parasites. Marine fishes parasitized by didymozoid trematodes and *Scolex polymorphus* cestodes are cosmopolitan in their distribution, and these offshore species may occasionally infect larval gobioids that eat infected intermediate hosts during the marine phase of their amphidromous life cycles. Thus, the rare and apparently random parasitization of these larval fishes results in their sporadic occurrence among streams with no obvious geographic patterns of distribution. Similarly, it is the vagility of piscivorous birds that accounts for other native parasites to be distributed among streams throughout the entire archipelago. Furthermore, it is likely that the pattern of visitation of streams and feeding on stream fishes by native birds such as the black crowned night heron that explains the sporadic distribution of these native helminths among streams.

Two species of native helminths are somewhat exceptional in that they are either more common than other native parasites or they display a different distributional pattern. Font (1997, 2003) showed that the native acanthocephalan *Southwellina hispida* is nearly as common as two of the most abundant introduced parasites and infects fish in approximately one-third of all Hawaiian streams that have been surveyed. Although over 50 of the bright orange cystacanths may be found in large specimens of hosts such as *Eleotris sandwicensis*, each parasite is about the size of a rice grain and has its attachment organ, the proboscis, retracted within its body. Cystacanths are non-

Table 1. Scientific and common names of native gobioid and exotic poeciliid fishes from Hawaiian streams that were examined for parasites.

Scientific Name	Common Name
Native Fishes	
Awaous guamensis	*ʻoʻopu nākea*
Lentipes concolor	*ʻoʻopu ʻalamoʻo*
Sicyopterus stimpsoni	*ʻoʻopu nōpili*
Stenogobius hawaiiensis	*ʻoʻopu naniha*
Eleotris sandwicensis	*ʻoʻopu ʻakupa*
Introduced Fishes	
Poecilia reticulata	guppy
Poecilia mexicana	shortfin molly
Xiphophorus helleri	green swordtail
Gambusia affinis	mosquitofish

motile; they attach to connective tissues along the outside of the intestinal track, and feed by absorption of nutrients. No host tissue damage is associated with these cysts and the amount of host nutrients that are absorbed is minimal, therefore pathogenicity seems unlikely at infection levels that have been recorded in native gobioids. This parasite, however, does have the potential for pathogenicity in fish-eating birds such as black crowned night herons that serve as definitive (= final) hosts in which the acanthocephalan becomes a sexually mature adult. An epizootic in Kanahā Pond near Kahului, Maui resulted in the deaths of black crowned night herons from which large numbers of adult *Southwellina* sp. were recovered with their spinous proboscises embedded in the intestinal mucosa (Thierry Work, USGS, National Wildlife Health Research, Honolulu, pers. comm.). Some effort, therefore, has been devoted to life cycle studies of *S. hispisda* in Hawaiʻi. Cystacanths are often found attached to the outside of the intestinal wall in the abdomen of crayfishes examined in my laboratory Louisiana. However, I have failed to find cystacanths in the red swamp crawfish *Procambarus clarkii* introduced into Hawaiian streams, nor have I found infections in *Atyoida bisulcata*, *Macrobrachium grandimanus,* or *M. lar* collected from streams where native gobioids harbor cystacanths. The search for the decapod crustacean host of this acanthocephalan in Hawaiian streams will continue. The discovery of rare infections of cystacanths in green swordtails and shortfin mollies, but only in the very largest specimens of these fishes, may provide a clue regarding the size of crustaceans that may host *S. hispida.*

The second parasite species that deviates from the pattern of rare occurrence and sporadic distribution of native stream species is the nematode *Spirocamallanus istiblenni*. This roundworm occurs commonly in many near shore coral reef fishes. Where estuarine conditions exist in stream mouths (*muliwai*), this parasite can become abundant in *E. sandwicensis* (Rigby & Font, 1997). In general, heaviest infections occur in larger estuaries where opportunities for transmission from marine fishes to stream fishes are greater. The discovery of *S. istiblenni* in blue-lined snappers or *taʻape* (*Lutjanus kasmira*) introduced into the Hawaiian archipelago from the Marquesas Islands caused Rigby & Font (1997) to question the native status of *S. istiblenni.* They considered the possibility that this parasite should be regarded as cryptogenic in the sense of Carlton (1996), that is, a species that is neither demonstrably native nor introduced. More recently, the introduced black tailed snapper or *toʻau* (*L. fulvus*) and native Hawaiian flagfin or *āholehole* (*Kuhlia xenura*) from the Wailoa River estuary in Hilo, occurring syntopically with infected *E. sandwicensis,* were found to harbor *S. istiblenni.*

One final species of native parasite that displays a distinct pattern of distribution and can attain large population sizes is the leech *Aestabdella abditovesiculata*. Originally misidentified by Font *et*

Table 2. Helminth parasites of stream fishes: sources, means by which fish host obtains its infection, life cycle stage that develops in the fish, and role of fish in parasite's life cycle (type of host).

Parasite	Source	Mode of infection of fish	Stage in fish	Type of host
Trematoda				
Ascocotyle tenuicollis	exotic; poeciliid fishes	cercaria from snail penetrates fish	metacercaria	intermediate
Centrocestus formosanus	exotic; poeciliid fishes	cercaria from snail penetrates fish	metacercaria	intermediate
Didymozoidae	native; marine fishes	infected copepod eaten by fish (?)	larva	intermediate
Strigeoidea	native; birds (?)	cercaria from snail	metacercaria	intermediate
Cestoda				
Bothriocephalus acheilognathi	exotic; poeciliid fishes	infected copepod eaten by fish	adult	definitive
Cyclophyllidea	native; birds (?)	unknown	cysticercus	intermediate
Scolex polymorphus	native; marine fishes	infected copepod eaten by fish	plerocercoid	intermediate
Monogenea				
Salsuginus sp.	exotic; poeciliid fishes	larva attaches to fish gill	adult	definitive
Nematoda				
Camallanus cotti	exotic; poeciliid fishes	infected copepod eaten by fish	adult	definitive
Spirocamallanus istiblenni	native; marine fishes	infected copepod eaten by fish	adult	definitive
Acanthocephala				
Southwellina hispida	native; birds	infected decapod eaten by fish (?)	cystacanth	intermediate
Hirudinea				
Aestabdella abditovesiculata	native; marine fishes	direct attachment	adult	definitive
Cystobranchus sp.	native; marine fishes (?)	direct attachment	adult	definitive
Myzobdella lugubris	exotic; fishes or crustaceans	direct attachment	adult	definitive

al. (2002) and assumed to be introduced, the correct identification was provided to me by Eugene Burreson (pers. comm.) and reported (Font, 2003) as a native species that commonly infects Hawaiian marine fishes. Similar to the distribution of *S. istiblenni,* the leech is especially common in extensive estuaries and parasitizes mainly *E. sandwicensis,* but also occurs on other syntopic native gobies. Examination of fishes in streams leading into estuaries has shown that these fishes lose their leech infestations when salinity levels drop to zero parts per thousand. Experimental confirmation of the inability of this leech to survive prolonged exposure to pure freshwater was accomplished by placing *E. sandwicensis* heavily infested with *A. abditovesiculata* in aquariums containing stream water collected far upstream (*mauka*) of estuaries and recording leech detachment and subsequent mortality.

Two factors make specific identification and elucidation of life cycles of the remaining native helminth parasites problematic. First, most of these helminths are incompletely developed, anatomically undifferentiated larval stages of parasites that become adults in fish eating birds or marine fishes. Lack of definitive anatomical characteristics permits identification only to higher taxonomic categories and an understanding of life cycles and means of transmission consistent only with the broad patterns displayed by these higher taxa. Because of the immaturity and rarity of these helminths, classical techniques for determining specific identity and life cycles of these species are unlikely to be fruitful. However, newer molecular techniques that compare nucleic acid sequences of these larvae in stream gobioids with adult helminths in birds and marine fishes may prove to be a valuable tool to provide this information. Fortunately, these problematic species, because of their rarity in streams, are not a high priority concern for the conservation and management of native stream fishes.

Introduced Parasites

Exotic freshwater fishes have been introduced into Hawaiian streams in association with anthropogenic activities. Devick (1991) has provided a thorough documentation with regard to the chronology, purpose, degree of success, and other details of these introductions. With regard to parasites that have been introduced by man into Hawaiian streams, the most important of these introductions are associated with poeciliid fishes. Among the several species of poeciliids that have been brought to the archipelago, the most widely distributed and abundant species are the guppy, *Poecilia latipinna,* the shortfin molly, *P. mexicana,* the green swordtail, *Xiphophorus helleri*, and the mosquitofish, *Gambusia affinis.* These live bearers were introduced into the majority of Hawaiian streams for control of mosquitoes and through aquarium releases. In association with the introduction of these fishes, the helminths that parasitized them at the time of their release into Hawaiian streams also became established in the streams. Furthermore, most of these helminths display broad host specificity and were able to infect the native gobioid fishes inhabiting these streams. What is not well appreciated is the fact that some of these parasites, in fact two of these species with the highest disease potential, are not only unnatural parasites of Hawaiian gobioids but do not occur naturally in these poeciliids either. The two parasites in question, the roundworm *Camallanus cotti,* and the tapeworm *Bothriocephalus acheilognathi* are native to Asia. The Asian fish tapeworm is a common parasite of carp in the Orient and has been introduced globally with carp aquaculture (Dove, 1998). Because of its broad host specificity, the tapeworm now infects many other fish hosts throughout its range of introduction. Similarly, the roundworm infects many fish throughout its native range in the Orient. The roundworm has been disseminated worldwide in association with the aquarium trade. Mark Rigby (pers. comm.) has established a web site for reports of *C. cotti* by freshwater aquarists, and has documented the occurrence of this parasite in aquarium fishes on all continents. There are also reports in the literature of populations of *C. cotti* that have escaped captivity and become established in nonaquarium habitats, e.g. Korea (Kim *et al.*, 2002). Because the poeciliids that now occur in Hawaiian streams are native to the New World, outside of the natural range of this roundworm and tapeworm, poeciliids did not bring these two parasites with them from North and South America. A more likely scenario is that the poeciliids introduced into Hawai'i first were parasitized by the roundworm and tapeworms in ponds on the archipelago where they were reared syntopically with parasitized Asian carp and other Oriental fishes. Only after acquiring these parasites in lentic habitats were the poeciliids introduced into Hawaiian streams where native gobioids became infected.

Many other freshwater fishes have been introduced into Hawaiian streams (Devick, 1991; Yamamoto & Tagawa, 2000; Staples & Cowie, 2001) but their role in parasite transmission is largely unknown. A preliminary survey has found black chin tilapia *Sarotherodon melanotheron* from streams along the windward coast of O'ahu infected with *C. cotti*. Exotic fish species are particularly prevalent in streams from urban areas (e.g., Mānoa Stream in Honolulu) but have not been surveyed for parasites. There is clearly a need for more research in this area to determine what other species of parasites occur in these exotic fishes, to document which of these parasites are capable of infecting native fish hosts, and to assess the potential for the spread of these parasites to other streams.

The six species of non-native parasites found in Hawaiian streams differ markedly from each other in their potential to cause disease in native stream gobioids, and thusly, differ in their importance to any program designed for the conservation and management of these fishes. Therefore, these parasites will be discussed in reverse order of their importance to fish conservation with the greatest disease threat associated with the roundworm *C. cotti* covered last.

Two of the helminths introduced into Hawaiian streams represent no conservation threat because they parasitize poeciliid fishes and do not infect the native stream gobioids. The first of these, *Salsuginus* sp., occurs on the gills of green swordtails. This helminth belongs to the Class Monogenea, Family Dactylogyridae, a taxon widely known for its extremely narrow host specificity. Syntopic gobioids taken from Hawaiian streams and even syntopic guppies and shortfin mollies do not harbor this parasite. The life cycle of this parasite is direct and involves fish to fish transmission. When infected green swordtails were maintained for 2 weeks in a 40 liter aquarium with *E. sandwicensis, Awaous guamensis, Lentipes concolor, Stenogobius hawaiiensis,* and *Sicyopterus stimpsoni*, none of the gobioids became infected.

Similarly, the heterophyid trematode *Centrocestus formosanus* infected only poeciliid fishes, with heaviest infections occurring in green swordtails. The species has broad host specificity and has been reported from reported from fishes belonging to several families including Eleotridae (Salgado-Maldonado *et al.*, 1995). Syntopic eleotrids and gobiids examined from localities where swordtails were infected have not revealed *C. formosanus* infections, although more specimens of native hosts will be examined as they become available.

In contrast, a second heterophyid trematode *Ascocotyle tenuicollis* previously reported only from the conus arteriosus of mosquitofish *G. affinis* was discovered in the conus of 3 of the 5 native gobioids that occurred syntopically with mosquitofish from Waiākea Pond in Hilo. The first intermediate host for this parasite in Hawai'i is the snail *Melanoides tuberculata* (Martin, 1958). Neither *L. concolor* nor *Si. stimpsoni* has been collected from habitats containing this snail; therefore their susceptibility to infection with *A. tenuicollis* is unknown. Because *M. tuberculata* has been introduced into so many Hawaiian streams, the potential for native species to be infected with this parasite is widespread. In Waiākea Pond where *M. tuberculata* is not abundant, native fishes harbored levels of infection too sparse to be considered pathogenic. However, in Louisiana, heavy infections in mosquitofish can result in the swelling of the conus arteriosus to several times its normal size. In Hawai'i, the potential for morbidity and mortality in native fishes is real wherever large populations of this snail occur.

The leech *Myzobdella lugubris* is unique among Hawaiian fish helminths in that it is the only parasite on native stream fishes that is known to both stream biologists and to the general public. Unlike other freshwater helminths in Hawai'i, leeches are ectoparasites, readily visible to the naked eye even upon casual inspection of fishes, and do not require dissection of the host or microscopical examination. The source of *M. lugubris* into Hawai'i is problematic. The leech is a common external parasite of many freshwater fishes throughout the southeastern United States, and also infests decapod crustaceans, grass shrimp *Palaemonetes pugio,* and blue crabs *Callinectes sapidus* from oligohaline estuaries of the Atlantic and Gulf of Mexico. Fishes are utilized by leeches as a source of nutrients, i.e., blood and tissue fluids, but crustaceans are used as hard substrate for leeches to deposit their egg cocoons in an environment consisting mostly of soft sediments (Overstreet, 1978). Because both freshwater fishes and blue crabs from the Southeast USA have been introduced into the archipelago, it is not known which of these represents the original source of introduction. Ironically, in Hawai'i the behavior of this leech apparently has become modified. Stream dwelling crustaceans are unparasitized, presumably because the leeches cement their cocoons to the rocky substrate of stream beds. It is also extremely rare to find a poeciliid harboring *M. lugubris* whereas stream dwelling gobioids are often heavily infested. A probable explanation for this differential parasitism is the fact that gobioids are demersal and are much more likely to come into direct contact with a leech attached to a rock than is a nektonic poeciliid swimming higher in the water column. Because of their visibility, these leeches provide stream biologists with an opportunity for quantifying levels of infections without removing gobioid hosts from streams. In my experience, two areas of caution are warranted if these stream surveys are to be conducted. First, some leeches may attach within the mouth or inside the opercula of fishes and not be readily visible. In addition, based upon my experience in using microscopy to perform dissections to detect internal parasites, very small specimens of *M. lugubris* on fish skin or fins may go unnoticed unless the fish is examined with a stereomicroscope. Secondly, in oligohaline Hawaiian estuaries, *M. lugubris* may co-occur with *A. abditovesiculata*. With experience and with foreknowledge of their possible co-occurrence, however, these two species can be distinguished anatomically and by color pattern. *Myzobdella lugubris* has oral and caudal suckers with diameters that are less than or equal to the body width and is yellow-green. *Aestabdella abditovesiculata* has a caudal sucker that is greater in diameter than the body width and is dark brown-black. The method of attachment with the caudal sucker by these two leeches differs and offers some insight with regard to potential for pathogenicity. Although the caudal sucker of *A. abditovesiculata* is much more robust, it attaches weakly to the fishes' surface, does not produce tissue damage, and is readily removed with forceps. The caudal sucker of *M. lugubris* is embedded deeply into the fin or flesh of its fish host and produces a very visible lesion. Often this wound is surrounded by a white cloudy patch, presumably representing a secondary bacterial or fungal infection. These white patches are often visible in the absence of an attached leech and indicate the leeches may repeatedly move from place to place on its host producing multiple lesions. In addition to the tissue damage caused by the caudal sucker of *M. lugubris*, both species use the oral sucker for feeding on fish blood, and the bodies of these leeches are often red with blood that can be seen through their distended epidermis.

Because the Asian fish tapeworm *B. acheilognathi* parasitizes carp, a species that is of importance to the aquaculture industry, many studies on the biology of this parasite have been published, with particular emphasis on its pathology. Cyclopoid copepods serve as intermediate hosts, and transmission to the definitive host occurs when fishes eat infected copepods. The cosmopolitan distribution of these copepods, including streams in Hawai'i, and the broad specificity of the tapeworm for fish hosts is responsible for the invasiveness of this parasite. It currently is found wherever carp have been introduced for aquaculture or where poeciliids, another group of susceptible hosts, have been released for mosquito control. The Asian fish tapeworm is regarded as the most widely distributed freshwater fish tapeworm in the world. Poeciliids in Hawaiian streams are infected with this tapeworm and are responsible for its transfer to native stream fishes. Adult tapeworms parasitize *E. sandwicensis* and *A. guamensis* but have not been found in the other 3 species of stream gobies. The trophic ecology of *Si.stimpsoni* provides insight that explains the tapeworm's absence from this potential host. This goby feeds mainly on blue-green algae and diatoms attached to the surface of rocks. The absence of copepods from the diet of *Si. stimpsoni* (Kido, 1996; Fitzsimons *et al.*, 2003) has been confirmed by me through the examination of intestinal contents of over 50 specimens in the course of parasitological examination. In the laboratory, several juvenile *Si. stimpsoni* and *A. guamensis* were placed in 2-liter containers of stream water to which a concentrated sample of the copepods *Macrocyclops albidus* from Alenaio Stream, Hilo were added. When examined microscopically after two days, the intestines of *Si.stimpsoni* were empty, but the intestines of *A. guamensis* each contained over 100 copepods. The absence of *B. acheilognathi* from *St. hawaiiensis* and *L. concolor*, both of which feed upon copepods, is presently unexplained and requires further investigation.

One of the most remarkable aspects of the biology of this tapeworm is the fact that this species that is found naturally in boreal climates is able to thrive in tropical waters. In the latter environment tapeworm reproduction is not restricted to the warmer months of the year as is the case in the northern parts of its native range or even in temperate climates such as North Carolina where the species has been introduced (Granath & Esch, 1983). This ecological release, decoupling reproduction from its natural seasonal constraints may, to some extent, account for the successful worldwide spread of this parasite.

Of the 14 species of helminths found in Hawaiian streams, by far the most prevalent, abundant, and widely distributed is the invasive nematode, *Camallanus cotti* (Font & Tate, 1994; Font 1997, 1998). Because of its abundance and associated potential for causing disease in native stream fishes, its biology is especially relevant when considering research programs focused on the conservation of native fishes. Several aspects of the biology of this roundworm are remarkably similar to that of the Asian fish tapeworm. First, both species are native to the Orient and both have been introduced into Hawaiian streams concomitant with the release of poeciliid fishes that are native to the New World. Both the roundworm and tapeworm use cyclopoid copepods as intermediate hosts (i.e., in spite of their taxonomic differences they have the same basic life cycle pattern and mode of transmission). Both are very broadly host specific, and infect several families of freshwater fishes. Both produce disease in their fish hosts.

Long term monitoring of parasite populations in Waiāhole and Waianu Streams has provided us with an opportunity to study the host specificity, population dynamics, and seasonal aspects of the reproduction of this species (Vincent & Font 2003a, b). Our most important findings from those studies included the observation that all exotic poeciliids in those streams are efficient reservoirs for infections of native fishes. All ontogenetic stages of these fishes were found infected and roundworms were very abundant even in juvenile fishes. The observations indicate that immature fishes even as small as 10 mm standard length should not be overlooked in any survey for this parasite. Female roundworms were found in both summer and winter. Reproduction occurs throughout the year, and recruitment of new infections into the fish community is continuous. Over the span of 4 years from 1995 to 1999, *C. cotti* showed no yearly differences in prevalence or abundance and indicated that the population is established, persistent, stable, and unlikely to diminish spontaneously through natural causes. The conservation implications of these data are that reduction or elimination of *C. cotti* from Hawaiian streams is unlikely to occur naturally, but will require human intervention.

Additional multi-year monitoring of *C. cotti* populations was conducted in Hakalau and Nanue Streams on the Big Island. Study sites were located approximately 4 km upstream from the ocean and in these habitats the only fish present was *L. concolor.* Over 80% of gobies with mature gonads were infected with roundworms. The roundworm population was both persistent and stable, in spite of climactic changes associated with El Niño events that resulted in highly variable amounts of rainfall from year to year. Small specimens of *L. concolor* with immature gonads, presumably new arrivals at the upstream site, were rarely infected with roundworms (prevalence <10%). Because worm burdens increased after arriving upstream, it can be concluded that fish were becoming infected at the site and not downstream where juvenile *L. concolor* passed through populations of poeciliids and other species of gobioids that were heavily infected with roundworms. In a previous study, Font & Tate (1994) showed that juvenile *L. concolor* newly recruited from the ocean and collected near the mouth of Hakalau Stream were not infected with roundworms, whereas syntopic *A. guamensis* were parasitized. At the upstream sites in both Hakalau and Nanue Streams, we searched waters above the population of infected *L. concolor* for the presence of poeciliids that might serve as sources of roundworms that might infect these gobies; however, we found none. The conclusion that can be drawn from this study has ominous implications for conservation and management of native stream fishes. Once *C. cotti* has become established in native fishes, the continued presence of poeciliids is no longer necessary. The roundworm can persist in native fishes even if all infected exotic fishes are removed from the stream.

Camallanus cotti infects 3 of the 5 species of native stream gobioids. Infections are absent from *Si. stimpsoni* for the same reason that this fish is also uninfected with *B. acheilognathi*; it does not feed upon copepods that serve as intermediate hosts for both the roundworm and tapeworm. Lack of physiological suitability may explain the absence of mature roundworms in *St. hawaiiensis.* Careful microscopical examination of the mucosa of the posterior intestine of this fish revealed the presence of dead, sclerotized mouthparts of *C. cotti* 3rd and 4th stage juveniles, devoid of the remainder of the body. It can be concluded that *St. hawaiiensis* can become infected, but that the roundworms cannot survive in that host.

A somewhat related observation was made when examining *C. cotti* infection data from *E. sandwicensis*. This fish harbored the largest numbers of the roundworm when compared with all other stream fishes; levels of infection often exceeded 100 worms. However, the vast majority of these specimens were 3rd or 4th stage juveniles. Rarely mature males and females were found in the intestine, but, among the thousands of *C. cotti* examined from this host, not a single gravid female worm was ever found. It may be concluded that *E. sandwicensis* represents a sink for *C. cotti* removing parasites from the stream without compensatory reproduction by the parasite.

One other facet of the population dynamics of the roundworm in *E. sandwicensis* that deserves attention is the finding of the heaviest worm burdens in the largest specimens of this host species. These large sleepers are unlikely to utilize copepods as a food source at this stage in their life history; rather they prey mainly on small fishes. It seems reasonable to conclude that they acquire specimens of *C. cotti* from infected small fishes that are acting as paratenic (= transport) hosts. Experimental confirmation of this hypothesis is required to document the first example of paratenic transmission by a member of the genus *Camallanus*.

Parasites in Estuaries

The ecology of estuarine ecosystems in Hawai'i is understudied, but likely plays an important role in the interactions that occur in stream ecosystems. More research effort needs to be directed toward the study of *muliwai* dynamics where faunal interchange between ocean and streams take place (Robert T. Nishimoto, pers. comm.). This research focus is particularly important in archipelagoes such as the Hawaiian Islands where most species comprising the native stream fauna are amphidromous (Fitzsimons *et al.*, 2002; McDowall, 2003). Encouraged by Nishimoto, an emphasis on parasite ecology and transmission dynamics of parasites of native fishes occupying the *muliwai* has recently been initiated by focusing on the parasites of *āholehole* or Hawaiian flagfin (*Kuhlia xenura*), *'ama'ama* or striped mullet (*Mugil cephalus*), and *'o'opu 'akupa* or sleeper, (*E. sandwicensis*). The most important preliminary finding is that *K. xenura* and *E. sandwicensis* can become infected with both of the two closely related nematodes, the exotic freshwater roundworm *C. cotti* and the presumed native marine roundworm *S. istiblenni*. Some individuals of both fishes simultaneously hosted both species of roundworms. Several aspects of these dual infections require further study, perhaps the most significant of which might be their use as biological indicators of fish movements between marine to freshwater habitats. *Mugil cephalus* was not infected with either species of roundworm. However, most specimens of striped mullet from Waiākea Pond estuary in Hilo harbored monorchiid trematodes conspecific with specimens from the introduced *kanda* or Marquesan mullet (*Valamugil engeli*). Whether these monorchiid trematodes are native to Hawai'i or introduced from the Marquesas is currently under investigation.

Future Directions

Parasites play important roles in Hawaiian stream ecosystems. In developing a plan for future research in stream biology in the archipelago, parasitology research should be included in any comprehensive program focusing on the biology of Hawaii's freshwater fauna. Although much basic parasitology research needs to be done in streams, I will emphasize studies that can be directly applied to the conservation and management of native stream fishes and invertebrates.

The most important questions involving the relationship of parasites and native Hawaiian stream fishes are questions of disease. Do the parasites known to occur in Hawaiian streams cause morbidity or mortality in native gobioids? Is fish fecundity reduced by these parasites as has been documented for other helminths infecting fishes (Heins *et al.*, 2004)? Do these parasites kill their fish hosts?

What seems like a simple question, "do parasites kill their hosts" is perhaps one of the most difficult for parasitologists to answer. Direct evidence of mortality is hard to obtain because only rarely are major fish kills caused by helminth parasites where many infected hosts all die simultaneously and the fish kill is witnessed by humans. If fish are killed by parasites individually, here and there, now and then, the detection of these parasite induced kills is unlikely. Furthermore, small fish are more vulnerable to parasitic disease than are large fish harboring the same number (or biomass) of helminths. Detection of small fish that die from parasites is even more unlikely. More problematic is the fact that parasites weaken fish and make them more vulnerable to predators like piscivorous birds or carnivorous fishes. The differential predation of these weakened parasitized fish is virtually impossible to document in natural ecosystems.

Additional approaches to the study of fish disease include histopathological and pathophysiological studies to determine the nature and extent of damage to parasitized fishes. Currently, the histopathological changes to naturally infected Hawaiian gobies caused by *C. cotti* are underway in my laboratory in collaboration with John Fournie of the Environmental Protection Agency. A related avenue of study is to experimentally infect fishes in the laboratory in order to quantify levels of infection, i.e., number of worms required to kill a fish host. Studies of this nature are dependent upon the ability to complete the parasite's life cycle in the laboratory in order to experimentally infect fish with known numbers of parasites.

Life cycle studies on one of the most important exotic parasites in Hawai'i, the roundworm *C. cotti* are currently in progress at Southeastern Louisiana University. Levsen (2001) reported that he was able to transmit *C. cotti* directly from fish to fish in contradiction to all previous life cycle studies on *Camallanus* spp. that have demonstrated the necessity of a copepod intermediate host for successful completion of the life cycle. Such an unusual report requires independent verification which was tested in the laboratory at Southeastern Louisiana University. In three carefully controlled replicate experiments, fish exposed directly to 1st stage juveniles of *C. cotti* did not become infected whereas all fish exposed to copepods containing 3rd stage juveniles of *C. cotti* did become infected. Either Levsen was working with a strain of *C. cotti* that departs from all known species of *Camallanus* or his experiments were flawed in that, undetected by him, his aquarium systems were contaminated with copepods. In addition to the significance of this finding with regard to our knowledge of the basic biology of camallanid nematodes, there is an important practical conclusion that can be drawn from this experiment regarding the laboratory breeding of gobioids such as *L. concolor*. The production of laboratory reared gobies as potential sources of fishes for stream restocking programs would be unlikely if *C. cotti* had a direct life cycle because the infection of juvenile fishes from infected parents could not be prevented. However, confirmation of an indirect life cycle for this parasite demonstrates that uninfected laboratory-reared gobies can be obtained simply by taking the precaution of removing copepods from breeding aquariums. This precaution could be achieved simply by filtering the water through a fine mesh plankton net.

Theoretical aspects of the evolutionary biology of pathogens must be better understood by conservation biologists and managers. It is incumbent upon parasitologists and pathologists to adequately explain the nature of disease. Parasitologists have done a poor job of explaining aspects of host-parasite relationships, particularly with regard to parasite-induced disease, to both the general public and to other biologists. The future of stream research in Hawai'i must have an education component. Both scientists and laypersons need to be informed regarding the biology of Hawaiian streams, and parasitologists must become more actively involved in this educational process.

One misconception held by many non-parasitologists is that a good parasite never kills its host. While there are some examples of host-parasite relationships that support this notion, there are many exceptions. If the death of a host enhances the likelihood of parasite transmission and increased fecundity, then selection for increased parasite virulence or modified behavior may evolve (Ewald, 1994). For example, *Diplostomum spathaceum* is a trematode that becomes sexually mature in gulls. Larval *D. spathaceum* preferentially infect the eyes and brains of fishes. These parasitized fishes have reduced visual acuity, swim near the water's surface, and make predation by gulls and transmission of the parasite to the next life cycle host more probable.

Another aspect of the "common wisdom" regarding parasites is that ancient host-parasite relationships evolve toward coexistence and new host-parasite relationships display greater pathogenicity. Although, once again, there is some support for this notion, there are many exceptions. As stated previously, the overriding factor regarding parasite pathogenicity is the whether increased virulence enhances or decreases parasite survival and fecundity. This concept is of particular relevance to exotic parasites in Hawaiian stream fishes. This "newness" argument is often stated, but is in fact irrelevant. *Camallanus cotti* and *B. acheilognathi* are not ancestral parasites of poeciliids and new parasites of Hawaiian gobioids. In fact, these Oriental parasites were newly acquired by New World poeciliids prior to their introduction in Hawaiian streams. If anything, the Hawaiian gobioids have amphidromous Indo-Pacific gobioids as ancestors, and indicate that for these two Oriental parasites they probably have an "older" relationship with Hawaiian gobioids than exotic poeciliids. Based upon this information, there is no reason to believe that the roundworm and tapeworm are more pathogenic to native Hawaiian stream fishes than to the exotic fishes that introduced them into the streams.

Helminth parasites of other exotic fishes represent a potential "Pandora's Box" for Hawaiian stream fauna. These alien fishes are virtually unstudied by parasitologists and may harbor helminths currently unknown from Hawai'i. It may be that these new parasites occur in exotic fishes in streams that are so degraded that they are presently uninhabitable by native gobioids. However, although these exotic parasites may not represent any current direct threat to native fishes, their presence in the archipelago may facilitate their transmission to streams where native fishes are present. In addition, the knowledge of the presence of these exotic parasites is instrumental to stream restoration if habitat quality of these degraded streams is to be improved in the future.

The concept of "parasite" is far broader than the current focus on helminth parasites in Hawaiian streams. In fact, in conducting helminth surveys, I have encountered several protistan parasites of native Hawaiian stream fishes. At least one of these protistans, *Ichthyophthirius multifilis* ("ich"), is a know fish pathogen. Perhaps even more importantly, no data have been collected with regard to viral, bacterial, and fungal pathogens in Hawaiian fishes. These taxa contain some of the most lethal of all fish parasites. It is strongly recommended that the parasitology research program for Hawaiian streams be expanded to include parasitologists who have the expertise to conduct research on other groups of parasites in addition to helminths.

Finally, the expansion of research by stream biologists to emphasize estuarine areas and stream mouths that represent the connection between stream and ocean is a clear need. This *muliwai* habitat is an integral component of the stream ecosystem and represents the focal point of faunal interchange between freshwater and the ocean. The leading researchers on amphidromy (Kinzie, 1991; Fitzsimons *et al.*, 2002; McDowall, 2003) have for years emphasized the importance of this key habitat, and it clearly is time that increased effort be devoted to its study.

Acknowledgments

I thank William S. Devick, J. Michael Fitzsimons, and Robert T. Nishimoto for generously sharing their knowledge of Hawaiian streams and especially for their enthusiastic support of my work on fish parasites. They helped me integrate my data on parasites into the "big picture" of the dynamics of Hawaiian stream ecosystems and the amphidromous species that occupy it. This research would not have been possible without the help of dedicated students from Southeastern Louisiana University and personnel from the Division of Aquatic Resources who assisted me in both the field and the laboratory. I am appreciative of the funding that I received the Hawai'i Division of Aquatic Resources and the Dingell - Johnson / Wallop – Breaux Sport Fish Restoration Project.

Literature Cited

Bush, A.O., K.D. Lafferty, J.M. Lotz & A.W. Shostak. 1997. Parasitology meets ecology on its own terms: Margolis *et al.* revised. *Journal of Parasitology* **83**: 575–583.

Carlton, J.T. 1996. Biological invasions and cryptogenic species. *Ecology* **77**: 1653–1655.

Devick, W.S. 1991. Patterns of introductions of aquatic organisms into Hawaiian fresh waters. *In*: New directions in research, management and conservation of Hawaiian freshwater stream ecosystems. *Proceedings of the 1990 Symposium on Freshwater Stream Biology and Fisheries Management.* State of Hawai'i, Department of Land and Natural Resources, Division of Aquatic Resources, Honolulu, p. 189–213.

Dove, A.D.M. 1998. A silent tragedy: parasites and the exotic fishes of Australia. *Proceedings of the Royal Society of Queensland* **107**: 109–113.

Ewald, P.W. 1994. *Evolution of infectious diseases*. Oxford University Press, New York. 298p.

Fitzsimons, J.M., J.E. Parham & R.T. Nishimoto. 2002. Similarities in behavioral ecology among amphidromous and catadromous fishes on the oceanic islands of Hawai'i and Guam. *Environmental Biology of Fishes* **65**: 123–129.

———., **M.G. McRae, H.L. Schoenfuss & R.T. Nishimoto.** 2003. Gardening behavior in the amphidromous Hawaiian fish *Sicyopterus stimpsoni* (Osteichthyes: Gobiidae). *Ichthyological Exploration of Freshwaters* **14**: 185–191.

Font, W.F. 1997. Improbable colonists: helminth parasites of freshwater fishes on an oceanic island. *Micronesica* **30**: 105–115.

———. 1998. Parasites in paradise: patterns of helminth distribution in Hawaiian stream fishes. *Journal of Helminthology* **72**: 307–311.

———. 2003. The global spread of parasites: what do Hawaiian streams tell us? *BioScience* **53**: 1061–1067.

———. & **D.C. Tate.** 1994. Helminth parasites of native Hawaiian freshwater fishes: an example of extreme ecological isolation. *Journal of Parasitology* **80**: 682–688.

———., **M. Wheeler & R.T. Nishimoto.** 2002. Discovery of a second species of introduced leech (Hirudinea: Piscicolidae) parasitizing native stream fishes on the island of Hawai'i. *Bishop Museum Occasional Papers* **69**: 46–48.

Granath, Jr., W.O. & G.W. Esch. 1983. Temperature and other factors that regulate the composition and infrapopulation densities of *Bothriocephalus acheilognathi* (Cestoda) in *Gambusia affinis* (Pisces). *Journal of Parasitology* **69**: 1116–1124.

Heins, D.C., B. Ulinski, J. Johnson & J.A. Baker. 2004. Effect of the cestode macroparasite *Schistocephalus pungitii* on the reproductive success of ninespine stickleback, *Pungitius pungitius*. *Canadian Journal of Zoology* **82**: 1731–1737.

Kinzie, R.A., III. 1991. Hawaiian freshwater ichthyofauna. *In: New directions in research, management and conservation of Hawaiian freshwater stream ecosystems.* Proceedings of the 1990 Symposium on Freshwater Stream Biology and Fisheries Management, State of Hawaii, Department of Land and Natural Resources, Division of Aquatic Resources, Honolulu, p. 18–39.

Kido, M.H. 1996. Diet and food selection in the endemic Hawaiian amphidromous goby, *Sicyopterus stimpsoni* (Pisces: Gobiidae). *Environmental Biology of Fishes* **45**: 199–209.

Kim, J-H., C. J. Hayward & G-J. Heo. 2002. Nematode worm infections (*Camallanus cotti*, Camallanidae) in guppies (*Poecilia reticulata*) imported to Korea. *Aquaculture* **205**: 231–235.

Levsen, A. 2001. Transmission ecology and larval behavior of *Camallanus cotti* (Nematoda, Camallanidae) under aquarium conditions. *Aquarium Sciences and Conservation* **3**: 315–325.

Maciolek, J.A. 1984. Exotic fishes in Hawaii and other islands of Oceania. *In*: W.R. Courtenay, Jr. & J.R. Stauffer, Jr. (eds.), *Distribution, biology, and management of exotic fishes.* The Johns Hopkins University Press, Baltimore, p. 131–161.

Martin, W.E. 1958. The life histories of some Hawaiian heterophyid trematodes. *Journal of Parasitology* **44**: 305–318.

McDowall, R.M. 2003. Hawaiian biogeography and the islands' freshwater fish fauna. *Journal of Biogeography* **30**: 703–710.

Overstreet, R.M. 1978. *Marine maladies? Worms, germs, and other symbionts from the northern Gulf of Mexico.* Blossman Printing, Inc., Ocean Springs, Mississippi. 140 p.

Pritchard, M.H. & G.O. Kruse. 1982. *The collection and preparation of animal parasites.* Technical Bulletin No. 1. The Harold W. Manter Laboratory. University of Nebraska Press, Lincoln, 141 p.

Rigby, **M**.**C**. **&** **W**.**F**. **Font**. 1997. Redescription and range extension of *Spirocamallanus istiblenni* Noble, 1966 (Nematoda: Camallanidae) from coral reef fishes in the Pacific. *Journal of the Helminthological Society of Washington* **64**: 227–233.

Rupp, **H**.**R**. 1996. Adverse assessments of *Gambusia affinis*: an alternate view for mosquito control practitioners. *Journal of the American Mosquito Control Association* **12**: 155–166.

Salgado-Maldonado, **G**., **R**.**V**. **Ma-Isabel** **&** **J**. **J**. **Campos-Perez**. 1995. Metacercariae of *Centrocestus formosanus* (Nishigori, 1924) (Trematoda) in freshwater fishes in Mexico and their transmission by the thiarid snail *Melanoides tuberculata*. *Studies on Neotropical Fauna and Environment* **30**: 245–250.

Staples, **G**.**W**. **&** **R**.**H**. **Cowie**. 2001. *Hawai'i's invasive species*. Mutual Publishing, Honolulu. 116 p.

Vincent, **A**.**G**. **&** **W**.**F**. **Font**. 2003a. Host specificity and population structure of two exotic helminths, *Camallanus cotti* (Nematoda) and *Bothriocephalus acheilognathi* (Cestoda), parasitizing exotic fishes in Waianu Stream, O'ahu, Hawai'i. *Journal of Parasitology* **89**: 540–544.

———. **&** **W**.**F**. **Font**. 2003b. Seasonal and yearly population dynamics of two exotic helminths, *Camallanus cotti* (Nematoda) and *Bothriocephalus acheilognathi* (Cestoda), parasitizing exotic fishes in Waianu Stream, O'ahu, Hawai'i. *Journal of Parasitology* **89**: 756–760.

Vitousek, **P**.**M**., **C**.**M**. **D'Antonio**, **L**.**L**. **Loope**, **M**. **Rejmanek** **&** **R**. **Westbrooks**. 1997. Introduced species: a significant component of human-caused global change. *New Zealand Journal of Ecology* **21**: 1–16.

Yamamoto, **M**.**N**. **&** **A**.**W**. **Tagawa**. 2000. *Hawai'i's native and exotic freshwater animals*. Mutual Publishing, Honolulu. 200 p.

Biology of Hawaiian Streams and Estuaries. Edited by N.L. Evenhuis & J.M. Fitzsimons. *Bishop Museum Bulletin in Cultural and Environmental Studies* 3: 171–181 (2007).

Hīhīwai (*Neritina granosa* Sowerby) Recruitment in ʻĪao and Honomanū Streams on the Island of Maui, Hawaiʻi

Skippy Hau
Hawaii Department of Land & Natural Resources, Division of Aquatic Resources, 130 Mahalani Street, Wailuku, Hawaiʻi 96793; email: skippy.hau@hawaii.gov

Abstract

Juvenile *hīhīwai* (*Neritina granosa*), endemic freshwater snails of Hawaiʻi, were collected from ʻĪao and Honomanū Streams on the island of Maui. Each stream has two or three diversions at various elevations which removes most of the stream flow before reaching the ocean. The lack of flow restricts *hīhīwai* to the estuary. Groundwater maintains freshwater habitats for their survival. Both sites were impacted occasionally by large ocean swells and extreme high tides that mixed salt water with fresh or formed berms that blocked stream flow to the ocean. *Hīhīwai* continue to migrate from the ocean. This paper describes their recruitment and reviews some of the conditions found in each stream. Monthly counts were from 0–576 in ʻĪao Stream (1999–2004) while Honomanū Stream counts were from 67–912 (2001–2004). Monthly shell lengths averaged from 1.5–6.1 mm in ʻĪao Stream and 2.2–9.1 mm in Honomanū. The persistence of juvenile *hīhīwai* recruitment confirms the possibility for restoring native stream populations if "natural flow" is restored. Stream restoration should be based on the needs of the slowest migrating animal such as *hīhīwai*. A slow-migrating species like *hīhīwai* may be a good indicator of the adequacy of stream flow during stream restoration programs.

Introduction

ʻĪao Stream is on the northeastern portion of the West Maui Mountains and Honomanū Stream is on the north flank of Haleʻakalā volcano (Fig. 1). Both streams are in windward watersheds. These streams are home to a variety of native fauna including *ʻoʻopu* (fish), *ʻōpae* (shrimp), and *hīhīwai* (snail). During increased stream flows, larvae hatch from eggs and are transported to the ocean. After developing in the ocean for several months, they return to freshwater and migrate upstream as post larvae (Lindstrom & Brown, 1994; Nishimoto & Kuamoʻo, 1997; Radtke *et al.*, 2001). In November 1997, after several weeks of stream flow, about 10 *hīhīwai* (*Neritina granosa*) were seen migrating upstream on the bottom of the concrete run in ʻĪao Stream. The *hīhīwai* traveled more than 500 meters from the ocean. Diversions have reduced stream flow in ʻĪao and Honomanū Streams and greatly reduced the migration success of native stream animals like *hīhīwai*. This paper will quantify the various size classes of *hīhīwai* found in each estuary.

Materials and Methods

Study streams

ʻĪao Stream is a second order stream more than 12 km in length. Puʻu Kukui is the highest point at 1764 m elevation and receives more than 900 cm annual rainfall. Part of ʻĪao Stream was modified with concrete channels for flood control since 1981. Two diversions are connected to Maniania Ditch & ʻĪao-Waikapu Ditch (240 m) and Waiheʻe Ditch (80 m). The third diversion to the Kama Ditch (~123 m) is no longer being used.

Honomanū Stream is a second order stream about 14 km long. Headwaters are about 2700 m elevation. Three diversions are connected to the Lower Kula Pipeline (936 m elevation), Koʻolau/Wailoa Ditch (400 m), and Spreckels Ditch (529 m).

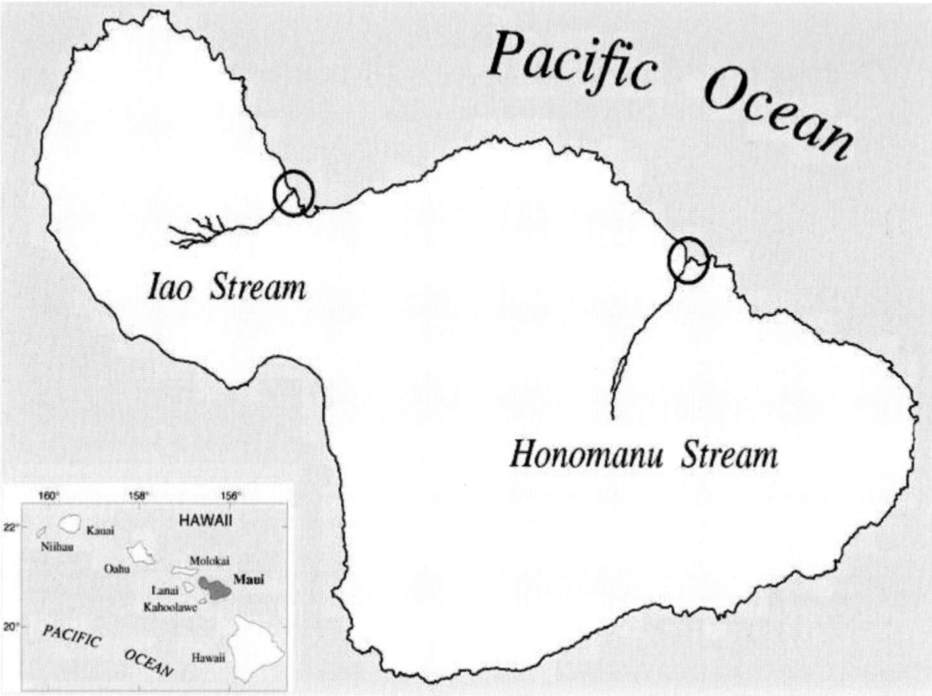

Figure 1. ʻĪao and Honomanū Streams on the island of Maui.

Most of the time, flow is absent in the lower sections of both streams. *Hīhīwai* sampling was initiated to document their presence in the ʻĪao (Fig. 2) and Honomanū (Fig. 3) estuaries. *Hīhīwai* were collected by hand from each stream mouth for up to one hour each month. Rocks were turned over and examined for *hīhīwai* attached to the bottom (Fig. 4). A mask and snorkel was used to collect *hīhīwai* in the deeper Honomanū estuary. A refractometer was used to measure salinity changes in the estuary. Fresh water was present as groundwater seeps at both stream mouths even when surface flow was absent.

Maximum shell lengths of snails were measured with a set of calipers to the nearest 0.1 mm. After *hīhīwai* were identified and measured, they were released in continuous parts of selected streams. Most of the *hīhīwai* collected from ʻĪao Stream were released in ʻĪao State Park (above the ʻĪao Valley intake).

Rainfall information for the Wailuku site came from a rain gauge (WUKH1) located at 55 m elevation while a neighboring Waikamoi rain gauge (#341) at 369 m elevation was used to approximate Honomanū rainfall. The Division of Water Resources Management, Department of Land & Natural Resources, provided monthly rainfall data.

Results

Monthly counts varied from 0 to 576 *hīhīwai* in ʻĪao Stream while Honomanū Stream counts ranged from 67 to 912 (Fig. 5). Monthly counts for both streams increased during June through August. Monthly shell lengths ranged from 1.5 to 6.1 mm for *hīhīwai* from ʻĪao Stream and 2.2 to 9.1 mm in Honomanū Stream (Fig. 6). Due to the intermittent nature of both streams, successful migration was restricted to the estuary. Most *hīhīwai* will not survive beyond the estuary because of dry stream beds and the lack of consistent stream flow.

Figure 2. ʻĪao Stream mouth collection site (groundwater-maintaining freshwater habitat).

Figure 3. Honomanū Stream collection site. *Hīhīwai* moved upstream during large swells and high tides.

Figure 4. *Hīhīwai* recruits (<8 mm shell length) attached to the bottom of a rock.

Size classes were compared annually for ʻĪao and Honomanū Streams (Figs. 7, 8). Most (93%) of the *hīhīwai* measured 5 mm shell length or less and are mostly represented in the first two size classes. Increased survival and limited growth was found in Honomanū Stream.

Total rainfall for January to March was correlated with maximum *hīhīwai* counts in ʻĪao Stream later in the year ($R^2 = 0.9289$). Total *hīhīwai* counts for July to September were plotted with mean monthly rainfall for January to March (Fig. 9). The increased rainfall at the beginning of each year results in increased larvae transported to the ocean. Juvenile *hīhīwai* return to stream mouths after several months of development and are reflected in the July to September counts for ʻĪao and Honomanū Streams.

Salinity in the ʻĪao estuary ranged from 0–2 ppt up to 14–28 ppt near the ocean during times of large incoming swells and high tides.

Discussion

Amphidromy and Stream Flow

Maciolek (1978) and Ford (1979) discussed the amphidromous life cycle that involves an obligatory period of larval development in the sea. Similar to fish larvae, there appears to be a limited window of opportunity for *hīhīwai* larvae to reach saltwater. Veligers held in fresh water died within six days (Ford, 1979). *Hīhīwai* larvae were found to occur twice a year in drift collections in Palauhulu Stream in Keʻanae (Hau *et al.*, 1992), about 2.3 km east of Honomanū Stream. *Hīhīwai* larvae was present three to five months prior to juvenile recruitment upstream (Unpublished).

The first major stream flow often triggers reproduction and the release of larvae for development in the ocean (Maciolek, 1978; Ford, 1979). During low flow conditions, *hīhīwai* populations are often found in pool and run areas. The first major freshet after a period of low flow often causes a sudden decrease in water temperature and an increase in dissolved oxygen. This scouring flow

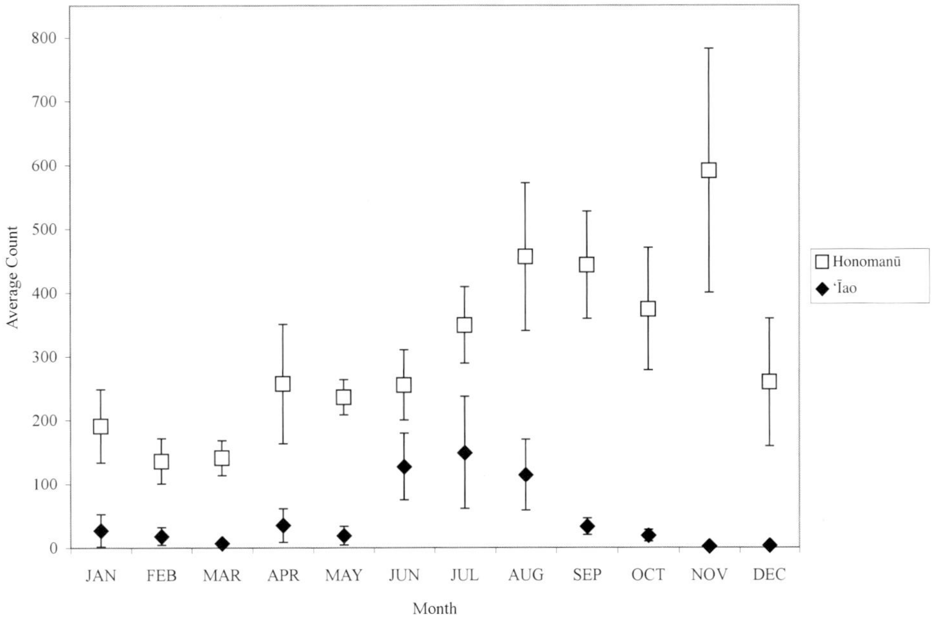

Figure 5. Monthly average counts for *hīhīwai* in ʻĪao (1999–2004) and Honomanū (2001–2004) Streams.

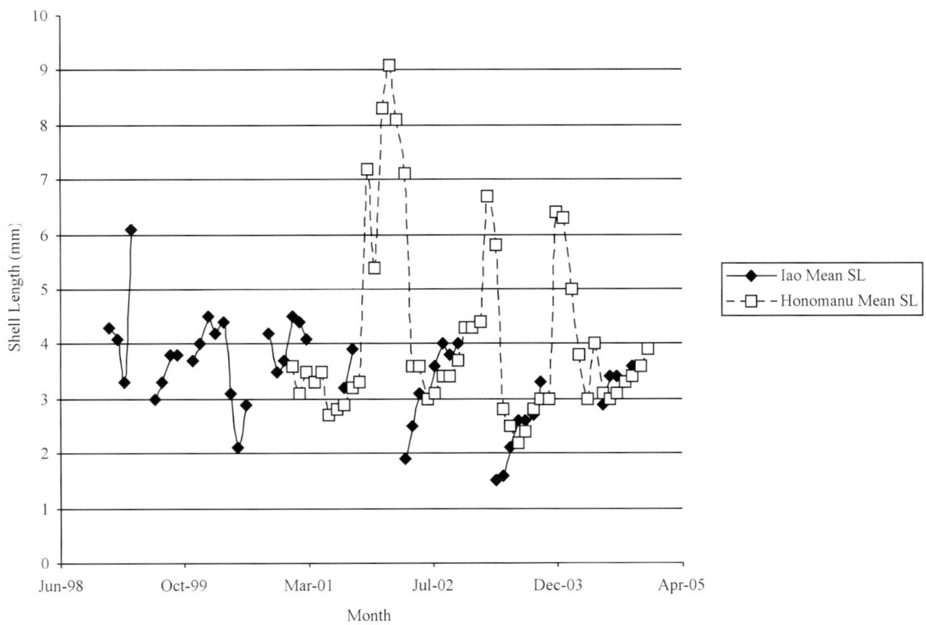

Figure 6. Comparison of mean monthly shell lengths of *hīhīwai* from ʻĪao and Honomanū Streams.

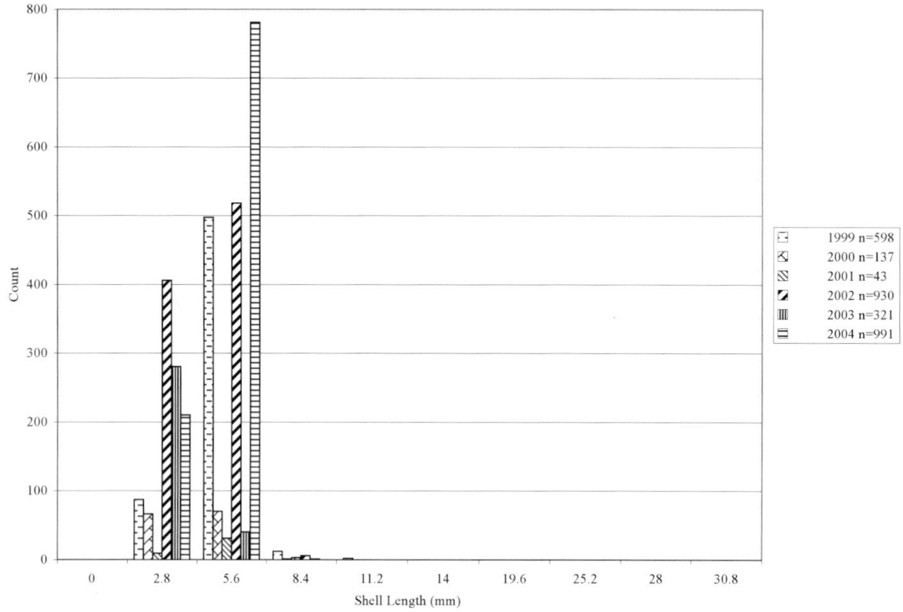

Figure 7. *Hīhīwai* size classes for ʻĪao Stream (1999 to 2004).

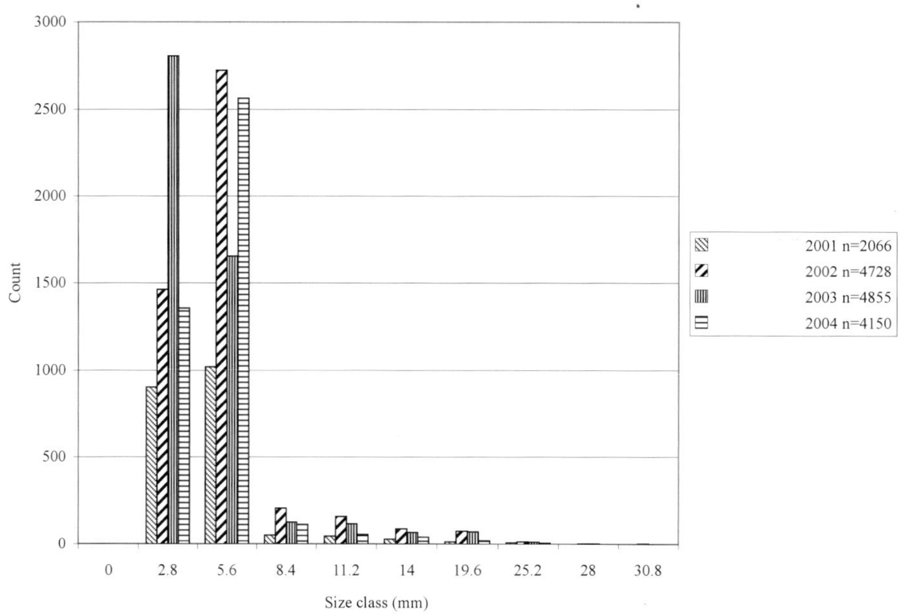

Figure 8. *Hīhīwai* size classes for Honomanū Stream (2001 to 2004).

provides optimum conditions for *hīhīwai* egg cases to be laid on hard substrate and the transport of hatched larvae to the ocean as quickly as possible.

The streams occasionally experience heavy rains and flash flooding which temporarily establishes the *mauka-makai* connection (from the mountain to the ocean) that is vitally important for amphidromous animals migrating between the ocean and fresh water. The connection is maintained intermittently after storms with flows that exceed diversion capacities. Many papers have documented post larval migration of hinana or juvenile fishes (Gobiidae) and *'ōpae* or shrimp (*Atyoida bisulcata*) in response to increased flow (Lindstrom & Brown, 1994; Nishimoto & Kuamo'o, 1997; Tate, 1997).

Hīhīwai still attempt to migrate into 'Īao and Honomanū estuaries even though both streams have been diverted for more than 100 years. Water collected by diversions is transported to agricultural lands by a comprehensive system of irrigation ditches and reservoirs (Wilcox, 1996). These diversions may also be carrying *hīhīwai* larvae away from the ocean.

Dewatering can inhibit upstream migration of postlarvae, which are critical to the life cycles of many native species (Brasher, 2003). The removal of water through diversions (Fig. 10) reduces stream discharge, lowers flow velocity, decreases water depth, and increases water temperature. The amount of run, riffle, and pool habitats are reduced for stream organisms like *hīhīwai*. An insect study in Kinihapai Stream (tributary of 'Īao Stream) suggest that torrential flows are a factor regulating habitat availability for *Telmatogeton torrenticola* and the reduced discharge could significantly reduce the amount of useable habitat for this and other stream fauna (Benbow *et al.*, 1996).

For diverted streams, the requirement for two or more flows are often overlooked and are needed to allow post larvae *hīhīwai* and other stream animals sufficient space and time to migrate upstream. Depending on the duration of the rainy season, these later flows may need to exceed diversion capacities and be able to break open a natural berm built up by large winter swells (>10 m) generated from the North Pacific. Substrate, which naturally moves downstream, blocks the stream from flowing into the ocean. On the other hand, with consistent rainfall, there is sufficient flow to prevent this build up and a continuous stream connection to the ocean is maintained.

The maintenance of median flow with prolonged periods of elevated discharge has been shown to be important for successful reproduction of *'o'opu alamo'o* (*Lentipes concolor*) (Way *et al.*, 1998). The diversion of Waikolu Stream on the island of Moloka'i dampened the natural seasonal discharge cycle, exacerbated natural low flow conditions, and increased the likelihood of prolonged periods of extremely low flow. Although *'o'opu alamo'o* appears to be capable of reproducing throughout the year, the species' gonadal activity is correlated with monthly periods of high water flow. Fish from Waikolu Stream had a 'boom or bust' reproductive pattern, and the population had reduced or no reproduction when stream flow conditions reached extreme low levels (Way *et al.*, 1998).

Groundwater

At both study sites, springs and groundwater becomes much more important in maintaining the freshwater stream habitat near the shore. *Hīhīwai* juveniles and adults are restricted to fresh waters and not found in brackish water environments (Ford, 1979). During November 1975, the population of neritids in the lower 90-m reach of Waiohue Stream on Maui was subjected to seawater inundation during periods of extremely high surf (Ford, 1979). The influence of large ocean swells and seasonal high tides reaching the estuary can increase salinity and cause *hīhīwai* to migrate upstream, away from brackish water. There could be an unknown mortality of juvenile *hīhīwai* caused by increased salinity. Depending on tides, *hīhīwai* was often the only snail migrating upstream while *hapawai (Neritina vespertina)* and *pipiwai* (*Theodoxus cariosus*) remained in the estuary (Maciolek, 1978; Titcomb, 1978). *Hīhīwai* was periodically absent at the mouth but located more than 50–75 m upstream in the 'Īao and Honomanū estuaries.

Groundwater emerging from several springs in the Honomanū basalt near the shoreline is estimated to be 1.4 mgd (2.2 cfs) of flow (Gingerich, 1999). The Honomanū estuary is much larger and covers over 200 m in length, over 10 m in width, and a maximum depth of 2 m. This larger area of stable aquatic habitat results in significantly higher *hīhīwai* counts and allows recruits to grow to larger sizes (>20 mm).

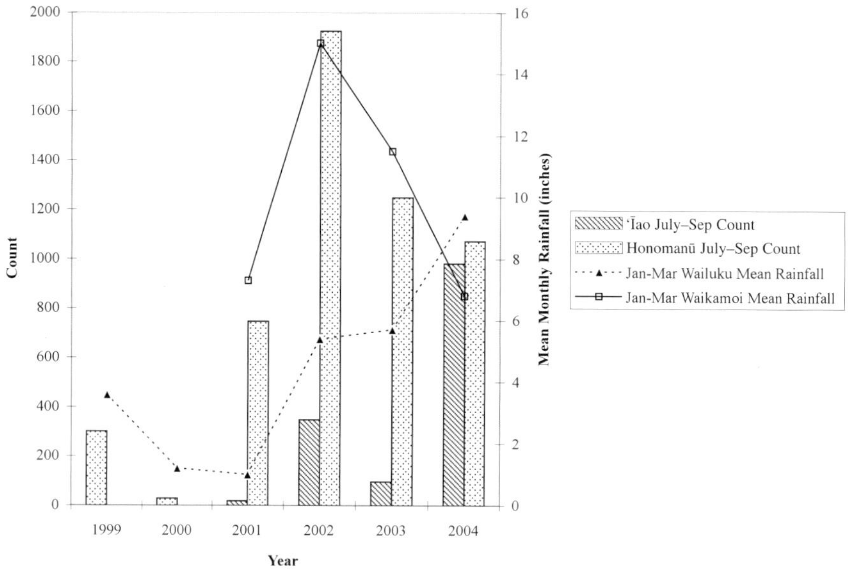

Figure 9. Total *hīhīwai* counts (July to September) in ʻĪao and Honomanū Streams plotted with mean monthly rainfall (January to March).

Flow is reduced in ʻĪao Stream from water tunnels and diversions. Lower ʻĪao Stream often lacks stream flow (Yamamoto & Tagawa, 2000). ʻĪao Stream is subject to prolonged dry periods, which reduces the amount of aquatic habitat to less than 100 m^2, and water depths are often less than 0.3 m. During rainy periods, the ʻĪao estuary can be greater than 100 m in length, more than 20 m in width, and about 0.5 m in depth. The concrete channel and flood control modifications help to shunt stream flow directly to the ocean resulting in shallow runs, riffles, and pools.

From July 1993 to June 2003, ʻĪao Stream discharge averaged 59.9 cfs (38.7 mgd) (U.S. Geological Survey Water-Data ʻĪao Stream Gauge No. 16604500). Based on water declarations, estimated stream flow was reduced from 80% in FY1993 to 96% in FY2003. The Spreckels Ditch, Kama Ditch, and ʻĪao Valley intakes were identified as major diversions. Together with private declarations a total of 129.6 cfs (83 mgd) were reported (Tagomori, 1991). Total water declarations exceed actual stream flow by more than two times.

Upstream Migration

The upstream movement of *hīhīwai* appears to be influenced by several factors. In general, there is limited upstream movement in diverted streams because of intermittent stream flows. Similar to migrating *hinana* and *ʻōpae, hīhīwai* will die after being stranded when flow stops and the streams dry.

Migration lines appear to be an efficient way to move en masse during times of increasing stream flow. The tendency of *hīhīwai* to aggregate and form lines and mucus trails was observed in both ʻĪao and Honomanū Streams. After a passing storm on 13 April 2004, I encountered snails starting to form lines (up to 7.5 cm) at the Honomanū Stream mouth. I also found several migrating *hīhīwai* carrying smaller ones. Small one-mm *hīhīwai* were found on larger ones with shells between 2 to 5 mm. This "hitchhiking" behavior was also reported for *Cochliopina tryoniana* migrating with *Neritina latissima* in the Rio Claro in Costa Rica (Schneider & Lyons, 1993); smaller *C. tryoniana* sometimes attached to larger *N. latissima*.

Longer migration lines appear in runs and riffles further upstream. On 8 May 2004, after heavy

Figure 10. `Īao Stream Diversion (below USGS Stream Gauge No. 16604500). Most stream flow is diverted into irrigation ditches.

rains, I found *hīhīwai* traveling along trails up to 60 cm long on the rocky streambed that is normally dry below the Honomanū highway bridge. A total of 315 *hīhīwai* measured from 2.3–5.9 mm and averaged 3.4 mm shell length (S.E. = 0.03). The mucus trail appears to reduce friction and "grease" the way for other *hīhīwai* to follow. These trails seem to reduce time spent in exploring other directions. As flow decreases in certain habitat conditions, some *hīhīwai* appear to independently stop migrating.

Dry streambeds, concrete channels, or multiple diversions impede and prevent upstream migration of *hīhīwai* beyond the estuary. During July 1991, Hodges (1992) found *hīhīwai* in wet gravel under large cobbles and boulders in the middle of the dry intermittent Honomanū streambed. On one occasion, he observed tens of thousands of young *hīhīwai* (up to 10 mm shell length) in a migrating column over a meter and downstream for 500 m. It appears this observation was during an exceptional rainy period with continuous stream flow that allowed *hīhīwai* to migrate upstream. When stream flow stops, similar to drought conditions, *hīhīwai* are forced initially into standing pools that eventually dries leaving moist areas under boulders and wet gravel.

In Pua'alu'u Stream, snails did not occur above 185 m in windward East Maui (Ford, 1979). The diversion at 390 m significantly restricted *hīhīwai* to a short reach immediately above the head of the estuary (Ford, 1979). This same condition exists for 'Īao and Honomanū Streams. On Maui, *hīhīwai* was recorded at 140 m in continuous 'Alelele Stream (Division of Aquatic Resources Surveys September 1994).

The majority of migrating *hīhīwai* found in 'Īao and Honomanū Streams are less than five mm shell length. When present, *hīhīwai* from 'Īao Stream appears to be represented by two size classes with few growing larger than 5 mm. The 'Īao estuary has been impacted by droughts at various times of the year. Honomanū Stream appears to be a better nursery with significantly more *hīhīwai* growing to larger size classes. The increased groundwater helps to insure a stable freshwater environment.

Water recharge estimates in the ʻĪao watershed have continued to decrease with the loss of agriculture (Meyer & Presley, 2001). More water recharge is necessary to sustain healthy groundwater flow. Water diversion projects have had enormous impacts on streams after the degrading of Hawaiian forests and the introduction of grazing animals. Perennial streams became intermittent and springs dried up (Wilcox, 1996). In the Wailuku watershed, increasing development and changes in land use activities including urbanization have increased impervious surfaces and drainage runoff.

ʻĪao and Honomanū Streams represent many other streams in Hawaiʻi that have been diverted. Restricted stream flows have resulted in smaller estuaries and prevented *hīhīwai* from migrating to higher elevations. Unless the animals reach adequate freshwater stream habitats, they are unable to grow into healthy reproducing populations. In Honomanū and ʻĪao Streams, the diversion of over 90% of the stream flow results in intermittent stream conditions, which limit the average growth of *hīhīwai* to less than 10 mm. The recruitment of *hīhīwai* and other amphidromous species requires consistent stream flows.

Groundwater plays a very important role in maintaining freshwater stream habitats near the ocean. Similar to *Neritina punctulata* in Puerto Rico which requires estuary connections for larval development (Pyron & Covich, 2003), the connections between the estuary and upstream reaches need to be maintained to avoid local extinctions of *hīhīwai*. The persistence of *hīhīwai* recruitment confirms the possibility for restoring native stream populations if "natural flow" is restored. Stream restoration should be based on the needs of the slowest migrating animal such as *hīhīwai*. A slow-migrating species like *hīhīwai* may be a good indicator of the adequacy of stream flow during stream restoration programs. Future attention should be focused on the flow requirements for sustaining reproducing populations of *hīhīwai* and other amphidromous species and not just on their mere presence. Native stream animals need successful spawning, larval development, and recruitment for each island (Radtke *et al.*, 2001).

Acknowledgments

Many thanks to Isabella A. Abbott, Darrell G.K. Kuamoʻo, Robert T. Nishimoto, Glenn R. Higashi, Wesley Hau, Lori L. Ohta, Norman Yoshioka, Matt Wong, Stephen B. Gingerich, Richard Fontaine, J. Michael Fitzsimons, David Higa, Neal D. Fujii, and Alison Sherwood who provided assistance, support, and kind words of encouragement. Special thanks to Ivor D. Williams & Thierry M. Work who reviewed my raw data and offered helpful comments. Thanks also to Luis E. Menoyo (USGS) for getting copies of their excellent maps showing my study sites. I especially wish to recognize two *kupuna*, Sarah Kaʻaaumo and Luther Kanae, Sr., who shared their stories and observations from earlier times before their passing.

Literature cited

Benbow, M.E., A. J Burky & C.M Way. 1996. Habitat specificity of the endemic Hawaiian midge, *Telmatogeton torrenticola* (Telmatogetoninae). *Bulletin of the North American Benthological Society* **13**: 136–137.

Brasher, A.M. 2003. Impacts of human disturbances on biotic communities in Hawaiian streams. *BioScience* **53**: 1052–1060.

Ford, J.I. 1979. Biology of a Hawaiian fluvial gastropod *Neritina granosa* Sowerby (Prosobranchia: Neritidae). M.S. Thesis, University of Hawaii, Honolulu, Hawaii: 94 p.

Gingerich, S.B. 1999. Ground-water occurrence and contribution to streamflow, Northeast Maui, Hawaii. *United States Geological Survey Water-Resources Investigations Report* 99-4090. 69p.

Hau, S., C.M. Way & A.J. Burky. 1992. *Life cycle of the endemic limpet* Neritina granosa *(Sowerby), in Palauhulu Stream, Maui. 43rd Annual Meeting of American Institute of Biological Sciences.* Ecological Society of America, 9–13 August, Honolulu, Hawaii (abstract).

Hodges, M.H.-D. 1992. Population biology and genetics of the endemic Hawaiian stream gastropod *Neritina granosa* (Prosobranchia: Neritidae): implications for conservation. Honors Thesis, Wildlife Biology Program University of Montana, Missoula, Montana, July 1992.

Lindstrom, D.P. & C.L. Brown. 1994. Early development and biology of the amphidromous Hawaiian stream goby *Lentipes concolor. Proceedings of the Fourth Indo-Pacific Fish Conference, Systematics and Evolution of Indo-Pacific Fishes*, 28 November–4 December, 1993, Bangkok, Thailand, p. 397–409.

Maciolek, J.A. 1978. Shell character and habitat of nonmarine Hawaiian neritid snails. *Micronesica* **14**: 209–214.

Meyer, W. & T.K. Presley. 2001. The response of the ʻĪao Aquifer to ground-water development, rainfall, and land-use practices between 1940 and 1998, Island of Maui, Hawaii: *United States Geological Survey Water-Resources Investigations Report* **00-4223**, 60 p.

Nishimoto, R.T. & D.G.K. Kuamoʻo. 1997. Recruitment of goby postlarvae into Hakalau stream, Hawaii Island. *Micronesica* **30**:41–49.

Pyron, M. & A.P. Covich. 2003. Migration patterns, densities, and growth of *Neritina punctulata* snails in Rio Espiritu Santo and Rio Mameyes, Northeastern Puerto Rico, *Caribbean Journal of Science* **39**: 338–347.

Radtke, R.L., R.A. Kinzie III & D.J. Shafer. 2001. Temporal and spatial variation in length of larval life and size at settlement of the Hawaiian amphidromous goby *Lentipes concolor. Journal of Fish Biology* **59**: 928–938.

Schneider, D.W. & J. Lyons. 1993. Dynamics of upstream migration in two species of tropical freshwater snails. *Journal of the North American Benthological Society* **12**: 3–16.

Tagomori, M. 1991. Division of Water Resource Management. [Letter to Mr. Henry Sakuda, Administrator, Division of Aquatic Resources. January 29, 1991.]

Tate, D.C. 1997. The role of behavioral interactions of immature Hawaiian stream fishes (Pisces: Gobiodei) in population dispersal and distribution. *Micronesica* **30**: 51–70.

Titcomb, M. 1978. Native use of marine invertebrates in old Hawaii. *Pacific Science* **32**: 325–375.

Way, C.M., A.J. Burky, J.M. Harding, S. Hau & W.K.L.C. Puleloa. 1998. Reproductive biology of the endemic goby, *Lentipes concolor*, from Makamakaʻole Stream, Maui and Waikolu Stream, Molokaʻi. *Environmental Biology of Fishes* **51**: 53–65.

Wilcox, C. 1996. *Sugar water: Hawaii's plantation ditches*. University of Hawaiʻi Press, Honolulu, Hawaii. 191 p.

Yamamoto, M.N. & A.W. Tagawa. 2000. *Hawaiʻi's native and exotic freshwater animals*. Mutual Publishing, Honolulu. 200 p.

Biology of Hawaiian Streams and Estuaries. Edited by N.L. Evenhuis & J.M. Fitzsimons. *Bishop Museum Bulletin in Cultural and Environmental Studies* 3: 183–193 (2007).

Why Sweat the Small Stuff: the Importance of Microalgae in Hawaiian Stream Ecosystems

Matthew L. Julius
Department of Biological Sciences, St. Cloud State University, St. Cloud, Minnesota 56301; email: mljulius@stcloudstate.edu

Abstract

Microalgae are well known for their importance in aquatic ecosystems and for their utility as environmental indicators. These attributes are emphasized here for microalgae, especially diatoms, occurring in Hawaiian stream systems. Attention is paid to the influences stream morphology has on the distribution of algae in streams on Hawai'i and Kaua'i. Maximum current velocities and flood frequency appear to be of particular importance in regulating community structure through re-initiation of the benthic algal successional cycle. The effects of variations in microalgal community structure on the grazing activities of native algivorous fishes are discussed, and suggestions are provided with regard to stream management strategies providing for the preservation of these fish.

Introduction

It is important to mention the background and perspective influencing the content of this manuscript. Discussion and examples are limited to observation and experimentation made over the last three plus field seasons on Hawai'i and Kaua'i. The observations and experimentation summarized here are appropriate for this paper's objectives. These are:

1. understanding the major differences between Hawaiian and temperate stream systems;
2. explaining the spatial distribution of microalgal (not visible with naked eye) communities in Hawaiian streams;
3. describing the role of microalgal communities in Hawaiian stream ecosystems; and
4. explaining how these factors relate to one another in context of making management decisions for conserving native fishes in streams and estuaries.

The two islands dealt with represent the youngest and oldest, as well as, the range of naturally occurring variation in Hawaiian stream systems making them good examples for objective 1. Additionally, these two islands represent the extremes of mechanistic variations influencing algal community structure in terms of objectives 2 and 3. The specific endpoints driving the text are also limited; gobioid fishes for freshwater systems and mullet species for marine systems, but again these specific fishery endpoints lend themselves to the objectives of the manuscript. The two fish groups (goby and mullet) used as examples in this article are appealing because of their cultural and economic importance. Each fish group includes native species of cultural and economic interest. The mechanistic factors influencing algal communities also influence the distribution of native fishes. Both fish groups also include algal grazing species. These components allow objective 4 to be achieved in an efficient and relevant context.

Nature and Composition of Hawaiian Stream Systems

Hawaiian vs. temperate streams

To understand the composition and development of algal communities in Hawaiian streams, researchers must comprehend the structure and mechanics of these aquatic systems and their surrounding environments. As with most investigations of freshwater ecology, studies of lotic ecosystems have disproportionately focused on temperate zone systems with study sites smaller than 10 m^2

and individual investigations lasting less than one year (Lodge *et al.*, 1998). This investigative and literature bias must be recognized to avoid applying generalizations appropriate for temporal zone systems to Hawaiian streams. Notable structural and environmental differences in Hawaiian streams compared to temperate systems include the minimal (perhaps nonexistent) hyporheic zone, narrow seasonal variation, and the smaller, steeper overall stream system. These structural differences in Hawaiian streams directly influence mechanistic factors impacting stream biota. Not the least of these factors include nutrient regimes, flow regimes, and residence time.

The minimized hyporheic zone is a result of the islands' volcanic origins and the limited amount of sediment covering this volcanic substrate. This streambed morphology limits the degree to which nutrient buffering can occur in the stream system and generally restricts nutrient inputs from inflow activities to surface sediment runoff. Minimized seasonal variation creates a situation in which temperature and light regimes for many species are always near optimal conditions and reduce seasonal succcessional patterns observed in the biota of temperate systems. Perhaps the most dramatic structural difference from temperate systems is the short, steep morphology of the overall stream system. Influences of this stream architecture on flow regimes and residence time are obvious.

Nutrients, specifically nitrogen and phosphorus, in stream systems of moist tropical environments are typically considerably higher than those in temperate systems with typical total nitrogen exports around 998 kg km^{-2} yr^{-1} and total phosphorus exports approximating 46 kg km^{-2} yr^{-1} (Lewis, 1986). These are approximately three to four times greater than values for temperate regions lakes, 300 kg km^{-2} yr^{-1} of total nitrogen and 10 kg km^{-2} yr^{-1} of total phosphorus (Schindler *et al.*, 1974; Jansson, 1979). Runoff from the streams' riparian zone almost certainly accounts for a high proportion of the nutrient values given little trickling and filtration occurs via soils before storm runoff enters the stream systems. These runoff events and the short, steep architecture of the stream systems also produce a much shorter residence time for these systems in comparison to those in temperate regions. Average values for temperate regions are approximately 16 days for unimpounded stream systems (Vörösmarty *et al.*, 1997) compared to nearly daily flushing in Hawaiian systems. Related to the short residence times are the relatively high maximum flow rates of the Hawaiian stream systems. Heavy rain events (>100mm per day) are not uncommon and result in major flash flood events (Kodama & Barnes, 1997; Kodama & Bursinger, 1998). Based upon particle size, rubble to boulder sized (64 to 128+ mm) (Einsele, 1960) particles, in the stream beds flood events have to achieve maximum velocities in excess of 200 cm s^{-1} (Morisawa, 1968; Allan, 1995). The strong currents and frequent flooding events maybe the features most useful in understanding the biotic interactions in the Hawaiian stream systems. With the relatively high nutrients and consistent temperature and light regimes, these variable hydrologic regimes have a strong effect on the streams' biota (Poff & Ward, 1989). Bright (1980) identified flood periodicity among the most influential mechanisms for restructuring the invertebrate stream community on Palau, another Pacific island of volcanic origin, and conditions appear similar for Hawaiian islands.

Streams on Hawai'i vs. streams on Kaua'i

Differences worth noting exist between the streams of Hawai'i and Kaua'i. The oldest island, Kaua'i, formed over 5 mya and rose to approximately 2600 m before eroding down to its present maximum elevation of 1598 m. In contrast, the youngest island, the Island of Hawai'i emerged less than 0.5 mya and has eroded little from its maximum height of 4600 m. As a result of differences in elevational gradient, these islands do not contain identical stream systems. The erosion of Kaua'i has resulted in broad meandering streams with large estuaries. These streams are only rarely interrupted by waterfalls, which are usually located in stream headwaters. In contrast, streams on Hawai'i have steep slopes resulting in higher velocity streams with few (if any) estuaries. Moreover, frequent waterfalls that can exceed 100 m in height and include many that terminate directly on the beach or in the ocean interrupt streams on Hawai'i. Furthermore, weather movements across Hawai'i precipitate out moisture and create the arid conditions on the Kona side of the island and moist conditions on the Hilo side of the island where flood events occur on a near monthly scale (Kodama & Barnes, 1997). This is not the case on Kaua'i where flood events are less frequent. The frequent waterfall

Figure 1. Scanning electron micrographs of benthic microalgae. **A:** the tuft forming diatom *Synedra acus*. **B:** the stalk forming diatom *Rhoicosphenia curvata*. **C:** epiphytic diatoms growing on a filamentous green algae. Scale bars equal 200 μm for A and C and 10 μm for B.

interruptions in stream systems on Hawai'i may limit the degree to which nutrient spiraling (Newbold *et al.*, 1983) can occur because of the inability of biota to move back upstream.

Distribution of microalgae in Hawaiian streams

The composition and mechanistic factors of Hawaiian streams discussed above directly influences the distribution of microalgae in the stream systems. The close association of microalgae with their surrounding environmental conditions dates to the earliest observations in aquatic ecology (Naumann, 1919), and they continue to be used as indicators of environmental health (Lowe & Pan, 1996). The nutrient regimes in the streams are conducive to algal growth with ratios of N:P averaging 22:1 (Lewis, 1986). This N:P ratio suggests phosporus is the limiting nutrient (Vallentyne, 1974) and results in eukaryotic microalgae being the dominant forms in Hawaiian stream systems. Studies of nutrient availability and periphyton on Oahu are consistent with this taxon/nutrient distribution pattern (Larned & Santos, 2000). The composition of this eukaryotic community is largely dominated by diatoms (Bacillariophyta), and further suggests that silica is in abundance in these streams. This is supported by Derry *et al.*'s (2005) observation that most of the silica released on Hawaiian islands is processed as biogenic silica.

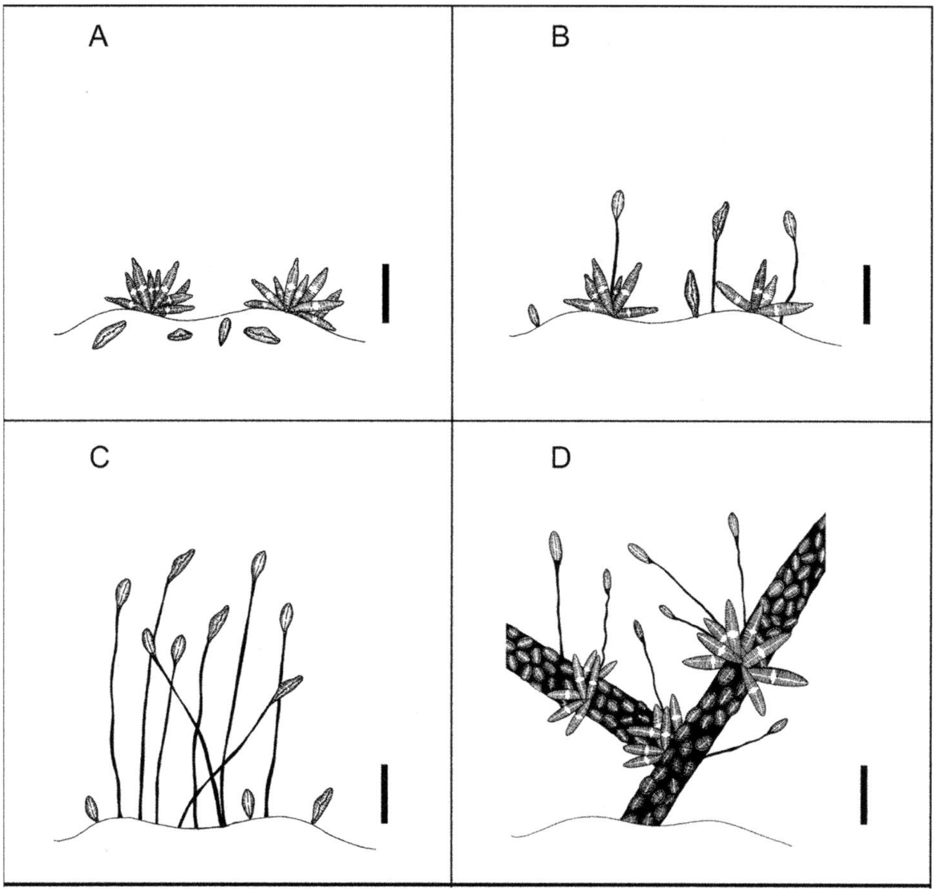

Figure 2. Hypothetical representation of benthic algal community succession, modified from Tuji (2000). **A:** First colonizers, characterized by colonies of tuft forming species and motile stalk growing forms. **B:** Early development of stalk forming species continuing. **C:** Dominant phase of stalk growing species. **D:** Community apex with macrophyte supporting numerous epiphytic species.

Among algae, the diatoms are the most widely used environmental indicators (Stoermer & Smol, 2000). The specimens observed from Hawaiian freshwater stream systems largely emphasize the environmental conditions of the streams described above. The diatom flora of the streams largely consists of benthic taxa occurring in one of two growth forms; tuft and stalk formers (Fig. 1A,B) (Julius *et al.*, 2005). Typical species assemblages comprise taxa associated with warm, nutrient-rich, hardwater, flowing systems with a pH ranging from 7.5 to 8.0 (Patrick & Reimer, 1966, 1975). Planktonic taxa are absent in the stream assemblages; these algae are a frequent component of most temperate systems (Stoermer & Julius, 2003). This is most likely a function of the absence of lakes draining into the Hawaiian stream systems.

Flow regimes are also important in structuring the overall algal community and directly influence when streams are dominated by filamentous macroalgae or by diatom microalgae. Peak algal biomass in high nutrient and high light environments occurs at modest stream velocities (30–60 cm s^{-1}) (Stevenson, 1996). Attached filamentous algae typically dominate in higher flow rates, and unattached filamentous forms persist in slower currents. Attached macroalgal filaments often act as a growth substrate for diatom microalgae (Fig. 1C), and unattached macroalagal filaments do not typically support diatom microalgal epiphytes. Only attached filamentous macroalgae and stalk forming diatoms are drag resistant and accumulate biomass in high flow environments (Reiter & Carlson, 1986; Dodds & Gudder, 1992).

During flash floods stream velocities become great enough to overturn the substrates of epilithic macro- and microalgae; based upon particle size discussed above these currents would exceed 200 cm s^{-1}. Flood events of this type will result in a clearing of the substrate surface initiating a distinct successional pattern (Hoagland *et al.*, 1982; Tuji, 2000; Fig. 2). Initially, a substrate is colonized by tumbling and motile species of diatoms, and short, tuft-like algal colonies develop (Fig. 2A). Stalk-forming species eventually begin to develop and coexist with tuft-forming taxa (Fig. 2B). As the stalk forming taxa become taller, the tuft-forming taxa are excluded via light competition (Fig. 2C). Eventually, macroalgae develop and exclude stalk-forming taxa from occurring directly on the rock substrate. At this point, stalk attachments shift from the rock to the stems of macrophyte algae, which are also colonized by tuft-forming and abundant adnate epiphytes (Fig. 2D). This last stage represents the development of peak biomass in benthic algal communities (Biggs, 1996). The entire successional process, from bare rock to epiphyte-supporting macroalgae, takes approximately one month to complete (Tuji, 2000). This timeframe has been corroborated with artificial substrate experiments in streams on Hawai'i (Julius, unpubl. data).

The frequency with which this successional pattern is initiated may well characterize the major differences in algal distribution between stream systems on Hawai'i and stream systems on Kaua'i. As previously mentioned, flash flood events on Hawai'i occur frequently. This is not the case for streams on Kaua'i where flash flood events are less frequent. The successional cycle is constantly being re-initiated in streams on Hawai'i. Periods of peak biomass are not persistent (Julius *et al.*, 2005) and microalgal taxa are in a near constant logarithmic growth phase (Biggs, 1996). In contrast, streams on Kaua'i are frequently found at peak biomass and the successional pattern is initiated only a few times per year. The best illustration of this dichotomy between streams on the two islands is abundance of the filamentous diatom *Hydrosera whampoensis* (Fig. 3). *Hydrosera* colonies are easily disturbed in strong currents and large colonies take several months to form. The taxon is extremely abundant in streams on Kaua'i, and form long filamentous colonies (hundreds of cells). In contrast, it is almost nonexistent in collections from streams on Hawai'i and, when present, occurs as a unicell or in short 2–5 cell filaments. This distribution contrasts streams on Hawai'i and Kaua'i and emphasizes the differences in periodicity of flood events.

Role of microalgal communities in Hawaiian stream ecosystems

Diatoms appear to be a particularly satisfactory food source for many aquatic animals, including fishes (Volkman *et al.*, 1989; Ahlgren *et al.*, 1990). This role as a quality food source is a function of the high energy lipid produced as a product of photosynthesis (Stoermer & Julius, 2003) like other members of the Stramenopile algal lineage, and contrasts with other major groups of algae which produce forms of starch as photosynthetic products. While discussions of microalgae as a food source frequently focus on members of the invertebrate community (Bott, 1996), the Hawaiian Islands provide some of the most spectacular examples of adult fish grazing on algal communities.

Gobioid fishes

The role of algae as a food source for the amphidromous gobioid fishes of Hawai'i is well documented (Kido, 1996a,b). Algae appear to be particularly important to climbing gobioid species, where grazing dietary habits appear to be reacquired with the development of climbing phylogenetically (Thacker, 2002). For at least one of these fishes, *Sicyopterus stimpsoni,* benthic diatoms appear to be the primary source of dietary sustenance. The adult jaw structure of the fish is highly modified for rasping and scraping benthic algal communities during early stages of their successional pattern (Julius *et al.* 2005). More dramatic is the evidence that this fish species can exist exclusively on thin coatings of epilithic diatoms exclusively (Schoenfuss *et al.*, 2004). This species also serves as an example illustrating the relationship between algal growth habit and availability to grazers. Steinman (1996) described limitations of stalked and tuft-forming taxa as a food source for invertebrate grazers (see Fig. 2A–C) and filamentous algae with epiphytes as a food source for another class of grazers (see Fig. 2D). This functional barrier appears to explain the coexistence of *Awaous guamensis* and *S. stimpsoni*, where *A. guamensis* feeds on filamentous algae and its epiphytes and *S. stimpsoni* feeds on the epilithic diatom taxa.

Native mullet

Preliminary investigation of the floristic composition of stomach contents from *Mugil cephalus* revealed that the contents consisted almost exclusively of diatoms. This result is not surprising, since Wells (1984) determined that algae comprised a major dietary source for this species. The floristic composition of this diatom assemblage for the Hawaiian fish, however, was more restricted than expected. A single diatom taxon, *Melosira moniliformis*, comprised the majority of the species identified. This taxon ranged from 70–90% of the total specimens recovered from the fish's stomachs. This diatom species distribution is not reflective of the diatom community composition in Hawaiian estuarine algal communities (Hustedt, 1942). This statement is accurate both in terms of total species numbers and overall abundance of *M. moniliformis*; and strongly suggests the fish is selectively feeding on this diatom species. The high abundance of *M. moniliformis* was present in the stomachs of all specimens observed from collections made in both Hawai'i and Kaua'i and percent composition of Hawai'i or Kaua'i collections were roughly equivalent. This seems to suggest a preference for the species by *Mugil cephalus* during foraging activities. The potential for this food preference is further emphasized by the fact that the other diatom species identified in the stomachs readily grow as epiphytes on *Melosira moniliformis*.

Although *Mugil cephalus* and *Melosira moniliformis* are both estuarine taxa, they are mentioned in this article because of the strong role the stream systems have in maintaining their interdependent ecologies. *Melosira moniliformis* is a colonial, centric diatom widely distributed throughout the entire southern Pacific and Atlantic seaboard. Like other *Melosira* species, *M. moniliformis* is tychoplanktonic, and exists both in benthic and planktonic communities. Typically, species are brought into planktonic habitats during storm events when waters are actively circulating through the benthic sediments of aquatic habitats due to the storm winds (Sicko-Goad *et al.*, 1986). Resuspension of *Melosira* typically breaks the colonies from longer, large cell number colonies into shorter, small cell number colonies. While in the plankton, the cells grow rapidly because light and nutrient conditions are more favorable than those encountered in benthic habitats. Increased light is a function of being higher in the water column, and increased nutrients are from resuspension of benthic sediments by storm events where nitrogen and phosphorus are trapped via sedimentation. As cell division proceeds in planktonic environments, *Melosira* colonies increase in length and become heavier. Weight is especially critical in diatom species because of the silica cell wall. Once colonies reach a certain length buoyancy can no longer be maintained, and the colonies settle into benthic environments. This peculiar ecology often results in large "blooms" of *Melosira* in post storm event waters. In the case of Hawaiian stream systems, blooms of *M. moniliformis* are regulated by flood events from the freshwater systems into near shore marine environments. These blooming events attract *Mugil cephalus*. This relation between the algae and the fish is so well understood that residents of Hilo actively harvest *Melosira moniliformis* after flood events to use as bait for *Mugil cephalus* (R. Nishimoto, pers. comm.).

Implications for Management of Native Fishes

Allan (2004) completed a review of the relationship between land use and stream health. Of the six factors mentioned in the article, hydrologic alteration appears to be the one of greatest concern in managing Hawaiian freshwater stream systems. Given the relationship between stream current, flood events, and biotic interactions described above there appears to be a distinct threshold or tolerance level beyond which hydrologic conditions should not be altered. Most significant is the requirement for periodic flooding events in the stream ecosystem; these could be on a near monthly scale for

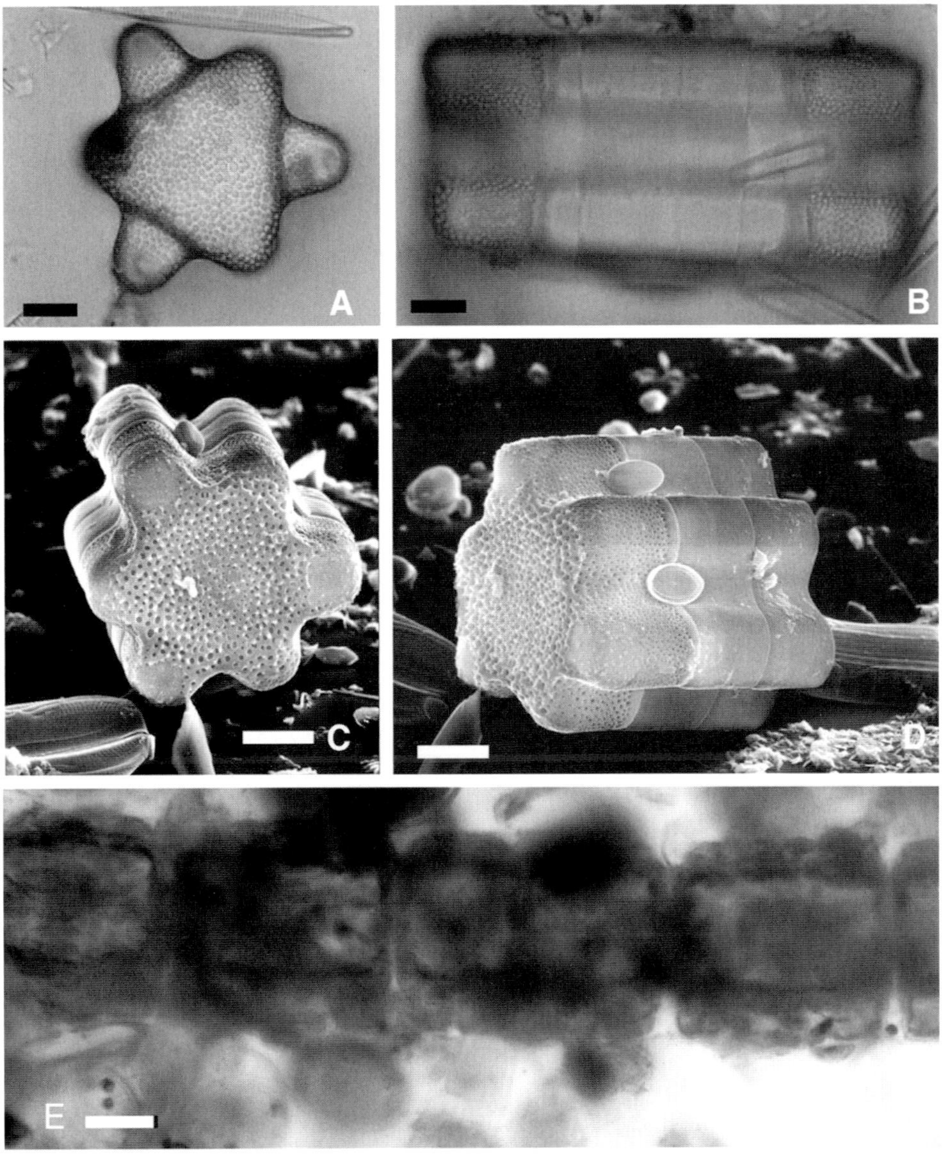

Figure 3. *Hydrosera whampoensis*. **A:** Light micrscope valve view of acid cleaned specimen. **B:** Light micrope girdle view of acid cleaned specimen.**C:** Scanning electon microscope oblique view. **D:** Scanning electron microscope girdle view. **E:** Light microscope view of live colony. Scale bars equal 10 μm.

streams on Hawai'i or semi-annual scale for streams on Kaua'i. Biotic responses to modifications in these threshold limited degredation events are non-linear (Allan, 2004), and once the threshold is broached the stream biota degrades rapidly.

Gobioid fishes

For *Sicyopterus stimpsoni* to continually gain access to its algal food, benthic algal succession must periodically be re-initiated through biological or physical disturbance events. Algal succession could be re-initiated through repeated grazing or periodic limitation of light, but it might also be re-initiated by substrate disruption via water column turbulence. These disturbance events may be critical to the maintenance of suitable food sources for *S. stimpsoni* in stream systems.

Field observations of feeding behavior in *S. stimpsoni* are consistent with the hypothesis that this species contributes to the disturbance of benthic algae and thereby continually re-initiates algal succession. Adult fish appear to "farm" large feeding rocks in upper stream reaches by continually scraping the same rock over a period of several days or weeks (Fitzsimons *et al.*, 2003). This activity effectively re-initiates benthic algal succession and prevents the development of long filamentous algae, which would make the rock difficult for these fish to feed from. Such grazing could also help to exclude other grazing fish species (Schoenfuss *et al.*, 2004), which typically feed on longer filamentous algae (Kido, 1996b).

In terms of maintaining this fish species on Hawai'i and Kaua'i, the conservation approaches may vary slightly based upon differences in streams on each island. For streams on Hawai'i periodic flood events must be maintained on a near monthly schedule to re-initiate the benthic algal successional pattern. This will ensure a constant food source for *S. stimpsoni* and emulate current stream conditions on the island in the most appropriate fashion. For streams on Kaua'i, appropriate sized feeding rocks provided to support *S. stimpsoni* would have to be within a specific size range. The feeding rock should be (1) large enough to allow development of early succession benthic algal taxa on some portions of the rock while feeding took place on other portions and (2) small enough to allow complete grazing of the rock before filamentous taxa became well developed. These expectations are consistent with the descriptions of feeding rocks provided by Fitzsimons *et al.* (2003). Rocks of the specific size described will provide an adequate feeding substrate for *S. stimpsoni* and accounts for the limited periodicity of flood events on Kaua'i.

In some instances the stream flow tolerance threshold for gobioid fishes has been exceeded. Numerous stream branches on Hawai'i have been impounded or diverted to such an extent that stream velocities are well below 30 cm s^{-1}. In these instances, the unattached filamentous algae *Spirogyra* dominate the stream segment. These filamentous algae do not support the growth of epiphytes, and these stream segments are devoid of both *Awaous guamensis* and *S. stimpsoni* (Schoenfuss, personal observation). These segments should be flooded to clear out *Spirogyra* and re-initiate the benthic algal successional cycle.

Native mullet

Use of the alga *Melosira moniliformis* as bait may have developed on and be limited to Hawai'i, particularly the windward side of the island, rather than on the other Hawaiian islands because of topography. Frequent flash floods in streams on Hawai'i directly affect the frequency of *Melosira moniliformis* blooms along the windward coast of the islands. As turbidity is increased by flood waters along the coast, *Melosira moniliformis* is actively resuspended into planktonic environments, and blooms are triggered. This scenario makes *Melosira moniliformis* readily available for harvest by anglers on the windward side of Hawai'i and less available to anglers at other locations where *Melosira moniliformis* blooms are less frequent and predictable, such as Kaua'i.

The recent introduction of the mullet species *Valamugil engeli* has caused great concern over appropriate management strategies for protecting the native mullet *Mugil cephalus*. The invasive species appears to be more abundant on Kaua'i than on Hawai'i, and this may be a result of stream systems on Kaua'i approaching their tolerance threshold for hydrologic alteration. *Valamugil engeli* appears to consume a wider variety of algal species in coastal Hawaiian ecosystems than the native fish *Mugil cephalus*. With diminished flooding events, blooms of *Melosira moniliformis* become less frequent. This may well provide the opportunity for the invasive mullet species to garner a competitive advantage over the native fish. In terms of management strategies, flood events should be maintained in stream systems on Hawai'i to provide an optimal feeding environment for *Mugil cephalus*. Resources can then be disproportionately applied to preventing the spread of *V. engeli* on Kaua'i where it is unreasonable to try to use flood regulation as a control mechanism.

Conclusions

Our ability to understand and manage the health of Hawaiian streams is directly linked to how well we understand the ecology of the biota and mechanics of the system. Unfortunately, the ability to compare both biota and mechanics of Hawaiian systems with that of other well-studied ecological systems is limited. It is also difficult to predict how the new challenges and problems for maintaining Hawaiian aquatic ecosystem health may manifest themselves. This does not mean that research should be funded without expectation of its results having some practical application. Researchers must make an effort to derive general principles from studying applied problems (Harris, 1994), and these applied problems may well provide interesting large-scale and long-term studies of great interest to basic research questions (Lodge *et al.*, 1998). If this relationship can be achieved between researcher and manager, maximum benefit per research dollar both in terms of application and knowledge expansion will be produced.

An area lending itself to this research relationship is a study focusing on interactions between Hawaiian freshwater streams and coastal marine environments to control the spread of *V. engeli*. This invasive mullet species moves between coastal marine and freshwater habitats as juveniles, and the regulation of the native mullet species food source appears to be linked with freshwater discharges into coastal marine environments. The following research objectives should be pursued in context of this problem:

- A floristic survey of the gut contents of *M. cephalus* and *V. engeli* should be conducted.
- The feeding mechanics of each fish should be examined and documented.
- Feeding trials and competition studies using varying food sources should be conducted.

Once these research objectives are complete, management decisions can be made to create environments favoring the behaviors and ecology of *M. cephalus*. This research will also expand the limited knowledge base concerning the feeding behaviors and mechanics of algivorous fish by relating specific algal taxa and growth forms with mechanistic constraints.

Acknowledgments

I express my appreciation to the Division of Aquatic Resources, Department of Land and Natural Resources, State of Hawai‘i for their support of this research. I am also indebted to my research colleagues on this project, Drs. Heiko Schoenfuss and Richard Blob. Finally, the friendship and collegiality of Bob Nishimoto, Mike Fitzsimons, Darrell Kuamo‘o, Bruce Kaya, Lance Nishiura, and Troy Shimoda is the most highly valued outcome of these research activities.

Literature Cited

Ahlgren, G., L. Lundstedt, M. Brett & C. Forsberg. 1990. Lipid composition and food quality of some freshwater phytoplankton for cladoceran zooplankters. *Journal of Plankton Research* **12**: 809–818.

Allan, J.D. 1995. *Stream ecology: structure and function of running waters*. Chapman & Hall, London. 388 p.

———. 2004. Landscapes and riverscapes: the influence of land use on stream ecosystems. *Annual Review of Ecology, Evolution, & Systematics* **35**: 257–284.

Biggs, B.J.F. 1996. Patterns in benthic algae of streams, p. 31–56. *In*: R.J. Stevenson, M.L. Bothwell & R.L. Lowe (eds.), *Algal ecology: freshwater benthic ecosystems*. Academic Press, San Diego.

Bott, T.L. 1996. Algae in microscopic food webs. *In*: R.J. Stevenson, M.L. Bothwell & R.L. Lowe (eds.), *Algal ecology: freshwater benthic ecosystems*. Academic Press, San Diego. p. 574–607.

Bright, G.R. 1980. Secondary production in a tropical island stream. *Limnology & Oceanography* **27**: 472–480.

Derry, L.A., A.C. Kurtz, K. Ziegler & O.A. Chadwick. 2005. Biological control of terrestrial silica cycling and export fluxes to watersheds. *Nature* **433**: 728–731.

Dodds, W.K. & D.A. Gudder. 1992. The ecology of *Cladophora*. *Journal of Phycology* **28**: 415–427.

Einsele, W. 1960. Die Strömungsgeschwindigkeit als beherrschender Faktor bei der limnologishcen Gestaltung der Gewasser. *Österreichs Fischerei* **2**: 1–40.

Fitzsimons, J.M., M.G. McRae, H.L. Schoenfuss & R.T. Nishimoto. 2003. Gardening behavior in the amphidromous Hawaiian fish *Sicyopterus stimpsoni* (Osteichthyes: Gobiidae). *Ichthyological Explorations of Freshwaters* **14**: 185–191.

Harris, G.P. 1994. Pattern, process, and prediction in aquatic ecology. A limnological view of some general ecological problems. *Freshwater Biology* **32**: 143–160.

Hoagland, K.D., S.C. Roemer & J.R. Rosowski. 1982. Colonization and community structure of two periphyton assemblies, with emphasis on the diatoms (Bacillariophyceae*). American Journal of Botany* **69**: 188–213.

Hustedt, F. 1942. Süsswasser-Diatomeen des indomalyayischen Archipels und der Hawaii-Inseln. *Internationalen Revue der Gesamten Hydrobiologie und Hydrographie* **42**: 1–252.

Jansson, M. 1979. Nutrient budgets and regulation of nutrient concentrations in a small subarctic lake in Northern Sweden. *Freshwater Biology* **9**: 213–231.

Julius, M.L., R. Blob & H.L. Schoenfuss. 2005. The survival of *Sicyopterus stimpsoni*, an endemic amphidromous Hawaiian gobiid fish, relies on the hydrological cycles of streams: evidence from changes in algal composition of diet through growth stages fish. *Aquatic Ecology* **39**: 473–484.

Kido, M.H. 1996a. Diet and food selection in the endemic Hawaiian amphidromous goby, *Sicyopterus stimpsoni* (Pisces: Gobiidae). *Environmental Biology of Fishes* **45**: 199-209.

———. 1996b. Morphological variation in feeding traits of native Hawaiian stream fishes. *Pacific Science* **50**: 184–193.

Kodama, K.P. & G.M. Barnes. 1997. Heavy rain events over the south-facing slopes of Hawaii: attendant conditions. *Weather & Forecasting* **12**: 347–367.

———. **& S. Businger**. 1998. Weather and forecasting challenges in the Pacific region of the National Weather Service. *Weather & Forecasting* **13**: 523–546.

Larned, S.T. & S.R. Santos. 2000. Light- and nutrient-limited periphyton in low order streams of Oahu, Hawaii. *Hydrobiologia* **432**: 101–111.

Lewis, W.M., Jr. 1986. Nitrogen and phosphorus runoff losses from a nutrient-poor tropical moist forest. *Ecology* **67**: 1275–1282.

Lodge, D.M., S.C. Blumenshine & Y. Vadeboncoeur. 1998. Insights and application of large-scale, long-term ecological observations and experiments, p. 202–235. *In*: W.J. Resettarits, Jr. & J. Dernardo (eds.), *Experimental ecology: issues and perspectives*. Oxford University Press, New York.

Lowe, R.L. & Y Pan. 1996. Benthic algal communities as biological monitors, p. 705–739. *In*: R.J. Stevenson, M.L. Bothwell & R.L. Lowe (eds.), *Algal ecology: freshwater benthic ecosystems*. Academic Press, San Diego.

Morisawa, M. 1968. *Streams: Their dynamics and morphology*. McGraw-Hill, New York. 175 p.

Naumann, E. 1919. Några Synpunkter Angående Limnoplanktons Ökologi med Särskild Hänsyn till Fytoplankton. *Svensk Botanisk Tidskrift* **13**: 129–163. [English Translation by the Freshwater Biological Association, publication #49.]

Newbold, J.D., J.W. Elwood, R.V. O'Neill & A.L. Sheldon. 1983. Phosphorus dynamics in a woodland stream ecosystem: A study of nutrient spiraling. *Ecology* **64**: 1249–1265.

Patrick, R. & C.W. Reimer. 1966. *The diatoms of the United States I*. Academy of Natural Sciences Philadelphia, Monograph **13**. 688 p.

———. **& C.W. Reimer**. 1975. *The diatoms of the United States II, part 1*. Academy of Natural Sciences Philadelphia, Monograph 13. 213 p.

Poff, N.L. & J.V. Ward. 1989. Impications of streamflow variability and predictability for lotic community structure: A regional analysis of streamflow patterns. *Canadian Journal of Fisheries and Aquatic Sciences* **46**: 1805–1818.

Reiter, M.A. & R.E. Carlson. 1986. Current velocity in streams and the composition of benthic algal mats. *Canadian Journal of Fisheries and Aquatic Sciences* **43**: 1156–1162.

Schindler, D.W., H.E. Welch, J. Kalff, G.J. Brunskill & N. Kritsch. 1974. Physical and chemical limnology of Char Lake, Corwallis Island (75? N Lat.). *Canadian Journal of Fisheries and Aquatic Sciences* **31**: 585–607.

Schoenfuss, H.L., M.L. Julius & R. Blob. 2004. Colonization of a recent, volcanically formed freshwater habitat: an example of primary succession. *Icthyological Exploration of Freshwaters* **15**: 83–90.

Sicko-Goad, L.M., E.F. Stoermer & G. Fahnenstiel. 1986. Rejuventation of *Melosira granulata* (Bacillariophyceae) resting cells from the anoxic sediments of Douglas Lake, Michigan, 1. Light microscopy and 14C uptake. *Journal of Phycology* **22**: 22–28.

Steinman, A. 1996. Effects of grazing on freshwater benthic algae, p. 341–347. In: R.J. Stevenson, M.L. Bothwell & R.L. Lowe (eds.), *Algal ecology: freshwater benthic ecosystems*. Academic Press, San Diego.

Stevenson, R.J. 1996. The stimulation and drag of Current. *In*: R.J. Stevenson, M.L. Bothwell & R.L. Lowe (eds.), *Algal ecology: freshwater benthic ecosystems*. Academic Press, San Diego. p. 321–340.

Stoermer, E.F. & J. Smol. 2000. *The diatoms: applications for earth and environmental sciences*. Cambridge University Press, London. 469 p.

———. **& M.L. Julius.** 2003. Centric diatoms. *In*: J.D. Wehr & R.G. Sheth (eds.), *Freshwater algae of North America: classification and ecology*. Academic Press, San Diego. p. 559–594.

Thacker, C.E. 2002. Molecular phylogeny of the gobioid fishes (Teleostei: Perciformes: Gobioidei). *Molecular Phylogenetics & Evolution* **26**: 354–368.

Tuji, A. 2000. Observation of developmental processes in loosely attached diatom (Bacillariophyceae) communities. *Phycological Research* **48**: 75–84.

Vallentyne, J.R. 1974. *The algal bowl: lakes and man*. Ottawa: Department of the Environmental, Fisheries, & Marine Services. 185 p.

Volkman, J.K., S.W. Jeffery, P.D. Nichols, G.I. Rogers & C.D. Garland.1989. Fatty acid and lipid composition of 10 species of microalgae used in mariculture. *Journal of Experimental Marine Biology & Ecology* **128**: 219–240.

Vörösmarty, C.J., K.P. Sharma, B.M. Gekete, A.H. Copeland, J. Holden, J. Marble & J.A. Lough. 1997. The storage and aging of continental runoff in large reservoir systems of the world. *Ambio* **26**: 210–219.

Wells, R. 1984. The food of the grey mullet (*Mugil cephalus* L.) in Lake Waahi and the Waikato River at Huntly, *New Zealand Journal of Marine and Freshwater Research* **18**: 13–19.

Biology of Hawaiian Streams and Estuaries. Edited by N.L. Evenhuis & J.M. Fitzsimons. *Bishop Museum Bulletin in Cultural and Environmental Studies* 3: 195–206 (2007).

Where Are We Now Regarding Hawaiian Stream Algal Systematics? (A Suspiciously Cosmopolitan Flora)

Alison R. Sherwood
Botany Department, 3190 Maile Way, University of Hawai'i, Honolulu, Hawai'i 96822 USA; email: asherwoo@hawaii.edu

Abstract

The last decade has witnessed a great advancement in our understanding of the Hawaiian stream algal flora. Several large surveys have produced comprehensive data sets for the four largest main Hawaiian islands (Kaua'i, O'ahu, Maui, and Hawai'i), which have resulted in a substantial increase of recorded taxa for the state. A compiled checklist is presented for the stream algal flora of Hawai'i, which includes 174 taxa. Overall, the Hawaiian stream algal flora is estimated to be only 2.3% endemic, which is very low compared to other groups of organisms in the state. The major drawback of most stream algal studies to date has been the reliance upon morphology-based identification techniques. As a result, most Hawaiian stream algal collections have had taxonomic names "force fit" to them from floras based on other geographical regions around the world. This has led to the recognition of a largely cosmopolitan Hawaiian stream algal flora with a very low rate of endemism, compared to other groups of organisms. Although it is recognized that the stream algal flora may indeed be more cosmopolitan than other groups of organisms in Hawai'i due to its high dispersal capability, the current estimate is regarded as suspiciously high. Artificial deflation of the degree of endemism may have arisen from a combination of "force-fitting" names from other regional floras, a reliance on morphological characters for identification, and a lack of taxonomic expertise. Estimates of the degree of endemism in the Hawaiian stream algal flora will likely increase with the inclusion of molecular- and genome-based characters in future taxonomic studies.

Introduction

"Across the world there is a prevailing view that freshwater algae are cosmopolitan. The notion has seldom been tested and is unlikely to be true in genetic terms....The degree of endemicity is probably masked by the 'force-fitting' of European names..." (Tyler, 1996, p.127)

Freshwater algae are largely viewed as cosmopolitan in their geographic distribution (Tyler, 1996). A number of inseparable factors likely contribute to this impression. First, since floristic works are only available for a limited number of geographical regions, phycologists have relied heavily on available floras as opposed to developing concepts of new taxa for each new area explored. For example, the most comprehensive sources of cyanobacterial taxonomy are based on the European flora, yet are used worldwide (e.g., Geitler, 1932; Anagnostidis & Komárek, 1985). Second, most phycologists still rely almost exclusively on morphological characters for routine identifications, especially for ecological studies that require the identification of many collections, and the use of these characters may lead to an underestimate of genetic diversity. This is especially true for algae that have very simple morphologies, with few defining characters (e.g., the crustose red alga, *Hildenbrandia* [Sherwood & Sheath, 2003]). Third, the paucity of well-trained taxonomists in the current generation leaves the door open for misidentification and over-reliance on the small number of available floras. These factors have all likely played a role in our understanding of the stream algal flora of the Hawaiian Islands and resulted in a floristic checklist that is comprised almost exclusively of cosmopolitan species, or at least those known to inhabit a broad tropical or subtropical range of geographical regions.

The evolutionary breadth of organismal life contained under the informal classification of "algae" can be misleading in some cases. Although a completely satisfactory definition for algae does not exist, in general they include photosynthetic organisms that are not members of the green plant lineage. Algae thus contain members of two of the three domains of life (Eubacteria and Eukaryota), and within the eukaryotes encompass a wide variety of quite distinct evolutionary lines. It should be recognized that the comparisons and statistics presented in the current paper are representative of an incredibly wide lineage diversity – a case which does not exist (to this degree) for any other group of organisms included in this symposium.

The first written records of Hawaiian stream algae were published approximately 130 years ago. The earliest known record dates back to Nordstedt (1876) on collections of freshwater algae and charophytes (henceforth included within the freshwater algae) from an expeditionary report to the islands. Since then, a number of papers have dealt with various aspects of the Hawaiian stream algal flora, from ecological perspectives of these primary producers (LaPerriere, 1995; Chong, 1996) to diatom (Fungladda *et al.*, 1983) and macroalgal taxonomy (Vis *et al.*, 1994; Filkin *et al.*, 2003), and phenology and phylogenetics of red algae (Sherwood *et al.*, 2004). A bibliographic checklist summarizing the non-marine algal records from Hawai'i was recently published (Sherwood, 2004) and included indications of the endemic versus non-endemic elements of the flora.

One of the most challenging questions regarding Hawaiian freshwater algae is the determination of the native flora, which is virtually impossible for several reasons. The historical records of freshwater algae in Hawai'i are patchy, and a very incomplete picture is available prior to the first Polynesian settlements (between 400–1100 AD) and European contact (1778 AD) (Zeigler, 2002). Additionally, freshwater algae played a very minor ethnobotanical role in the diet of early Hawaiians (Abbott, 1984), and thus cultural records of the stream algal flora are also lacking. Another factor is that the earliest identifications (from 1876 onward) were by researchers working with identification resources developed for vastly different regions of the world. Given this, how close can we come to determining the native stream algal flora of the Hawaiian Islands? Two different approaches are available: algae in relatively pristine areas could be assumed to represent the native flora, or molecular tools could be used to assess the degree of divergence with respect to representatives in other areas (Sherwood, 2006). However, it will be almost, if not entirely impossible to discover what the original floristic composition was, as it is almost certain that alterations occurred following human settlement to the islands (Sherwood, 2006).

The present paper is a summary and critical assessment of the Hawaiian stream algal literature. Trends are presented from published reports of stream algae in Hawai'i. An emphasis is placed on how the future use of molecular characters may influence our understanding of the origins and the level of endemism of the Hawaiian stream algal flora.

Materials and Methods

Summary of previously published literature

Stream algae records from previously published literature reports were gathered and summarized. The list includes only peer-reviewed publications, or those currently under review in the primary literature. Only records from stream habitats are included (i.e., this analysis excludes standing water habitats such as taro patches, reservoirs, fish ponds, and other brackish-water areas). However, much of the older literature combines records from a variety of habitats under a single category, and these could not be separated in the current analysis. The list is largely derived from the stream algal records contained in Sherwood (2004), with the addition of recent literature (Sherwood, 2006). Only the soft algae are included in the present analysis (i.e. the diatoms are excluded). Both microalgae (those algae requiring light microscopy for visualization of the entire algal thallus) and macroalgae (benthic algae with a thallus construction visible to the naked eye) are included in the comparative list. Taxa endemic to the Hawaiian Islands are marked with an asterisk.

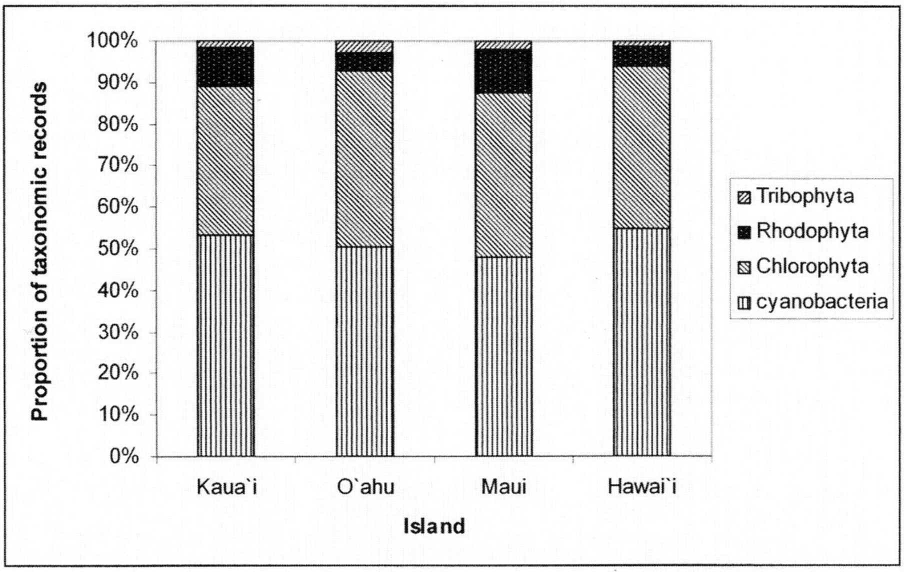

Figure 1. The proportion of stream algae records from each of the four main Hawaiian Islands belonging to each broad taxonomic category (cyanobacteria, Chlorophyta, Rhodophyta and Tribophyta).

Analyses of published records

Taxonomic records were sorted by broad taxonomic category: cyanobacteria (or "blue-green algae"), Chlorophyta (green algae), Rhodophyta (red algae), Dinophyta (dinoflagellates), Chrysophyta (chrysophytes), and Tribophyta (yellow-green algae or xanthophytes), and compared across islands to determine if similar proportions of taxa from each category were identified from each island. The Chrysophyta and Dinophyta were excluded from the analysis since only a single taxon was identified from each group. Similarly, only records from the four main Hawaiian Islands were included in the analysis since these were the only islands for which enough data were available to make meaningful comparisons. A similar analysis previously presented for stream macroalgae of the Hawaiian Islands (Sherwood, 2006) allows a comparison between the macroalgal stream algal flora and the entire stream algal flora.

Published molecular analyses of a broad taxonomic range of freshwater algae were evaluated for reported levels of genetic divergence across different geographical regions. These data were used to explore the implications of including molecular or genetic data for re-interpretation of the taxonomy of the algal groups included in the current study.

Results and Discussion

Summary of the Hawaiian stream algal flora

The non-diatom component of the stream algal flora of the Hawaiian Islands is comprised of an estimated 174 taxa, including 83 cyanobacteria, 77 Chlorophyta, 7 Rhodophyta, 1 Dinophyta, 1 Chrysophyta, and 5 Tribophyta (Table 1). These records were drawn from a total of 22 literature reports, ranging in publication date from the late 1800's (Nordstedt, 1876) to the present (Sherwood, 2006). Of these 174 taxa, only four, or 2.3% of the total, are believed to be endemic to the Hawaiian Islands. This value is extremely low when compared to the estimate of 5.0% for the entire non-marine algal flora (Sherwood, 2004), not to mention the estimated values for the marine red algal

Table 1. Stream algal taxa, including both macroalgae and microalgae, recorded in the literature for the Hawaiian Islands. Taxa believed to be endemic to Hawai'i are indicated with an asterisk (*). Island names are abbreviated as follows: L = Laysan Island, Ni = Ni'ihau, K = Kaua'i, O = O'ahu, Mo = Moloka'i, M = Maui and H = Hawai'i. Reference numbers correspond with those in the Literature Cited section.

Taxon	Distribution	References
Cyanobacteria		
Anabaena catenula (Kütz.) Bornet et Flahault	K, O	7, 9, 19
Anabaena sp.	O	8, 12, 17
Aphanothece naegelii Wartm. ex Rabenh.	O	7, 8, 9, 17, 18, 19
Calothrix braunii (A. Br.) Bornet et Flahault	M, H	12
C. fusca (Kütz.) Bornet et Flahault	M	6, 7, 9, 12, 19
Chamaesiphon elongatus (Nordst.) Kann	K, O	6, 10, 13
syn. *C. curvatus* var. *elongatum* Nordst.		
Cylindrospermum catenatum Ralfs	O	7, 9, 19
C. stagnale (Kütz.) Bornet et Flahault	K, O, H	3, 7, 9, 12, 19
Cylindrospermum sp.	K, O, M, H	3, 8, 10, 12, 17
Dichothrix baueriana (Grun.) Bornet et Flahault	M	12
Fischerella ambigua (Nägeli) Gomont	O	7, 8, 9, 19
Geitlerinema splendidum (Grev. ex Gom.) Anagn.	O, H	3, 12
Gloeocapsa magma (Bréb.) Komárek et Anagn.	O, H	6, 7, 8, 9, 10, 19
G. polydermatica Kütz.	O, H	7, 9, 19
G. quarternata Kütz.	O, H	7, 8, 9, 17, 18, 19
Heteroleibleinia distincta (Schmidle) Anagn. et Komárek	O, H	7, 9, 19
syn. *Lyngbya distincta* (Nordst.) Schmidle		
Hydrocoleum meneghinianum Kütz.	H	11, 12
Leptolyngbya angustissima (W.West et G.S.West) Anagn. et Komárek	O	3, 12
L. foveolarum (Mont. ex Gomont) Anagn. et Komárek	O	11, 12
L. tenuis (Gomont) Anagn. et Komárek	K, M	13, 15, 20
syn. *Phormidium tenue* (Menegh.) Gomont		
**Lyngbya cladophorae* Tilden	O, H	7, 9, 19
L. major Menegh.	O	12
L. martensiana Menegh.	O, H	7, 9, 18, 19
Lyngbya sp.	O, H	8, 17
Microchaete uberrina N.Carter	K	3, 12
Microcoleus lacustris (Rabenh.) Farl.	O, H	3, 12, 15
M. vaginatus (Vaucher) Gom.	O, H	2
Nostoc commune Vaucher	O	6, 7, 9, 12, 17, 18, 19
N. foliaceum Moug.	O, H	7, 9, 18, 19
N. linckia (Roth) Bornet ex Bornet et Thur.	K, O	3, 6, 7, 9, 12, 19
syn. *N. piscinale* Kütz.		
N. paludosum Kütz.	M, H	6, 7, 9, 12, 19
N. pruniforme C. Ag.	K, O, M, H	11, 12
N. verrucosum (L.) Vaucher	O, M, H	7, 9, 12, 17, 18, 19
Nostochopsis lobatus Wood emend. Geitler	K, O, M	3, 12
N. radians Bharadwaja	K, O	3, 12
Oscillatoria limosa C.Agardh ex Gom.	K, O, H	3, 12
O. princeps Vauch. ex Gom.	H	12
O. sancta Kütz.	K, O, H	7, 8, 9, 12, 18, 19
O. subbrevis Schmidle	K	12
Oscillatoria sp.	K, O, H	3, 8, 10, 17
Phormidium ambiguum Gomont	K, M	11, 12
P. amoenum (Kütz. ex Gomont) Anagn. et Komárek	K	3, 12
P. autumnale C.Ag. ex Gom.	M	12
P. corium (C.Agardh) Gomont	K, O	3, 7, 8, 9, 12, 18, 19
syn. *P. papyraceum* Gomont		
P. favosum Gomont	O	7, 9, 18, 19

Table 1 (continued)

Taxon	Distribution	References
P. formosum (Bory ex Gomont) Anagn. et Komárek syn. *Oscillatoria formosa* Bory	O, H	7, 8, 9, 12, 19
P. inundatum Kütz.	K, O, M, H	12, 20
P. pachydermaticum Frémy	M	12
P. putealis (Mont. ex Gomont) Anagn. et Komárek	H	12
P. retzii (C.Agardh) Gomont	K, O, M, H	2, 3, 12, 13, 16, 20
P. splendidum (Grev. ex Gomont) Anag. et Komárek syn. *Oscillatoria splendida* Grev.	O, H	3, 19
P. stagninum C.B.Rao	O	11, 12
P. subfuscum Kütz.	K, O, M	12, 20
P. subincrustatum Fritsch et Rich	O, H	11, 12
P. tinctorium Kütz.	K, O	11, 12
P. truncicola Ghose	K, O	12
Phormidium spp.	K, O, M, H	8, 10, 12, 17
Planktothrix planctonica (Elekin) Anag. et Komárek syn. *Microcoleus lyngbyaceus* (Kütz.) P.L.Crouan et H.M.Crouan	L, K, H	2
Pseudanabaena sp.	O	3, 12
Rivularia haematites (DC.) C.Agardh	K, H	12, 20
R. minutula (Kütz.) Born. et Flah.	K, H	12
Schizothrix calcicola (C.Agardh) Gomont	Ni, K, O, M, H	2
S. friesii (C.Agardh) Gomont	K, O, M, Mo, H	2, 20
S. lacustris A.Braun ex Gomont	K	20
S. rivulis (Wolle) F.E.Drouet	K, H	12, 13
Scytonema arcangelii Bornet et Flahault	O, M, H	12, 20
S. chiastum Geitler	K	3, 12
S. coactile Mont. ex Kütz.	K	3, 12
S. crispum (C.Agardh) Bornet	K, O, M, H	6, 7, 9, 10, 12, 16, 18, 19
S. fritschii S.L.Ghose	H	12, 15
S. guyanese (Mont.) Bornet et Flahault	O	6, 7, 9, 19
S. myochrous (Dillwyn) C.Agardh	O, M	20
S. ocellatum Lyngb.	O	6, 7, 8, 9, 19
S. rivulare Borzí	O, M	7, 8, 9, 12, 18, 19
S. tolypothricoides Kütz.	K, M, H	3, 12, 20
S. varium Kütz.	O	6, 7, 8, 9, 19
Scytonema sp.	O, H	8, 17
Spirulina maior Kütz.	O, H	7, 8, 9, 19
Stigonema mamillosum (Lyngb.) C.Agardh ex Bornet et Flahault	H	12
Tolypothrix distorta (O.F.Müll.) Kütz.	O, H	7, 8, 9, 12, 18, 19
T. nodosa Bharadw.	H	11, 12
T. tenuis Kütz. emend. A.W.F.Schmidt syn. *T. lanata* (Desv.) Wartmann	O, H	3, 9, 12, 20
Xenococcus kerneri Hansg.	O	6, 7, 8, 9, 19
Chlorophyta		
Basicladia chelonum (Collins) W.E.Hoffmann et Tilden	O	3, 12, 14
Binuclearia tectorum (Kütz.) S.Berger ex Wichmann syn. *B. tatrana* Wittr.	K	12, 13, 14
Bulbochaete sp.	K, O, H	7, 8, 10, 12, 13
Chaetophora elegans (Roth) C.Agardh	K, O	3, 12
Chaetosphaeridium globosum (Nordst.) Kleb. syn. *Herposterion globosum* Nordst.	H	6, 9, 10
Chara braunii C.C.Gmelin	O, M	12, 16
C. zeylanica var. *diaphana* f. *armata* (F.J.F.Meyen) Zaneveld	K, O	6, 7, 9, 10, 22
Characium groenlandicum Richter	Mo	6, 8, 9
Cladophora fracta (O.F. Müller ex Vahl) Kütz.	O	3, 6, 7, 9, 12

Table 1 (continued)

Taxon	Distribution	References
C. fracta var. *rigidula* (Kütz.) Rabenh.		21
C. glomerata (L.) Kütz.	K, O, M, H	3, 12, 13, 18, 20
syn. *C. canalicularis* (Roth) Kütz.		
C. glomerata var. *crassior* (C.Agardh) C.Hoek	K, M, H	12, 20
**C. longiarticulata* var. *valida* F.Brand	O	1, 10
C. rivularis (L.) C.Hoek	O	20
C. sericea (Huds.) Kütz.	O	7, 8, 9
syn. *C. nitida* Kütz.		
Cladophora sp.	K, O, M	5, 8, 12, 13, 17
Cloniophora macrocladia (Nordst.) Bourr.	K, O, H	4, 6, 7, 8, 9, 10, 12
syn. *Draparnaldia macrocladia* Nordst., *Stigeoclonium macrocladium* (Nordst.) Schmidle		
C. plumosa Kütz. emend. Bory	K, O, M, H	3, 4, 12, 13, 20
C. spicata Schmidle emend. A.K.Islam	K, O, M, H	12, 13, 14
Closteriopsis longissima Lemmerm.	O	6, 7, 8, 9
Coleochaete irregularis Pringsh.	O, H	6, 7, 9, 10
C. orbicularis Pringsh.	O, H	6, 7, 9, 10
Cylindrocystis sp.	H	12
Dactylococcus infusionum var. *minor* Nordst.	O	6, 7, 9, 10
Desmidium sp.	O, M, H	12
Dictyosphaerium pulchellum H.C.Wood	O	6, 7, 8, 9
Geminella minor (Nägeli) Heering	M, H	12
Haematococcus pluvialis Flot.	throughout	6, 9
Klebsormidium fluitans (Gay) Lokhorst	H	11, 12
syn. *K. rivulare* (Kütz.) Morison et Sheath		
K. subtile (Kütz.) Tracanna ex Tell	M	6, 7, 9, 11, 12, 16
Microspora pachyderma (Wille) Lagerh.	K	3, 12
Monoraphidium convolutum var. *pseudosabulosum* Hindák		8
syn. *Dactylococcus infusionum* Nägeli		
Mougeotia capucina (Bory) Nordst.	O, H	6, 7, 9, 10
Mougeotia spp.	K, O, M, H	3, 8, 12, 16
Nitella gracilis ssp. *havaiensis* (Nordst.) R.D.Wood	O, H	6, 7, 8, 9, 10, 22
syn. *N. havaiensis* Nordst.		
Oedogonium crispum var. *hawiense* Nordst.	O, H	6, 7, 9, 10
O. globosum Nordst.	O	6, 7, 9, 10
O. undulatum (Bréb.) A.Br.	O	12
Oedogonium spp.	K, O, M, H	3, 5, 7, 8, 12, 13, 17
Pediastrum boryanum (Turpin) Menegh.	H	6, 7
P. duplex var. *reticulatum* Lagerh.	O	6, 9
syn. *P. duplex* var. *clathratum* A.Braun		
P. integrum var. *braunianum* (Grunov) Nordst.	H	6, 9, 10
P. tetras (Ehrenb.) Ralfs	O	6, 9, 10
Pediastrum sp.		8
**Pithophora macrospora* F.Brand	H	1
Raphidium polymorphum Fresen.	throughout	6, 7, 8, 9
Rhizoclonium crassipellitum W.West et G.S.West	O	11, 12
R. hieroglyphicum (C.Agardh) Kütz.	K, O, M, H	12, 13, 20
R. hieroglyphicum var. *hosfordii* (Wolle) Collins	K,O	14
Schizomeris leibleinii Kütz.	O, M	12, 14
Schroederia setigera (Schröd.) Lemmerm.	O	6, 7, 8, 9
Spirogyra dictyospora C.C.Jao	M	11, 12, 16
S. dubia Kütz.	M	11, 12
S. elegantissima Ling et Zheng	O, H	11, 12
S. fallax (Hansgirg) Wille	H	11, 12
S. fluviatilis Hilse	H	12, 18

Table 1 (continued)

Taxon	Distribution	References
Spirogyra spp.	throughout	3, 5, 7, 8, 9, 10, 12, 13, 16
Stigeoclonium amoenum var. *novizelandicum* Nordst.	O	18
S. falklandicum Kütz.	O	6, 7, 9, 10, 18
S. fasciculare Kütz.	K, H	11, 12
S. flagelliferum Kütz.	H	12
S. lubricum (Dillw.) Kütz.	K, O, M	12
S. nudiusculum Kütz.	O	18
S. pachydermum Prescott	K	11, 12
S. segarare A.K.Islam	K	3, 12
S. setigerum Kütz.	O, H	3, 12
S. stagnatile (Hazen) Collins	O, M, H	3, 12, 20
S. subsecundum Kütz.	K, O, H	3, 12, 20
S. tenue (C.Agardh) Kütz.	K, O	7, 9, 20
S. variabile Kütz.	O	12
Stigeoclonium sp.	O	8, 17
Triploceras sp.		8
Ulothrix minulata Kütz.	O	7, 9
U. tenerrima Kütz.	M	12
U. tenuissima Kütz.	H	3, 12
Ulothrix spp.		5, 8
Zygnema spp.	K, O, M, H	8, 10, 12, 13, 20
Rhodophyta		
Audouinella chalybea (Roth) Bory	H	12
A. eugenea (Skuja) C.C.Jao	K, O, M	12, 20
A. pygmaea (Kütz.) Weber Bosse	K, O, H	3, 12, 20
Audouinella sp.	K, O, M	12
**Batrachospermum spermatiophorum* Vis et Sheath	M	12, 16, 20
Compsopogon coeruleus (Balb.) Mont. syn. *Compsopogonopsis leptoclados* (Mont.) Krishnamurthy	K, O, M, H	3, 12, 16, 20
Hildenbrandia angolensis Welw. ex W.West et G.S.West	K, O, M, H	3, 12, 13, 20
Dinophyta		
Hemidinium nasutum F.Stein	O	6, 8
Chrysophyta		
Dinobryon sertularia Ehrenb.	Mo	6, 8, 9
Tribophyta		
Tribonema aequale Pascher	H	12
T. affine (G.S.West) G.S.West	O	20
T. elegans Pasch.	K	12
Tribonema sp.	O	12
Vaucheria sp.	O, M	11, 12

flora (19.5%; Abbott, 1999), the angiosperm flora (89%; Wagner & Funk, 1995), and the insect fauna (99%; Wagner & Funk, 1995) of the Hawaiian Islands.

An examination of the number of stream algal records from each broad taxonomic category, by island, indicates that approximately equal proportions of taxa are represented on each island (Fig. 1). This trend is very similar to that previously observed for the stream macroalgal flora, in that the cyanobacteria and Chlorophyta are most strongly represented, followed by the Rhodophyta, with a much smaller proportion represented by the Tribophyta (Fig. 1; Sherwood, 2006).

The "suspiciously cosmopolitan" nature of the Hawaiian stream algal flora

With an estimated 2.3% level of endemism, and a large number of taxa known from a broad range of geographic regions, the Hawaiian stream algal flora is indeed suspiciously cosmopolitan. This is especially true in light of estimates for other groups of the Hawaiian biota (see above section). The cosmopolitan nature of the freshwater algal flora, however, is tightly linked to the species concept employed for the organisms and how it practically relates to species identification. A morphological species concept is most commonly used for freshwater algae (Ichimura, 1996), and, in the vast majority of cases, the taxonomic boundaries are not tested according to the biological species concept (but see McCourt & Hoshaw, 1990 as an example of a rare exception). This can be explained through reasons of practicality – the smallness of size, and either infrequency or outright lack of sexual reproduction for many taxa means that the biological species concept can not be reasonably applied in such studies.

It has been argued that rare taxa may potentially be mistaken for endemics, for example, if insufficient information is available regarding their general distribution (Tyler, 1996). However, given that only four out of 174 taxa in the Hawaiian stream algal flora are believed to be endemic, this is likely not heavily biasing the estimate of endemism for Hawaiian stream algae, at least in a positive direction.

Quickly evolving lineages (relative to dispersal rates) are likely to be characterized by a high degree of endemism (Hoffmann, 1996). Conversely, truly cosmopolitan taxa can only arise if their rate of speciation is slower than their dispersal rate (Hoffmann, 1996). A case could be made for a cosmopolitan freshwater algal flora, based on the information available regarding their dispersal abilities. In general, freshwater algae are believed to take advantage of a number of dispersal means, including water (e.g., within a watershed), other organisms (vectors such as waterfowl and other birds, insects, and mammals, including humans), or air transport (long-distance dispersal by wind) (Kristiansen, 1996). If freshwater algae, or their propagules and reproductive products, are indeed employing all of the above mechanisms for dispersal on a frequent basis, then it is possible that little opportunity exists for reproductive isolation and subsequent speciation. However, one of the hallmarks of the Hawaiian Archipelago is the extreme geographical isolation of the island chain (Juvik & Juvik, 1998). Although freshwater algae may have a remarkably high dispersal capability, the fact remains that Hawai'i is located more than 3,500 km from the nearest continental landmass, and thus it represents a low-probability destination via the above-described dispersal mechanisms. It is almost inevitable, then, that Hawaiian stream algae are on unique evolutionary trajectories, and the signposts of these trajectories may be most easily revealed using molecular techniques. These evolutionary diversions will be interesting to explore since they may give clues as to how the selective pressures operating on Hawaiian stream algae differ from those acting on algae in other geographical locales.

Insights from molecular data – case studies from the literature

In general, researchers do not test their working definitions of taxa through direct application of the biological species concept. However, a number of published studies have examined the molecular diversity patterns for individual taxa of freshwater algae. Such patterns can aid in the discrimination between taxonomic lineages that are truly widespread in geographical distribution, and those that contain previously unsuspected molecular diversity. Some examples are presented below.

The only known study to date to employ molecular markers for taxonomic investigations of Hawaiian stream algae used DNA sequence analysis of the *rbc*L gene (coding for the large subunit of the Rubisco enzyme) to confirm the endemicity of the freshwater red alga, *Batrachospermum spermatiophorum* Vis et Sheath (Sherwood *et al.*, 2004). This distinctive alga was first described and recognized as endemic to the Hawaiian Islands on the basis of its unique morphology and reproductive anatomy (Vis *et al.*, 1994). In this instance, comparisons of *rbc*L gene sequence data from a Hawaiian representative to other sequences of the well-studied and globally distributed genus *Batrachospermum* (e.g., Vis *et al.*, 1998; Vis & Entwisle, 2000) revealed yet another line of evidence to support the endemic status of this alga.

One of the most convincing cases for the inclusion of molecular genetic assessments of individual taxa stems from a study of a freshwater cyanobacterium. The molecular diversity of the cosmopolitan freshwater alga, *Phormidium retzii* (C.Ag.) Gomont, was recently examined by using random amplified polymorphic DNA (RAPD) markers and DNA sequence analysis of the 16S rDNA gene (Casamatta *et al.*, 2003). On the basis of morphological characteristics, this species is one of the most broadly delimited within the cyanobacteria, and it is also widespread throughout Hawaiian stream systems. The levels of genetic diversity found in *P. retzii* samples from throughout the North American range of the taxon were, in some cases, very high for a single species. For example, sequence similarities for the 16S rRNA sequence data comparisons ranged from 88.4-98.4% (Casamatta *et al.*, 2003). Other studies have reported that 16S rRNA sequence similarities of less than 97% are indicative of comparisons between different species (Stackebrandt & Goebel, 1994) and show that the current circumscription of *P. retzii* likely includes a number of different taxa.

As in the previous cyanobacterial case study, some eukaryotic stream algae have also been demonstrated to possess high molecular diversity that is masked by an inconspicuous morphology. For example, the freshwater red alga, *Hildenbrandia angolensis*, was examined as part of a global survey of the order Hildenbrandiales (Sherwood & Sheath, 2003). *Hildenbrandia angolensis* is common in Hawaiian stream systems on all four main Hawaiian Islands, but is commonly overlooked due to its diminutive, crustose morphology (Sherwood, 2006). Comparisons of both the *rbc*L and 18S rRNA genes for samples throughout its range in North America, as well as a single sample from the Philippines, revealed a large amount of genetic diversity within this taxon, compared with other red algae (Sherwood & Sheath, 2003). However, since the pattern of genetic diversity could not be tied to morphological or anatomical characters, taxonomic changes were not effected. Continued study of other characters of this alga, however, such as life history variations, biochemical features, or fine structure, may yield useful taxonomic features that can be used to further discriminate taxa along lines of molecular variation.

Several members of the volvocacean green algae have also been subjects of biogeographic molecular diversity studies. For example, internal transcribed spacer (ITS) DNA comparisons of isolates from widespread collections of *Gonium pectorale* Mueller indicate that only 7% of the DNA sequence positions were variable - an extremely low value (Coleman *et al.*, 1994; Coleman, 1996). Additionally, no cline was found to significantly correlate with the geographic origin of the samples. Thus, *G. pectorale* may indeed be a cosmopolitan taxon. By comparison, the authors used the same technique to examine 12 worldwide isolates of *Pandorina morum* Bory belonging to one of two morphologically indistinguishable "subspecies" or "syngens", based on interbreeding capabilities (Coleman, 1996). In this case the ITS sequence data suggested that far more DNA sequence diversity was present in *P. morum* than *G. pectorale*, and that syngens are good taxonomic groupings insofar as mating types are more strongly associated than close geographical isolates (Coleman, 1996). Even within the concept of a species, then, it is possible to have morphologically indistinguishable, reproductively isolated lineages with different biogeographic distributions.

Conclusions and Recommendations for Future Research Directions

Based on current understanding of the Hawaiian stream algal flora, most taxa are cosmopolitan in distribution, and very few are endemic to the Hawaiian Islands. However, a true estimate of the endemism of this flora will not be available until molecular methods are employed on a routine basis to confirm or refute the endemic or cosmopolitan nature of the collections. For many studies, however, such as general ecological investigations, a fine taxonomic scale may not be necessary for the kinds of conclusions being drawn. In any study of stream algae, the taxonomic structure used for identification should be noted so that investigators will be aware of the taxonomic limitations of that study.

Molecular techniques for stream algal systematics research have thus far been largely based on comparative DNA sequence analysis. For conspicuous and character-rich taxa, such as the red alga *Batrachospermum spermatiophorum*, such analyses may not be as critical to include, but, for those "less-charismatic" taxa that comprise the majority of the flora; these methods may uncover revealing patterns of genetic diversity. These patterns may correspond to morphological characters provided sufficient time and effort is spent searching for these connections, for example, at the electron microscopic level. In any case, increased use of molecular data in stream algal taxonomic studies will aid our understanding of the breadth of geographic distribution and the evolutionary diversification of these organisms.

Acknowledgments

I thank the many people who have helped in the field over the last several years: H. Juergen Boehmer, Anne Brasher, Wayne Chiasson, James Dasinger, Ron Englund, Mike Fitzsimons, Skippy Hau, Don Heacock, Glenn Higashi, Michael Kido, Darrell Kuamo'o, Charlie MacDonald, Robert Nishimoto, Gernot Presting, David Preston, Sonia Stephens, Morgan Vis, and Reuben Wolff. The Hawai'i Division of Aquatic Resources has been the primary supporter of stream algae research. A.R.S. was partially supported on a Natural Sciences and Engineering Research Council of Canada postdoctoral fellowship during her first two years of stream algae work in Hawai'i (2001–2002).

Literature Cited

(Numbered references are used in Table 1)

Abbott, I.A. 1984. *Limu: An ethnobotanical study of some Hawaiian seaweeds*. National Tropical Botanical Garden, Lawa'i, Kaua'i, Hawai'i. 39 p.

———. 1999. *Marine red algae of the Hawaiian Islands*. Bishop Museum Press, Honolulu, Hawai'i, 477 p.

Anagnostidis, K. & J. Komárek. 1985. Modern approach to the classification system of cyanophytes. 1. Introduction. *Archiv für Hydrobiologie* Supplement **71**: 291–302.

1. **Brand, F.** 1905. Anheftung der Cladophoraceen und über verschiedene polynesische Formen dieser Familie. *Beihefte zum Botanischen Centrablatt* **18**: 165–193.

Casamatta, D.A., M.L. Vis & R.G. Sheath. 2003. Cryptic species in cyanobacterial systematics: a case study of *Phormidium retzii* (Oscillatoriales) using RAPD molecular markers and 16S rDNA sequence data. *Aquatic Botany* **77**: 295–309.

Chong, C. 1996. Effect of flow regimes on productivity in Hawaiian stream ecosystems. *In:* Will stream restoration benefit freshwater, estuarine, and marine Fisheries? Proceedings of the October 1994 Hawai'i Stream Restoration Symposium. *DLNR Technical Report* **96-01**: 152–157.

Coleman, A.W. 1996. Are the impacts of the earth's history discernable in the current distributions of freshwater algae? *Hydrobiologia* **336**: 137–142.

———., **A. Suarez & L.J. Goff**. 1994. Molecular delineation of species and syngens in Volvocacean green algae. *Journal of Phycology* **30**: 80–90.

2. **Drouet, F.** 1968. *Revision of the Classification of the Oscillatoriaceae*. Academy of Natural Sciences of Philadelphia, Monograph **15**, 370 p.

3. **Filkin, N.R., A.R. Sherwood & M.L. Vis**. 2003. Stream macroalgae of the Hawaiian Islands: 23 additional collection sites. *Pacific Science* **57**: 421–431.

Fungladda, N., I. Kaczmarska & S.R. Rushforth. 1983. A contribution to the freshwater diatom flora of the Hawaiian islands. *Bibliotheca Diatomologica* **2**: 1–103.

Geitler, **L**. 1932. Cyanophyceae. *In*: *Rabenhorst's kryptogamen-flora von Deutschland, Österreich und der Schweiz.* Vol. 14. Akademie Verlagsgeschichte, Leipzig. 1196 p.

Hoffmann, **L**. 1996. Geographic distribution of freshwater blue-green algae. *Hydrobiologia* **336**: 33–40.

Ichimura, **T**. 1996. Genome arrangement and speciation in freshwater algae. *Hydrobiologia* **336**: 1–17.

4. **Islam**, **A**.**K**.**M**.**N**. 1961. The genus *Cloniophora* Tiffany. *Revue Algologique* **6**: 7–32.

Juvik, **S**.**P**. & **J**.**O**. **Juvik**. 1998. *Atlas of Hawai'i.* 3rd Edition. University of Hawai'i Press, Honolulu. 333 pp.

5. **Kido**, **M**.**H**. 1997. Food relations between coexisting native Hawaiian stream fishes. *Environmental Biology of Fishes* **49**: 481–494.

Kristiansen, **J**. 1996. Dispersal of freshwater algae – a review. *Hydrobiologia* **336**: 151–157.

LaPerriere, **J**.**D**. 1995. Riffle algal ecology of small streams during the rainy season: islands of Hawai'i, Maui and O'ahu. *Tropical Ecology* **36**: 59–72.

6. **Lemmerman**, **E**. 1905. Die Algenflora der Sandwich-Inseln. Ergebnisse einer Reise nach dem Pacific. H. Schauinsland 1896/97. *Englers Botanische Jahrbücher für Systematik, Pflanzengeschichte, und Pflanzengeographie* **34**: 607–632.

7. **MacCaughey**, **V**. 1917. The phytogeography of Manoa Valley, Hawaiian Islands. *American Journal of Botany* **4**: 561–603.

8. ———. 1918a. Algae of the Hawaiian Archipelago I. *Botanical Gazette* **65**: 42–57.

9. ———. 1918b. Algae of the Hawaiian Archipelago II. *Botanical Gazette* **65**: 121–149.

McCourt, **R**.**M**. & **R**.**W**. **Hoshaw**. 1990. Noncorrespondence of breeding groups, morphology, and monophyletic groups in *Spirogyra* (Zygnemataceae, Chlorophyta) and the application of species concepts. *Systematic Botany* **15**: 69–78.

10. **Nordstedt**, **C**.**F**.**O**. 1876. De Algae aquae dulcis et de Characeis ex insulis Sandvicensibus a Sv. Berggren 1875 reportatis. *Kongliga Fysiografiska Sällskapet i Lund* **7**: 1–24.

Sheath, **R**.**G**. & **K**.**M**. **Cole**. 1992. Biogeography of stream macroalgae in North America. *Journal of Phycology* **28**: 448–460.

11. **Sherwood**, **A**.**R**. 2004. New records of freshwater macroalgae and diatoms from the Hawaiian Islands. *Bishop Museum Occasional Papers* **79**: 1–8.

———. 2004. Bibliographic checklist of the nonmarine algae of the Hawaiian Islands. *Bishop Museum Occasional Papers* **80**: 1–23.

12. ———. 2006. Stream macroalgae of the Hawaiian Islands: a floristic survey. *Pacific Science* **60**: 191–205.

13. ———. & **M**.**H**. **Kido**. 2002. Watershed-scale comparisons of algal biodiversity in high quality proximate Hawaiian stream ecosystems. *Pacific Science* **56**: 431–440.

14. ———., **M**.**H**. **Kido** & **C**.**W**. **Morden**. 2001a. New records of freshwater macroalgae from the Hawaiian Islands. 1. Chlorophyta. *Bishop Museum Occasional Papers* **74**: 1–3.

15. ———., **M**.**H**. **Kido** & **C**.**W**. **Morden**. 2001b. New records of freshwater macroalgae from the Hawaiian islands. 2. Cyanophyta. *Bishop Museum Occasional Papers* **74**: 3–5.

———. & **R**.**G**. **Sheath**. 2003. Systematics of the Hildenbrandiales (Rhodophyta): gene sequence and morphometric analyses of global collections. *Journal of Phycology* **39**: 409–422.

16. ———., **M**.**L**. **Vis**, & **R**.**G**. **Sheath**. 2004. Phenology and phylogenetic positioning of the Hawaiian endemic freshwater alga *Batrachospermum spermatiophorum* (Rhodophyta, Batrachospermales). *Phycological Research* **52**: 193–203.

Stackebrandt, **E**. & **B**.**M**. **Goebel**. 1994. Taxonomic note: a place for DNA-DNA reassociation and 16S rRNA sequence analysis in the present species definition in bacteriology. *International Journal of Systematic Bacteriology* **44**: 846–849.

17. **Tilden**, **J**.**E**. 1901. Algae collecting in the Hawaiian Islands. *Postelsia* **1**: 135–175.

18. ———. 1902. Collection of algae from the Hawaiian Islands. *[Thrum's] Hawaiian Annual* **1902**: 106–113.

19. ———. 1910. Minnesota Algae, Volume 1. The Myxophyceae of North America and adjacent regions including Central America, Greenland, Bermuda, the West Indies and Hawaii. *Report of the Survey Botanical Series VIII.* Minneapolis, Minnesota. 328 p.

Tyler, P.A. 1996. Endemism in freshwater algae – with special reference to the Australian region. *Hydrobiologia* **336**: 127–135.

Vis, M.L. & T.J. Entwisle. 2000. Insights into the phylogeny of the Batrachospermales (Rhodophyta) from *rbc*L sequence data of Australian taxa. *Journal of Phycology* **36**: 1175–1182.

———., **G.W. Saunders, R.G. Sheath, K. Dunse & T.J. Entwisle.** 1998. Phylogeny of the Batrachospermales (Rhodophyta) inferred from *rbc*L and 18S ribosomal DNA gene sequences. *Journal of Phycology* **34**: 341–350.

20. ———., **R.G. Sheath, J.A. Hambrook & K.M. Cole.** 1994. Stream macroalgae of the Hawaiian Islands: a preliminary study. *Pacific Science* **48**: 175–187.

Wagner, W.L. & V.A. Funk. 1995. Preface, p. ix–xi. *In*: Wagner, W.L. & V.A. Funk (eds.), *Hawaiian biogeography: evolution on a hot spot archipelago*. Smithsonian Institution Press, Washington, D.C. and London.

21. **Wille, N.** 1915. Süßwasseralgen von den Samoainseln, Hawaii, den Salomonsinseln und Ceylon, gesammelt von Dr. K. Rechinger. *Denkschriften der Kaiserlichen Akademie der Wissenschaften* **91**: 141–213.

22. **Wood, R.D. & K. Imahori.** 1965. *A revision of the Characeae.* Volume 1. J. Cramer, Weinheim. 904 pp.

Zeigler, A.C. 2002. *Hawaiian natural history, ecology, and evolution*. University of Hawai'i Press, Honolulu. 477 p.

Biology of Hawaiian Streams and Estuaries. Edited by N.L. Evenhuis & J.M. Fitzsimons. *Bishop Museum Bulletin in Cultural and Environmental Studies* 3: 207–232 (2007).

Aquatic Insect Taxa as Indicators of Aquatic Species Richness, Habitat Disturbance, and Invasive Species Impacts in Hawaiian Streams[1]

Ronald A. Englund
Hawaii Biological Survey, Bishop Museum, 1525 Bernice Street, Honolulu, Hawai'i 96817, USA: email: englund@bishopmuseum.org

Mark G. Wright
Department of Plant and Environmental Protection Sciences, University of Hawai'i at Mānoa, Honolulu, Hawai'i, 96822, USA

Dan A. Polhemus[2]
Department of Entomology, National Museum of Natural History, Smithsonian Institution, Washington, D.C. 20705, USA

Abstract

In this study we provide a synthesis of numerous stream assessments in the Hawaiian Islands that began in the early 1990s and have continued to the present. Data from numerous sites within the five major high Hawaiian Islands with flowing streams (excluding Lāna'i, which lacks flowing waters) were used to assess native and introduced aquatic insect communities, the impacts of various invasive freshwater species and the threats from habitat disturbance. The primary objective of this study was to provide the first comprehensive analysis of aquatic insect populations in various urbanized and virtually pristine stream reaches on the five major Hawaiian Islands, and to assess if various suites of introduced aquatic species may be impacting aquatic insect populations.

We were also interested in assessing the suitability of native aquatic insects as key indicator, flagship, or umbrella species regarding the overall health of Hawaiian aquatic ecosystems. If key indicator species can be found, then aquatic habitats with high native biodiversity can be identified and management efforts can be made to ensure this high level of biodiversity persists. These indicator species could also be used for monitoring future rehabilitation programs on disturbed streams.

Introduction

Detailed distribution and abundance data for invertebrates such as aquatic insects are lacking for most tropical regions, and this lack of basic knowledge hinders the development of conservation planning efforts. The Hawaiian Islands are an exception to this rule because of a long history of entomological collections starting in the 1800s, and the infrastructure of a major museum and large university in close proximity to a wide range of aquatic habitats. Because of its extreme isolation, Hawai'i has the greatest percentage of unique fauna in the world with an estimated 98% endemicity rate for the 5,368 described insect species (Eldredge & Evenhuis, 2003). Most research efforts in the Hawaiian Islands have been focused on the amazing adaptive radiations and ecological adaptations found within the terrestrial insect fauna, with far fewer resources devoted to studying insects found within freshwater habitats. In aquatic systems, the insect group historically receiving the greatest attention has been the Odonata (damselflies and dragonflies), with other taxa such as aquatic flies (Diptera) or true bugs (Heteroptera) being assessed at various levels of intensity. While most

1. Contribution 2007-004 to the Hawaii Biological Survey.
2. Present Address: Hawaii State Department of Land and Natural Resources, Division of Aquatic Resources, Honolulu, Hawai'i, USA.

of the early research involved taxonomic descriptions of new species, some early pioneers such as F.X. Williams conducted life history and basic ecological studies on the Hawaiian aquatic insect fauna (Williams, 1936).

Although life history and limited ecological studies have been conducted on a small number of Hawaiian aquatic insect species, this study is the first to examine broad scale patterns of entire communities found within individual watersheds, islands, or different islands. While various authors have demonstrated the impacts of specific introduced aquatic species on native Hawaiian freshwater species (Englund & Polhemus, 2001; Englund, 1999; Font, 1998; Font & Tate, 1994), a quantitative examination of the potential suitability of different aquatic insect taxa as indicator species representing the ecological health of a particular Hawaiian aquatic ecosystem has not previously been attempted. For the purposes of this study we define ecological health as an intact Hawaiian watershed containing greater numbers of native species than an urbanized and highly disturbed watershed.

In this study we provide a synthesis of numerous stream assessments in the Hawaiian Islands that started in the early 1990s and continue to the present. Data from numerous sites within 5 of the major high Hawaiian Islands with flowing streams (excluding Lāna‘i) were used to assess native and introduced aquatic insect communities, the impacts of various invasive freshwater species and the threats from habitat disturbance (see Figs. 1–5 for site maps).

The primary objective of this study was to provide the first comprehensive analysis of aquatic insect populations in various urbanized and virtually pristine stream reaches on the five major Hawaiian Islands, and to assess how various suites of introduced aquatic species may be impacting these aquatic insect populations. Additionally, given that one of the major goals for conservation biologists is maintaining biodiversity in highly endemic areas such as in Hawaii, we were also interested in assessing the suitability of native aquatic insects as key indicator, flagship, or umbrella species regarding the overall health of Hawaiian aquatic ecosystems. If key indicator species can be found, then aquatic habitats with high native biodiversity can be more readily identified and management efforts can be undertaken to ensure this high level of biodiversity persists. These indicator species can also be used for monitoring future rehabilitation programs on disturbed streams.

Many species of native Hawaiian aquatic insects are now threatened with extinction because of reduced ranges resulting from habitat loss and invasive species (Liebherr & Polhemus, 1997; Englund, 1999, 2001, 2002). Preserves for threatened and endangered species are often designed to protect habitats that permit the maximum number of species to be conserved, often by using surrogate species that are believed to represent the needs of other threatened species using the same habitat (Simberloff, 1998; Andelman & Fagan, 2000; Rubinoff, 2001). Three classes of surrogate species have been identified and include: (1) flagship species, or charismatic species attracting public support, (2) umbrella species, or species requiring large areas of habitat needing protection thereby also providing protection for other species, and (3) biodiversity indicators, or species whose presence indicates areas with high species richness (Andelman & Fagan, 2000).

In the present study we make the first attempt to assess the sensitivity of both native fish and aquatic insect species to introduced species and to other major watershed perturbations such as diversions or concrete channelization. This was done by collecting from a wide variety of aquatic insect habitats ranging from heavily urbanized and channelized streams, to pristine sections of watersheds accessible only by helicopter. A holistic evaluation of Hawaiian streams requires not only the assessment of the five native species of freshwater fish and several large species of easily observed invertebrates (i.e., crustaceans), but also the 300–400 estimated species of native Hawaiian aquatic insects. Unlike aquatic vertebrates, many aquatic insects have narrow habitat tolerances meaning they can only live in certain flowing water microhabitats, for example seeps or cascade splash-zones. These narrow habitat preferences also increase the vulnerability of aquatic insects to stream disturbances such as stream channelization, dewatering, sedimentation, and alien species introductions. Because native Hawaiian aquatic insects are much less flexible in their habitat requirements than aquatic vertebrates, it then follows that insects may provide a better monitoring and stream assessment tool than vertebrates. Stream macrofauna such as the native fish, crustaceans, and neritid snails are migratory and are not necessarily co-evolved to a specific stream system, unlike many Hawaiian aquatic insects. This study therefore makes a first attempt at integrating the various factors that appear to be presently limiting the distributions of native aquatic insects in Hawaii, or factors that make habitats suitable for the survival of endemic species.

Materials and Methods

Streams on Kauai, O'ahu, Moloka'i, Maui, and Hawai'i islands (Table 1, Figs. 1–5) were surveyed for both native and introduced species in a wide range of aquatic habitats, ranging from coastal lowlands at sea level to high elevation reaches only accessible by helicopter, thus covering the entire gradient of habitats available in the islands. The highest elevation sampled in a particular stream reach was recorded and determined with a combination of USGS topographic maps and handheld altimeters. Efforts were made to standardize insect collections at each sample site as similar habitats and collecting techniques were used at each station.

Aquatic Insects

Collections of both immature and adult specimens were made with yellow pan traps, aerial sweep nets, aquatic dip nets, kick-netting, and Surber (benthic) samplers around all aquatic habitats at each study site. Visual observations of aquatic insects were also conducted above and around the stream. Sampling of damselflies and dragonflies (Odonata) was also emphasized, because six Hawaiian species are currently considered Candidate Species by the U.S. Fish & Wildlife Service.

Benthic sampling centered on kick-netting and involved vigorously disturbing the substrate upstream of a fine meshed aquatic net to displace any aquatic invertebrates inhabiting the stream substrate. The use of frequent kick-netting allowed for a greater sample size and resulted in increased effort for invertebrate collections. Benthic sampling also included collecting individual variously sized rocks and then using a toothbrush or forceps to remove immature insects. Above and below water visual observations for aquatic insects were also conducted as we hiked between sampling stations. Sampling effort was focused on all suitable aquatic habitats such as splash zones around riffles and cascades, wet rock faces associated with springs and seeps, waterfalls, nearby wetland areas associated with the streams, and variously-sized stream substrates. All aquatic habitats were sampled. All insect specimens were stored in 95% ethanol for curation and identification and voucher specimens are currently housed in the Bishop Museum and Smithsonian Institution collections.

Freshwater Fish, Introduced Crustaceans, and Amphibians

One of the primary objectives of this study was to assess where specific suites of aquatic organisms have been introduced into a particular Hawaiian watershed. Thus, observations and limited collections of freshwater fish, crustaceans, and amphibians were undertaken to verify species identities. Fish and introduced crustaceans and amphibians were either collected with nets and hand seines, or identified underwater while snorkeling. Many of these aquatic insect surveys were jointly conducted with biologists from the Hawaii Division of Aquatic Resources (HDAR) assessing native and introduced fish populations, thus we have integrated the results of their findings with our aquatic insect findings. HDAR fish collection data was accessed from their stream survey website at: [http://www.hawaii.gov/dlnr/dar/streams/stream_data.htm].

Statistical Analysis

Multiple-species data are notoriously difficult to analyze in a clear and meaningful manner. Multivariate statistical analysis of community data offer a means of detecting patterns in similarity of species composition of sample sites, and a means of identifying species associated with specific environmental conditions. Canonical correspondence analysis is an analytical method that can be used to unravel patterns in complex ecological data sets (Leps & Smilauer, 2003).

Presence/absence data for the insect species was subjected to canonical correspondence analysis (CCA), a direct gradient analysis method, which summarizes relationships between response variables (in this case, insect species assemblages in 39 study sites) and environmental variables

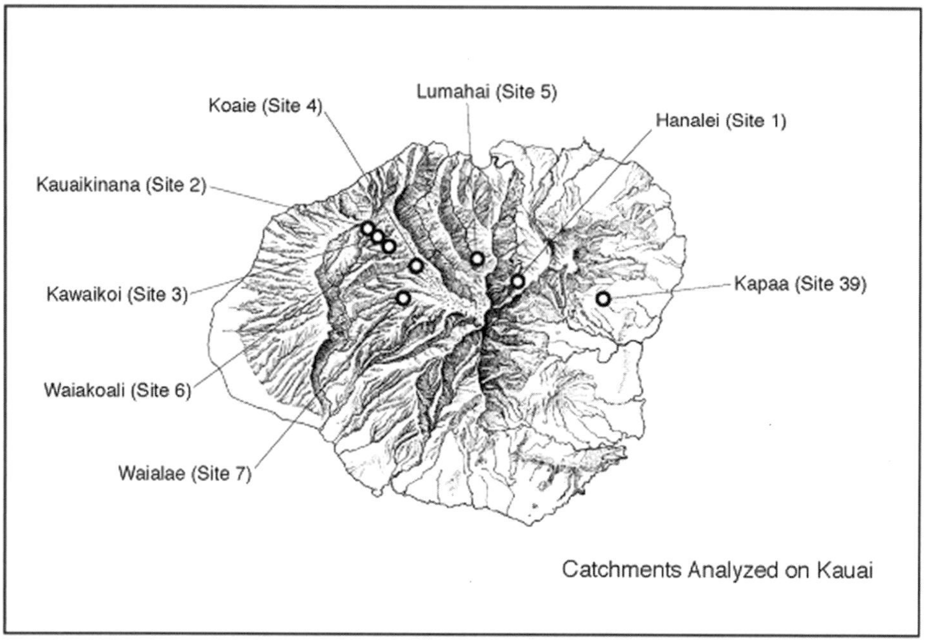

Figure 1. Streams sampled for aquatic biota during this study on the island of Kaua'i.

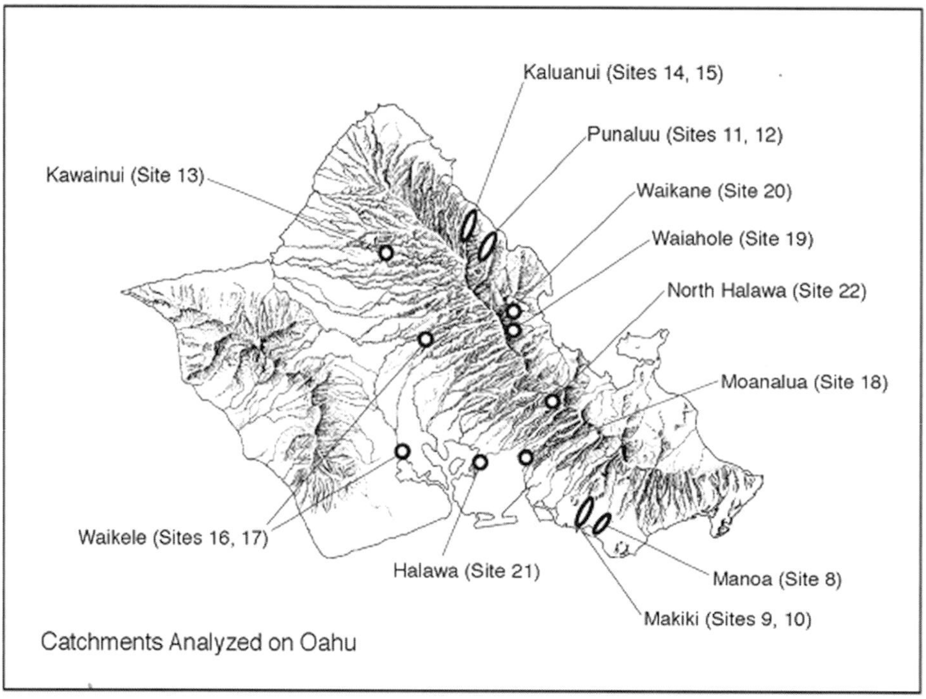

Figure 2. Streams sampled for aquatic biota during this study on the island of O'ahu.

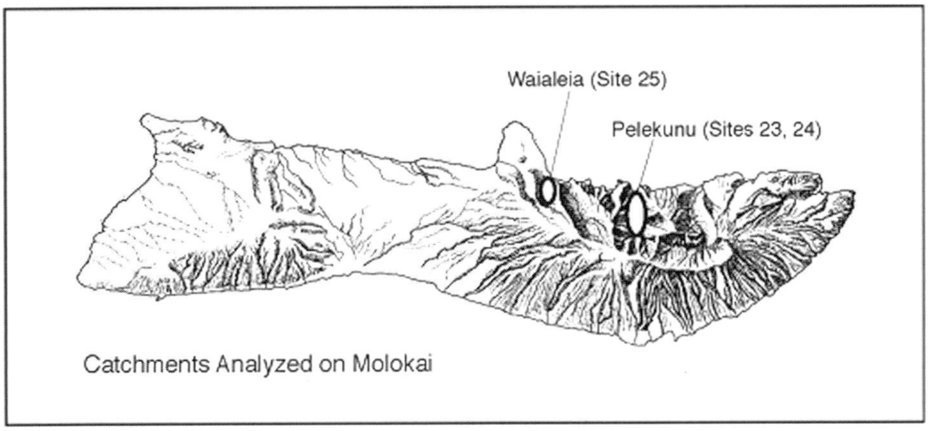

Figure 3. Streams sampled for aquatic biota during this study on the island of Moloka'i.

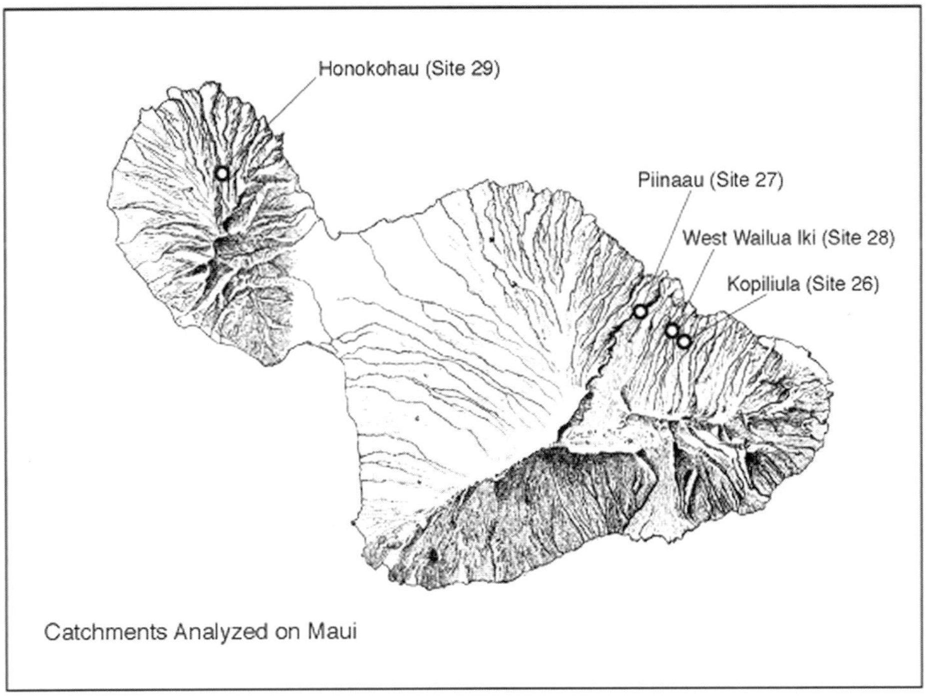

Figure 4. Streams sampled for aquatic biota during this study on the island of Maui.

(Leps & Smilauer, 2003). The analyses were conducted using CANOCO 4.5 and CanoDraw software (Ter Braak & Smilauer, 2002). CANOCO performs multivariate ordination on species data, calculating chi-square distance between samples, and plotting sample and species scores these on canonical (constrained) axes, determined by correlations between specified environmental variables and species scores. Plots of ordinations are generated by CanoDraw (Ter Braak & Smilauer, 2002).

The ordinations were initially done for all species, and then broken down by insect family. Families were analyzed separately as each has different ecological characteristics, and meaningful

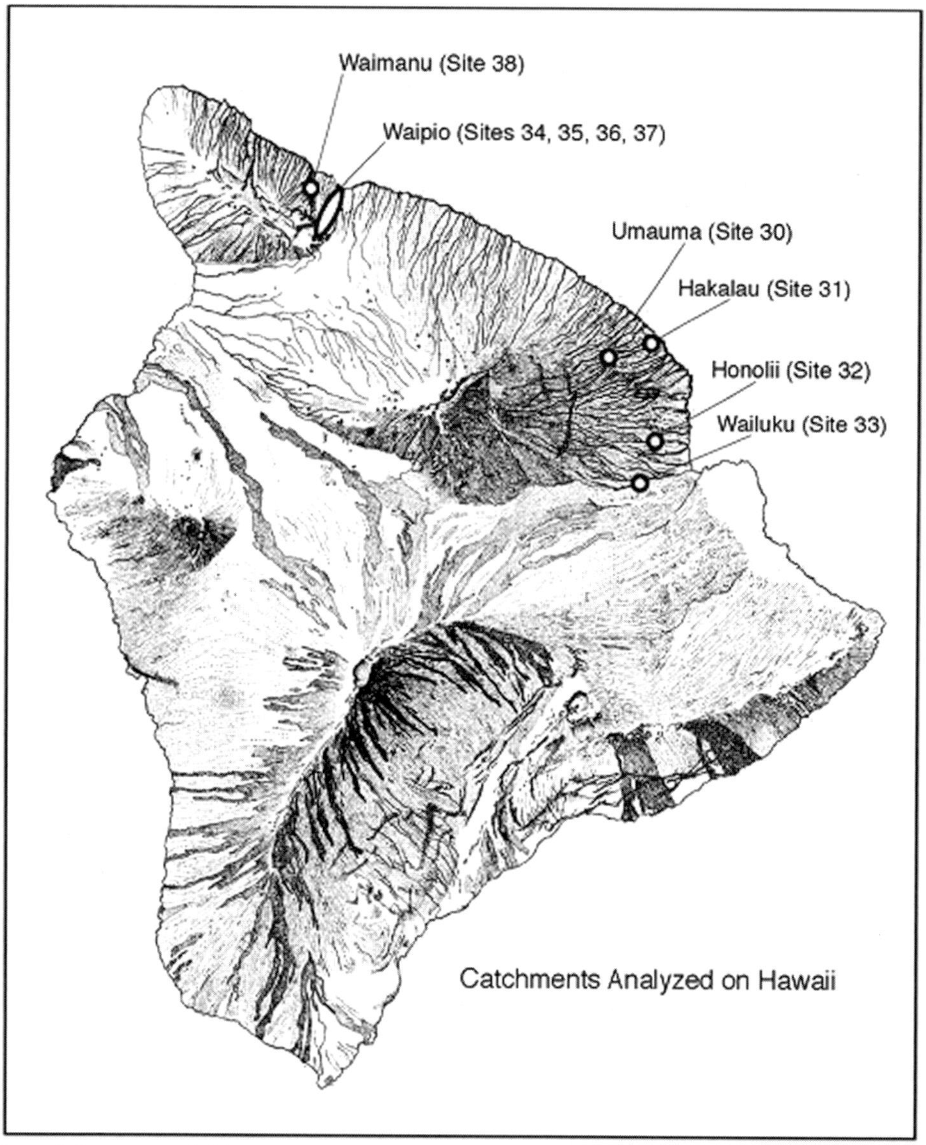

Figure 5. Streams sampled for aquatic biota during this study on the island of Hawai'i.

graphical analyses could be presented with the reduced data sets. Environmental variables that were selected were: island (coded as 1–5, Kaua'i =1; Hawai'i = 5); elevation (m.a.s.l.); type of stream (coded as 1 = undiverted, not channelized; 2 = concrete channel; 3 = channelized no concrete; 4 = diverted, below diversion but not channelized); presence or absence of indigenous and exotic fish species; and presence or absence of exotic frogs. Exotic fish species and frogs were included as environmental variables because they may impact indigenous insects negatively, or in some cases, they may be associated with either positive or negative environmental conditions that are suitable for certain communities of aquatic insects. The former situation is the case for poeciliid fish that are often the only fish species found in concrete channelized Hawaiian streams, while the latter situation is

Table 1. Island, stream sampling sites, sampling date, and stream type assessed for native and introduced aquatic insects, fishes, crustaceans, and amphibians in the Hawaiian Islands. The Hawaii Division of Aquatic Resources stream database was also consulted for fish species composition. Stream type at sample reach: 1 = undiverted, not channelized, 2 = concrete channel, 3 = channelized no concrete, 4 = diverted, below diversion.

Site No.	Stream (elevation surveyed- m)	Stream Type	Date(s) Sampled	Reference
Kaua'i				
1	Hanalei (380 m)	1 (Undiverted)	Nov 1994	Polhemus (1995)
2	Kauaikinana (1035 m)	1 (Undiverted)	Aug 1997, Aug 1998, Jan 1999	Englund and Polhemus (2001)
3	Kawaikoi (1035 m)	1 (Undiverted)	Aug 1997, Aug 1998, Jan 1999	Englund and Polhemus (2001)
4	Koaie (1175 m)	1 (Undiverted)	Aug 1997, Jan 1999, July 2000	Englund and Polhemus (2001)
5	Lumahai (430 m)	1 (Undiverted)	Nov 1994	Polhemus (1995)
6	Waiakoali (1035 m)	1 (Undiverted)	Aug 1997, Aug 1998, Jan 1999	Englund and Polhemus (2001)
7	Waialae (1095 m)	1 (Undiverted)	Jan 1999	Englund and Polhemus (2001)
O'ahu				
8	Manoa (5-75 m)	2 (Concrete Channel)	April 2004	Englund and Arakaki (2004)
9	Makiki (2 m)	2 (Concrete Channel)	April 2004	Englund and Arakaki (2004)
10	Makiki (45 m)	2 (Concrete Channel)	April 2004	Englund and Arakaki (2004)
11	Punaluu (31 m)	1 (Undiverted)	Nov 2002	Englund *et al.* (2003a)
12	Punaluu (275 m)	1 (Undiverted)	Nov 2002	Englund *et al.* (2003a)
13	Kawainui (Anahulu trib) (305 m)	1 (Undiverted)	Apr 2003	Englund *et al.* (2003a)
14	Kaluanui (100 m)	1 (Undiverted)	Nov 2002	Englund *et al.* (2003a)
15	Kaluanui (762 m)	1 (Undiverted)	Jan 1994	Englund *et al.* (2003a)
16	Waikele (0-1 m)	3 (Channelized, no concrete)	Mar 1993, Dec 1997, Aug 1998	Englund and Filbert (1999), Englund *et al.* (2000)
17	Waikele (381 m)	1 (Undiverted)	Mar 1993	Englund (1993)
18	Tripler Stream (79 m)	1 (Undiverted)	Mar 1995-Jan 2005	Evenhuis *et al.* (1995), Englund 2001
19	Waiahole (Waianu trib) (60 m)	1 (Undiverted)	Feb-Aug 1995, Feb, May, Nov 2002, Apr 2003	Filbert and Englund (1995), Englund *et al.* (2003b)
20	Waikane (210 m)	1 (Undiverted)	Feb-Aug 1995, Feb, Nov 2002, Mar 2003	Filbert and Englund (1995), Englund *et al.* (2003b)
21	Halawa (0-1 m)	2 (Concrete Channel)	Nov, Dec 1997, Mar-Aug 1998	Englund *et al.* (2000)
22	N. Halawa (300 m)	1 (Undiverted)	Jan 1991-Feb 1994	Polhemus (1994)

Table 1. (continued)

Site No.	Stream (elevation surveyed- m)	Stream Type	Date(s) Sampled	Reference
Moloka'i				
23	Pelekunu (0-1 m)	1 (Undiverted)	Jan 1991, Apr 2000, May 2001, May 2002	Englund & Arakakii (2003)
24	Pelekunu (182-237 m)	1 (Undiverted)	Apr 2000, May 2001, May 2002	Englund & Arakaki (2003)
25	Waialeia (0-60 m)	1 (Undiverted)	Nov 1998	Englund, unpublished data
Maui				
26	Kopiliula (610 m)	1 (Undiverted)	Jan 2003	Englund *et al.* (2003a)
27	Piinaau (731 m)	1 (Undiverted)	Jan 2003	Englund *et al.* (2003a)
28	W. Wailua Iki (493)	1 (Undiverted)	Jan 2003	Englund *et al.* (2003a)
29	Honokohau (450 m)	1 (Undiverted)	Jan 2003	Englund *et al.* (2003a)
Hawai'i				
30	Umauma (713 m)	1 (Undiverted)	Mar 2003	Englund *et al.* (2003a)
31	Hakalau (0-10 m)	1 (Undiverted)	Dec 1993	Polhemus (1995)
32	Honolii (536-640 m)	1 (Undiverted)	Feb 2002, Mar 2003	Englund *et al.* (2002), Englund *et al.* (2003a)
33	Wailuku (670 m)	1 (Undiverted)	Mar 2003	Englund *et al.* (2003a)
34	Wailoa (Waipio) (0-1 m)	4 (Diverted)	Mar 2001, 2003-2005 (quarterly)	Englund *et al.* (2001)
35	Hiilawe (Waipio) (15 m)	1 (Undiverted)	Mar 2003-2005 (quarterly)	Englund *et al.* (2001)
36	Wailoa (Waipio) (190 m)	4 (Diverted)	Oct 1996, Nov 1998	Englund & Filbert (1997), Englund & Preston (1999)
37	Kawainui (Waipio) (425 m)	1 (Undiverted)	Oct 1996	Englund & Filbert (1997)
38	Waimanu (90 m)	1 (Undiverted)	Dec 1998	Englund & Preston (1999)
Kaua'i				
39	Kapa'a (80-120 m)	1 (Undiverted)	Nov 1994	Polhemus (1995)

also true for the indigenous fish species included. Their association with certain insect species, demonstrated by their correlation as "environmental variables" may serve as a surrogate for true environmental variables that were not directly measured. For each analysis, environmental variables that significantly affected variation in community structure were selected by Monte Carlo simulation (499 permutations), with the six best predictors selected automatically by CANOCO.

The results of these analyses provide extensive information about the communities analyzed (Leps & Smilauer, 2003). Sample sites are arranged in space (the ordination) based on similarity of insect communities; species are similarly arranged, and their proximity to sample sites and other species in the ordination are indicative of their association with sites, and other species. Environmental variables are plotted as vectors on the ordinations, each indicating the relative contribution it makes toward defining each axis plotted. The longer the vector, the greater the effect is has in explaining an environmental gradient; the smaller the angle between a vector and an axis, the more closely correlated that variable is with the gradient of points plotted. Finally, canonical correspondence axes (CCA) can be viewed as linear combinations of environmental variables along which insect community data are plotted according to similarity of species composition.

A primary objective of this study was to determine what insect species are typically associated with pristine or disturbed habitats. The availability of presence / absence data for fish and frogs in these habitats allowed us furthermore to assess the contribution that they might have on shaping insect communities, and also to determine whether any are specifically associated with pristine habitats.

Results

A list of species collected during this study can be found in the Appendix, along with a code number for each species as shown in Figs. 6–13. The ordination of all species in the data set, from all sites, showed that there were patterns along gradients, but these could not be clearly explained from the full data set (which produced a complex graph with many overlaid points, and no distinct patterns). To better understand the patterns within the data, each family of native insects was analyzed separately. The cumulative percentage of variation on species composition and species environment relationship for families and selected genera is shown in Table 2. Higher variance and species environment relationship accounted for with large, diverse taxonomic assemblages (e.g., Dolichopodidae) provide more robust and meaningful results when compared to smaller taxonomic assemblages such as Telmatogeton.

Coenagrionidae

Figure 6 shows the ordination for the native Coenagrionidae (*Megalagrion*) and the three introduced damselfly species (numbered 16, 17, and 18 on Fig. 6), where 84.5% of the species-environment relation was explained (Table 2) by the first three correspondence axes. This ordination clearly defines the sample sites along a gradient defined by "island" and "elevation". Elevation was autocorrelated with stream type, and they are thus largely functionally equivalent in these analyses. The Kauai samples were grouped in a clearly defined cluster along CCA1, with the Hawai'i samples at the other end of that spectrum, for a loosely defined cluster. O'ahu, Moloka'i and Maui are distributed along the gradient (Fig. 6). The second axis (CCA 2) was defined by introduced Mexican molly *Poecilia mexicana*, and the introduced bullfrog *Rana catesbeiana* and elevation. This may be interpreted as the Kauai sites being associated with highest elevation and absence of *P. mexicana* and *R. catesbeiana*; clearly there is a negative correlation of the presence of these two species and the absence of native *Megalagrion* damselflies. Of particular interest was the fact that the three introduced damselfly species *Ischnura posita*, *Ischnura ramburii*, and *Enallagma civile*, (numbers 16-18 on Fig. 6) were also closely clustered around axis of the introduced *P. mexicana* and the disturbed streams and sites associated with this fish species. These results were also encouraging as it indicates that our CCA results were sensitive at delineating communities of introduced taxa, even though all of these introduced damselflies are commonly caught with native *Megalagrion* damselflies.

It is interesting to note that O'ahu streams at elevations lower than 200 m.a.s.l. were clustered

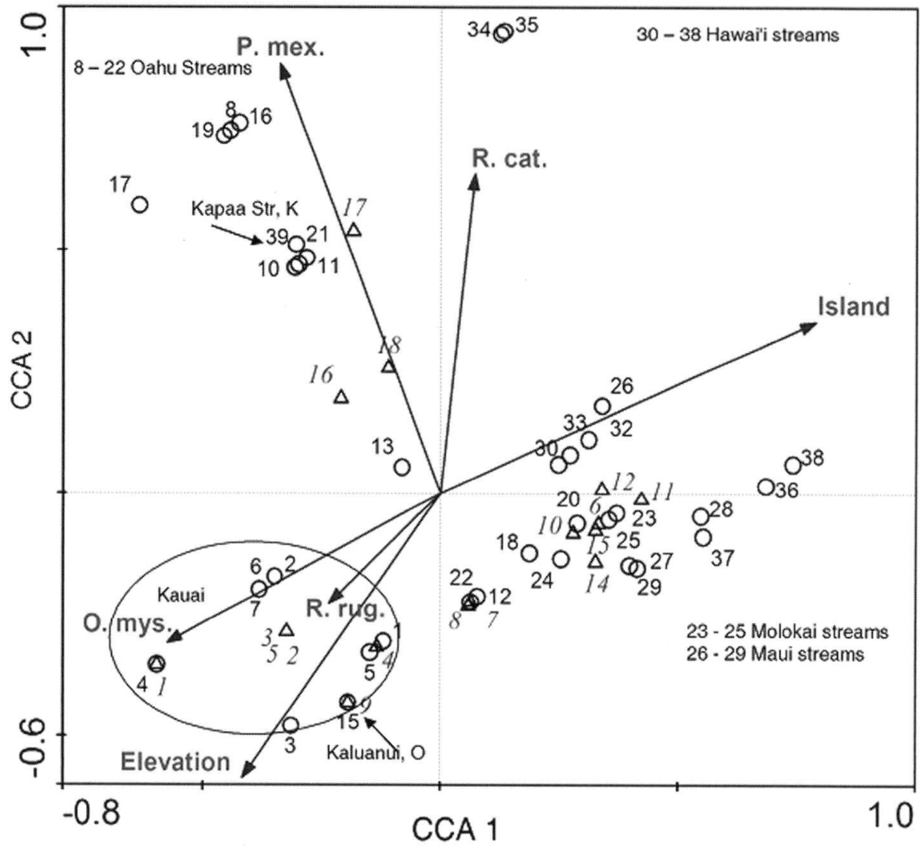

Figure 6. Canonical correspondence (CCA) ordination of sites and species-environment relationship for native Coenagrionidae (*Megalagrion*).

in a distinct group, and that Kapa'a Stream, Kaua'i (site 39) was grouped with them, rather than with the other Kaua'i streams. In contrast, Kaluanui Stream (O'ahu, 762 m.a.s.l.) was grouped with the Kaua'i sample sites (Fig. 6). Indeed, most higher elevation sites from O'ahu, such as Kawainui (upper Anahulu) (site 13), Waikane (site 20), and North Halawa (site 22), had greater similarity with less disturbed islands than low elevation sites on O'ahu. A number of species such as *Megalagrion eudytum*, *M. heterogamias*, *M. oresitrophum*, *M. orobates*, and *M. vagabundum* (all Kauai endemics) were closely associated with the pristine sites on Kaua'i.

Dolichopodidae

"Island" and elevation were the major determinates of CCA 1 for these aquatic flies (Fig. 7) with the high elevation Kaua'i sites forming a distinct group, O'ahu also distinct, and the other islands showing a spread along the axis. CCA 2 was largely a function of the presence of indigenous fish, depending on their presence or absence. While not as high damselflies, 65% of the cumulative variance in species composition (Table 2) was explained by the first three correspondence axes. This ordination clearly identifies Dolichopodidae as being effective indicators of stream quality, for example, there are certain species associated with the Kaua'i sites that could perform such a function. What was especially striking, was the occurrence of indigenous fish species being correlated with certain Dolichopodidae species, particularly for the Moloka'i, Maui, and Hawai'i sample sites. Of great

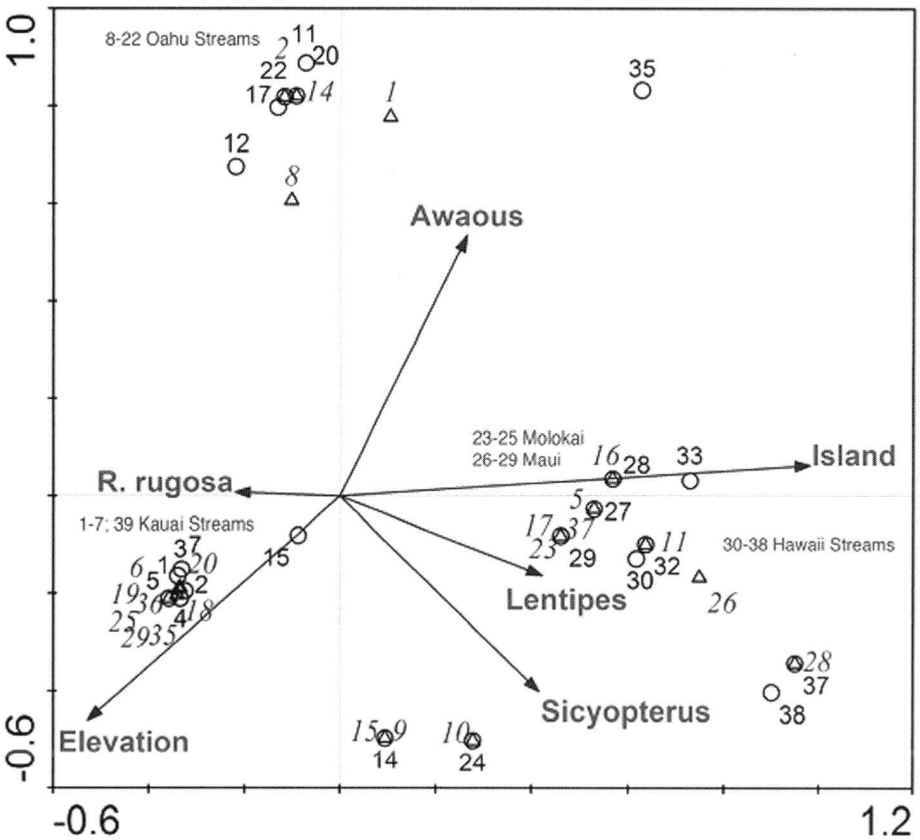

Figure 7. Native Dolichopodidae (all taxa) sites and species-environment relationship using CCA (CCA1 vs. CCA2).

interest was that the top three fish species associated with the Dolichopodidae were the 3 native stream species; *Lentipes concolor*, *Awaous guamensis*, and *Sicyopterus stimpsoni* (Figure 7).

Chironomidae

The ordination for this family of aquatic flies provided 70.4% explanation of variability by the first two axes, with all environmental variables retained (Fig. 8, Table 2). The resolution of this ordination is relatively high; however, it shows strong associations of these flies with alien taxa (Fig. 8). If the analysis was reduced to only the genus *Telmatogeton*, and excluding crustaceans as environmental variables, yet adding indigenous fish, the ordination (Fig. 9) shows clear separation of samples by "island", and strong associations of *Telmatogeton abnormis*, *T. fluviatilis*, *T. hirtus*, and *T. williamsi* with indigenous fish species (e.g. *Lentipes concolor*).

Ephydridae

The ordination (CCA1 vs. CCA2) (Fig. 10) was severely skewed by sample site 21 (Halawa Stream at Pearl Harbor) and the native *Atissa oahuensis* (species 99) (Fig. 11); CCA 2 and 3 accounted for 83.1% of the variation (Table 2), with the samples forming groups defined primarily by "island". However, it would appear that the Ephydridae may be less responsive to the environmental variables we examined. Site 39 (Kapa'a Stream, an impacted, low elevation stream), for example, is included with other Kaua'i sites, which from an overall indicator species perspective offers little in terms of

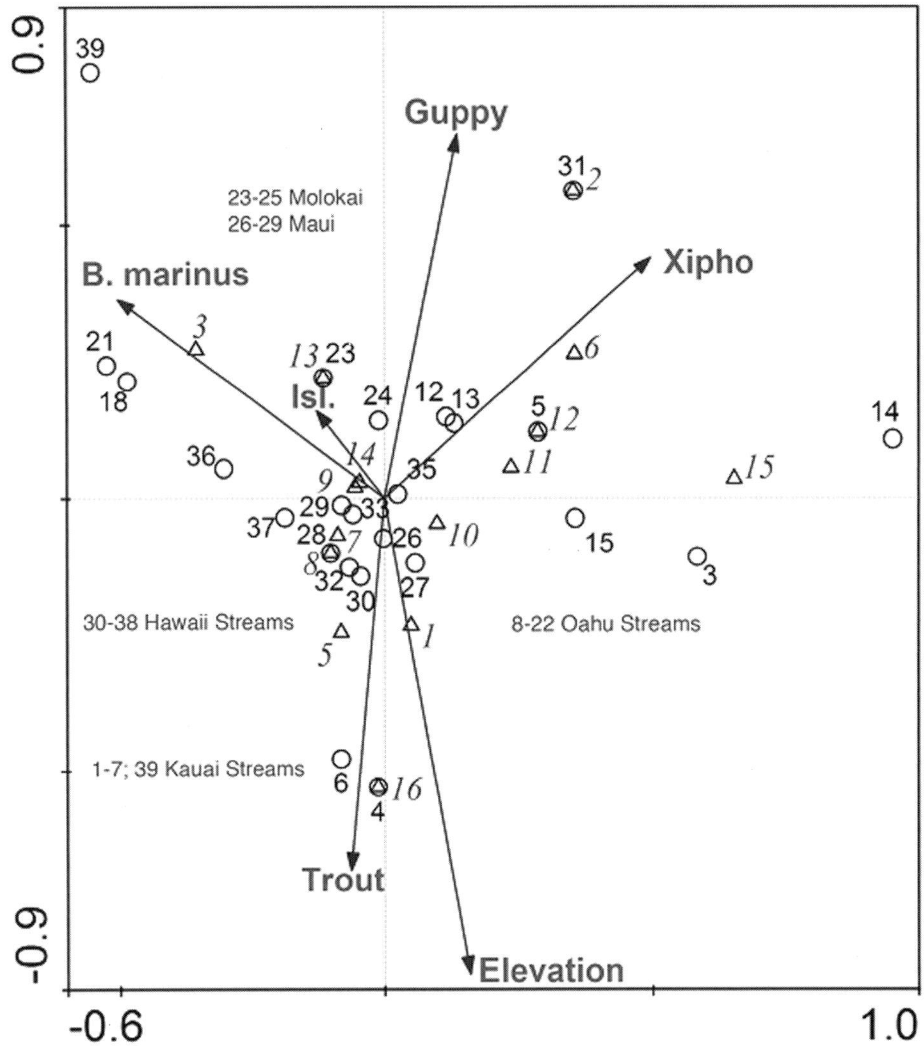

Figure 8. Native Chironomidae (all taxa) sites and species-environment relationship using CCA (CCA1 vs. CCA2).

identifying impacted habitats. It was also of interest to examine the genus *Scatella* because it is one of most dominant native aquatic insect groups in Hawaiian streams. CCA 2 and 3 accounted for 82.8% of the variation (Fig. 11, Table 2), thus *Scatella* by itself is responsive to environmental variables. They were, however, most strongly associated with alien fish and amphibian species, and low elevation native fish species. In contrast to the native species where environmental associations were not always clear, certain introduced ephydrids were clearly associated with disturbed environments, such as *Placopsidella marquesana*, *Scatella stagnalis*, and *Donaceus nigronotatus*.

Canacidae

Because of their association with torrenticolous habitats it was hypothesized that the endemic genus *Procanace* would be sensitive to disturbed habitats or introduced aquatic taxa. The ordination for this family provided 86.5% explanation (Table 2) of variability by the first two axes, with all environ-

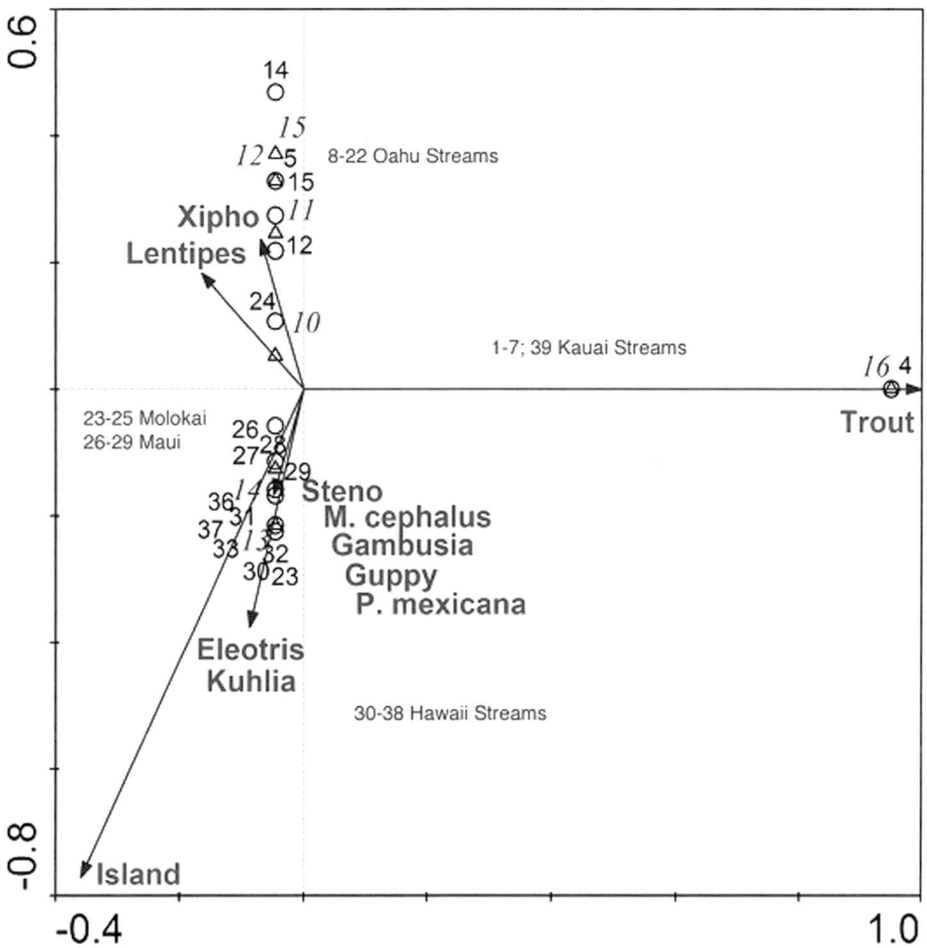

Figure 9. Native Chironomidae (*Telmatogeton* spp. only) sites and species-environment relationship using CCA (CCA1 vs. CCA2).

mental variables retained (Fig. 12). Running CCA for only the genus *Procanace* increased the level of variability to 92.3% (Fig. 13, Table 2), with good resolution by island, and accounted for relatively strong associations with indigenous fish species.

Amphibian Impacts

Hawai'i currently has three species of introduced aquatic amphibians, *Bufo marinus*, *Rana catesbeiana*, and *R. rugosa*. Of greatest concern according to CCA analysis was *R. catesbeiana*, with the other two species showing little impact in regard to native insect taxa. This is because *B. marinus* is found in mainly highly disturbed low elevation areas, while *R. rugosa* is found in high elevation areas and is often co-associated with endemic aquatic insects. As shown on Fig. 6, *R. catesbeiana* was associated with *Poecilia mexicana* in degraded habitats, while *R. rugosa* was by contrast usually found in high quality habitats on Kaua'i and O'ahu (see Figs. 6 or 12), and likely because of its small size is having little impact on dolichopodid Diptera or native damselflies, except perhaps to exclude certain species of the latter from preferred fast water breeding sites with its gelatinous egg masses.

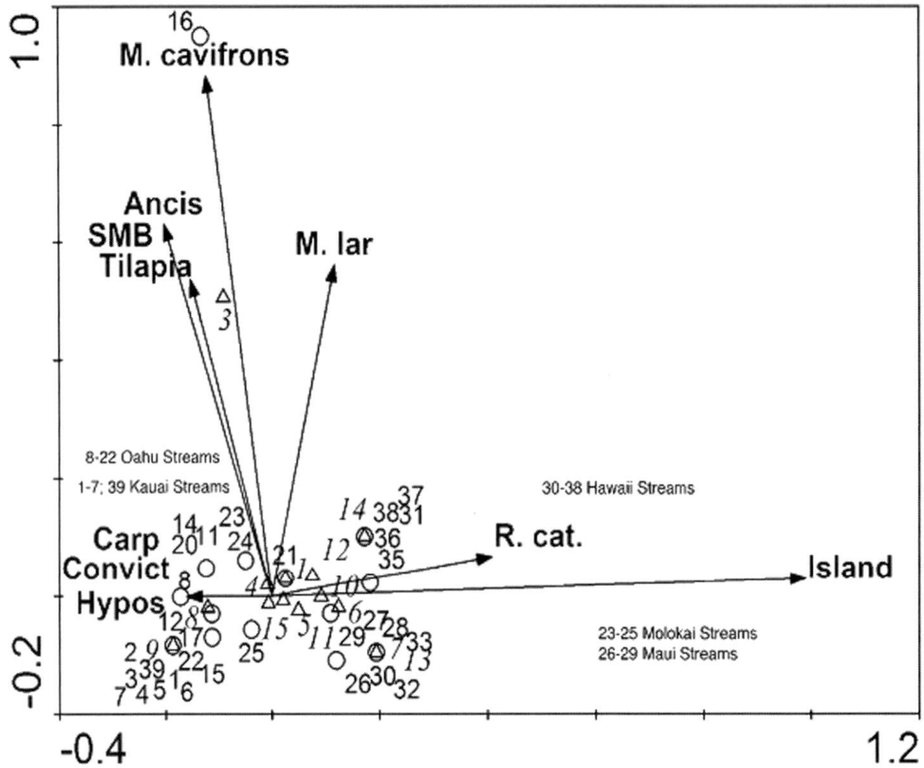

Figure 10. Native Ephydridae site and species-environment relationship using CCA (CCA2 vs. CCA3).

Discussion

These findings represent the first attempt at elucidating statistical associations of native aquatic insect faunas with environmental variables such as alien fish species, elevation, and stream disturbance. Our results have also allowed us to explore relationships between native aquatic insects and indigenous stream fish. The primary objective of this study was determine what, if any, species of aquatic insects are associated with pristine or disturbed habitats. A significant finding was that at least two groups, the native *Megalagrion* damselflies and dolichopodid flies, exhibited statistical relationships that appear to reflect correlations with disturbed and undisturbed environments (Figs. 6 and 7). Several aquatic insect families also exhibited obvious groupings, with sites from Kaua'i and O'ahu often clustered together, while Maui and Moloka'i sites often grouped together with the Hawai'i sites, as shown in Figs. 6 and 7. That these patterns may reflect the evolutionary history of the *Megalagrion* and dolichopodid species is of great future research interest; as is the fact that these patterns also show consistency in identifying sites with similar levels of impact among the different islands. *Telmatogeton* spp. was another assemblage of taxa showing clear separation by islands and strong associations with native fish taxa such as *Lentipes concolor*.

These findings then lend credence that *Megalagrion* damselflies, dolichopodid flies, *Procanace* spp., and *Telmatogeton* spp. (giant Hawaiian midges) are all suitable as indicator species for diverse aquatic habitats worthy of preservation and conservation attention. Ephydridae also had high resolution, but this associated with disturbance rather than with pristine conditions. At the family level, dam-

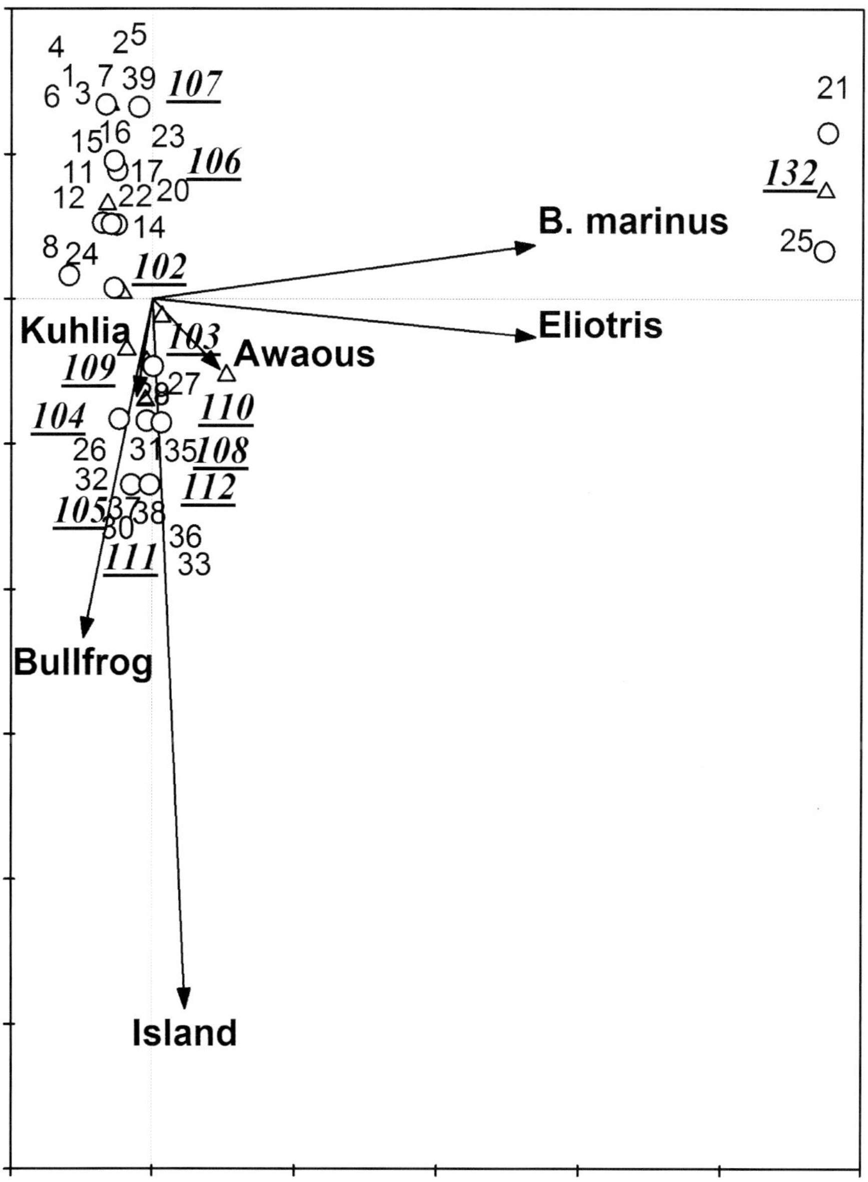

Figure 11. Native Ephydridae (*Scatella* spp. only) species-environment relationship using CCA.

selflies and canacid flies received the highest species-environment relationship score (84.5%) for the CCA. Because Hawaiian damselflies have a larger species assemblage than canacids (18 vs. 12 species analyzed here) their results are more meaningful than canacids, suggesting that damselflies have the most easily detected sensitivity of the aquatic insect taxa we assessed, and show the clearest patterns in community composition and responses to environmental factors. Odonata are well known to the public because of their large size and stunning appearance, and because of this would certainly qualify as the most charismatic of the aquatic insects in Hawaii, and thus could also be considered a flag-

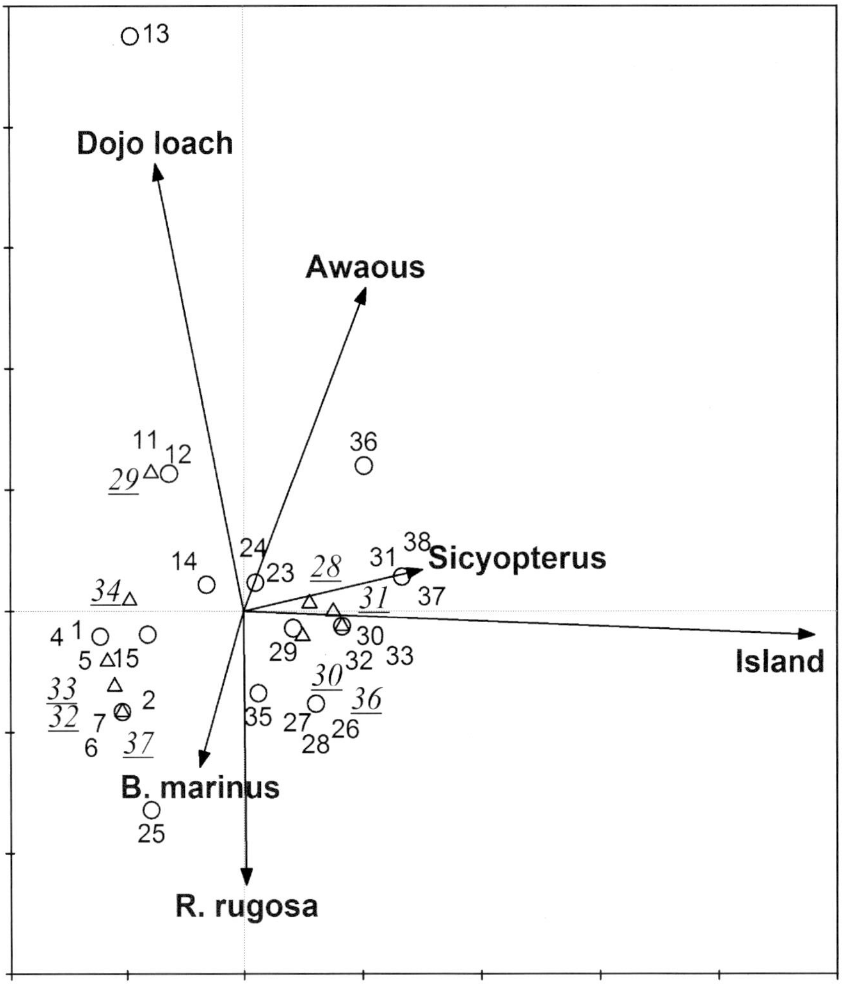

Figure 12. Native Canacidae (all taxa) site and species-environment relationship using CCA (CCA1 vs. CCA2).

ship species (Andelman & Fagan, 2000). On a more controversial note, our data suggest that Hawaiian damselflies would fall under the dual role of an umbrella species (Andelman & Fagan, 2000), or species defined as requiring such large areas of habitat that their protection might simultaneously protect other aquatic species. Because native damselflies will only be found in areas with little disturbance, this would in turn lead to healthy populations of native stream fish species being found in the same area. In contrast, ephydrids and all chironomids had well defined axes and groupings associated with disturbed habitats in our analyses (Figs. 10–13), suggesting these species are more resistant to both a disturbed environment and alien aquatic species, and are thus not good indicator candidates for pristine conditions. One of the weaknesses of the current study, which used presence / absence data rather than abundance data. The availability of abundance data would make the CCA considerably more robust. Nonetheless, the analyses provide credible characterizations of the streams surveyed.

Field observations indicate that *Telmatogeton* spp. are now found only in exceedingly pristine, high volume, and high water quality environments. Because of these requirements and the prevalence of water diversions on Hawaiian streams *Telmatogeton* are now difficult to find in the

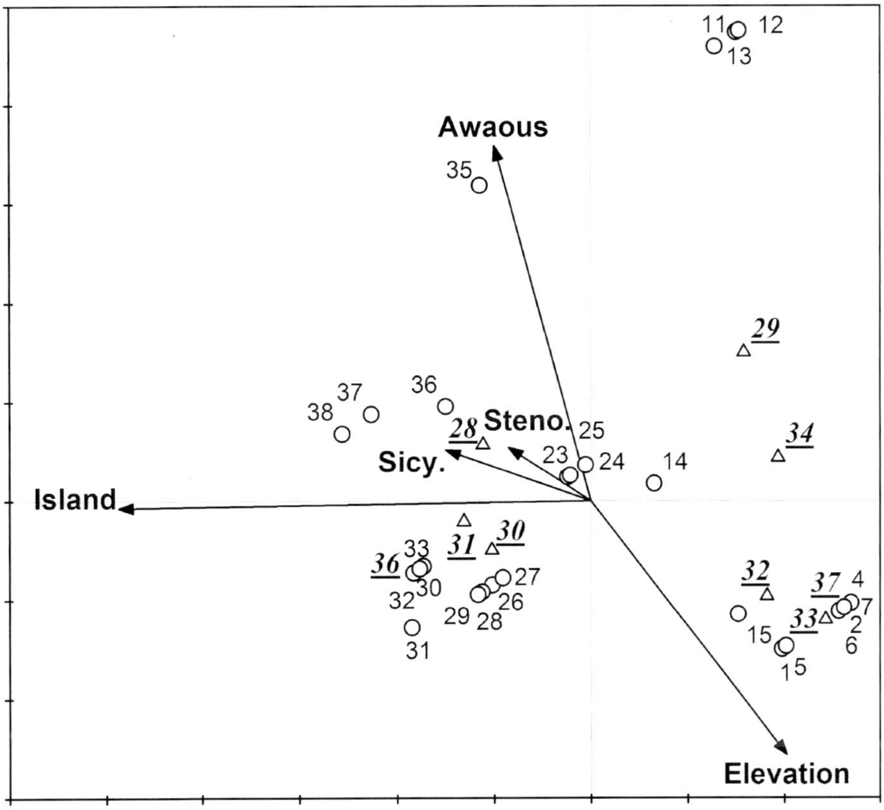

Figure 13. Native Canacidae (*Procanace* spp. only) species-environment relationship using CCA.

Hawaiian islands and are becoming increasingly rare, and for example, this genus is now found in only 4 of 57 streams on O'ahu (Englund & Polhemus, unpubl. data). The current rarity and naturally low species richness (7 spp.) in the genus *Telmatogeton* resulted in an inflated degree of variance accounted for; rarity of species in this genus precludes them from being an effective indicator species. These giant Hawaiian chironomids may not be as charismatic as the Hawaiian damselflies, they are easy for untrained observers to identify in the field because of their large and distinctive white larval cases on stream boulders, and hence could make an ideal suite of indicator species if they were more common.

The conservation community has recently had heated debates on the various conceptual and practical values of indicator, umbrella, flagship, and keystone species (Simberloff, 1998; Andelman & Fagan, 2000; Rubinoff, 2001) when it comes to the assessment and preservation of biodiversity. In the Hawaiian Islands there has also been some degree of controversy, with inappropriate attempts to use Index of Biotic Integrity (IBI), developed for continental salmonid streams, to rank and assess Hawaiian streams (Parham, 2005). The shortcomings of the use of IBI in tropical insular streams with low natural fish diversity were well recently documented (Parham, 2005), but further problems exist with IBI as used in Hawai'i in that native aquatic insects are excluded from the metrics. Thus, in Hawai'i the dominant component of native aquatic biodiversity, the 400+ species of native aquatic insects, have been overlooked. Our findings that certain native insect taxa such as the *Megalagrion* damselflies, canacid, and dolichopodid flies are correlated with the presence of native

Table 2. Cumulative percentage of variation in species-environment and species composition explained by correspondence axes 1–3 by family or genus.

Insect family / taxon (# spp. included in analysis)	% Variance account for by first three axes	
	Spp.-environment	**Species only**
Coenagrionidae (18)	84.5	30.4
Dolichopodidae (38)	65.0	22.9
Chironomidae (16)	70.4	26.3
Canacidae (12)	84.5	32.7
Ephydridae (15)	83.1	40.9
Telmatogeton (genus level) (7)	78.9	54.3
Procanace (genus level) (10)	92.3	29.5
Scatella (genus level, native only) (14)	82.8	40.1

indigenous stream fish indicates that any assessment of native streams should necessarily be conducted in a more holistic fashion than has been practiced with IBI in Hawai'i (e.g., Parham, 2005).

While the indicator species concept has received considerable criticism because it is both difficult to determine which species are the best indicators, or even what a species should indicate (Simberloff, 1998), we feel the indicator concept still has value for Hawaiian streams, especially in light of our findings from the present study indicating certain native aquatic insect taxa are sensitive to physical disturbance and alien species. For example, with funding for habitat conservation measures likely to remain at a low level, these findings can be used to identify taxa and stream areas that have high conservation value, thus prioritizing allocation of resources. In this case, we define areas of high conservation value as Hawaiian streams and adjacent wetlands with high biodiversity of native aquatic taxa. The presence of native species from the highly diverse groups such as dolichopodid and *Megalagrion* damselflies in a Hawaiian stream indicates that the stream has not been greatly disturbed by alien species or physically altered. In addition, if these two groups of taxa are present it usually means that many endemic and indigenous species will be co-associated with them, and that there will often be healthy populations of native stream fish as well.

We therefore conclude that for the highly endemic and diverse aquatic insect fauna in Hawaiian streams the indicator species concept still has value. Until now, most attention and resources have been focused on freshwater fish as indicators (Parham, 2005), but our results indicate the nearly exclusive use of native Hawaiian stream fish as indicator species in models such as is the current practice with IBI in Hawai'i should be re-examined.

Our results indicate that there are certain advantages to using certain aquatic insect taxa as indicators for highly diverse Hawaiian aquatic habitats, and streams that maintain these indicator species should have a high conservation priority. Although the use of aquatic insects as indicator species in Hawaiian streams has both advantages and drawbacks (Table 3) as compared to native fish, advantages include greater specificity and increased sensitivity to external disturbances.

While data for this research of necessity was collected in a species presence or absence format, future directions in Hawaiian aquatic insect research could focus on developing techniques to further quantify specific aquatic insect populations. This study is the first to shed light on the fact that Hawaiian aquatic insects and native stream fish populations are closely linked, yet we are only just beginning to understand the relationships between different groups of native aquatic insects, let alone the interactions between stream fish and insects. Two major obstacles remain in obtaining quantitative data on native Hawaiian aquatic insect populations, taxonomic and ecological. Most of the taxonomic descriptions and illustrations of native aquatic insect taxa have been from the adult aerial stage, and few systematic larval descriptions exist for most taxa. Even some of the well-studied groups such as the genus *Megalagrion* have numerous undescribed larval stages. Very few descriptions exist for the other aquatic insect groups, and some taxa such as the diverse Chiro-

Table 3. Summary of advantages and disadvantages of using native aquatic insects versus native freshwater fish as species for monitoring the health of an aquatic ecosystem.

Taxa	Advantages	Disadvantages
Fish	1. Easily identifiable 2. Charismatic species 3. Culturally important	1. Open system: impacts outside watershed have great influence 2. Broad habitat preferences (less sensitive to disturbance) 3. Found only at lower elevations (900 m max) 4. Usually not above diversions/dams 5. Migratory: impacts outside watershed influence Population 6. Only 5 species
Insects	1. Closed system: impacts outside of watershed have no influence 2. Certain groups easily identifiable 3. Charismatic species (a few) 4. Narrow habitat Preferences 5. More sensitive to disturbance 6. Found above diversions 7. Occurrence correlated with indigenous fish 8. 400+ species	1. Many groups difficult to identify (taxonomic knowledge required) 2. 400+ species

nomidae are taxonomically quite difficult in the larval stage. With the exception of the *Megalagrion* damselflies, most native aquatic insects evolved in wave-swept marine habitats and have secondarily invaded and radiated into freshwater habitats (Howarth & Polhemus, 1991). These native insects are almost exclusively then found in torrenticolous riffle and cascade habitats, which are difficult to quantify with benthic enumeration devices such as a Surber or Hess sampler. Future research should be directed at further refining quantitative sampling methods for such taxa. For instance, new technologies such as DNA extraction from larval aquatic insects, statistically sound methods of collecting adults, and new methods to sample torrenticolous habitats would increase our knowledge of this highly endemic fauna, thus helping to ensure its ultimate preservation.

Acknowledgments

Funding for these surveys was provided by the Hawaii Division of Aquatic Resources, Sport Fish Restoration Program, the U.S. Fish & Wildlife Service, and the Hawaii DLNR Natural Areas Reserve System. Additional support for the Moloka‘i surveys came from The Nature Conservancy Hawai‘i: Moloka‘i Office, Tina Lau, Steph Loo, Brian Naeole. We would like to thank the following people from Hawaii Division of Aquatic Resources/DLNR NAR for assisting in these studies: Bill Devick, Bob Nishimoto, Bill Puleloa, Glenn Higashi, Darrell Kuamo‘o, John Kahiapo, Skippy Hau, Mike Yamamoto, and Betsy Gagné. Bishop Museum personnel providing assistance included: David Preston, Frank Howarth, Neal Evenhuis, Myra McShane, and Tracie Mackenzie. We especially thank Mike Fitzsimons and Bob Nishimoto for chairing this symposium and providing the impetus for this paper to be written.

Literature Cited

Andelman, S.J. & W.F. Fagan. 2000. Umbrellas and flagships: efficient conservation surrogates or expensive mistakes? *Proceedings of the National Academy of Sciences* **97**: 5954–5959.

Eldredge, L.G. & N.L. Evenhuis. 2003. Hawaii's biodiversity: a detailed assessment of the number of species in the Hawaiian Islands. *Bishop Museum Occasional Papers* **76**: 1–28.

Englund, R.A. 1993. A survey of the fish and aquatic insect fauna of the Waikele/Kipapa Streams, Oahu, Hawaii. BHP Environmental Technologies report prepared for Halekua Development Corp., Honolulu. 20 pp.

———. 1999. The impacts of introduced poeciliid fish and Odonata on endemic *Megalagrion* (Odonata) damselflies on O'ahu Island, Hawaii. *Journal of Insect Conservation* **3**: 225–243.

———. 2001. Long-term monitoring of one of the most restricted insect populations in the United States, *Megalagrion xanthomelas* (Selys-Longchamps), at Tripler Army Medical Center, O'ahu, Hawaii (Zygoptera: Coenagrionidae). *Odonatologica* **30**: 255–263.

———. 2002. The loss of native biodiversity and continuing nonindigenous species introductions in freshwater, estuarine, and wetland communities of Pearl Harbor, O'ahu, Hawaiian Islands. *Estuaries* **25**: 418–430.

———. **& Arakaki, K.T.** 2003. Report on long-term aquatic insect monitoring by Hawaii Biological Survey, Bishop Museum in Pelekunu Valley, Moloka'i, Hawaii. Report prepared for TNCH Moloka'i Office. 10 pp.

———. **& Arakaki, K.T.** 2004. Rapid biological inventories of streams in the Ala Wai watershed, O'ahu Island, Hawai'i. Hawaii Biological Survey Report prepared for Oceanit Laboratories, Inc., Honolulu, Hawai'i. 16 p.

———., **Arakaki, K.T., D.J. Preston, N.L. Evenhuis and M.K.K. McShane.** 2003a. Systematic Inventory of Rare and Alien Aquatic Species in Selected O'ahu, Maui, and Hawai'i Island Streams. Final report prepared for Hawaii Department of Land and Natural Resources, Division of Aquatic Resources, Honolulu. 14 pp.

———. **& R.B. Filbert.** 1997. Native and exotic stream organisms study in the Kawainui, Alakahi, Koiawe, and Lalakea Streams, Lower Hamakua Ditch watershed project, County of Hawaii. USDA-NRCS Contract No. 53-9251-6-275. 71 pp.

———. **& R.B. Filbert.** 1999. Flow restoration and persistence of introduced species in Waikele Stream, Oahu. *Micronesica* **32**: 143–154.

———. **& Polhemus, D.A.** 2001. Evaluating the effects of introduced rainbow trout (*Oncorhynchus mykiss*) on native stream insects on Kauai Island, Hawaii. *Journal of Insect Conservation* **5**: 265–281.

———. **& Preston, D.J.** 1999. Biological Assessment of the Lower Hamakua Ditch on the Hawaiian Stream Fly (*Sigmatineurum meaohi*) and other aquatic insects. Report. Bishop Museum. 31 pp.

———., **Preston, D.J., K. Arakaki & M.K.K. McShane.** 2003b. Aquatic insect surveys of four windward O'ahu stream systems impacted by the Waiāhole Ditch. Hawaii Biological Survey Report prepared for Hawaii Department of Land and Natural Resources, Division of Aquatic Resources, Honolulu. 8 pp.

———., **Preston, D.J., N.L. Evenhuis, R.H. Cowie, C. Imada, C. Puttock, & K. Arakaki.** 2001. Native and exotic organism study. Lower Wailoa River, Waipi'o, County of Hawaii. Report prepared for USDA Natural Resources Conservation Service, Honolulu, Hawaii. 57 p.

———., **Preston, D.J., R. Wolff, S. L. Coles, L. G. Eldredge & K. Arakaki.** 2000. Biodiversity of freshwater and estuarine communities in lower Pearl Harbor, O'ahu, Hawai'i with observations on introduced species. *Bishop Museum Technical Report* **16**: 1–166.

Evenhuis, N.L, D. Polhemus, S. Swift, K. Arakaki, & D. Preston. 1995. A study of the biology of the orangeblack Hawaiian damselfly (*Megalagrion xanthomelas*), with special reference to conservation of the population at Tripler Army Medical Center, Oahu. *Bishop Museum Technical Report* **8**, 81 pp.

Filbert, R. & R.A. Englund. 1995. Waiāhole ditch water contested case: biological assessments of windward and leeward streams. Pacific Aquatic Environmental consultants report prepared for Kamehameha Schools/Bishop Estate. 140 pp.

Font, W.F. 1998. Parasites in paradise: patterns of helminth distribution in Hawaiian stream fishes. *Journal of Helminthology* **72**: 307–311.

———. **& D.C. Tate.** 1994. Helminth parasites of native Hawaiian freshwater fishes: an example of extreme ecological isolation. *Journal of Parasitology* **80**: 682–688.

Howarth, F. G. & Polhemus, D.A. 1991. A review of the Hawaiian stream insect fauna. In: *New directions in research, management, and conservation of Hawaiian freshwater stream ecosystems, Proceedings of the 1990 Symposium on Freshwater Stream Biology and Management, State of Hawaii*, p. 40–51. Hawaii Division of Aquatic Resources, Honolulu.

Leps, J. & Smilauer, P. 2003. *Multivariate analysis of ecological data using CANOCO.* Cambridge University Press. pp. 269.

Liebherr, J. & D.A. Polhemus. 1997. Comparisons to the century before: the legacy of the R.C.L. Perkins and Fauna Hawaiiensis as the basis for a long-term ecological monitoring program. *Pacific Science* **51**: 490–504.

Nishida, G.M. 2002. Hawaiian terrestrial arthropod checklist. Fourth edition. *Bishop Museum Technical Report* **22**, 313 pp.

Parham, J.E. 2005. Survey techniques for freshwater streams on oceanic islands: important design considerations for the PABITRA Project. *Pacific Science* **59**: 283–291.

Polhemus, D.A. 1994. The aquatic insect fauna of North Halawa, Oahu, Hawaii. *In* F.G. Howarth *et al.*, A natural history survey of North Halawa Valley, Oahu, Hawaii. Bishop Museum report to Hawaii State Department of Transportation, H-3 Project. pp. 75–85. Bishop Museum, Honolulu, Hawaii. 112 pp.

———. 1995. A survey of the aquatic insect faunas of selected Hawaiian streams. Hawaii Biological Survey Report for the Commission on Water Resource Management, Department of Land and Natural Resources, State of Hawaii. Department of Natural Sciences, Bishop Museum. 128 pp.

Rubinoff, D. 2001. Gnatcatcher as umbrella species for insects. *Conservation Biology* **15**: 1374–1383.

Simberloff, D. 1998. Flagships, umbrellas, and keystones: is single-species management passé in the landscape era? *Biological Conservation* **93**: 247–257.

Ter Braak, C.J.F. & Smilauer, P. 2002. CANOCO reference manual and ConoDraw for Windows users guide: Software for Canonical Community Ordination (version 4.5). Microcomputer Power, Ithaca, New York. 500 pp.

Williams, F.X. 1936. Biological Studies in Hawaiian water-loving insects. Part 1. Coleoptera or beetles. Part 2. Odonata or dragonflies. *Proceedings of the Hawaiian Entomological Society* **9**: 235–345.

Yamamoto, M.N. & A.W. Tagawa. 2000. *Hawai'i's native and exotic freshwater animals.* Mutual Publishing, Honolulu. 200 pp.

APPENDIX

Biota found during this study and their native or introduced status. [Status taken from Yamamoto (2000) and Nishida (2002).]

Taxa	Species (Ind = Indigenous; End = Endemic; Int = Introduced)	Species Number on Figures 6–13
Native (Endemic)		
Aquatic Insects		
Odonata		
Aeshnidae	*Anax strenuus* (End)	1
Libellulidae	*Nesogonia blackburni* (End)	2
Coenagrionidae	*Megalagrion eudytum* (End)	3
	Megalagrion heterogamias (End)	4
	Megalagrion oresitrophum (End)	5
	Megalagrion orobates (End)	6
	Megalagrion vagabundum (End)	7
	Megalagrion hawaiiense (End)	8
	Megalagrion leptodemas (End)	9
	Megalagrion nigrohamatum nigrolineatum (End)	10
	Megalagrion oceanicum (End)	11
	Megalagrion xanthomelas (End)	12
	Megalagrion blackburni (End)	13
	Megalagrion calliphya (End)	14
	Megalagrion nesiotes (End)	15
	Megalagrion nigrohamatum nigrohamatum (End)	16
	Megalagrion oceanicum (End)	17
Heteroptera		
Nabidae	*Nabis gagneorum* (End)	18
	Saldula exulans (End)	19
Saldidae	*Saldula oahuensis* (End)	20
	Saldula procellaris (End)	21
Veliidae	*Microvelia vagans* (End)	22
Coleoptera		
Dytiscidae	*Rhantus pacificus* (End)	23
Hydrophilidae	*Limnoxenus semicylindricus* (End)	24
Lepidoptera		
Cosmopterigidae	*Hyposmocoma* sp. (End)	25
	Hyposmocoma sp. nr *montivolans* (End)	26
	Hyposmocoma sp. nr *saccophora* (End)	27
Diptera		
Canacidae	*Procanace acuminata* (End)	28
	Procanace bifurcata (End)	29
	Procanace confusa (End)	30
	Procanace constricta (End)	31
	Procanace nigroviridis (End)	32
	Procanace quadrisetosa (End)	33
	Procanace wirthi (End)	34
	Procanace new sp. 1 (End) (Oahu - Rare Alien Survey)	35
	Procanace new sp. 1 (End) (Hawaii koa timber survey)	36
	Procanace sp. (End)	37

Taxa	Species (Ind = Indigenous; End = Endemic; Int = Introduced)	Species Number on Figures 6–13
Native (Endemic)		
Aquatic Insects		
Ceratopogonidae	*Dasyhelea digna* (End)	38
	Dasyhelea hawaiiensis (End)	39
	Dasyhelea sp. (not *hawaiiensis*) (End)	40
	Dasyhelea platychaeta (End)	41
	Dasyhelea sp. (End)	42
	Forcipomyia hardyi (End)	43
	Forcipomyia kaneohe (End)	44
	Forcipomyia sp. (End)	45
Chironomidae	*Chironomus* sp. (End)	46
	Chironomus hawaiiensis (End)	47
	Clunio sp. nr. *vagrans* (End)	48
	Micropsectra sp. (End)	49
	Micropsectra hawaiiensis (End)	50
	Orthocladius sp. (End)	51
	Orthocladius grimshawi (End)	52
	Pseudosmittia paraconjuncta (End)	53
	Telmatogeton abnormis (End)	54
	Telmatogeton fluviatilis (End)	55
	Telmatogeton hirtus (End)	56
	Telmatogeton japonicus (End)	57
	Telmatogeton torrenticola (End)	58
	Telmatogeton williamsi (End)	59
	Telmatogeton sp. (End)	60
Dolichopodidae	*Campsicnemus brevipes* (End)	61
	Campsicnemus gloriosus (End)	62
	Campsicnemus labilis (End)	63
	Campsicnemus lepidochaites (End)	64
	Campsicnemus longitibia (End)	65
	Campsicnemus nigricollis (End)	66
	Campsicnemus modicus (End)	67
	Campsicnemus miritibialis (End)	68
	Campsicnemus patellifer (End)	69
	Campsicnemus ridiculus (End)	70
	Campsicnemus tibialis (End)	71
	Campsicnemus nr. *truncatus* (End)	72
	Campsicnemus williamsi (End)	73
	Campsicnemus sp. (End)	74
	Campsicnemus new sp. 1 (End) (Oahu- Rare Alien Surveys)	75
	Campsicnemus new sp. 2 (End) (Maui - Rare Alien Surveys)	76
	Campsicnemus new sp. 3 (End) (Maui- Rare Alien Surveys)	77
	Campsicnemus lawakua (End) Kokee	78
	Eurynogaster mediocris (End)	79
	Major minor (End)	80
	Elmoia multispinosa (End)	81
	"*Eurynogaster*" sp. (End)	82
	"*Eurynogaster*" new sp. (End) (Maui - Rare Alien Surveys)	83
	"*Eurynogaster*" new sp. (End) (Koa Timber survey -Hawaii)	84
	Paraliancalus metallicus (End)	85
	Sigmatineurum englundi (End)	86
	Sigmatineurum iao (End)	87
	Sigmatineurum meaohi (End)	88
	Sigmatineurum napali (End)	89
	Sigmatineurum nigrum (End)	90

Taxa	**Species** (Ind = Indigenous; End = Endemic; Int = Introduced)	**Species Number on Figures 6–13**
Native (Endemic)		
Aquatic Insects		
Dolichopodidae (cont.)		
	Sigmatineurum parenti (End)	91
	Sigmatineurum mnemogagne (End) (Hanawi)	92
	Sigmatineurum omega (End)	93
	Sigmatineurum puleloai (End) (Moloka'i)	94
	Sigmatineurum n. sp. 1 (End) Koke'e	95
	Sigmatineurum n. sp. 2 (End) Koke'e	96
	Sigmatineurum n. sp. (End) (Maui - Rare Alien Surveys)	97
	Thambemyia acrostichalis (End) (formerly *Paraphrosylus* sp.)	98
Ephydridae	*Atissa oahuensis* (End)	99
	Hydrellia tritici (End)	100
	Notiphilia insularis (End)	101
	Scatella cilipes (End)	102
	Scatella clavipes (End)	103
	Scatella (*Apulvillus*) *femoralis* (End)	104
	Scatella fluvialis (End)	105
	Scatella hawaiiensis (End)	106
	Scatella kauaiensis (End)	107
	Scatella (*Apulvillus*) *mauiensis* (End)	108
	Scatella oahuense (End)	109
	Scatella warreni (End)	110
	Scatella williamsi (End)	111
	Scatella new sp. (End) (Waimanu)	112
Muscidae	*Lispe* sp. (End)	113
	Lispocephala new sp. 1 (End) (Maui - Rare Alien Surveys)	114
	Lispocephala new sp. 2 (End) (Maui - Rare Alien Surveys)	115
Psychodidae	*Psychoda* sp. (End)	116
	Trichomyia hawaiiensis (End)	117
	Trichomyia sp. (End)	118
Limoniidae	*Dicranomyia hawaiiensis* (End)	119
	Dicranomyia grimshawi (End)	120
	Dicranomyia jacoba (End)	121
	Dicranomyia kauaiensis (End)	122
	Dicranomyia nigropolita (End)	123
	Dicranomyia perkinsi (End)	124
	Dicranomyia swezeyi (End)	125
	Dicranomyia stygipennis (End)	126
	Dicranomyia sp. (End)	127
	Gonomyia sp. (End)	128
Native (Indigenous)		
Insects		
Odonata		
Aeshnidae	*Anax junius* (Ind)	129
Libellulidae	*Pantala flavescens* (Ind)	130
Diptera		
Chironomidae	*Chironomus esakii* (Ind)	131
Ephydridae	*Scatella sexnotata* (Ind)	132
Tethinidae	*Dasyrhicnoessa insularis* (Ind)	133
	Dasyrhicnoessa (Ind)	134
	Dasyrhicnoessa vockerothi (Ind)	135

Taxa	Species (Ind = Indigenous; End = Endemic; Int = Introduced)	Species Number on Figures 6–13
Introduced Aquatic Insects		
Odonata		
Libellulidae	*Tramea abdominalis* (Int)	136
	Tramea lacerata (Int)	137
	Crocothemis servilia (Int)	138
	Orthemis ferrugenia (Int)	139
Coenagrionidae	*Ischnura posita* (Int)	140
	Ischnura ramburii (Int)	141
	Enallgma civile (Int)	142
Heteroptera		
Mesoveliidae	*Mesovelia amoena* (Int)	143
	Mesovelia mulsanti (Int)	144
Notonectidae	*Notonecta indica* (Int)	145
	Buenoa pallipes (Int)	146
Coleoptera		
Dytiscidae	*Rhantus guttulosus* (Int)	147
	Copelatus parvulus (Int)	148
Hydrophilidae	*Tropisternus lateralis* (Int)	149
Diptera		
Canacidae	*Procanace williamsi* (Int)	150
	Canaceioides angulatus (Int)	151
Ceratopogonidae	*Forcipomyia* sp. (Int)	152
	Atrichopogon jacobsoni (Int)	153
Chironomidae	*Cricotopus bicinctus* (Int)	154
	Polypedilum nubiferum (Int)	155
Dixidae	Dixa longistyla (Int)	156
Dolichopodidae	*Condylostylus longicornis* (Int)	157
	Chrysosoma globiferum (Int)	158
	Chrysotus longipalpus (changed from *pallidipalpus*) (Int)	159
	Chrysotus sp. 1 (Int)(Waipio)	160
	Dolichopus exsul (Int)	161
	Pelastoneurus lugubris (Int)	162
	Syntormon flexibile (Int)	163
	Tachytrechus angustipennis (Int)	164
	Thinophilus hardyi (Int)	165
Empididae	*Hemerodromia stellaris* (Int)	166
Ephydridae	*Brachydeutera ibari* (Int)	167
	Ceropsilopa coquilletti (Int)	168
	Discocerina mera (Int)	169
	Hecamede granifera (Int)	170
	Hydrellia williamsi (Int)	171
	Donaceus nigronotatus (Int)	172
	Lytogaster gravida (Int)	173
	Ochthera circularis (Int)	174
	Paratissa pollinosa (Int)	175
	Placopsidella marquesana (Int)	176
	Typopsilopa sp. (Int)	177
	Scatella stagnalis (Int)	178
Muscidae	*Lispe assimilis* (Int)	179

Taxa	**Species** (Ind = Indigenous; End = Endemic; Int = Introduced)	**Species Number on Figures 6–13**
Introduced Aquatic Insects		
Diptera (cont.)		
Psychodidae	*Clogmia albipunctata* (Int)	180
Sciomyzidae	*Sepedon aenescens* (Int)	181
Tethinidae	*Tethina variseta* (Int)	182
Limoniidae	*Dicranomyia advena* (Int)	183
	Erioptera bicornifer (Int)	184
Trichoptera		
Hydropsychidae	*Cheumatopsyche analis* (Int)	185
Hydroptilidae	*Hydroptila icona* (Int)	186
	Hydroptila potosina (Int)	187
	Oxyethira maya (Int)	188
Fish	*Lentipes concolor* (End)	189
	Sicyopterus stimpsoni (End)	190
	Awaous guamensis (Ind)	191
	Stenogobius hawaiiensis (End)	192
	Eleotris sandwicensis (End)	193
	Mugil cephalus (Ind)	194
	Kuhlia xenura (End)	195
	Gambusia affinis (Int)	196
	Poecilia reticulata (Int)	197
	Poecilia mexicana (Int)	198
	Poecilia latipinna (Int)	199
	Limia vittata (Int)	200
	Poecilia (misc) spp. (Int)	201
	Xiphophorous helleri (Int)	202
	Micropterus dolomieui (Int)	203
	Oncorhynchus mykiss (Int)	204
	Misgurnus anguillicaudatus (Int)	205
	Mugilogobius cavifrons (Int)	206
	Tilapia/Oreochromis spp. (Int)	207
	Cichlasoma managuense (Int)	208
	Archocentrus (*Cichlasoma*) *nigrofasciatus* (Int)	209
	Hemichromis elongatus (Int)	210
	Melanochromis johanni (Int)	211
	Hypsophrys nicaraguensis (Int)	212
	Amphilophus citrinellum (Int)	213
	Ancistris temminicki (Int)	214
	Hypostomous watwata (Int)	215
	Cyprinus carpio (Int)	216
Amphibians	*Bufo marinus* (Int)	217
	Rana catesbeiana (Int)	218
	Rana rugosa (Int)	219
Crustaceans	*Procambarus clarkii* (Int)	220
	Macrobrachium lar (Int)	221

Biology of Hawaiian Streams and Estuaries. Edited by N.L. Evenhuis & J.M. Fitzsimons. *Bishop Museum Bulletin in Cultural and Environmental Studies* 3: 233–246 (2007).

Biology Recapitulates Geology: the Distribution of *Megalagrion* Damselflies on the Ko'olau Volcano of O'ahu, Hawai'i[1]

DAN A. POLHEMUS[2]
Department of Entomology, MRC 105, National Museum of Natural History PO Box 37012, Smithsonian Institution, Washington, DC 20013-7012 USA; email: bugman@bishopmuseum.org

Abstract

Populations of endemic *Megalagrion* damselflies breeding in upland streams have exhibited a progressive decline in both range and numbers on O'ahu since the late 1970s, based on analysis of historical collection data and results of recent surveys. In particular, conservation status surveys conducted from 1991 onward determined that the 4 upland stream-breeding taxa on the island, 3 of which are endemic to O'ahu, have disappeared from many catchments they formerly inhabited, particularly in the Wai'anae Mountains and on the leeward side of the Ko'olau Mountains. The remaining populations on the island are now disproportionately concentrated on windward slopes of the latter range, where they are clustered around exposures of the core dike complex of the Ko'olau volcano. This geological formation traps groundwater in dike-segregated compartments, thereby producing permanent surface flow in the upper reaches of many windward Ko'olau gulches, coupled with naturally interrupted midreaches immediately below that block the upstream migration of invasive species. The discovery of this correlation between geology and damselfly distributions has allowed predictive location of additional colonies by using geological maps, permitting future surveys to be more accurately targeted, and providing an objective basis for the delimitation of habitat critical to the survival of these species.

Introduction

Endemic damselflies in the genus *Megalagrion* were formerly a common component of Hawaiian stream and wetland biotas at elevations ranging from sea level to at least 2000 m. In response to ecological opportunities offered by the topographically complex and geographically isolated Hawaiian archipelago, specialized species of *Megalagrion* evolved to exploit a wide variety of aquatic ecosystems as breeding habitats, ranging from anchialine pools and stream terminal reaches in the lowlands to rheocrenes and rushing streams in the highlands. Many of these habitat specialists also evolved as single island endemics, particularly on the geologically older main islands of Kaua'i and O'ahu. As of result of their bright colors, insular endemism, fascinating ecological specializations, and relatively large size in relation to most other Hawaiian insects, these damselflies have attracted continuing attention from researchers in taxonomic entomology, systematics, and ecology (Williams, 1936; Zimmerman, 1948; Polhemus, 1997; Jordan *et al.*, 2003).

With the advent of European contact and subsequent colonization, many *Megalagrion* species experienced significant range contractions or even extirpation on particular islands as a result of physical habitat modifications and introduction of invasive species (Polhemus & Asquith, 1996; Polhemus, 1996; Liebherr & Polhemus, 1997). Such impacts were particularly severe on O'ahu, the most heavily developed and populated island in the archipelago. Beginning in 1991, a comprehensive survey program for native damselflies, funded in large part by the U.S. Fish and Wildlife Service, was undertaken on O'ahu by staff from the Bishop Museum and other partner organizations, eventually covering 150 sites on the island (Fig. 1). These surveys demonstrated that at least one

1. Contribution No. 2007-005 to the Hawaii Biological Survey.
2. Present address: Hawaii State Department of Land and Natural Resources, Division of Aquatic Resources, 1151 Punchbowl St., Honolulu, Hawai'i 96813, USA; email: bugman@bishopmuseum.org.

Table 1. Distribution of stream-breeding *Megalagrion* damselfly populations on O'ahu. Populations are defined as demes separated by significant topographic barriers sufficient to preclude regular gene flow, and may include separate branches of the same overall stream catchment.

	NUMBER OF REMAINING POPULATIONS				
TAXON	Leeward Wai'anae	Windward Wai'anae	Leeward Ko'olau	Windward Ko'olau	Total
M. oceanicum	0	0	0	8	8
M. leptodemas	0	0	2	2	4
M. n. nigrolineatum	0	0	6	15	21
M. hawaiiense	2	2	2	10	16

lowland species, *M. pacificum*, had been completely extirpated from O'ahu, and that another lowland species, *M. xanthomelas*, was reduced to a single remnant population (Liebherr & Polhemus, 1997). Both of these species bred in lentic habitats or stream terminal reaches, which experienced significant modification for agriculture beginning as early as the 19th century, and sustained similarly early impact from alien fishes imported for food sources by Asian immigrant laborers (Polhemus & Englund, 2003).

Similar declines, however, were also apparent by the late 20th century for three upland species breeding in lotic habitats: *M. oceanicum*, *M. leptodemas*, and *M. nigrohamatum nigrolineatum*. All three of these taxa are endemic to O'ahu, and formerly occurred in both major mountain ranges on the island, the Ko'olau and Wai'anae (Polhemus & Asquith, 1996). By the late 1990s (and probably decades earlier) they had been completely extirpated from the Wai'anae Mountains, and persisted to varying degrees only in streams draining from the Ko'olau volcano. In addition, one other upland stream-associated species, *M. hawaiiense*, managed to retain a few small populations in the Wai'anae Mountains and remained locally abundant in certain Ko'olau catchments. The comprehensive surveys of the 1990s revealed a further interesting pattern in regard to these remnant distributions: they were to a large degree clustered on the windward slope of the Ko'olau Mountains (Table 1), particularly in the northern windward quadrant, where deeply incised catchments reveal exposures of the core dike complex of the Ko'olau volcano. It thus became apparent that in addition to biological determinants, the distribution of the remaining populations of O'ahu-endemic stream-breeding damselflies was also being dictated to some extent by geomorphological and petrological factors. In the present paper, this possible relationship is examined, beginning with a review of the geology of O'ahu and the Ko'olau volcano, followed by an analysis of distribution patterns for the taxa in question.

Geology and Geomorphology of the Ko'olau Volcano

The island of O'ahu is the third largest in the Hawaiian chain, with an area of 1574 km^2, and lies in the eastern Pacific Ocean at approximately 21°30'N and 158°00'W. The bulk of the island consists of the heavily eroded remnants of a pair of shield volcanoes formed over the Hawaiian hot spot: the Wai'anae volcano in the west, and the Ko'olau volcano in the east (Fig. 2). The Wai'anae volcano is the older of the two, with its earliest rocks dating from 3.7 My, while the Ko'olau volcano is over a million years younger, with oldest rocks dating to 2.6 My (Carson & Clague, 1995). At their peak, the elevations of the Wai'anae and Ko'olau volcanoes are estimated to have reached 2200 m and 1900 m, respectively (Carson & Clague, 1995), but both have now been greatly lowered by erosion, with the current summit of the Wai'anae Range (Mt. Ka'ala) attaining 1231 m, and the summit of the Ko'olau Range (Kōnāhuanui) standing at only 960 m. Despite its lower height, the Ko'olau volcano is the wetter of the two, due to the prevailing trade wind patterns, with its windward flank receiving up to 7000 mm of rainfall a year, while the wettest secton of the Wai'anae Mountains, the

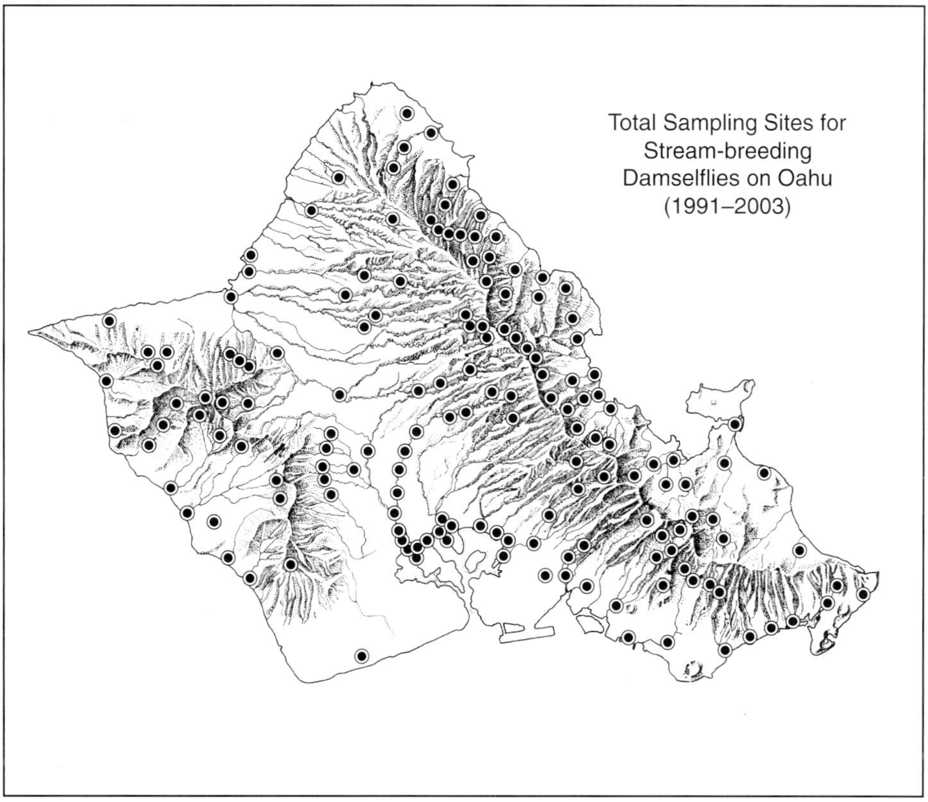

Figure 1. Sites on O‘ahu sampled for *Megalagrion* damselflies from 1991 to 2003 inclusive.

Mt. Ka‘ala massif, receives less than 2000 mm (Giambelluca *et al.*, 1986). This difference in precipitation regime has a marked effect on the number of perennial streams draining from the two mountain ranges; of the 57 perennial streams recognized on O‘ahu, the Ko‘olau volcano has 48, while Wai‘anae has only 9 (Hawaii Stream Assessment, 1990).

Erosion has also heavily reduced the flanks of both volcanoes, particularly on their seaward faces. Such erosion is strikingly asymmetrical, with the interior slopes of the volcanoes being gradual and sloping inward to a broad central valley formed in the dihedral between their two abutting lava shields, and the outer seaward slopes in large part precipitous, with tall cliffs and steep slopes. It was initially thought that these steep seaward slopes were simply the result of headward erosion by the combined action of wind, rain and sea, but bathymetric mapping initiated in the 1960s revealed enormous submarine debris fields offshore of the seaward faces of both the Wai‘anae and Ko‘olau volcanoes, leading to a hypothesis by Moore (1964) that these steep seaward flanks were the result of catastrophic slope failures associated with giant landslides. This hypothesis has been subsequently supported by more sophisticated bathymetric mapping techniques, including the GLORIA multibeam side-scan sonar system of the 1980s, and the JAMSTEC multibeam surveys combined with GPS navigation that were employed in the late 1990s (Moore & Clague, 2002).

The realization that giant landslides have been a significant factor in the erosional history of the Ko‘olau volcano allows a reassessment of the geomorphology of the Ko‘olau Mountains, and of the influence of such landslides on the development and bed profiles of stream catchments within the range. Based on most recent evidence, it appears that one major event, the Nu‘uanu landslide, was responsible for most of the mass wasting experienced by the windward Ko‘olau Mountains. This

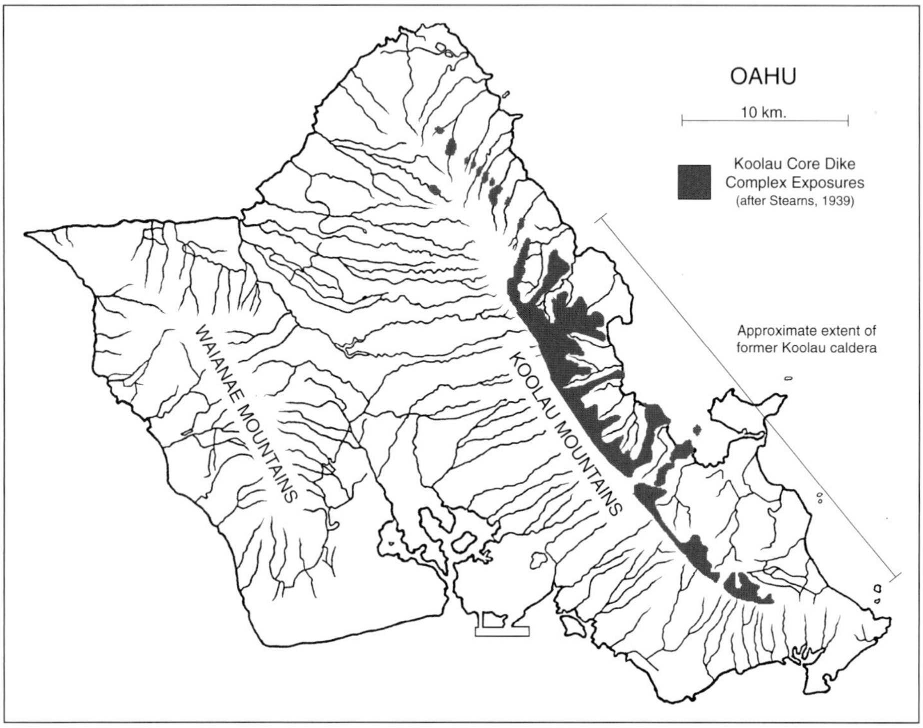

Figure 2. Map of O'ahu showing location and extent of exposures of the core dike complex of the Ko'olau Volcano and approximate former extent of the Ko'olau caldera.

landslide occurred prior to 1.5 my (Clague *et al.*, 2002), and involved the entire northeastern flank of the Ko'olau caldera, which formerly extended from Waimanalo to near Punalu'u (Fig. 2), a distance of approximately 28 km (Macdonald *et al.*, 1983). This is the sector of the windward Ko'olau Mountains that now exhibits bowl-shaped valleys with extremely steep headwater reaches, open midreaches, and long terminal reach estuaries. By contrast, the section of the windward Ko'olau Mountains lying northwest of the old caldera, referred to subsequently as the northern windward Ko'olau Mountains, laying beyond the rim of the caldera, did not experience giant landslides and has streams with very different catchment structure.

Unlike the bowl-shaped valleys formed in the old caldera, the streams of the northern windward Ko'olau Mountains from Kaluanui Stream northward have perched headwater reaches with well integrated drainage patterns occupying open, hanging valleys modestly incised into the tholeiitic basalts of the Ko'olau Volcanic Series that represent the remaining outer windward slope of the original volcano. These hanging catchments make abrupt transitions into narrow canyons up to 500-m deep entrenched into this old windward surface, and eventually debouch onto relatively short terminal reaches lacking extensive estuaries. The boundary between the perched headwaters and the midreach canyons in most of these northwest Ko'olau systems (particularly Kaluanui, Ma'akua, Kaipapa'u and Koloa) is usually marked by one or more high waterfalls that form a significant break in the stream bed profile. In addition, the canyon bottoms are heavily infilled with coarse alluvium, creating a deep hyporheic zone into which the base flow sinks except during spates, making these stream ecosystems "naturally interrupted" according to the classification of Polhemus *et al.* (1992).

Examination of geological maps reveals that the transition from perched headwaters to midreach canyons in these northern windward Ko'olau systems also correlates with the points at which these streams have eroded downward into the upper margin of the volcano's core dike com-

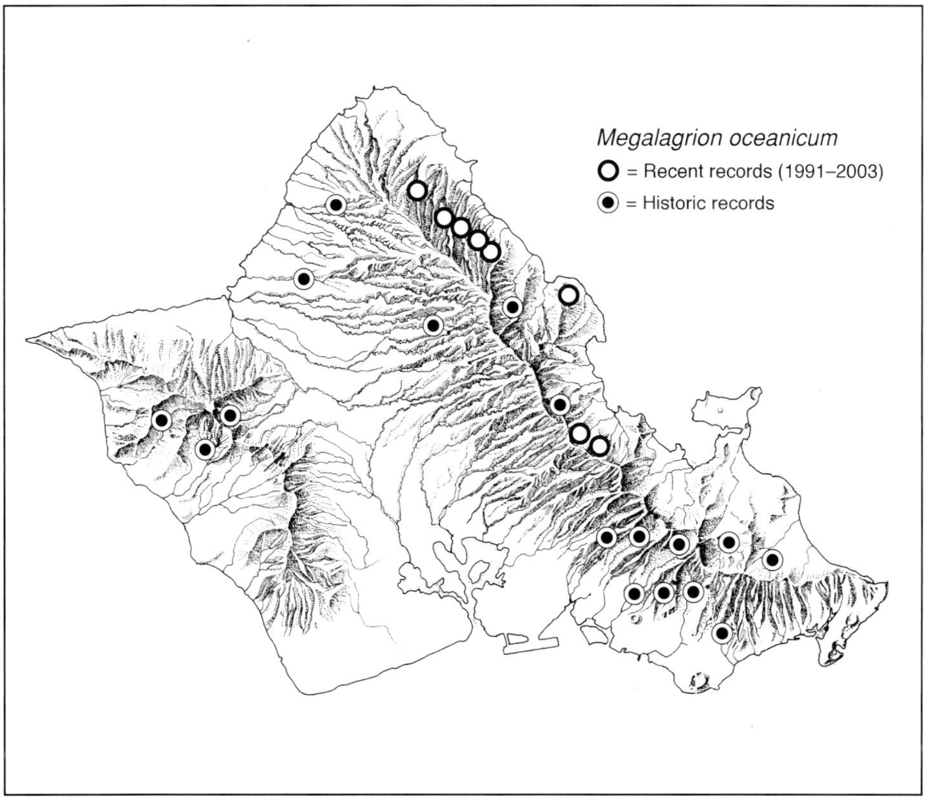

Figure 3. Former and current distribution of *Megalagrion oceanicum* on O'ahu.

plex (Stearns, 1939). This dike complex consists of numerous parallel, vertically-oriented dikes of hard, dense, microcrystalline basalt, in some cases up to 400/km (Macdonald *et al.*, 1983), that represent the remains of the lava conduits that fed the volcano's main rift zone, and are tightly clustered along the remains of this rift zone that forms the southeast to northwest longitudinal axis of the Ko'olau Range (Takasaki *et al.*, 1969). The exact definition of the "core dike complex" has varied among authors; Wentworth (1951) considered it to be the zone containing over 100 dikes per mile, while Takasaki *et al.* (1969) defined it as areas where dikes constitute more than 5 percent of the total rock volume (see Fig. 8 in this latter work for a graphic representation of dike densities in the upper Waiāhole and Kahana systems). Because of their structure, the dike rocks are relatively impermeable, and therefore trap groundwater in the more porous exposures of the Ko'olau Volcanic Series that intervene between them. As noted by Takasaki *et al.* (1969), the dikes generally retard the movement of groundwater in a seaward direction, instead forcing it to move laterally to points of discharge where the dike complex is cut by stream valleys. The dikes are also harder and less susceptible to erosion than the surrounding Ko'olau volcanics, and create vertical waterfalls where streams encounter them in the course of downcutting.

In the bowl-like valleys from Waimānalo to Punalu'u the core dike complex crops out broadly amid vertical cliffs along the valley heads at elevations between 180 and 240 meters (Takasaki *et al.*, 1969), having been exposed by the combined forces of the Nu'uanu landslide and subsequent headward erosion. In the entrenched catchments from Kaluanui to Mālaekahana, the core dike complex is by contrast exposed only to limited extents at or near the heads of the midreach canyons (Fig. 2) at elevations ranging from 200–400 m. These points of core dike complex exposure in turn correlate with the presence of high waterfalls, and the point at which base flow in these systems makes a tran-

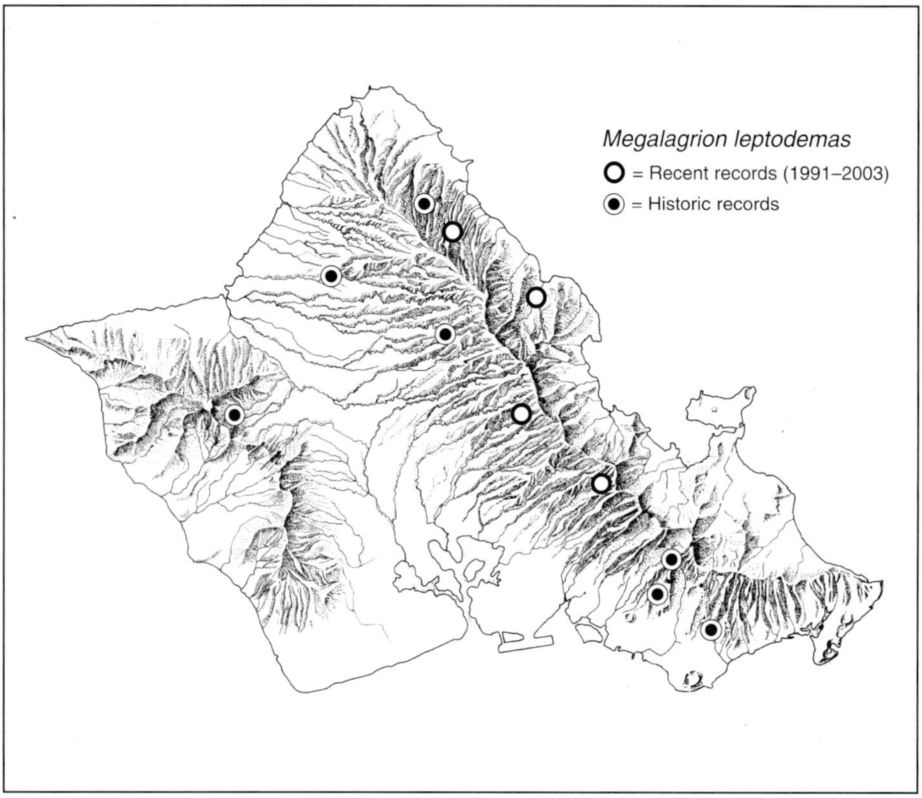

Figure 4. Former and current distribution of *Megalagrion leptodemas* on O'ahu.

sition from perennial surface flow to hyporheic, suballuvial flow. These correlations have in turn had a significant influence on the distribution of the remaining populations of endemic stream-breeding damselflies on O'ahu.

Distribution of Damselfly Populations on the Ko'olau Volcano

The island of O'ahu is known to have historically supported 8 damselfly species in the endemic Hawaiian genus *Megalagrion*. Of these, one, *M. oahuense*, breeds terrestrially beneath banks of *uluhe* ferns on upland ridges; another, *M. koelense*, breeds in the phytotelmata of the climbing pandanus (*Freycinetia arborea*) in similar habitats; and two more, *M. xanthomelas* and *M. pacificum*, breed in lentic habitats or slow stream terminal reaches (Polhemus, 1996). Of these four species, only *M. oahuense* is endemic to O'ahu. These four are not considered further in the current paper.

The native damselfly biota of upland stream habitats, by contrast, consists of three species endemic to O'ahu, plus one species more widespread across the Hawaiian chain. These species, which formerly co-occurred in many catchments throughout the island, are as follows:

Megalagrion oceanicum

This large orange-and-black species is endemic to O'ahu and breeds in stream riffles and rapids, or on waterfall faces. Historic collection records show that *M. oceanicum* was originally present in the windward and leeward drainages of both the Ko'olau and Wai'anae mountain ranges, although it is now extirpated from the latter, with the last specimens from the Wai'anae Range having been taken

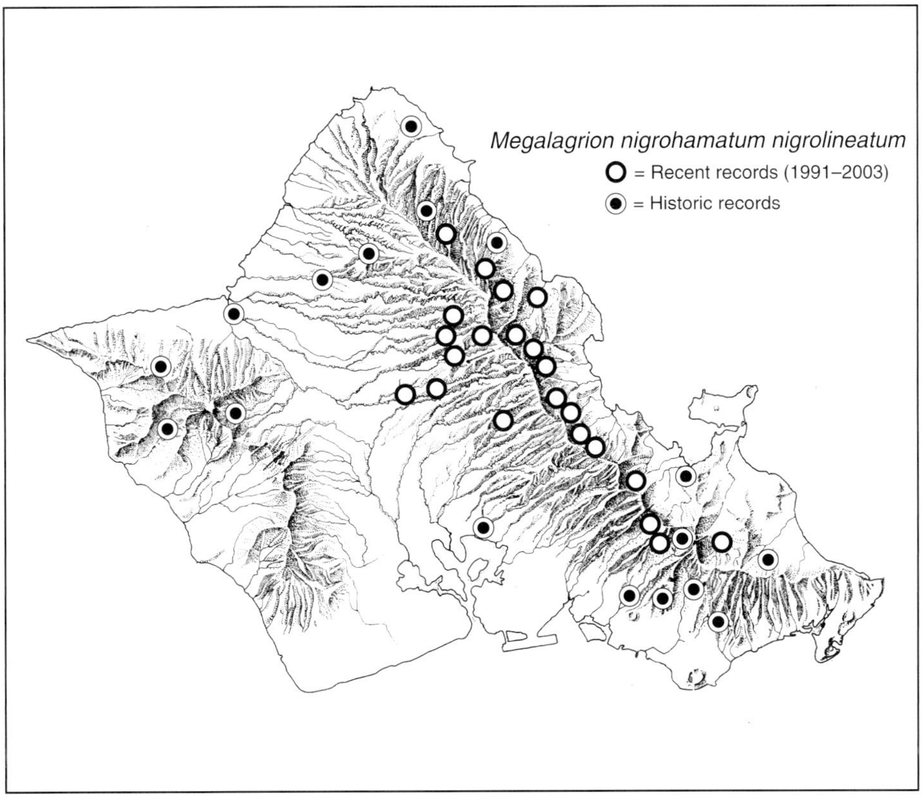

Figure 5. Former and current distribution of *Megalagrion nigrohamatum nigrolineatum* on O'ahu.

above Wai'anae in 1948. Current surveys indicate that the species has also been completely extirpated from the leeward Ko'olau Mountains, and now occurs only as a series of scattered populations on the windward side of the Ko'olaus, occupying 8 stream catchments from Kahalu'u Stream in the southeast to Kahamainui Stream in the northwest (Fig. 3). The persistence of the two southernmost populations on branches of the 'Āhuimanu system (Waihe'e and Kahalu'u) is in fact in doubt; neither has been reconfirmed in nearly a decade, despite several recent visits by the author to Kahalu'u, and intensive scientific study of the Waihe'e catchment from 1999–2001 by researchers from the Biological Resources Division of the U.S. Geological Survey (Brasher *et al.*, 2004).

Megalagrion leptodemas

This slender, predominantly bright red species is endemic to O'ahu and breeds in headwater and upper midreach stream pools. Historic collection records indicate that this species occurred in certain windward drainages of the Wai'anae Mountains (Hale'au'au), and on both the windward and leeward sides of the Ko'olau Mountains. Current surveys have found that this species is now extirpated in the Wai'anae Mountains, and survives in the Ko'olau Mountains only as widely scattered colonies on both the windward and leeward sides of that range. Only four populations are known to exist at present, occupying the North Hālawa, Waiawa, Ma'akua and Kahana drainages (Fig. 4).

Megalagrion nigrohamatum nigrolineatum

This endemic O'ahu subspecies of *M. nigrohamatum* is characterized by broad panels of yellow, orange or reddish coloration on the pterothorax, uninterrupted by lateral stripes, and breeds in headwater and midreach stream pools. Historic collection records show that this species was originally

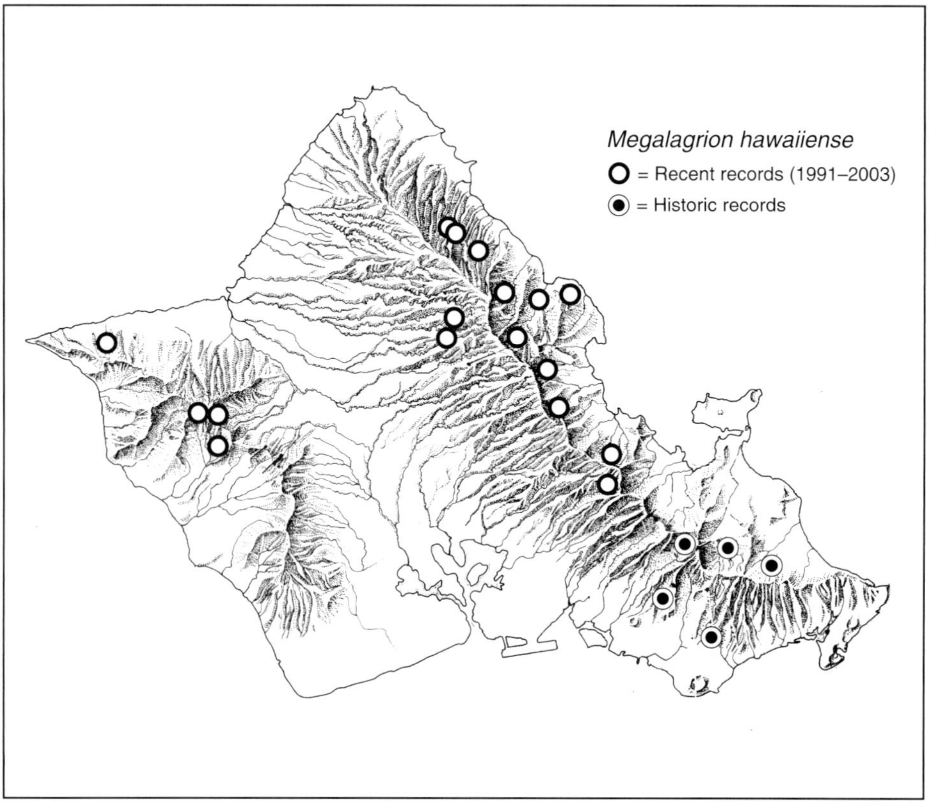

Figure 6. Former and current distribution of *Megalagrion hawaiiense* on O'ahu.

present in the windward and leeward drainages of both the Ko'olau and Wai'anae ranges, but current surveys indicate that the species was completely extirpated in the Wai'anae Mountains sometime after 1980 (the last record being from Pāhole Gulch in that year). The species persists in a modest number of drainages on both the windward and leeward sides of the Ko'olau Mountains (Fig. 5). Some of the remaining populations are robust, and the species shows a degree of tolerance for physically disturbed habitats, provided these are free of introduced poeciliid fishes (Englund, 1999).

Megalagrion hawaiiense

This is a widespread species, occurring on all major high islands in the Hawaiian archipelago except Kaua'i, and breeding on rheocrenes. Across its range it exhibits considerable color polymorphism; although the typical color pattern of males is orange-and-black, variants include aqua-and-black males from the Hāna coast of Maui, and a dark blue-and-black form from the windward Ko'olau Mountains. On O'ahu, historic collection records show that this species occurred in both the Wai'anae and Ko'olau mountain ranges at intermediate to high elevations. Of two color forms present on O'ahu, the typical orange-and-black form occurs in the Wai'anaes and northern leeward Ko'olaus, while the unusual dark blue-and-black form is found in the southern and windward Ko'olaus (Fig. 36 in Polhemus & Asquith, 1996). Recent surveys indicate that this species has been extirpated in the Wai'anae Mountains south of Kolekole Pass (the last record from that section of the range being from Palikea Peak in 1957), but persists north of the pass in the Mt. Ka'ala summit bogs and in the upper drainages of the Wai'anae Kai area below Pu'u Kalena. In the Ko'olau Mountains the dark form of this species persists as numerous small populations associated with headwall seeps

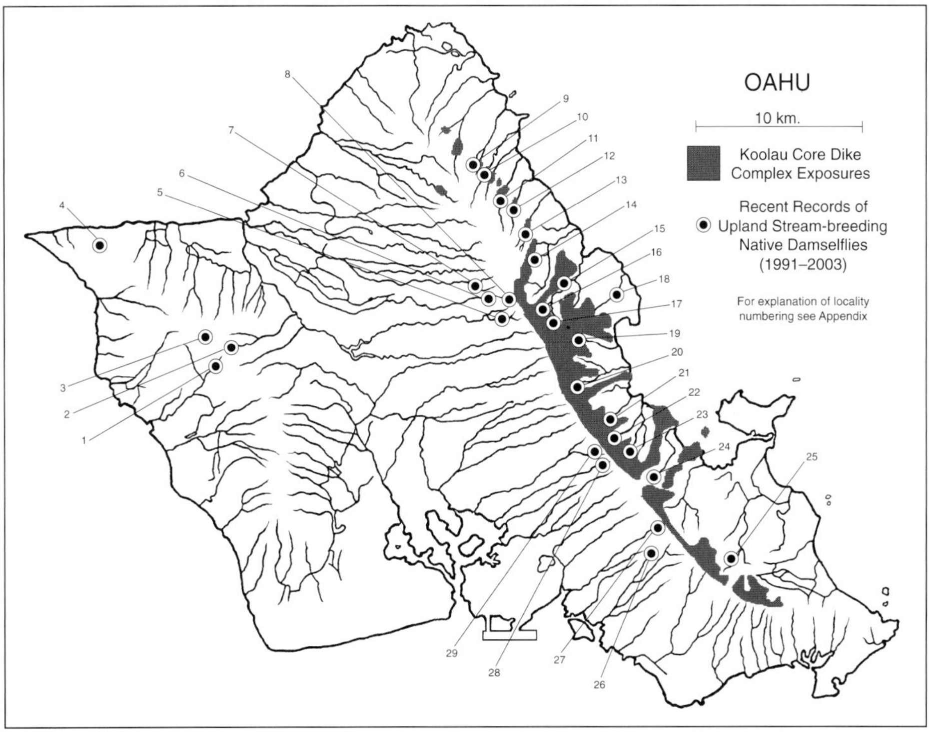

Figure 7. Distribution of currently known populations of upland stream-dwelling *Megalagrion* damselfly species on O'ahu in relation to exposures of the Ko'olau core dike complex. Site numbers correspond to those used in the Appendix, which provides details for each locality.

and riparian rheocrenes, particularly in the windward drainages (Fig. 6).

In examining the distributions of these stream dwelling damselflies on O'ahu (Figs. 3–7), it becomes clear that by far the majority of the remaining populations are confined to the windward drainages of the Ko'olau Mountains (Table 1, and see data presented in Englund, 1999). In addition, overall species richness is clearly greatest in the incised catchments of the northern windward Ko'olau Mountains, and correlates with exposures of the core dike complex. In particular, stream midreaches traversing the core dike complex in this section of the island are almost the only localities in which *M. oceanicum* may still be found. The open, bowl-like valleys of the old caldera from Waikāne southeastward are particularly deficient in native damselflies in comparison to the windward drainages from Kualoa Point northwestward, supporting only scattered populations of *M. nigrohamatum nigrolineatum* and *M. hawaiiense* (except for very small remnant populations of *M. oceanicum* in the 'Āhuimanu system which may already be extirpated—see previous discussion), and southeast of the 'Āhuimanu supporting only two populations of *M. nigrohamatum nigrolineatum*. All of these remaining populations in the southeast sector of the range are concentrated near the broad exposures of the core dike complex at the valley headwalls. Although it may be possible that additional populations of both *M. nigrohamatum nigrolineatum* and *M. hawaiiense* persist near some of the headwall springs behind Waimānalo, the fact remains that even on the windward flank of the Ko'olau Mountains, which represents the last refuge for O'ahu's endemic stream-breeding damselflies, the bowl-like valleys of the old caldera have lost nearly all of their original native damselfly populations, except those associated with the core dike complex exposures.

Discussion

The definition of critical habitat for insect species at risk is one of the more difficult and contentious issues in conservation biology. In some cases, such as insects restricted to particular host plant species, the matter is superficially straightforward, but the exercise then becomes definition of critical habitat for the host plant rather than the insect, and leads to questions of minimum viable patch size. In cases where the insect taxon is a generalist predator or ranges widely across a spectrum of local ecosystems, the matter can become even more complex. Aquatic insect taxa are conceptually more tractable in regard to delineation of critical habitat, since they occupy ecosystems with discrete boundaries, but even here there are differences between lentic habitats such as lakes, ponds and springs, which are discrete landscape features with uniform attributes, and lotic habitats such as streams, which are linear and vary in their characteristics along the length of a catchment.

The current study demonstrates that critical habitat for a particular group of aquatic insects at risk, native Hawaiian *Megalagrion* damselflies breeding in upland streams on the island of O'ahu, can be accurately and predictably defined by evaluation of the geological structure of the catchments they inhabit. In particular, it is clear that colonies of four *Megalagrion* species breeding in upland streams are concentrated around exposures of the core dike complex of the Ko'olau volcano (Fig. 7), and that by using geological maps to locate such exposures it is possible to effectively focus survey efforts, and unambiguously define habitats essential to the continued survival of these species.

The correlation between the remaining populations of native, stream-breeding damselflies on O'ahu and exposures of the core dike complex of the Ko'olau volcano is striking, and probably results from several factors:

1. *Precipitation* - Due to prevailing wind patterns in the northeastern Pacific, the windward sides of the Hawaiian Islands receive significantly more precipitation than the leeward sides. This in turn has an effect on the distribution of perennial streams within islands, and their associated aquatic biota, including *Megalagrion* damselflies. On O'ahu, all remaining populations of upland, stream-breeding *Megalagrion* species are now found in areas that lie within the 2000 mm median annual isohyet (Giambelluca *et al.*, 1986). Even so, the 2000 mm median annual isohyet also covers large areas of the leeward Ko'olau and lower windward Ko'olau slopes that once supported native damselfly populations but now do not, and the remaining damselflies are distinctly clustered and discontinuous along stream reaches that lie within this precipitation zone. Additional factors must thus be involved in determining their current distribution.

2. *Catchment flow regime* - The core dike complex has been known for over a century to be an important aquifer, supplying significant groundwater inflow to collection systems such as the Waiāhole Ditch (Stearns & Vaskvik, 1935; Takasaki *et al.*, 1969; Hirashima, 1971). Exposures of this formation thus represent areas of predictable spring outflow for dike-confined groundwater at the headwalls of the bowl-like valleys in the southeastern Ko'olau Mountains, creating rheocrene habitats that are exploited by *M. hawaiiense*. Due to the dense and impermeable nature of the core dike complex rocks, they also create a zone of both augmented base flow and minimal hyporheos along the incised streams which cross them in the northern windward Ko'olau Mountains, and bring nearly the entire base flow to the surface over a certain distance. It is this feature which is apparently of greatest importance in correlating with the distributions of *M. oceanicum*, *M. leptodemas*, and *M. nigrohamatum nigrolineatum*. In the incised canyons of the north, the stream reaches lying immediately below exposures of the core dike complex make a sudden transition from bedrock to alluvial beds, and base flow makes a similar sudden transition from surface to hyporheic. This hyporheic base flow frequently remains subsurface for 1–2 km below the core dike complex contact, and surfaces again only near the seaward terminus where it floats on top of the inland-percolating marine water table. These naturally interrupted catchments therefore also produce a zone of relatively unsuitable habitat downstream of the core dike complex exposures, and confine damselfly populations to the vicinity of the core dike complex itself or to the headwater reaches upstream of it (see Fig. 4 in Englund, 1999).

3. *Invasive species* - The natural dichotomy of surface flow encountered below exposures of the core dike complex in the canyons from Kaluanui northwestward creates a barrier to the upstream migration of invasive aquatic species, particularly poeciliid fishes, which have been circumstantially implicated in predation on native damselfly populations. Englund (1999), in a study including the majority of windward Ko'olau catchments, demonstrated a strong negative correlation between the presence of poeciliids and that of native damselflies, and in the same paper provided figures that

detailed the elevational distribution of *Megalagrion* species in individual catchments at elevations now known to correspond with those of core dike complex exposures. The dry channels of the naturally interrupted midreaches below the core dike complex also appear to be effective barriers to the upstream migration of invasive amphibians and non-native damselflies. In the bowl-shaped valleys of the southeastern windward Koʻolaus, by contrast, no such zones of natural flow interruption originally existed due to higher base flows fed from more extensive exposures of the core dike complex at the valley heads, thus allowing more effective upstream penetration of invasive species.

4. *Urbanization* - The bowl-like valleys of the southeastern Koʻolau Mountains have open floors that are highly suitable for urban and suburban development, or for agriculture. As a result, they have been heavily altered in the past 200 years by the growth of towns such as Kāneʻohe, Kailua, and Waimānalo, coupled with extensive nursery and pasturing in the latter area extending up to 5 km inland. These land uses have both altered the physical habitat, through the channelization of stream terminal and midreaches, and promoted introduction of invasive species for mosquito control or via releases from home aquaria (Polhemus & Englund, 2003). In addition, the headwalls of the bowl-like valleys have been more readily bored for water development tunnels and trans-basin diversions, resulting in significant losses of base flow (up to 50%) in catchments such as Kahana, Waikāne, Waiāhole, Waiheʻe, and Kahaluʻu (Hirashima, 1963, 1965. 1971; Takasaki *et al.*, 1969). The incised catchments northwest of Kahana Bay, by contrast, have much smaller areas of open, relatively level land along their terminal reaches, and have consequently been less affected by urbanization and agriculture. The few towns that do exist in this area, Hauʻula and Lāʻie, are both set close to the coast, and residential or agricultural developments extend at most 1 km inland. The incised valleys also lack extensive ditch and tunnel systems, because their dike exposures were limited and difficult to utilize. The topography of these catchments cut into the intact windward slopes that were never subjected to giant landslides has thus served to buffer their damselfly populations from a wide spectrum of human impacts.

The hypothesis that the core dike complex exposures correlate with current damselfly distributions is testable to some extent, because not all catchments on Oʻahu containing exposures of the this formation have yet been sampled for native damselflies. Several moderately incised but incompletely surveyed gulches remain in the northern windward quadrant of the range, particularly ʻIhiʻihi and Wailele [the record of *M. oceanicum* from this latter catchment in Englund (1999) was in error]. Of equal interest, a single exposure of the core dike complex is also exposed leeward of the Koʻolau crest in the upper Waimea system, and would certainly repay examination. At the same time, not all streams with core dike complex exposures harbor native damselflies: although the complex crops out along the upper midreach of Mālaekahana Stream, surveys here have indicated an absence of *Megalagrion* species, possibly because this stream valley is relatively dry for a windward location, not deeply entrenched, and has therefore been easily colonized by invasive species and formerly subjected to sugarcane cultivation in its lower section. Finally, the geological factors correlating with native damselfly distributions on Oʻahu may also have analogs on other islands in the Hawaiian chain. On eastern Maui, for instance, the old, hard flows of the Honomanū basalts have been exposed in certain deeply incised catchments on the windward flank of Haleakalā, such as Hanawī, and these areas are also known to support populations of increasingly rare *Megalagrion* species.

Clearly, some form of protection and management for the remaining core dike complex exposures and their proximal stream reaches is an essential step in efforts to retain the endemic stream-dwelling damselfly fauna of Oʻahu. Even so, these localities now appear to represent the "last of the last" in regard to populations of retreating species such as *M. leptodemas* and *M. oceanicum*, and it is an open question as to whether these few particularly optimal catchments constitute sufficient minimum core range to ensure the long-term survival of these species. The same considerations also apply to other endemic, stream-dwelling species on Oʻahu, particularly native Diptera in the genera *Telamatogeton* and *Procanace*, the remaining populations of which are also largely confined to stream reaches in proximity to exposures of the core dike complex.

In summary, geomorphology and petrology appear to be acting as significant factors in determining the distribution of the remaining populations of endemic, stream-breeding damselflies and other lotic insects on the island of Oʻahu, and probably elsewhere in the Hawaiian Islands. As such, they should be taken into account in any future effort to delineate critical habitat for these increasingly endangered species.

Acknowledgments

This paper represents the aggregate results of 15 years of field work on O'ahu, and could not have been completed without the help of numerous colleagues who assisted with stream surveys, museum support, and guidance with geological literature. Among those I particularly thank are Ron Englund, Bishop Museum; Kevin Johnson, University of Hawai'i; Adam Asquith, formerly of the U.S. Fish and Wildlife Service; Bill Devick, formerly of the Division of Aquatic Resources, Hawaii State Department of Land and Natural Resources; and Randall Filbert, whereabouts unknown.

Ron Englund, J. Michael Fitzsimmons, and an anonymous reviewer provided comments and corrections that greatly improved the quality of the final manuscript, and their efforts are gratefully acknowledged.

The distributional surveys of *Megalagrion* damselflies that form the core of this report were funded by an initial grant from the U.S. Fish and Wildlife Service, and by subsequent grants from the State of Hawai'i Division of Aquatic Resources. I thank these agencies for their continued support of research on Hawaiian damselflies and other native stream biota.

Literature Cited

Brasher, A.M.D., R.H. Wolff & C.D. Luton. 2004. Associations among land use, habitat characteristics, and invertebrate community structure in nine streams on the island of O'ahu, Hawaii, 1999–2001. *United States Geological Survey Water Investigations Report* **03-4256**. viii + 47 p.

Carson, H.L. & D.A. Clague. 1995. Geology and biogeography of the Hawaiian Islands, p. 14–29. *In*: W.L. Wagner & V.A. Funk (eds.), *Hawaiian biogeography: evolution on a hot spot archipelago*. Smithsonian Institution Press, Washington, D.C.

Clague, D.A., J.G. Moore & A.S. Davis. 2002. Volcanic breccia and hyaloclastite in blocks from the Nuuanu and Wailau landslides, Hawaii, p. 279–296. *In*: E. Takahashi, P.W. Lipman, M.O. Garcia, J. Naka & S. Aramaki (eds.), *Hawaiian volcanoes: deep underwater perspectives*. Geophysical Monograph 128, American Geophysical Union.

Englund, R.A. 1999. The impact of introduced poeciliid fish and Odonata on the endemic *Megalagrion* (Odonata) damselflies of O'ahu Island, Hawai'i. *Journal of Insect Conservation* **3**: 225– 243.

Giambelluca, T.W., M.A. Nullet & T.A. Schroeder. 1986. *Rainfall atlas of Hawaii*. State of Hawaii Department of Land & Natural Resources, Honolulu. vi + 267 p.

Hawaii Stream Assessment. 1990. *A preliminary appraisal of Hawaii's stream resources*. Report **R-84**. Hawaii Cooperative Park Service Unit, Western Region Natural Resources & Research Division. Honolulu. 294 p.

Hirashima, G.T. 1963. Influence of water-development tunnels on streamflow-groundwater relations in Haiku-Kahaluu area, Oahu, Hawaii. *State of Hawaii Division of Water & Land Development Circular* **C21**, 11 p.

———. 1965. Effects of water withdrawals by tunnels, Waihee Valley, Oahu, Hawaii. *State of Hawaii Division of Water & Land Development* Report **R28**: 38 p.

———. 1971. Tunnels and dikes of the Koolau Range, Oahu, Hawaii, and their effect on storage depletion and movement of ground water. *Geological Survey Water-Supply Paper* **1999-M**. United States Government Printing Office, Washington. iv + 21 p.

Jordan, S., C. Simon & D.A. Polhemus. 2003. Molecular systematics and adaptive radiation of Hawaii's endemic damselfly genus *Megalagrion* (Odonata: Coenagrionidae). *Systematic Biology*: **52**: 89–109.

Liebherr, J.K. & D.A. Polhemus. 1997. Comparisons to the century before: the legacy of R.C.L. Perkins and the *Fauna Hawaiiensis* as the basis for a long-term ecological monitoring program. *Pacific Science* **51** (4): 490–504.

Macdonald, G.A., A.T. Abbott & F.L. Peterson. 1983. *Volcanoes in the sea. The geology of Hawaii*. Second edition. University of Hawaii Press, Honolulu. x + 517 p.

Moore, J.G. 1964. Giant submarine landslides on the Hawaiian Ridge. *United States Geological Survey Professional Paper* **501-D**: 95–98.

———. **& D.A. Clague.** 2002. Mapping the Nuuanu and Wailau landslides in Hawaii, p. 223–244. *In*: E. Takahashi, P.W. Lipman, M.O. Garcia, J. Naka & S. Aramaki (eds.), *Hawaiian volcanoes: deep underwater perspectives*. Geophysical Monograph 128, American Geophysical Union.

Polhemus, D.A. 1993. Damsels in distress: a review of the conservation status of Hawaiian *Megalagrion* damselflies (Odonata: Coenagrionidae). *Aquatic Conservation: Marine and Freshwater Ecosystems* **3**: 343–349.

———. 1996. The orangeblack Hawaiian damselfly, *Megalagrion xanthomelas* (Odonata: Coenagrionidae): clarifying the current range of a threatened species. *Bishop Museum Occasional Papers* **45**: 30–53.

———. 1997. Phylogenetic analysis of the Hawaiian damselfly genus *Megalagrion* (Odonata: Coenagrionidae): implications for biogeography, ecology and conservation biology. *Pacific Science* **51**(4): 395–412.

———. **& A. Asquith.** 1996. *Hawaiian Damselflies: a field identification guide.* Hawaii Biological Survey Handbook No. 1. Bishop Museum Press, Honolulu. x + 122 p.

———. **& R.A. Englund.** 2003. Predictably sequential pathways for invasions on tropical Pacific islands, p. 58–60 *In*: L.A. Meyerson, K.A. Ciruna, A. Gutierrez & E. Watson (eds.), *Ecological and socio-economic impacts of invasive alien species on inland water ecosystems: report of experts' consultation.* Global Invasive Species Programme, Washington, D.C.

———., **J. Maciolek & J. Ford**. 1992. An ecosystem classification of inland waters for the tropical Pacific islands. *Micronesica* **25**(2): 155–173.

Stearns, H.T. 1939. *Geologic map and guide of the island of Oahu, Hawaii.* Territory of Hawaii, Department of Public Lands, Division of Hydrography, Bulletin 2. x + 76 p.

———. **& K.N. Vaskvik.** 1935. *Geology and ground-water resources of the island of Oahu, Hawaii.* Territory of Hawaii, Department of Public Lands, Division of Hydrography, Bulletin 1. xx + 479 p.

Takasaki, K.J., G.T. Hirashima & E.R. Lubke. 1969. *Water resources of windward Oahu, Hawaii.* Geological Survey Water-Supply Paper 1894. United States Government Printing Office, Washington. viii + 119 p., 3 maps.

Wentworth, C.K. 1951. *Geology and ground-water resources of the Honolulu-Pearl Harbor area, Oahu, Hawaii.* Board of Water Supply, Honolulu, Hawaii. 111 pp.

Williams, F.X. 1936. Biological studies in Hawaiian water-loving insects. Part I. Coleoptera or beetles. Part II. Odonata or Dragonflies. *Proceedings of the Hawaiian Entomological Society* **9**(2): 235–349.

Zimmerman, E.C. 1948. *Insects of Hawaii.* Volume 2. Apterygota to Thysanoptera inclusive. University of Hawaii Press, Honolulu. 475 p.

APPENDIX

Localities on O‘ahu at which native upland stream-breeding damselflies have been recorded from 1991–2003.

Locality numbers correspond to those in Figure 7 and represent sites at which one or more of the following species have been collected or unequivocably observed: *Megalagrion leptodemas*, *M. oceanicum*, *M. nigrohamatum nigrolineatum*, or *M. hawaiiense* (for detailed data on species occurring along particular streams and reaches see Englund, 1999). Dates in parentheses indicate the most recent year in which any of the preceding 4 species has been recorded at a given locality. Elevations are given in meters with foot equivalents in order to allow interface to USGS topographic maps.

1. Wai‘anae Mtns., Kānewai Stream (trib. to Honua Stream), upper Wai‘anae Valley, 425 m [1400 ft] (1994)
2. Wai‘anae Mtns., Mt. Ka‘ala, headwater tributary of Haleanau Gulch at edge of summit bog, 1160 m [3800 ft] (1991)
3. Wai‘anae Mtns., headwaters of Mākaha Stream below Mt. Ka‘ala summit bog, 1160 m [3800 ft] (1991)
4. Wai‘anae Mtns., small stream S. of Dillingham airfield, 150 m [500 ft] (1998)
5. Ko‘olau Mtns., upper Kaukonahua Stream, 460–520 m [1500–1700 ft] (1996)
6. Ko‘olau Mtns., Poamoho Stream, headwater reach, 520 m [1700 ft] (1996)
7. Ko‘olau Mtns., upper Helemano Stream, 490 m [1600 ft] (1996)
8. Ko‘olau Mtns., lake along upper Poamoho Stream, S. of Poamoho trail summit, 700 m [2300 ft] (1995)
9. Ko‘olau Mtns., upper Kahamainui Gulch, above Lā‘ie, 390 m [1280 ft] (1993)
10. Ko‘olau Mtns., Koloa Gulch, above Lā‘ie, from forks to first major waterfalls on each branch, 275–365 m [900-1200 ft] (1994)
11. Ko‘olau Mtns., Kaipapa‘u Stream, below first major waterfall, 305 m [1000 ft] (1994)
12. Ko‘olau Mtns., Ma‘akua Gulch, above Hau‘ula, at first waterfall, 150–215 m [500–700 ft] (1994)
13. Ko‘olau Mtns., upper Kaluanui Stream, above Sacred Falls, 670–760 m [2200-2500 ft] (1993)
14. Ko‘olau Mtns., Punalu‘u Stream, Waiho‘i Springs, north fork, 180 m [600 ft] (2001)
15. Ko‘olau Mtns., Kahana Stream, midreach, 30–270 m [100–880 ft] (1995)
16. Ko‘olau Mtns., Kahana Valley, headwaters of north branch of Kahana Stream at Intake 1 on Waiāhole Ditch tunnel, 245 m [800 ft] (1995)
17. Ko‘olau Mtns., Kahana Valley, headwaters of Kahana Stream at Intake 19 on Waiāhole ditch tunnel, 245 m [800 ft] (1995)
18. Ko‘olau Mtns., Makaua Gulch, above Ka‘a‘awa, 245 m [800 ft] (1994)
19. Ko‘olau Mtns., headwaters of Waikāne Stream at Waikāne Camp (site), 245 m [800 ft] (1995)
20. Ko‘olau Mtns., Uwau Stream (trib. to Waiāhole), headwaters, 245 m [800 ft] (1995)
21. Ko‘olau Mtns., Ka‘alaea Stream, near bridge on road, 25–30 m [80–100 ft] (1995)
22. Ko‘olau Mtns., Kahalu‘u Stream, headwaters of north branch, 150 m [500 ft] (2000)
23. Ko‘olau Mtns., Waihe‘e Stream, NW of Kāne‘ohe, waterfall and rocky stream at headwall, 215 m [700 ft] (2003)
24. Ko‘olau Mtns., He‘eia Stream in Ha‘ikū Valley, at Board of Water Supply pumphouse, 135 m [440 ft] (1995)
25. Ko‘olau Mtns., ‘Ōma‘o Stream, Maunawili Valley, 95 m [320 ft] (1995)
26. Ko‘olau Mtns., headwaters of Mo‘ole Stream (trib. to Nu‘uanu Stream) below Mt. Lanihuli, 520–610 m [1700–2000 ft] (1996)
27. Ko‘olau Mtns., headwaters of Kālihi Stream, above Wilson Tunnel, 305–365 m [1000–1200 ft] (1994)
28. Ko‘olau Mtns., North Hālawa Valley, tributary above highway tunnel, 305 m [1000 ft] (2000)
29. Ko‘olau Mtns., Waiwa Stream, upstream of rain gauge, 215–245 m [700–800 ft] (1995)

Biology of Hawaiian Streams and Estuaries. Edited by N.L. Evenhuis & J.M. Fitzsimons. *Bishop Museum Bulletin in Cultural and Environmental Studies* 3: 247–260 (2007).

Blue *hawaiiense* and Beyond: Conservation Genetics and Comparative Phylogeography of Four Hawaiian *Megalagrion* Damselfly Species (Odonata: Coenagrionidae)

STEVE JORDAN, EMILIE BARRUET, MARK OLAF
Department of Biology, Bucknell University, Lewisburg, Pennsylvania 17837, USA; email: sdjordan@bucknell.edu

BARBARA PARSONS
University of Kansas, Department of Ecology and Evolutionary Biology, Lawrence, Kansas 66045, USA

CHRIS SIMON
Department of Ecology and Evolutionary Biology, University of Connecticut, Storrs, Connecticut 06269, USA

Abstract

Hawaii's endemic *Megalagrion* damselflies are rivaled in their beauty and diversity only by the degree of threat posed to them by anthropogenic disturbance. In this preliminary study of phylogeography and conservation genetics, we have sequenced about 660 base pairs (bp) of the mitochondrial COII gene from 191 damselflies from four species, including 31 populations that span a gradient of endangerment. We applied phylogeographic analyses in order to understand their biogeographic history. Unlike *Megalagrion xanthomelas* and *M. pacificum*, *M. calliphya* and *M. hawaiiense* rarely share haplotypes between populations and between islands, even within the larger Maui Nui super-island, suggesting that these latter two species do not disperse as well across land or water. Their phylogenies also better reflect the geological history of the islands. We applied conservation genetic analyses in order to understand their genetic health. Under a conservation genetic paradigm, populations with low genetic diversity are generally considered to be at greater risk of decline and extinction than populations with high genetic diversity. Applying this principle to *Megalagrion* populations gave both expected and surprising results. Expected results included measurements of high diversity in most populations of *M. calliphya*, *M. hawaiiense*, and Hawai'i Island populations of *M. xanthomelas*. Low diversity was observed in populations known to be bottlenecked or relictual, including O'ahu and Maui *M. xanthomelas*, and *M. pacificum*. Surprising results included low genetic diversity in O'ahu Ko'olau and Hawai'i Onomea *M. hawaiiense*, Moloka'i *M. pacificum*, and West Maui *M. calliphya*. We propose that these latter three populations be monitored and managed to maximize their long-term genetic health.

Introduction

Damselflies of the endemic Hawaiian genus *Megalagrion* are arguably Hawaii's most visible native stream insects. Found on all the main high islands, this genus has long fascinated biologists with its beauty and diversity (McLachlan, 1883; Perkins, 1899; Kennedy, 1928; Williams, 1936; Zimmerman, 1948; Polhemus, 1997). In fact, the approximately 23 species of this genus exploit the full known range of damselfly habitats, from ponds, anchialine pools, and low gradient streams, to fast mountain streams, acidic bogs, plant leaf axils, and even terrestrial habitats. *Megalagrion* appears to have arrived in Hawai'i about 11 million years ago, colonizing high islands that are now eroded away (Jordan *et al.*, 2003). It is an outstanding example of adaptive radiation, with bursts of speciation correlated with morphological innovations allowing the exploitation of broadly diverse and novel habitats (Jordan *et al.*, 2003). Fourteen of its species are endemic to single islands. Nine are found on more than one island, and two, *M. xanthomelas* and *M. pacificum* are known historically from the entire chain.

Extensive surveys carried out by the Hawai'i Biological Survey (HBS) in the 1990s led to a broad modern understanding of the range and abundance of *Megalagrion* species, and how these attributes have changed since the early days of Hawaiian entomology (e.g., Polhemus, 1993; Englund, 1999a). Most species, including some long believed extinct, were located. However, not all species and populations were found to be healthy, and some were merely remnants of once-thriving communities described by the early Hawaiian entomologists (see, for example, Englund, 2001). Because of this work, six species or subspecies of *Megalagrion* are currently considered Candidate Species under the U. S. Endangered Species Act (USFWS, 2004).

Polhemus (1993) highlighted the threats posed to *Megalagrion* by invasive species (fish, frogs, and other invertebrates) and habitat destruction correlated with human activity. Polhemus (1997) went further and posited two general classes of threats to Hawaiian damselflies: 1) phylogenetically related ecological vulnerability and 2) "biogeographic misfortune". *Megalagrion xanthomelas* and *M. pacificum* illustrate the first case. They are sister species that breed in coastal habitats and lower to middle stream reaches. This puts them in direct conflict with most human activity on the islands. Three other species (*M. leptodemas*, *M. oceanicum*, and *M. nigrohamatum nigrolineatum*) that are not closely related to each other illustrate the second case. They have the misfortune of being native to O'ahu, which has experienced more anthropogenic ecological disturbance than any other Hawaiian island. This has taken a heavy toll on Oahu's native insects.

Although few formal quantitative surveys of *Megalagrion* have been published (Englund,1999a; Englund, 2001), workers from the HBS (R. Englund, N. Evenhuis, D. Preston, D. Polhemus, pers. comm.) and the U.S. Geological Survey (D. Foote, pers. comm.) have spent thousands of hours monitoring *Megalagrion* populations across the Hawaiian Islands, including many remote sites, within the context of their specific research goals. From this work, we have an excellent qualitative feel for the health of various *Megalagrion* populations based on ranges, abundances, ease of capture and sighting frequencies. In many cases, these observations span more than 15 years, and changes in population health over time have been observed. While a quantitative synthesis of this work is needed, our general knowledge of the health of some *Megalagrion* populations is excellent (e.g., see Polhemus, 2007). And while we have seen the exciting rediscovery of some *Megalagrion* populations (Evenhuis & Cowie, 1994; Englund, 1999b; Polhemus *et al.*, 1999), information on the health of others is disquieting.

The two main goals of conservation genetics are 1) the preservation of genetic diversity and the processes that foster it so that populations can meet the demands of a changing environment, and 2) the avoidance of inbreeding depression (Reed & Frankham, 2003). Inbreeding is a decrease in genetic diversity resulting from matings between related individuals. While many factors can decrease genetic diversity [e.g., certain life history and ecological strategies, natural selection, molecular structure of enzyme products, and many others—see Avise (2004)], inbreeding seems to trump all of these. That is, no matter what else is occurring, if inbreeding occurs, genetic diversity (heterozygosity and related measures) will decrease. This decrease in genetic diversity is often accompanied by inbreeding depression, a decrease in growth, survival or fertility (Avise, 2004).

Here we seek to quantify the genetic health of populations from four *Megalagrion* species by analyzing mtDNA sequences with phylogenetic and conservation genetic techniques. In particular, we discuss the relationship between genetic diversity at a mitochondrial locus and the abundance, range, and known demographic history of populations from four species of *Megalagrion*. Genetic data can serve at least three purposes toward the goal of *Megalagrion* conservation. First, phylogenetic analysis of genetic data can help us to identify species and their genetically unique subunits that merit conservation attention (Moritz, 1994; Crandall *et al.*, 2000). Second, genetic diversity is an indicator of the health of the population. Although Lande (1988) maintained that demographic factors have a greater influence on population endangerment and extinction than genetic factors, recent meta-analyses of hundreds of individual taxa have shown that may not be true in many cases (DeSalle, 2005). These studies indicate that population health is often influenced by low genetic diversity (Spielman *et al.*, 2004) and the loss of heterozygosity can have a deleterious effect on population fitness (Reed & Frankham, 2003). Finally, genetic diversity can give us an idea of whether and how recent perturbations have affected populations. For example, populations that have seen reductions in population size due to anthropogenic causes are expected to have lower genetic diversity than those that have not, due to inbreeding.

The four species analyzed here include populations of known and unknown health. These species were selected for the following reasons. First, all of them are widespread, being found on at least four separate islands. Second, *M. xanthomelas* and *M. pacificum*, which have recently been the subject of a more detailed phylogeographic analysis (Jordan *et al.*, 2005), are candidates for listing under the U.S. Endangered Species Act. Extensive work on these species by the HBS has given us an excellent qualitative sense of the health of their sampled populations. In particular, the *M. xanthomelas* population on O'ahu is known to have experienced recent severe bottlenecks (Englund, 2001). Because of this, we expect genetic diversity to be low in it, allowing us to use it as a reference point for comparison with results from other populations. *Megalagrion pacificum* is one of the most endangered of the Hawaiian damselflies, having been extirpated on Kaua'i and O'ahu. Third, we chose *M. hawaiiense* because it is morphologically variable across its range, and has been proposed for splitting into at least two separate species (Daigle, 2000). In particular, *M. hawaiiense* males display color variability. Most males of the species are red, but blue males occur in the Ko'olau Mountains of O'ahu, and males with red abdomens and aqua thoraces are found near waterfalls along the Hana coast of Maui. There has been disagreement in the literature about the taxonomic status of these color variants (Polhemus & Asquith, 1996; Daigle, 2000). Although *M. hawaiiense* has never been considered to be of conservation concern, its morphological variability suggests that certain populations may be unique, meriting specific conservation action. This species has not been monitored as much as the other three, and so we have less information about its health. Genetic data can give us a starting point for its conservation. Fourth, we chose *M. calliphya* to serve as a sort of positive control. It has never been considered threatened in any way, and its populations seem quite healthy across its range. In fact it has successfully colonized many human-created freshwater sources at Volcano Village on Hawaii Island, including rainwater cisterns and ditches (e.g., the moat around a USGS greenhouse described in the stream conference paper presented by Idelle A. Cooper & David Foote, "Response of *Megalagrion calliphya* to simulated drought events"). We therefore expect it to display higher levels of genetic diversity than populations that have been negatively impacted by human activities.

We have initiated this preliminary study because knowledge of population genetic diversity can contribute to an understanding of the conservation needs of *Megalagrion* damselflies. Here, we use DNA sequence data to identify conservation units within each species and discuss conservation, taxonomy, and biogeography. We also quantify genetic diversity across a gradient of population health levels, including populations known to be at risk and populations known to be healthy. Correlations of population health and genetic variability are then used to assess the health of poorly known populations. Inspection of results from this comparison suggest that several populations may need conservation attention. Our results from this preliminary study will guide us as we expand the scope of the project in the future by including more loci, individuals, populations, and analyses.

Materials & Methods

A total of 191 damselflies representing four species were sampled from 31 populations (Table 1, Fig. 1) over a 14-year period of time. In many cases, population sample sizes were limited by the difficulties of collecting rare organisms over diverse and rugged terrain. These sample sizes are smaller than would be optimal for a conservation genetics study, and haplotype frequencies estimated from them are thus less reliable than those from large populations. However, because of the exploratory nature of this study and the importance of the questions being addressed, we have chosen to retain these samples in order to maximize the information available to decision makers. DNA was extracted from thoracic or leg muscle using a Qiagen DNEasy kit.

Approximately 660 base pairs of the mitochondrial cytochrome oxidase II gene were amplified

Table 1. Populations, sample sizes, and haplotype distributions for four damselfly species included in the current study. X = *M. xanthomelas*, P = *M. pacificum*, H = *M. hawaiiense*, C = *M. calliphya*. Vertically adjacent numbers in boldface represent samples from neighboring locales that were aggregated for the population genetic analyses. Haplotype codes correspond to Figs. 2, 3, 4. Numbers following colons represent the number of individuals bearing a particular haplotype in a particular population.

	Sample size				
Locality sampled	*X*	*P*	*H*	*C*	**Haplotypes: Frequency**
O'ahu					
Tripler Army Medical Center (TAMC)	20				Xa:20
Mount Ka'ala			7		HWa:6, HWc:1
Ko'olau Mtns. Waikāne Stream			6		HWa:2, HWb:4
Moloka`i					
Kalaupapa Peninsula		**6**			Pa:6
Waikolu Valley			**8**	4	HWc:7, HWm:1 – Ca:2, Cb:2
Pelekunu Valley & taro patch near Keawenui Stream	14				Xa:5, Xb:6, Xc:2, Xe:1
Wailau Valley		**9**			Pa:6, Pc:1, Pe:1, Pf:1
Kamakou TNC Preserve			**1**	9	HWh:1 – Cb:3, Ci:1, Cj:2, Ck:1, Cm:1, Cn:1
Lāna'i					
Pipeline seepage, Maunalei Gulch	**6**				Xa:4, Xc:2
Kō'ele Lodge Ponds	**4**				Xa:2, Xb:2
Maui					
East Maui, Cape Hanamanioa Region	4				Xb:1, Xd:3
East Maui, Haipua'ena Stream		4	**2**		Pb:3, Pd:1 – HWf:2
East Maui, East Wailua Iki Stream			**3**	**6**	HWf:2, HWj:1 – Cf:2, Cg:1, Ch:2, Cp:1
East Maui, Upper Hanawī			**1**	**4**	HWk:1 – Ch:2, Cs:1, Ct:1
West Maui, Olowalu Stream			**2**		HWg:1, HWl:1
West Maui, Mt. 'Eke			**1**	4	HWg:1 – Ce:3, Cu:1
Hawai'i					
Kohala Mtns., upper Waikoloa Stream, N. of Waimea				**1**	Cd:1
Kohala Mtns. SE of Pu'u Pili				**1**	Cr:1
Kohala Mtns., upper Alakahi Stream at head of Waipi'o Valley			**3**		HWd:2, HWi:1
Kohala Mtns., Honokāne Nui Stream			**3**		HWd:3
Waikaumalo Co. Park	12				Xf:3, Xg:3, Xh:1, Xj:3, Xm:1, Xr:1
Onomea Str. above garden			4		HWe:4
Hawai'i Volcanoes National Park, Ola'a Pu'u Unit				11	Cc:6, Cd:2, Cg:1, Co:1, Cq:1
Kāwā Springs	15				Xf:7, Xh:6, Xp:1, Xq:1
Kaloko Springs	16				Xa:1, Xb:1, Xf:6, Xh:2, Xi:4, Xk:1, Xn:1
Total	91	19	41	40	

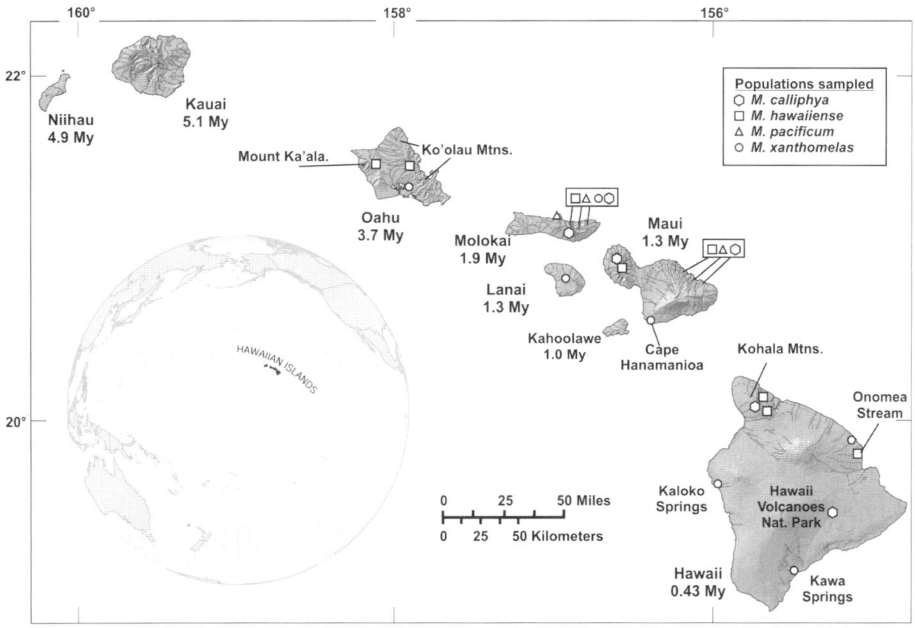

Figure 1. Map of the high Hawaiian Islands, showing sample sites for *Megalagrion calliphya*, *M. hawaiiense*, *M. pacificum*, and *M. xanthomelas*. Island ages based on K-Ar dating are shown (Clague & Dalrymple, 1987).

for each individual damselfly using the polymerase chain reaction (PCR) and primers C2-J-3102 (aaatggcaacatgagcacaayt) and TK-N-3773 (gagaccagtacttgctttcagtcatc) from Jordan *et al.* (2003). Standard PCR was carried out with the AmpliTAQ gold enzyme under the following conditions: 10 minutes at 95 °C followed by 35 cycles of 30 sec. at 95 °C, 30 sec. at 50 °C, and 60 sec. at 72 °C. PCR products were purified by using a Qiagen Qiaquick purification kit. Cycle sequencing using each of the PCR primers was carried out following the protocols from the manufacturer of the dye terminators (ABI or Beckman Coulter). Sequencing products were purified by using Sephadex spin columns or ethanol precipitation. DNA sequences were visualized on either an ABI 377, 3100, or Beckman Coulter CEQ 8000 instrument.

Individual primer sequences were compared and corrected for base-calling errors by using Sequencher 3.0 (Genecodes Corp., Ann Arbor, Michigan) or CodonCode Aligner (CodonCode Corp., Dedham, Massachusetts). Final alignments were generated using ClustalW (Thompson *et al.* 1994), and were adjusted by eye using MEGA (Kumar *et al.*, 2001).

Two types of phylogenetic analysis were performed on unique haplotypes from each data set: 1) maximum likelihood (ML) with bootstrapping and 2) Bayesian. Initial data exploration (not shown) demonstrated that, without fail, outgroup taxa were connected to the ingroup by relatively long branches, and that this could affect hypothesized ingroup relationships. Therefore, all analyses were performed without outgroups. Three separate phylogenetic analyses were performed. *Megalagrion xanthomelas* and *M. pacificum* are closely related and were analyzed together, and the other two species were analyzed separately. In order to correct for multiple substitutions at nucleotide sites and for variation in substitution rates among sites, we selected the simplest ML model that did not differ significantly from the most complex (best fitting) model (GTR+I+G) for each data set. This was done using the likelihood ratio test procedure of Frati *et al.* (1997), the Akaike information criterion method of Buckley *et al.* (2002) and PAUP* 4.10b (Swofford, 1998). This simplest adequate model reduces computing time and the variance of estimated parameters. We used this appropriate model to perform heuristic ML searches in PAUP* with 10 random addition sequence replicates and TBR branch swapping. Parameter estimates for the appropriate model were fixed after being estimated on an initial

Table 2. Population and sampling parameters for four species of Hawaiian damselflies. Haplotype and nucleotide diversity and their sampling variances were calculated for entire species, without subdivision. The genetic distance shown is the maximum value calculated between haplotypes within each species using the model indicated.

	N	Number Haplotypes	Aligned Sequence Length (bp)	Number Variable Sites/Pars. inform.	ML Model	Maximim Corrected Genetic Distance	Haplotype Diversity +/-V	Nucleotide Diversity +/-V
M. calliphya	40	20	653	27/19	HKY+I	0.028	0.95 +/- 0.017	0.0085 +/- 0.0046
M. hawaiiense	41	13	643	26/12	GTR+I	0.025	0.90 +/- 0.023	0.0105 +/- 0.0056
M. pacificum	19	6	660	32/20	HKY+G	0.0031	0.60 +/- 0.12	0.0011 +/- 0.0009
M. xanthomelas	91	16	660	32/20	HKY+G	0.032	0.82 +/- 0.03	0.0116 +/- 0.0060

neighbor joining tree. The parameter values were re-estimated on the resulting ML tree, and the heuristic search was performed again. If the results of the two heuristic searches agreed, we searched no further. ML bootstrap analysis was performed under the appropriate model with parameter values fixed based on estimates from the ML tree. We performed 200 pseudoreplicates in PAUP* by using a heuristic search with one random addition sequence replicate and retaining one tree.

Bayesian analysis was performed by using MrBayes 2.01 (Huelsenbeck & Ronquist, 2001). Unique haplotypes for each data set were analyzed using the Metropolis coupled Markov chain Monte Carlo method. Four chains, three heated and one cold, were run for 1 million generations each. Only sample points occurring after stationarity of the negative log likelihood score (-lnL) was achieved were used in calculating posterior probabilities. Stationarity was assessed by graphing the -lnL for every 100th generation, and looking for the point where these values leveled off near their optimal value. The burnin value (generations discarded) was different for each data set (2000–12000 generations).

Population genetic parameters were estimated using the Arelequin software package (Schneider *et al.*, 2000). We estimated both gene (haplotype) and nucleotide diversity for 22 populations. In some cases, neighboring locales were aggregated to increase sample sizes (Table 1). Gene diversity is roughly equivalent to the expected heterozygosity of diploid data. It represents the probability that two randomly selected haplotypes from a population will be different. Nucleotide diversity is the probability that two randomly selected homologous nucleotides from the population will be different. Estimates of these parameters become less reliable as sample sizes decrease.

Results

Aligned sequence lengths, the number of variable sites, and appropriate ML nucleotide substitution models are shown by species in Table 2. All sequences generated are available from GenBank. Please contact the first author for accession numbers and alignments. Alignments ranged from 643–660 bp in length, and were unambiguous. There were no insertions or deletions. Maximum corrected genetic distances within species (Table 2) ranged from 0.0031 for *M. pacificum* to 0.032 for *M. xanthomelas*, a ten-fold difference. Total numbers of haplotypes found per species varied from 6 for *M. pacificum* to 20 for *M. calliphya* (Table 2).

Phylogeographic patterns varied between species. We found 3 main clades for *M. xanthomelas* and *M. pacificum*, corresponding to *M. xanthomelas* predominantly from O‘ahu and Maui Nui, *M. pacificum*, and *M. xanthomelas* from Hawai‘i (Fig. 2). Many haplotypes were shared between populations and even between islands. On the other hand, *M. hawaiiense* populations rarely shared haplotypes, and only once was a haplotype found on 2 islands (HWc, Fig. 3). We found support for three clades, one comprised mainly of Maui and Hawai‘i haplotypes, another of O‘ahu haplotypes, and a third of mainly Moloka‘i haplotypes. Moloka‘i haplotypes were intermediate between the other two clades, in essence bridging the gap between O‘ahu and Maui in a manner consistent with the geological history of the islands. Finally, the 4 individuals from Onomea Stream on Hawai‘i all bore a haplotype that differed by a minimum of 11 and a maximum of 14 substitutions (0.019– 0.025 substitutions/site) from the rest of the species. Phylogeography of *M. calliphya* also shows little sharing of haplo-

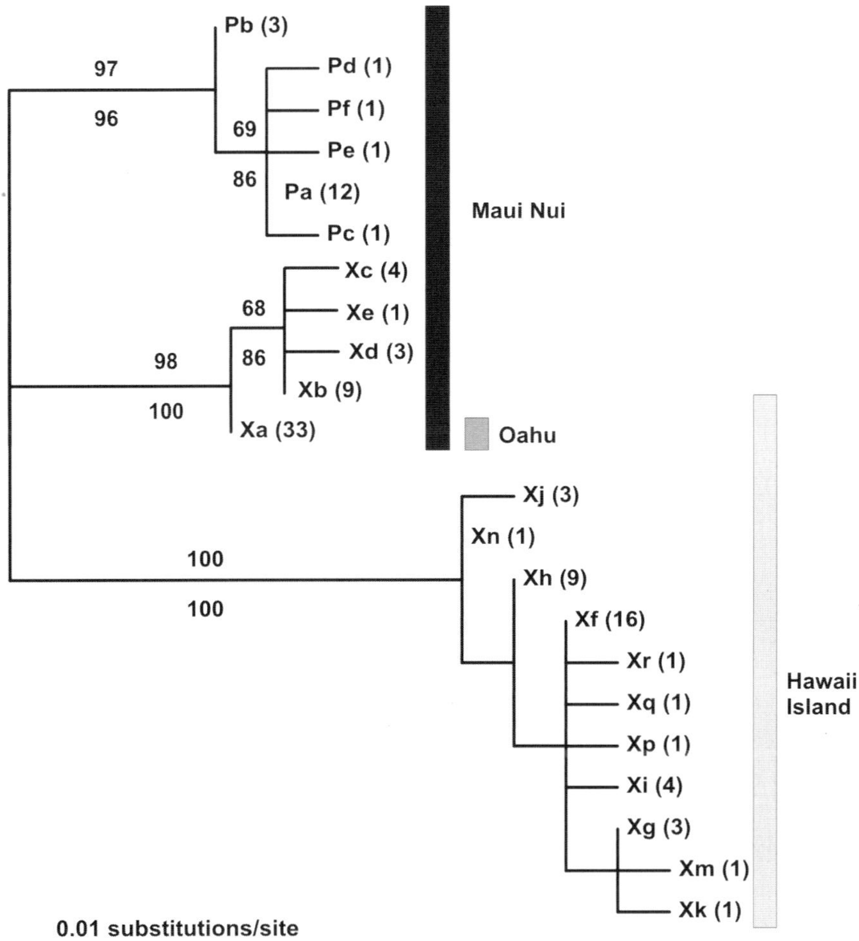

Figure 2. Maximum likelihood (ML) phylogram of 16 haplotypes from 91 individuals of *Megalagrion xanthomelas* and 6 haplotypes from 19 individuals of *M. pacificum*, generated using the HKY+G model in both maximum likelihood and Bayesian analyses. Numbers above the line are ML bootstrap values (200 pseudoreplicates) and numbers below the line are Bayesian posterior probabilities (one million generations). Due to the widespread nature of these haplotypes, identification of specific populations on the tree is not feasible. See Table 1 for specific haplotype distributions.

types between islands, but the distribution of two haplotypes serves to unite the islands in a manner consistent with geological history: haplotype Ca, from Moloka'i, shows little difference from Maui haplotypes, and haplotype Cg was found on East Maui and Hawai'i (Fig. 4).

Genetic diversity calculations for the entire species were telling (Table 2). In terms of both haplotype and nucleotide diversity, *M. pacificum* scores the lowest, with *M. xanthomelas* next in haplotype diversity. The other 2 species generally displayed the most genetic diversity in the study. Genetic diversity calculations for individual populations were much more variable (Fig. 5). In general, populations of *M. calliphya* displayed the most haplotype and nucleotide diversity, while *M. xanthomelas* displayed high haplotype diversity with lower nucleotide diversity. Once again, *M. pacificum* showed the least genetic diversity by either measure. *Megalagrion hawaiiense* was surprising, in that while its nucleotide diversity was comparable to *M. xanthomelas*, its haplotype diversity was generally lower.

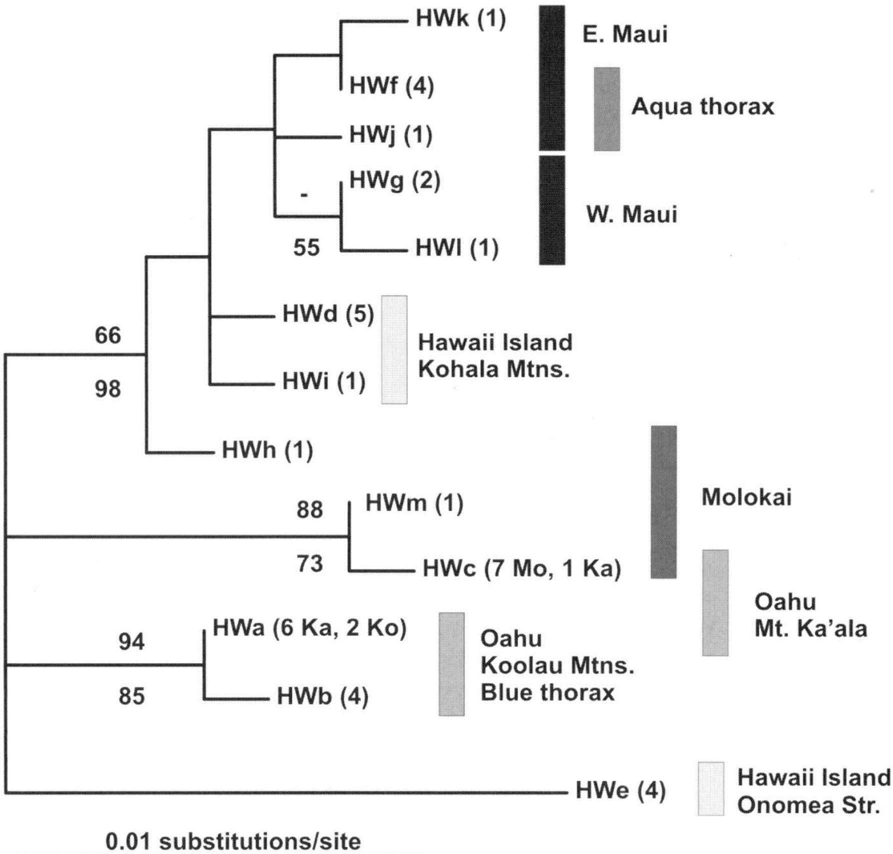

Figure 3. ML phylogram of 13 haplotypes from 41 individuals of *M. hawaiiense* generated as in Fig. 2, but using a GTR+I model. The occurrence of the blue and aqua color polymorphisms are noted.

Discussion

Comparative phylogeography

Megalagrion xanthomelas and *M. pacificum* share many more haplotypes between populations and islands than either of the other two species. Jordan *et al.* (2005) used phylogeographic analysis to propose that *M. xanthomelas* and *M. pacificum* do not often cross ocean channels, and that haplotype sharing seems to have occurred within the boundaries of contiguous or once contiguous islands (Hawai'i and Maui Nui). Populations of the two additional species analyzed here, *M. calliphya* and *M. hawaiiense* have many fewer shared haplotypes than *M. xanthomelas* and *M. pacificum*, suggesting that they have not often crossed ocean channels, but also that they do not experience as much gene flow within islands as *M. xanthomelas* and *M. pacificum*.

Megalagrion xanthomelas and *M. pacificum* show the greatest genetic differentiation between the Hawai'i Island clade and those from O'ahu and Maui Nui (Fig. 2). Two north island haplotypes found on Hawai'i are in low frequency and seem to be the result of limited immigration. This general pattern is consistent with the geological history of Hawai'i Island, which has never been connected to Maui Nui. *Megalagrion calliphya* and *M. hawaiiense*, however, both show much less differentiation between Hawai'i Island haplotypes and those found on Maui Nui (Figs. 3, 4). Many elements of their phylogeographic patterns roughly recapitulate the chronological history of the island

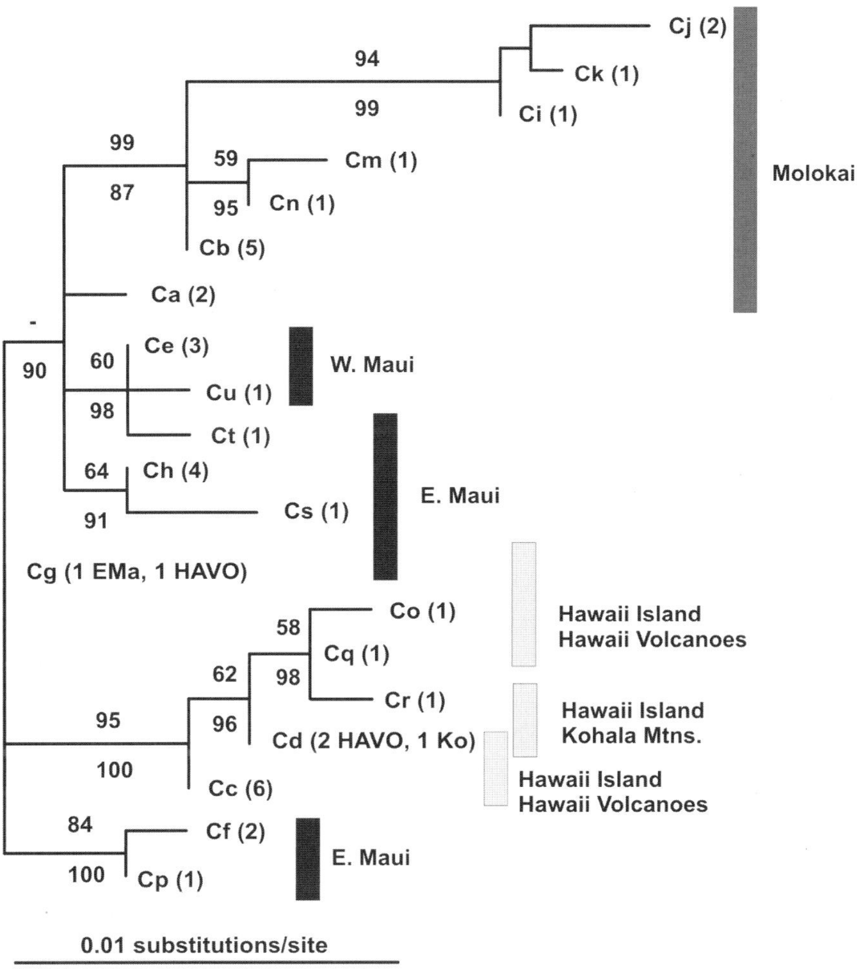

Figure 4. ML phylogram of 20 haplotypes from 40 individuals of *M. calliphya* generated as in Fig. 2, but using the HKY+I model.

creation (e.g., clustering of Moloka'i and O'ahu haplotypes, and Maui and Hawai'i Island haplotypes). These geological signatures differ from those of *M. xanthomelas* and *M. pacificum*, where more recent and local processes (e.g., hybridization and gene flow) are thought to be responsible for genetic patterns within Maui Nui and Hawai'i (Jordan *et al.*, 2005).

High levels of mtDNA divergence in *M. xanthomelas* led Jordan *et al.* (2005) to explore the possibility of incipient speciation of Hawai'i Island populations. This idea was ultimately rejected in light of both nuclear DNA sequences and morphology that support the monophyly of *M. xanthomelas*. Jordan *et al.* (2005) also noted that genetic distances between Hawai'i Island and O'ahu/Maui Nui populations of *M. xanthomelas* appear higher than they should based on the geological age of Hawai'i Island. They hypothesized that this might be due to an ancient event of introgressive hybridization between Maui Nui *M. pacificum* and *M. xanthomelas*. This hypothesis is supported by our observation of low genetic distances between Hawai'i Island and Maui Nui populations of *M. hawaiiense* and *M. calliphya* (Figs. 3, 4), which may better reflect the baseline inter-island genetic distance for *Megalagrion* in the absence of introgression.

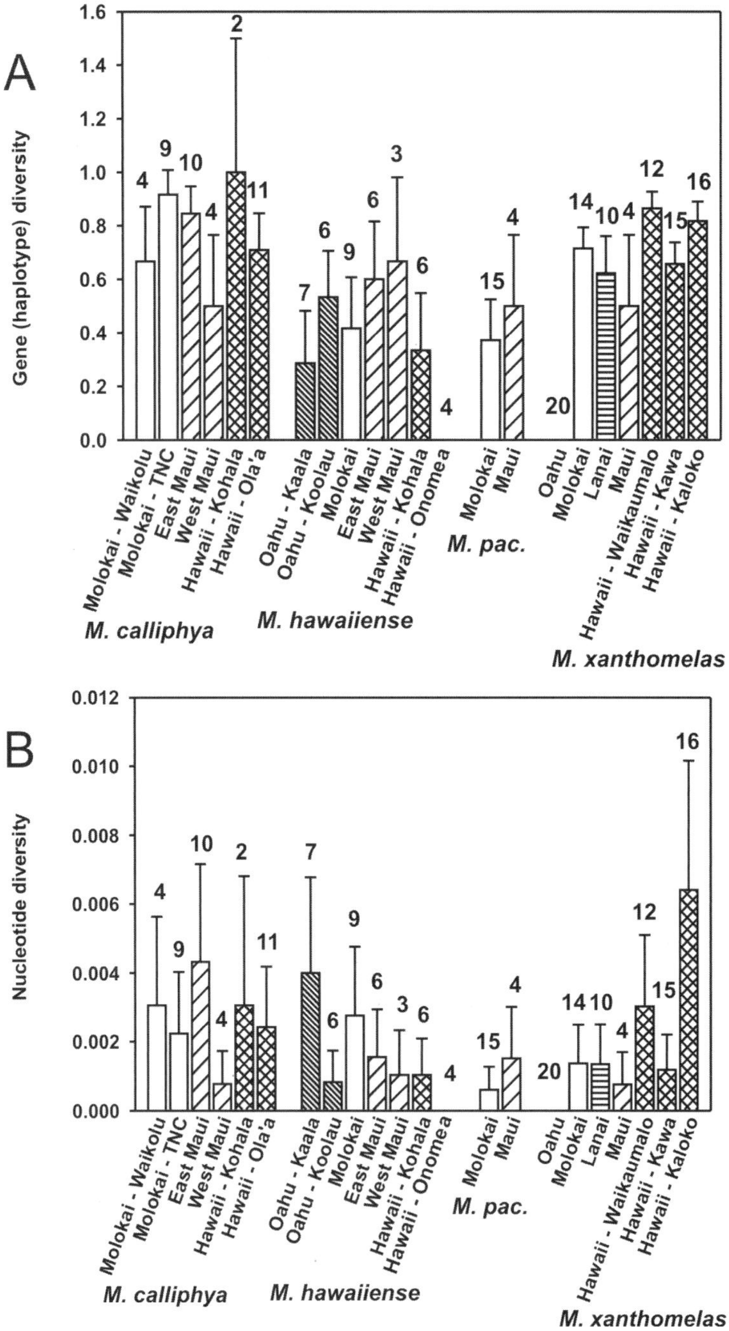

Figure 5. Comparison of population genetic parameters calculated for all populations of all species sampled using the Arlequin software package (Schneider *et al.*, 2000). Numbers above bars represent sample sizes. Whiskers are the sampling variance in each case. **A:** Gene or haplotype diversity, which is roughly equivalent to heterozygosity for diploid data; **B:** Nucleotide diversity, which is the probability that two homologous nucleotides sampled at random from the data set will be different.

Although we do not support species-level recognition for the Hawai'i Island populations of *M. xanthomelas*, we do feel that these genetically unique populations should be managed as a separate entity within this species. Of the 43 *M. xanthomelas* sampled from Hawai'i Island, 2 individuals from Kaloko bore haplotypes common on O'ahu and/or Maui Nui. The rest bore haplotypes that can be diagnosed from Maui Nui and O'ahu haplotypes at 11 nucleotide sites. Although the presence of the 2 northern haplotypes at Kaloko technically means that Hawai'i Island is not diagnosable from the other islands, the 11 diagnosable sites in the other 41 individuals represent a great deal of genetic uniqueness and we recommend that this be taken into account as Hawai'i Island populations are managed. Diagnosable island subunits were not observed within *M. pacificum*, *M. calliphya,* or *M. hawaiiense*. Significantly, this includes the finding of no genetically distinguishing characteristics for the O'ahu blue and Maui aqua populations of *M. hawaiiense* (Fig. 3), countering the argument that these color variants represent unique species (Daigle, 2000).

Conservation genetics

Long-term monitoring of several populations included in this study has shown them to be relatively healthy. Damselflies in these populations are abundant, the geographical area occupied is not unnaturally restricted (as *M. xanthomelas* is on O'ahu), and no significant declines in population size have been observed. Chief among these are populations of *M. calliphya* from Hawai'i Volcanoes National Park (HAVO) and East Maui. Other populations are known to be but small remnants of larger populations, or populations that were perhaps never very large. Included in these are *M. xanthomelas* populations from O'ahu and East Maui, which were both described as abundant by early Hawaiian entomologists (e.g., Perkins, 1913). *Megalagrion pacificum* is generally rare throughout its range, though it is locally abundant in populations on North Moloka'i (R. Englund, pers. comm.).

Results of the genetic diversity analyses confirm some of our previous knowledge (e.g., Englund, 2001) and offer surprises in some cases and important warnings in others. First the confirmations: Hawai'i Island *M. xanthomelas* show a great deal of diversity, although in spite of a large sample size, the Kawa population has less nucleotide diversity than Waikaumalo or Kaloko. The Kaloko nucleotide diversity is especially high because of the presence of 2 individuals there from the northern *M. xanthomelas* clade. Likewise, genetic diversity in *M. calliphya* is relatively high. In particular, the Moloka'i, East Maui and HAVO populations of *M. calliphya* display high nucleotide diversity, confirming the relative health of this species. This is not surprising to those who have spent time at Volcano, and have observed *M. calliphya* successfully using human and pig-created habitats.

O'ahu and Maui *M. xanthomelas* populations are thought to be relictual and their low genetic diversity is thus not surprising. The *M. xanthomelas* population in Honolulu, however, is an extreme case of genetic inbreeding, consistent with documented demographic bottlenecks (Englund, 2001). Although our sample size here was high (20 individuals), they all bore the same mtDNA haplotype, suggesting a lack of genetic diversity for meeting the challenges of a changing environment (Reed & Frankham, 2003). The future health of *M. xanthomelas* on O'ahu may depend on translocations of individuals from Moloka'i populations that also harbor the O'ahu mitochondrial haplotype. In contrast, *M. hawaiiense* from Moloka'i also appear to be quite genetically diverse.

Results of conservation genetic analysis for some populations were surprising. Among the pleasant surprises was *M. pacificum*, which, although rare on East Maui, appears to harbor more genetic diversity there than on Moloka'i. This is all the more surprising since our calculations here did not include an extremely divergent mtDNA haplotype from the Kīpahulu Valley (Jordan *et al.*, 2005). O'ahu Mt. Ka'ala populations of *M. hawaiiense* are also quite diverse in spite of the fact that extensive habitat has been lost in this region. Results of concern include those from *M. hawaiiense* populations from the Ko'olau Mountains of O'ahu, the Kohala Mountains of Hawai'i, and the West Maui Mountains, which show very little genetic diversity. This species has not previously been on the radar of conservationists, but the genetic data suggest that perhaps it should be. Although not diagnosable, its population at Onomea Stream is genetically quite distinct from the rest of the species and merits further conservation attention.

Finally, *Megalagrion pacificum* and *M. xanthomelas*, which enjoy some of their highest numbers on Moloka'i, show relatively low genetic diversity there. These populations, which have been considered the most healthy in the species, occur in some of the most pristine streams in Hawai'i (Polhemus, 1993). Nevertheless, in spite of a sample size of 15, the Moloka'i *M. pacificum* show a stark lack of genetic diversity. This suggests that the long term outlook for those populations may not be good, although their low genetic diversity could also be the result of historical bottlenecks caused by sea level fluctuations, and therefore present no cause for alarm (Jordan *et al.*, 2005). In any case, these populations should be carefully monitored, stream flows should be assured and any detection of invasive fish in these streams should be followed by aggressive eradication efforts.

An alternative perspective on the data can be gained by considering the diversity results by island (Fig. 5). Patterns among species within islands are not constant. Rather than all species doing well on relatively pristine Moloka'i, or poorly on O'ahu, results vary. This in part may be due to local extirpations: O'ahu has entirely lost *M. pacificum*, so its genetic diversity cannot be considered. *Megalagrion calliphya* does not occur on O'ahu, but its sister species, *M. leptodemas* is in desperate straits there (Polhemus, 1993). And yet, *M. hawaiiense* on Mt. Ka'ala has high nucleotide diversity. Moloka'i populations of *M. calliphya* and *M. hawaiiense* are quite healthy, but *M. xanthomelas* and *M. pacificum* show disappointingly low genetic diversity levels there. Hawai'i Island populations generally show high diversity, but *M. hawaiiense* bucks this trend.

It is worth noting that the results of this paper may represent a best-case scenario, as the data included in it are from damselflies sampled up to 14 years ago. The sampled populations may have declined in the meantime. In May 2005, for example, we failed to find any *M. xanthomelas* at Waikaumalo State park on Hawai'i Island. The stream was teeming with introduced fish and frogs. This example emphasizes the need for steps to be taken to immediately secure the genetic diversity that remains in this important endemic Hawaiian insect genus. Such efforts should begin with centralized record keeping and data sharing between all the agencies that have stewardship over, or interest in, *Megalagrion*. Populations with low genetic diversity should be managed in ways that increase gene flow and population sizes. Both of these factors are directly correlated with genetic diversity. These goals could be accomplished by ensuring adequate water flow, excluding and removing invasive species, and, in some cases, translocation.

In summary, we have assessed levels of genetic diversity in *Megalagrion* populations of known health. We have then estimated genetic diversity in a variety of *Megalagrion* populations and inferred their health. Based on these estimates, we propose that several populations may merit renewed conservation attention. We have shown that genetic diversity can be low in populations that have otherwise been considered healthy. Many examples in the literature suggest that low genetic diversity may be a sign of population endangerment and decreased fitness (Reed & Frankham, 2003; Spielman *et al.*, 2004). We have also shown that measuring genetic diversity may be a quick and reasonable method for assessing the health of populations that have not been extensively monitored. This approach shows promise in *Megalagrion*, in spite of small some sample sizes and analysis of a single locus. In the future, the expansion of this approach to remaining *Megalagrion* populations will be beneficial, and will improve on estimates of genetic diversity by sampling many more loci such as microsatellites and more individuals per population.

Acknowledgments

Specimens and field assistance were kindly provided by Ron Englund, Neal Evenhuis, Dan Polhemus, David Preston, Adam Asquith, Forrest Starr, Kim Martz, Bill Puleloa, Jen Morris, and David Foote. Thanks to Gordon Nishida and the Bishop Museum for maintaining and loaning valuable specimens. We are grateful for the extensive and insightful comments of an anonymous reviewer. This work was funded by the U.S. National Science Foundation Doctoral Dissertation Improvement program (Grant no. 0073314 to SJ), the U.S. Geological Survey Biological Resource Division, the UConn Department of Ecology and Evolutionary Biology, the Connecticut State Museum of Natural History's Penner and Slater Funds, the Canon National Parks Science Scholars Program, Sigma Xi, and the Bucknell University Department of Biology. We express our sincere gratitude to the Hawai'i Division of Aquatic Resources for organizing and funding the 2005 Stream Symposium at which these results were first presented and to J. Michael Fitzsimons and Neal Evenhuis for editing these conference proceedings. This paper is dedicated to the children of the *Kanu 'o ka 'Āina* Charter School, who have the privilege of sharing air, water, land, and time with these marvelous insects. Through them it will always be so.

Literature Cited

Avise, J.C. 2004. *Molecular markers, natural history, and evolution.* 2nd edition. Sinauer Associates, Sunderland, Massachusetts.

Buckley, T.R., P. Arensburger, C. Simon & G.K. Chambers. 2002. Combined data, Bayesian phylogenetics, and the origin of the New Zealand cicada genera. *Systematic Biology* **51**: 4–18.

Clague, D.A. & G.B. Dalrymple. 1987. The Hawaiian-Emperor volcanic chain. *In*: R.W. Decker, T.L. Wright & P.H. Stauffer (eds.), *Volcanism in Hawaii.* United States Geological Survey professional paper 1350. Washington, District of Columbia, United States Government Printing Office.

Crandall, K.A., O.R.P. Bininda-Emonds, G.M. Mace & R.K. Wayne. 2000. Considering evolutionary processes in conservation biology. *Trends in Ecology and Evolution* **15**: 290–295.

Daigle, J.J. 2000. The distribution of the Odonata of Hawaii. *Bulletin of American Odonatology* **6**: 1–5.

DeSalle, R. 2005. Genetics at the brink of extinction. *Heredity* **94**: 386–387.

Englund, R.A. 1999a. The impacts of introduced poeciliid fish and Odonata on the endemic *Megalagrion* (Odonata) damselflies of O'ahu Island, Hawaii. *Journal of Insect Conservation* **3**: 225–243.

———. 1999b. New records and range extensions of native Odonata (Coenagrionidae) and introduced aquatic species in the Hawaiian Islands. *Bishop Museum Occasional Papers* **59**: 15–19.

———. 2001. Long-term monitoring of one the most restricted insect populations in the United States, *Megalagrion xanthomelas* (Selys-Longchamps), at Tripler Army Medical Center, O'ahu, Hawai'i (Zygoptera: Coenagrionidae). *Odonatologica* **30**: 255–263.

Evenhuis, N.L. & Cowie, R.H. 1994. A survey of the snails, insects and related arthropods in the grounds of the Tripler Army Medical Center, Honolulu, Hawaii. *Bishop Museum Technical Report* 3, 21 + [2] p.

Frati, F., C. Simon, J. Sullivan & D.L. Swofford. 1997. Evolution of the mitochondrial cytochrome oxidase II gene in Collembola. *Journal of Molecular Evolution* **44**: 145–158.

Huelsenbeck, J.P. & F. Ronquist. 2001. MrBayes: Bayesian inference of phylogenetic trees. *Bioinformatics* **17**: 754–755.

Jordan, S., C. Simon, D. Foote & R.A. Englund. 2005. Phylogeographic patterns of Hawaiian *Megalagrion* damselflies (Odonata: Coenagrionidae) correlate with Pleistocene island boundaries. *Molecular Ecology* **14**(11): 3457–3470.

———., **C. Simon & D. Polhemus.** 2003. Molecular systematics and adaptive radiation of Hawaii's endemic damselfly genus *Megalagrion* (Odonata: Coenagrionidae). *Systematic Biology* **52**: 89–109.

Kennedy, C.H. 1928. The origin of the Hawaiian odonate fauna and its evolution within islands. *Proceedings of the Fourth International Congress of Entomology* **1**: 978–981.

Kumar, S., K. Tamura, I.B. Jakobsen & M. Nei. 2001. MEGA2: Molecular Evolutionary Genetics Analysis software. *Bioinformatics* **17**: 1244–1245.

Lande, R. 1988. Genetics and demography in biological conservation. *Science* **241**: 1455–1460.

McLachlan, R. 1883. Neuroptera of the Hawaiian Islands.—Part I. Pseudo-Neuroptera. *Annals and Magazine of Natural History* (5) **12**: 226–240.

Moritz, C. 1994. Defining 'Evolutionarily Significant Units' for conservation. *Trends in Ecology & Evolution* **9**: 373–375.

Polhemus, D.A. 2007. Biology recapitulates geology: the distribution of *Megalagrion* damselflies on the Ko'olau Volcano of O'ahu, Hawai'i. *In*: Evenhuis, N.L. & J.M. Fitzsimons (eds), Biology of Hawaiian streams and estuaries. *Bishop Museum Bulletin in Cultural and Environmental Studies* **3**: 335–346.

Perkins, R.C.L. 1899. Neuroptera. *Fauna Hawaiiensis* **2**(3): 63–77.

———. 1913. Introduction. *Fauna Hawaiiensis* **1**(6)**:** xv–ccxxviii.

Polhemus, D.A. 1993. Damsels in distress: a review of the conservation status of Hawaiian *Megalagrion* damselflies (Odonata: Coenagrionidae). *Aquatic Conservation: Marine & Freshwater Ecosystems* **3**: 343–349.

———. 1997. Phylogenetic analysis of the Hawaiian damselfly genus *Megalagrion* (Odonata: Coenagrionidae): implications for biogeography, ecology, and conservation biology. *Pacific Science* **51**: 395–412.

———. **& A. Asquith**. 1996. *Hawaiian damselflies: a field identification guide.* Bishop Museum Press, Honolulu. 122 p.

———., **H. Oppenheimer, F. Starr & K. Martz**. 1999. Notable rediscoveries of *Megalagrion* species on Maui (Odonata: Coenagrionidae). *Bishop Museum Occasional Papers* **59**: 27–29.

Reed, D.H. & R. Frankham. 2003. Correlation between fitness and genetic diversity. *Conservation Biology* **17**: 230–237.

Schneider, S., D. Roessli & L. Excoffier. 2000. *Arlequin ver. 2.000: A software for population genetics data analysis.* Genetics and Biometry Laboratory, Geneva, Switzerland.

Spielman, D., B.W. Brook & R. Frankham. 2004. Most species are not driven to extinction before genetic factors impact them. *Proceedings of the National Academy of Science of the United States of America* **101**: 15261–15264.

Swofford, D.L. 1998. *PAUP*. Phylogenetic analysis using parsimony (*and other methods).* Version 4. Sunderland, Massachusetts, Sinauer Associates.

Thompson, J.D., D.G. Higgins & T.J. Gibson. 1994. CLUSTAL W: Improving the sensitivity of progressive multiple sequence alignment through sequence weighting, position-specific gap penalties and weight matrix choice. *Nucleic Acids Research* **22**: 4673–4680.

United States Fish & Wildlife Service. United States Department of the Interior. 2004. Review of species that are candidates or proposed for listing as Endangered or Threatened. *Federal Register* **69**: 24876–24904. [http://endangered.fws.gov/candidates/ 2003_CNOR.pdf]

Williams, F.X. 1936. Biological studies in Hawaiian water-loving insects. Part I. Coleoptera or beetles. Part II. Odonata or dragonflies. *Proceedings of the Hawaiian Entomological Society* **9**: 235–349.

Zimmerman, E.C. 1948. *Insects of Hawaii.* Volume 2. Apterygota to Thysanoptera inclusive. Honolulu, University of Hawaii Press.

Biology of Hawaiian Streams and Estuaries. Edited by N.L. Evenhuis & J.M. Fitzsimons. *Bishop Museum Bulletin in Cultural and Environmental Studies* 3: 261–276 (2007).

Translocation and Monitoring Efforts to Establish a Second Population of the Rare *Megalagrion xanthomelas* (Sélys-Longchamps) on O'ahu, Hawai'i (Zygoptera: Coenagrionidae)[1]

David J. Preston, Ronald A. Englund & Myra K.K. McShane
Hawai'i Biological Survey, Bishop Museum, 1525 Bernice Street, Honolulu, Hawai'i 96817, USA; email: davidp@bishopmuseum.org

Abstract

The last remaining population of *Megalagrion xanthomelas* (Sélys-Longchamps) resides in a 100 meter reach of stream located on the grounds of Tripler Army Medical Center, O'ahu. Because actions may be taken that might jeopardize this only known O'ahu population, it has been considered imperative to establish a second population to prevent *M. xanthomelas* from going extinct on O'ahu. An attempt to establish this species at a stream in the Dillingham area of O'ahu was made in 1998, but unfortunately was unsuccessful. Because the Tripler population is so small and restricted in distribution, a second effort at translocation was attempted at a new location. We estimated the population size of *M. xanthomelas* at the Tripler site in 1997 and again in 2003 by mark-recapture and concluded that the Tripler population was stable and could withstand the removal and translocation of a small number of adults and larvae. A stream site located in Makiki Valley was selected for its lack of alien predators such as crayfish, prawns, and mosquito fish, and a number of adults and immatures were translocated to the Makiki site in August 2004. Monitoring of the Tripler and the Makiki sites is ongoing and an additional translocation of *M. xanthomelas* to Makiki is planned. Future conservation plans should also include the assistance of from the general public through avenues such as stocking of backyard ponds with *M. xanthomelas*.

Introduction

The native endemic damselfly *Megalagrion xanthomelas* (Sélys-Longchamps) inhabited lowland coastal wetland and the lowest reaches of streams on all the main Hawaiian Islands. *Megalagrion xanthomelas* is now severely restricted in range because lowland aquatic habitats over the past century have undergone major physical and biotic transformations. In contrast to some immature *Megalagrion* species having dagger-shaped or streamlined gills and being found in cascades and riffles, this species has round, paddle-shaped gills restricting it to areas with little or virtually no water current (Polhemus & Asquith,1996). Throughout Hawai'i, habitats containing still waters are generally found in the low elevation regions and are now quite limited.

Historic and current range of *Megalagrion xanthomelas*

Prior to 1993, it was speculated that *Megalagrion xanthomelas* existed on all the islands (Perkins, 1899), yet at that time there were no collections from Kaua'i, Lāna'i, or Kaho'olawe. Several specimens of *M. xanthomelas* collected from Kaua'i and Ni'ihau in the early part of the 20th century have been found in the Bishop Museum collection. It was not until 1993 that specimens of *M. xanthomelas* were collected on Lāna'i at two sites with unnatural water features (Polhemus, 1993). On O'ahu, this species was widely distributed and found in low elevation areas throughout the island, with historical collections starting in 1892 and ending in 1977 (Fig. 1) (Bishop Museum collection data). Surveys conducted by D.A. Polhemus in the 1990s found *M. xanthomelas* wide spread on Hawai'i Island where they are locally common in coastal wetlands in Puna, Ka'ū, and Kona districts (Polhemus, 1995). In 1997, *M. xanthomelas* was collected in 2 separate low-elevation locations on East and West Maui (Polhemus *et al.*, 1999).

1. Contribution No. 2007-006 to the Hawaii Biological Survey.

Megalagrion xanthomelas is now restricted to the islands of Hawai'i, Moloka'i, Lāna'i, Maui, and O'ahu. There are 4 populations known from Moloka'i, and it was reported to be abundant in artificial golf course ponds on Lāna'i (Polhemus, 1995). The last collection of this species from Lāna'i was made in 1998 at the golf course ponds of Koele Lodge. However, in March 2005 *M. xanthomelas* was not sighted during a return visit to those ponds. The only other known population on Lāna'i, located in Maunalei Gulch, was not visited during the present study and its status is currently unknown. Although once abundant on O'ahu, *M. xanthomelas* is now only known from one population located at Tripler Army Medical Center (TAMC). Polhemus (1995) stated "The population of *M. xanthomelas* occurring at TAMC appears to be a remnant of a much larger and more continuous population that formerly occupied the wetlands along the inner margin of Pearl Harbor". The current populations on O'ahu and Lāna'i are supported by artificial water augmentation. In the case of TAMC, natural stream flow is augmented by water supplied by a hose 100 m above the stream. At Maunalei Gulch, Lāna'i, the water was provided by vandals drilling holes in irrigation pipes, and presumably this population was originated by colonists from the ponds at Koele Lodge. Both the TAMC and Maunalei Gulch populations are at risk of extinction should these artificial water sources be cut off.

Conservation *in situ* of the remaining O'ahu population

Believed to be extinct on O'ahu since at least 1977, *Megalagrion xanthomelas* was rediscovered in 1994 by Neal Evenhuis of the Bishop Museum's Hawai'i Biological Survey. A population was found residing in a small section of an unnamed tributary to Moanalua Stream at TAMC, adminstered by the U.S. Department of Defense (Fig. 1). Soon afterwards, attempts were made to locate additional populations on O'ahu, but these efforts were unsuccessful. As this endemic species was formerly "a common insect in Honolulu gardens and lowland districts" (Perkins, 1913), it was imperative to preserve one of most conspicuous and beautiful components of the native Hawaiian aquatic fauna. Without the exceptional efforts of earliest researchers such as Perkins, later entomologists such as F.X. Williams (Williams, 1937), and more recently D.A. Polhemus (Polhemus, 1996), we would have had no idea that *M. xanthomelas* was threatened with extinction on O'ahu.

Because *M. xanthomelas* was now found only in approximately 100 m of habitat, concerns were raised about the continued existence of the Tripler population and possible loss of this species as a result of planned construction activities upslope from the site. To help ensure the survival of the damselfly population against possible watershed disturbances, it was decided to build 4 mitigation ponds just upslope of the stream (Fig. 2). Adult and larval damselflies were then transferred into the ponds from the stream to create a sub-population separate from but in close proximity to the stream. In October 1995, one month after the completion and filling of the mitigation ponds, a catastrophic flood eliminated the Tripler stream damselfly population, which survived only in the newly-filled mitigation ponds (Englund, 2001). The damselfly survived in the newly created ponds, and eventually recolonized the stream. Construction upslope of the stream then caused it to go dry, with as much as 40% of the 100-m Tripler stream channel having no water by May 1997 (Englund, 1998). The Veterans Administration then augmented water flow with a small garden hose, maintaining a low but fairly stable stream flow since 1997, which lead to a corresponding rebound in the Tripler *M. xanthomelas* population (Englund, 1998).

Difficulties of *in situ* conservation at TAMC

Examples abound regarding difficulties in preserving the remnant native Hawaiian flora and fauna (Banko *et al.*, 2001), and efforts in conserving the O'ahu *M. xanthomelas* population have also been problematic. The mitigation ponds were not designed properly because they did not closely resemble the natural stream habitat, except to provide a fish-free body of water with water plants to provide substrate for egg laying. There was no attempt to provide vegetation around the ponds to allow for hiding and foraging by the females, as is their natural behavior at the stream site. Shade-cloth was added in order to simulate the overhead canopy found along the stream. The shade-cloth also lowered the pond's water temperature, which prior to that time was much higher than the adjacent stream temperatures, and this likely hindered larval damselfly development and subsequent recruitment. As a result of very slow water flow in the mitigation ponds, they soon became infested by large alien dragonflies such as *Tramea* spp., and a large notonectid bug predator *Notonecta indica*, which proliferated and almost certainly preyed on some *M. xanthomelas* larvae. In addition to the non-native predators, the very slow water flow and high light levels, due to insufficient shading, lead to excessive algal growth filling the ponds.

An additional threat to the entire stream population was discovered when *Tilapia* fish were purposely introduced into the mitigation ponds by the TAMC groundkeeping staff apparently to create their own private fishpond. This was a serious concern as the damselflies at the Tripler stream were located only 200 m directly downstream of these ponds. In March 2000, due to these continuing problems and after assurances from the Veterans Administration that all efforts to maintain the stream and flow would be undertaken, the decision was made to permanently drain the ponds (Englund, 2001). Once again, it was fortuitous that the Tripler damselflies survived, because a large storm caused the ponds to fill and overflow down slope into the stream soon after the ponds were drained and the *Tilapia* removed. If the *Tilapia* had washed into the stream and become established, they would have decimated the larvae and ultimately caused the extinction of *M. xanthomelas* on O'ahu.

Conservation by relocation

Efforts focused on restoring portions of the endemic flora and fauna of the Hawaiian Islands have had a mixed record of success. In general, efforts in establishing additional populations to restore native plant species have been somewhat more successful than efforts to restore animals. Many examples of establishing new populations of endangered plants can be found, with the Mauna Kea silversword (*Argyroxiphium sandwicense*) a notable success story (Robichaux *et al.*, 2000). However, establishment of additional populations of rare bird or insect species in Hawai'i has had a mixed record. There have been only a few notable successes such as the Nene (*Branta sandvicensis*), but mostly failures such as *'alalā* (*Corvus hawaiiensis*), or *palila* (*Loxioides bailleui*) (Banko *et al.*, 2001). By far the vast majority of resources have been aimed at restoring bird populations, and with the exception of *M. xanthomelas*, we are not aware of any attempts in Hawai'i to establish additional populations of rare insects in order to preserve a species. Disease, low population sizes, and invasive species are just some of the major hurdles involving conservation efforts intended to restore populations in Hawai'i.

Because of the high continuing risk to *M. xanthomelas* at Tripler it was necessary to locate an alternate site suitable for translocation and establishment of a second population. The first attempt to translocate this damselfly species on O'ahu, or anywhere in Hawai'i, was undertaken in 1998 (Johnson, 2001) when damselflies were translocated from the Tripler site to an unnamed stream near the Dillingham Airfield, O'ahu. Although this effort was unsuccessful because of a crayfish introduction shortly after translocation was attempted (Johnson, 2001), it laid the groundwork for more successful future translocation attempts. After searching for a suitable site lacking invasive aquatic species, one was found in an area of upper Makiki Valley, O'ahu. On 18 August 2003, a small number of adults and larvae were moved to an unnamed tributary of Makiki Stream. A comparison of the arthropods and riparian plants found at the Dillingham and Makiki translocation sites is given in Tables 1 and 2.

Study objectives

The objectives of the damselfly mark-recapture sampling efforts at TAMC were to 1) document recruitment of new individuals to the population between sampling efforts; 2) assess the relative abundance of damselflies between monitoring periods using a standardized methodology; 3) provide a quick means of determining if the TAMC *M. xanthomelas* population was threatened by disturbance or stream dewatering; and 4) determine if the TAMC population would be impacted by removing a subset of the population during translocation efforts to establish an additional population.

Table 1. Comparison of plant species found at TAMC, Dillingham Stream, and Makiki.

PLANT SPECIES	COMMON NAME	FAMILY	STATUS	Dillingham	Makiki	TAMC
Abutilon grandifolium	hairy abutilon	Malvaceae	Introduced			+
Adiantum raddianum	maidenhair fern	Pteridaceae	Introduced	+		
Aleurites moluccana	*kukui*	Euphorbiaceae	Introduced	+	+	+
Arthrostema ciliatum	everblooming lavender	Melanstomataceae	Introduced			+
Asystasia gangetica	Chinese violet	Acanthaceae	Introduced	+		
Carica papaya	papaya	Caricaceae	Introduced	+		
Caryota urens	white palm	Palmaceae	Introduced			+
Cedrela odorata	toona, Spanish-cedar	Meliaceae	Introduced		+	
Cestrum nocturnum	night cestrum	Solanaceae	Introduced		+	
Christella parasitica	downy wood fern	Thelypteridaceae	Introduced	+	+	
Cinnamomum burmannii	Padang cassia	Lauraceae	Introduced		+	
Clidemia hirta	Koster's curse	Melastomataceae	Introduced		+	
Coccinia grandis	ivy gourd	Cucurbitae	Introduced			+
Colocasia esculenta	*taro, kalo*	Araceae	Introduced	+	+	
Commelina diffusa	*honohono*	Commelinaceae	Introduced	+	+	+
Cordyline fruticosa	*ti*	Agavaceae	Introduced	+	+	+
Cycas circinalis	Sago palm	Cycadacae	Introduced			+
Cyperus papyrus	papyrus	Cyperaceae	Introduced			+
Digitaria insularis	sourgrass	Poaceae	Introduced			+
Dracena fragrans	happy plant	Agavaceae	Introduced			+
Dryopteris sp.	Driopteris fern	Dryopteridaceae	Endemic	+		
Erythrina sandwicensis	*wili wili*	Fabaceae	Endemic	+		
Ficus benghalensis	India banyan	Moraceae	Introduced			+
Ficus microcarpa	banyan	Moraceae	Introduced	+	+	+
Haematoxylum campechianum	bloodwood tree	Fabaceae	Introduced			+
Impatiens wallerana	impatiens	Balsaminaceae	Introduced		+	
Justicia betonic	white shrimp plant	Acanthaceae	Introduced		+	+
Leucaena leucocephala	*koa haole*	Mimosaceae	Introduced	+		+
Mangifera indica	mango	Anacardiaceae	Introduced		+	+
Momordica charantia	bitter melon	Cucurbitae	Introduced			+
Musa xparadisiaca	banana	Musaceae	Introduced		+	
Nephrolepis cordifolia	sword fern	Nephrolepidaceae	Indigenous	+		
Nephrolepis exaltata	common sword fern	Nephrolepidaceae	Endemic	+	+	

Table 1. (continued)

PLANT SPECIES	COMMON NAME	FAMILY	STATUS	Dillingham	Makiki	TAMC
Odontonema tubiforme	fire spike	Acanthaceae	Introduced		+	
Oplismenus hirtellus	basketgrass	Poaceae	Introduced		+	
Paederua scandens	*Maile pilau*	Rubiaceae	Introduced			+
Panicum maximum	Guinea grass	Poaceae	Introduced			+
Passiflora edulis	passion fruit	Passifloraceae	Introduced		+	+
Phlebodium aureum	hare's foot fern	Polypodiaceae	Introduced	+		
Philodendron lacerum	philodendron	Araceae	Introduced		+	+
Pimenta dioica	all spice	Myrtaceae	Introduced		+	
Pipturus albidus	*mamaki*	Urticaceae	Endemic		+	
Plumbago zeylanica	leadwort	Plumbaginaceae	Introduced	+		
Pouteria sandwicensis	*`ala`a, aulu*	Sapotaceae	Endemic		+	
Psidium_guajava	common guava	Myrtaceae	Introduced		+	+
Ricinus communis	castor bean	Euphorbiaceae	Introduced			+
Rivina humilis	coral berry	Phytolaccaceae	Introduced			+
Sapindus oahuensis	*`aulu*, soapberry	Sapindaceae	Endemic	+		
Schefflera actinophylla	octopus tree	Apiaceae	Introduced		+	+
Schinus terebinthifolius	Christmas berry	Anacardiaciae	Introduced		+	
Senna surratensis	*kalamona*	Fabaceae	Introduced			+
Solanum americanum	popolo berry	Solanaceae	Introduced			+
Solanum seaforthianum	potato vine	Solanaceae	Introduced			+
Spathodea campanulata	African tulip tree	Bignoniaceae	Introduced	+	+	+
Synadenium grantii	African milk bush	Euphorbiaceae	Introduced			+
Syngonium sp.	arrowhead vine	Araceae	Introduced			+
Syzygium cumini	Java plum	Myrtaceae	Introduced	+		+
Terminalia catappa	false kamani	Combretaceae	Introduced			+
Thunbergia fragrans	white thunbergia	Acanthaceae	Introduced			+

Table 2. Insects recorded or observed from Tripler Army Medical Center compared to translocation sites. (NSR = new state record)

TAXA	TAMC	Dillingham Stream	Makiki
INSECTA			
BLATTARIA			
Blaberidae	*Pycnoscelus indicus* (Fabricius)	+	+
Blattellidae	*Periplaneta americana* (Linnaeus)	+	+
COLEOPTERA			
Aglycideridae	*Proterhinus* sp.		+
Canthridae	*Caccodes oceaniae* (Bourgeois)	+	
Cerambycidae	*Sybra alternans* (Wiedemann)	+	
Cerylonidae	Ceryloninae sp.		+ (NSR)
Ciidae	*Cis* sp.		+
Coccinellidae	*Curinus coeruleus* Mulsant	+	
	Olla abdominalis (Say)	+	
Corylophidae	*Corylophodes suturalis* (Sharp)	+	
Curculionidae	*Dryophthorus* sp.		+
	Neotrichus latisculus (Fairmaire)		+
	Scotytinae sp. 1		+
	Scotytinae sp. 2		+
	Scotytinae sp. 3		+
Histeridae	*Baccanius* sp.		+
Elateridae	*Conoderus exul* (Sharp)	+	+
Nitidulidae	*Stelidota geminata* (Say)		+
	Undetermined genus sp.		+
Scarabeidae	*Adoretus sinicus* Burmeister	+	+
	Protaetia fusca (Herbst)	+	
Staphylinidae	*Neosorius rufipes* (Motschulsky)		+
	Osorius rufipes Motschulsky	+	
	Pselaphinae sp.		+ (NSR)
Tenebrionidae	*Amarygmus morio* (Fabricius)		+
	Undetermined genus sp.		+
COLLEMBOLA (springtails)			
Entomobryidae	*Entomobrya* sp.	+	+
	Lepidocyrtus sp.	+	
	Tomocerus minor (Lubbock)	+	+
Neelidae	*Neelus* (*Megalothorax* sp.)	+	
Sminthuridae	*Dicyrtoma* (*Papirioides*) *dubia* (Folsom)	+	+
DIPTERA (flies)			
Calliphoridae	*Chrysomya megacephala* (Fabricius)	+	
Ceratopogonidae	*Atrichopogon jacobsoni* (de Meijere)	+	+
	Forcipomyia hardyi Wirth & Howarth	+	+
Chironomidae	*Chironomus hawaiiensis* Grimshaw	+	
Chloropidae	*Rhodesiella* sp.	+	
Culicidae	*Aedes albopictus* (Skuse)	+	+
Dolichopodidae	*Chrysosoma globiferum* (Wiedemann)	+	
	Chrysotus pallidipalpus Van Duzee	+	
	Chrysotus parthenus Hardy & Kohn	+	
	Pelastoneurus sp.	+	
	Syntormon flexibile (Becker)	+	
Drosophilidae	*Drosophila* sp. prob. *melanogaster* Meigen	+	+
Ephydridae	*Brachydeutera hebes* Cresson	+	
	Brachydeutera ibari Ninomiya	+	
	Scatella bryani (Cresson)	+	

Table 2 (continued). (NSR = new state record)

TAXA	TAMC	Dillingham Stream	Makiki
INSECTA			
DIPTERA (flies)			
Lauxaniidae	*Poecilominettia* sp.	+	
Phoridae	?*Spiniphora* sp.	+	
Psychodidae	*Psychoda* sp.	+	+
Sphaeroceridae	*Leptocera brevivenosa* Tenorio	+	
	Poecilosomella punctipennis (Wiedemann)	+	
Syrphidae	*Allograpta exotica* (Weidemann)	+	+
	Allograpta oblique (Say)	+	
	Simosyrphus grandicornis (Macquart)	+	
Tachinidae	*Trichopoda pilipes* (Fabricius)	+	
Tephritidae	*Bactrocera cucurbitae* (Coquillett)	+	
Tipulidae	*Limonia advena* (Alexander)	+	+
EMBIIDINA (web-spinners)			
Oligotomidae	*Oligotoma saundersii* (Westwood)	+	
HETEROPTERA (true bugs)			
Lygaeidae	*Cligenes marianensis* Usinger	+	
Notonectidae	*Buenoa pallipes* (Fabricius)	+	
	Notonecta indica Linnaeus	+	
Pentatomidae	*Brochymena quadripustulata* (Fabricius)	+	
	Nezara viridula (Linnaeus)	+	
Reduviidae	*Zelus renardii* Kolenati	+	+
Tingidae	*Leptodictya tabida* (Herrich-Schaeffer)	+	
HOMOPTERA			
Agallidae	*Agalliopsis* sp.	+	
Cicadellidae	*Homalodisca coagulata* (Say)	+	
Clastopteridae	*Clastoptera* sp.	+	+
Flatidae	*Melormenis basalis* Walker	+	+
Psyllidae	*Heteropsylla cubana* Crawford	+	
Oligotimidae	*Aposthonia oceania* (Ross)	+	
HYMENOPTERA			
Agaonidae	*Odontofroggattia* sp.	+	
Anthophoridae	*Xylocopa sonorica* Smith	+	
Apidae	*Apis mellifera* Linneaus	+	+ +
Aphelinidae	Undetermined genus sp.	+	
Braconidae	*Apanteles trifasciatus* Muesebeck	+	
Eulophidae	Undetermined genus sp.	+	
Eurytomidae	*Eurytoma* sp.	+	
	Sycophobia sp.	+	
Formicidae	*Anoplolepis longipes* (Jerdon)	+	+
	Camponotus sp.	+	
	Cardiocondyla wroughtoni (Forel)	+	
	Leptogenys falcigera Rogers	+	
	Pseudomyrmex gracilis mexicanus Rogers	+	
	Solenopsis geminata (Fabricius)	+	
	Tapinoma melanocepohalum (Fabricius)	+	
Proctotrupidae	Undetermined genus sp.	+	
Sphecidae	*Scelipheron caementarium* (Drury)	+	
Trichogrammatidae	Undetermined genus sp.	+	
Vespidae	*Delta curvatum* (Saussure)	+	+

Table 2 (continued). (NSR = new state record)

TAXA		TAMC	Dillingham Stream	Makiki
INSECTA				
ODONATA				
Coenagrionidae	*Ischnura posita* (Hagan)	+		
	Megalagrion hawaiiense (McLachlan)		+	
	Megalagrion xanthomelas (Sélys-Longchamps)	+		
Libellulidae	*Nesogonia blackburni* (McLachlan)		+	
	Orthemis ferruginea (Fabricius)	+		
	Pantala flavescens (Fabricius)	+		
	Tramea abdominalis (Rambur)	+		
	Tramea lacerata Hagan	+		
ORTHOPTERA				
Acrididae	*Oxya japonica* (Thunberg)			+
	Schistocerca nitens (Thunberg)	+		
Gryllidae	*Laupala* spp.			+
Tettigoniidae	*Conocephalus saltator* (Saussure)	+		+
	Elimaea punctifera (Walker)	+		+

Site descriptions

TAMC —The stream is located on leeward O‘ahu at 79 m elevation (Fig. 1), and flows for 95 m through a forest of introduced plants (Table 1). The wetted width of the stream is less than 1 m wide for most of its length, and has several pools that are 15–20 cm deep, though most of the stream is less than 10 cm in depth. A detailed description and map of the TAMC Stream study area can be found in Evenhuis *et al.* (1995), Polhemus (1996), and Pangelinan (1997). The TAMC stream now requires augmented water flow because construction in 1995 of a large Veterans Administration facility up slope of the TAMC stream disrupted the stream hydrology. The cement-lined mitigation ponds were constructed in 1995 drained in 2000, and were located approximately 200 m northwest of the TAMC stream. The ponds measured 7.3 m long by 3 m wide and had an average water depth of 0.6 m. These ponds contained cobble substrate brought from the TAMC stream and aquatic plants such as algae, water lily (*Nymphaea* sp.), water lettuce (*Pistia stratiotes*), and the large aquatic sedge (*Cyperus alternifolius*).

Makiki —The upper reaches of Makiki Stream are quite low-flowing, but it is unclear how historical water diversions may have impacted the current stream flow because virtually all upper tributary reaches of Makiki Stream were diverted at one time. Makiki Stream and its associated tributaries maintain a permanent, albeit low, flow. The upstream Herring Springs found at upper Moleka Stream was apparently completely diverted by several Board of Water Supply box culverts until 1994. This area, prior to it being diverted, was an important collecting site in the 1890s and early 1900s by R.C.L. Perkins and others for several now extinct native Oahu damselfly species (Englund, 1999). An unnamed spring and short section of stream found between Kānealole and Moleka Springs in the upper (>290 m) Makiki watershed was chosen as the translocation site for *M. xanthomelas*. This is the only upper elevation stream area lacking alien fish species in the watershed. This is also the only section of stream found to lack the introduced shrimp *Neocaridina denticulata sinensis* and contain the native shrimp *Atyoida bisulcata*. A 6-m slanted waterfall appeared to be a barrier to introduced fish and crustaceans, which were observed at the bottom of the waterfall, approximately 15 m downstream of the relocation site. All other areas of Makiki Stream examined during this study were dominated by alien species. Feral pig activities (e.g., wallows and vegetation damage) around upper Makiki tributaries (e.g., Moleka Springs, Kānealole Stream) appeared to contribute sediment into the stream.

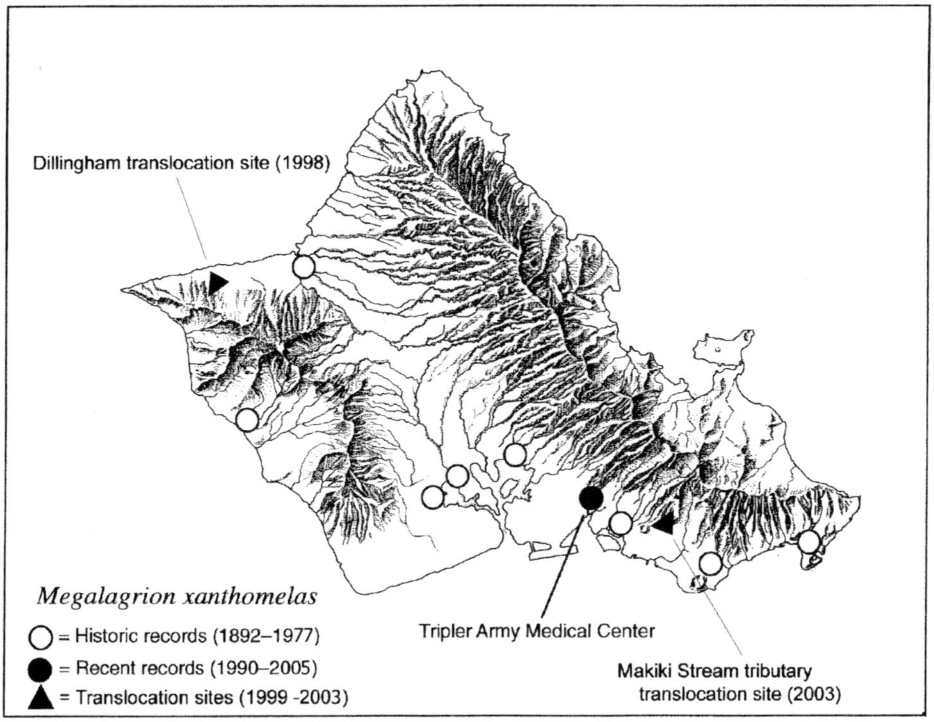

Figure 1. Map of O'ahu, Hawai'i showing locations of current and historic records for *Megalagrion xanthomelas* (modified from Evenhuis *et al.*, 1995).

Methods

The entire length of wetted stream was sampled starting at the upstream end of the man-made culvert where stream flow originates. Methods used were identical to previous research conducted on the *M. xanthomelas* TAMC population (Pangelinan, 1997; Englund, 1998). For this study, damselflies were marked and recaptured from May–July 1997, and March–August 2003. At least two observers collected adult damselflies while slowly walking down the stream. Each observed damselfly was netted and its wings marked with a permanent black extra fine felt tip marker. The number was recorded if a captured damselfly had been previously marked. After completing the slow downstream walk, which would average 3 hours, we returned slowly upstream to collect and mark any previously unmarked damselflies. Collection and observation times were consistent during each monitoring event to standardize sample effort. Only netted, marked individuals were counted as recaptures. New captures were marked, sexed, measured, and held at the stream sections where they were collected. All individuals were released later the same day, at the conclusion of the survey. Individual numbers were recorded with their sex noted and abdomen lengths measured, and recaptures were recorded and tandem pairs were noted as to whether any were previously marked. Population estimates were obtained by using the Petersen estimate of population size (Caughley, 1980). Although the Petersen estimate will in the long run result in an overestimation of population size (Caughley, 1980), we use it here because our mark-recapture studies in 1997 and 2003 were relatively short in duration, and they provide a relatively good estimate of damselfly population size.

Pond sampling (from 1997 until there draining in March 2000 was identical to Englund (2001). This involved thirty-minute damselfly counts being conducted at each of the four concrete mitigation ponds, with damselfly sex and behavioral activity was recorded during each thirty-minute count.

Figure 2. TAMC Mitigation ponds prior to drainage, February 2000.

Individual adults were not counted unless they were captured and the wings marked with a number. Quantitative aquatic net samples were taken at the ponds starting during the November 1997 monitoring. Three aquatic net sweeps approximately 1.25 m in length were taken at the surface, middle, and bottom of each pond. The net contents were then placed in a 500-m sieve and inspected for immature damselfly naiads (Englund, 2001).

TAMC Population Study Results

From May to July 1997, 355 damselflies were marked, with 129 recaptures made during the study period (Table 3). *Megalagrion xanthomelas* population estimates ranged from a low of 18 to a high of 221 individuals in 1997 (Fig. 3). Over a period of 148 days between March and August 2003, 1167 individual damselflies were captured and marked, with 742 recaptures during this period (Table 4). A high of 147 individuals were collected on 21 June 2003 and a low of 11 captured on 8 July 2003. As a result of more stable and higher stream flows, population estimates in 2003 increased on average more than in 1997 and ranged from 75–535 individuals (Fig. 4). Capture success was determined by the weather with results being better on clear, sunny days. In 2003, the population estimate was quite consistent with the exception of a marked decrease in July 2003 when numbers were exceptionally low. The July 2003 decrease was attributed to poor weather conditions that resulted in low damselfly capture rates, and not a decrease in real population size. Population estimates increased again to between 224–293 in August 2003 with the return of sunny sampling conditions.

Males were collected in higher numbers than females because their bright orange color made them easier to see and they were more inclined to fly when approached (Fig. 5). Males defended oviposition sites and stayed closer to the water, perching on rocks, twigs, and branches in the stream. Females are more cryptically colored and tend to sit still on vegetation or rocks when approached. Single females were not readily seen unless flying about, however tandem pairs were very noticeable, thus we collected most females when they were in tandem with a male. Single females tended to be at greater distances from the water than single males and only came to the water to mate and lay eggs.

Search and selection of suitable sites for translocation

In conjunction with monitoring of the TAMC damselfly population, an extensive search was carried out on O'ahu to locate potential relocation sites for *Megalagrion xanthomelas*. Suitable sites needed to be free of alien predators, particularly poeciliid fish, prawns, and crayfish. During this study only two O'ahu sites were located that met these requirements. One site was located on private lands

Table 3. Sample dates and recaptures during the 1997 mark-recapture study at TAMC stream and mitigation ponds.

Date	No. recapture / day (stream + ponds)	No. new captures / day (stream + mitigation ponds)	Total captures / day
11-May-1997	0	17	17
17-May-1997	12	3	15
18-May-1997	17	4	21
22-May-1997	15	10	25
13-Jun-1997	13	26	39
6-Jul-1997	8	42	50
10-Jul-1997	21	47	68
13-Jul-1997	43	77	120
Total	129	226	355

Table 4. Sample dates and recaptures during the 2003 mark-recapture study at TAMC stream only.

Date	No. recapture / day	No. new captures / day	Total captures / day
21-Mar-2003	0	111	111
25-Mar-2003	30	86	116
1-Apr-2003	39	93	132
8-Apr-2003	59	82	141
15-Apr-2003	58	74	132
23-Apr-2003	48	58	106
29-Apr-2003	54	57	111
6-May-2003	45	64	109
13-May-2003	43	73	116
20-May-2003	41	77	118
27-May-2003	75	62	137
3-Jun-2003	67	80	147
10-Jun-2003	79	62	141
1-Jul-2003	14	42	56
8-Jul-2003	6	5	11
29-Jul-2003	11	69	80
8-Aug-2003	26	72	98
15-Aug-2003	47	62	109
Total	742	1229	1971

and could not be used because of access issues. Another site free of alien fish was also located in Makiki Valley, and was located at an unnamed tributary of Makiki stream, roughly 75 m below Herring Springs and was also remote enough to discourage human disturbance. The lack of predators resulted from a natural barrier created by an approximate 6-m waterfall located downstream of the site. The Makiki site was assessed for water quality, stream corridor vegetation, and available food resources. Stream bank vegetation, algae, and detritus, all used for oviposition, were comparable to the TAMC site. The Makiki site had denser vegetative over-story. The streambed at Makiki was sufficiently varied both in vegetation and benthic substrate and should provide adequate usable habitat for the larvae and adults. It should be noted that the Japanese wrinkled frog (*Rana rugosa*) was abundant at the Makiki site and was of some concern. However, this frog is common in most

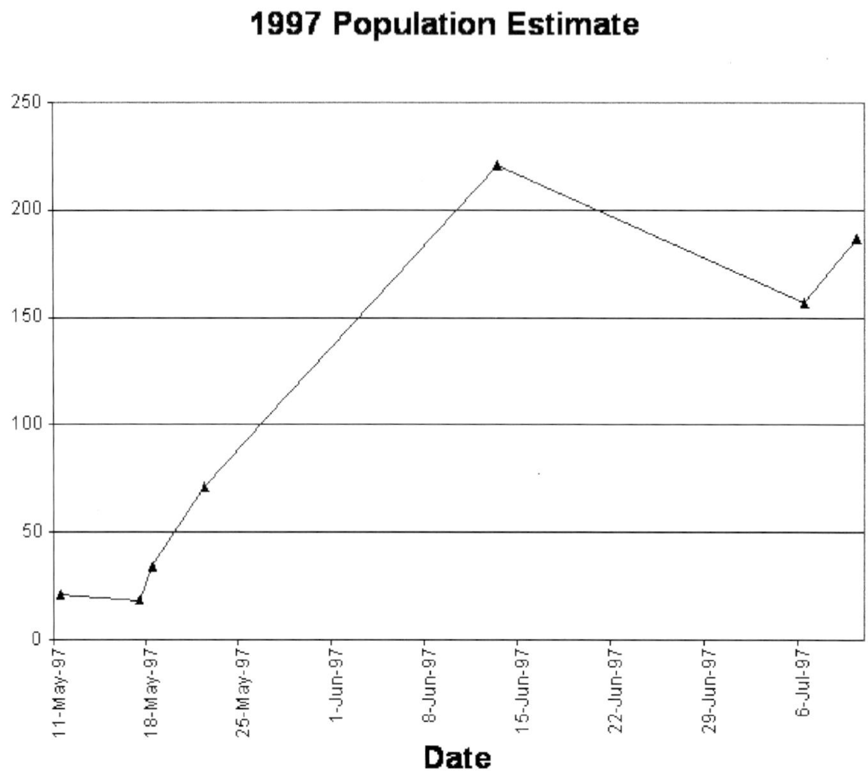

Figure 3. Population estimates from 1997 mark-recapture study at TAMC Stream and mitigation ponds.

streams where damselflies still occur and does not seem to be a significant threat (Englund, 1999).

Translocation of *Megalagrion xanthomelas*

Damselflies are most active during sunny and warm periods of the day thus collecting was planned for good weather. On the morning of 18 July 2003, 35 adult (21♂, 14♀) *M. xanthomelas* and 30 larvae were collected from TAMC. Lone adults were numbered and placed singly into plastic snapcap vials while tandem pairs were placed together in the vials. Late instar damselfly larvae were collected from the TAMC Stream and placed singly in water containers. The adults and larvae were kept in a small soft-sided cooler containing a small amount of ice which kept the adults and larvae still and also prevented overheating during transportation to the Makiki site. Damselfly larvae were released at several pools within the Makiki site, and 5 larvae were released in each pool. Pools were assessed to make sure they contained enough macroalgae and a varied substrate to allow the larvae to find cover and search for prey. Adults were released at points between adjacent pools where the larvae were placed. A second collection of 33 adults (17♂, 16♀) was made at TAMC on 25 August 2003, of which 9 pairs were collected in tandem or were mating when captured. Larvae were not collected for this second phase of the translocation. Our population estimates clearly indicate that removing these individuals from the TAMC Stream did not negatively impact the damselfly population here. For example, the damselfly population estimates increased between mark-recapture studies conducted between 29 July and 8 August 2003 (Fig. 4), several weeks after the first translocation efforts conducted on 18 July 2003 when 35 adults and 30 late instar larvae were removed from the TAMC Stream.

Three visits to the Makiki site after the translocation in August, September, and November of 2003 failed to detect any recruitment of *M. xanthomelas*. Unusually heavy flooding in this region on

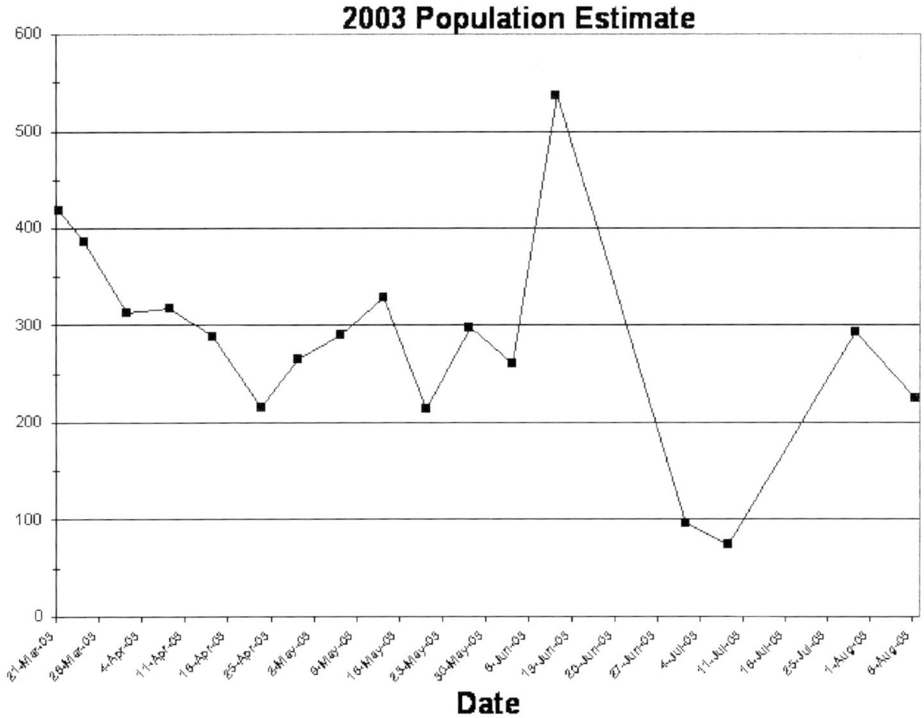

Figure 4. Population estimates from 2003 mark-recapture study at TAMC Stream only.

30 October 2004 may have hindered damselfly recruitment. Although damselfly recruitment was not documented at the Makiki translocation site, a very important finding of long-term adult survival of the translocated damselfly adults was documented during this study. One individual marked adult male damselfly translocated from the TAMC Stream on 18 August 2003 was resighted and recaptured on 25 August 2003, though no larvae were observed. Because adult *Megalagrion* damselflies will not survive without food for more than a few days, this finding indicates that adequate forage existed at the Makiki site for this damselfly to persist for at least a one-week period. Unfortunately funding did not allow for long-term monitoring of the Makiki site but monitoring of the TAMC site continues. Mark-recapture population studies at the TAMC Stream have not been conducted since 2003, but we have determined through visual observations that the TAMC population remains stable at this time. Future translocations are planned for June of 2005 when the TAMC population should be at a high enough level to withstand a 10–20% loss of adults to a translocation site.

Discussion

The only O'ahu population of *M. xanthomelas* was monitored since 1997, with intensive mark-recapture studies conducted in 1997 and 2003. The conclusion here is that the size of the TAMC *M. xanthomelas* population is a direct result of usable habitat size, and as this habitat is extremely limited, another population must be established elsewhere. The continued survival of the *M. xanthomelas* population has been due to an astonishing series of events, but stochastic events such as large rainstorms or chemical spills will eventually eliminate the TAMC damselfly population because it is so severely restricted and small in size.

The population estimates from 1997 and 2003 indicate an increase in damselfly numbers since

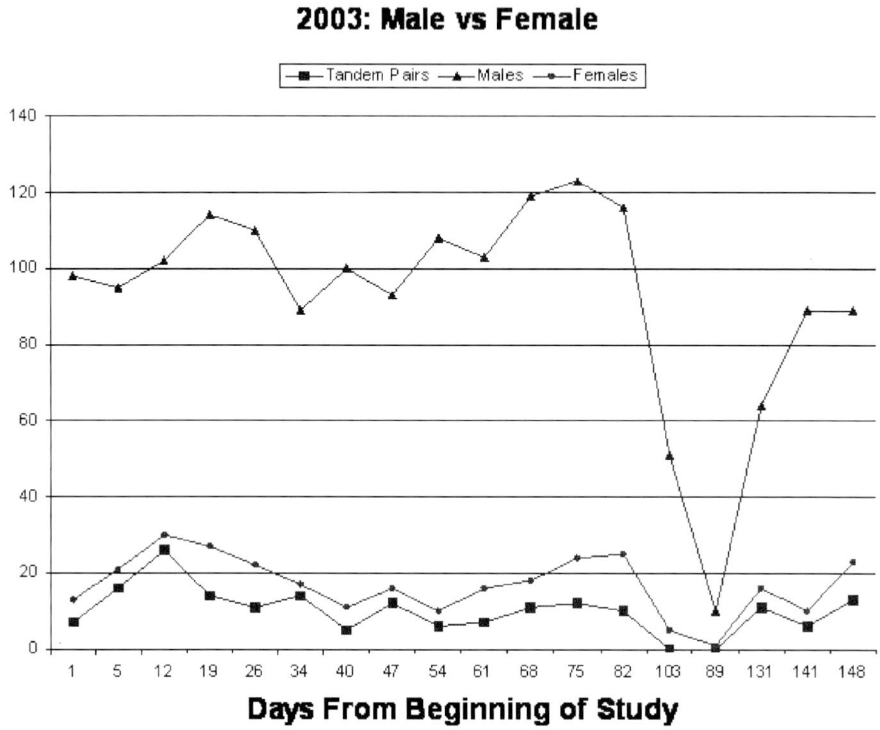

Figure 5. Number of individual males, females, and tandem pairs captured in 2003 at Tripler stream.

nearly 40% of the TAMC Stream habitat was lost from dewatering in early 1997. Excluding the July population estimate data that were taken during suboptimal cloudy and rainy conditions, the 2003 estimates indicate a relatively stable population ranging from 250–350 adults. Negative impacts from removing a small portion of the TAMC stream population were not observed, even though approximately 10–20% of the estimated adult population (and a corresponding number of larvae) were removed from the TAMC stream for translocation to the Makiki site on two separate occasions.

The TAMC stream is an extreme case of a restricted habitat, where a small section of usable damselfly habitat is bounded by channelization at the lower end and by drainage culverts at the upper end, with habitat regulated by water availability. If the area of useable habitat were to be reduced, the larvae might become crowded and start to cannibalize each other thus reducing the adult numbers. This reduction in adult numbers would then be reflected as a reduction of eggs being produced and therefore an overall reduction in population size.

Because low-elevation natural stream and wetland areas lacking alien fish and crustaceans species on O'ahu are virtually nonexistent, another option that should be pursued is the use of artificial habitats for translocation sites. *Megalagrion xanthomelas* used to be "a common insect in Honolulu gardens and in lowland districts" (Perkins, 1913), and it once again could be found in urban habitats and gardens of Honolulu. Areas such as backyard ponds, golf course waterways, and even hotel or landscaping waterfalls or water gardens could be stocked with damselfly larvae. It would be necessary to make sure these habitats were free from alien fish or crustacean species, but this could be encouraged with small grants or technical assistance in the form of aid to remove the aliens prior to damselfly stocking. There are precedents for this in the Hawaiian Islands, and a similar technique is currently being used where private citizens are encouraged and allowed to propagate endangered plants in their backyards. School groups or other educational institutions could also be encouraged to propagate native damselflies, in ways similar to the Honolulu City and County "Adopt-A-Stream" program that encourages regular cleaning of urban streams. The continued existence of the TAMC population of *M. xanthomelas* in <100 m of stream habitat illustrates the fact that large swaths of pristine habitat are not necessary for this species' continued existence. It is of great urgency that actions are taken prior to another major disturbance to the last remaining O'ahu population.

Acknowledgments

We thank Dan Polhemus for providing the damselfly distribution map and Mark Wright for assisting with population estimates. We also thank Lorena Wada and the U.S. Fish and Wildlife Service for funding and encouraging this study. Keith Arakaki and Anna Scott provided field technical assistance. We are grateful to the following mark-recapture collectors: Marguerite Butler, Brett Harper, Lei Panui, Ray Hoy, Melinda McKinney, and Anna Scott. Frank Howarth and Dan Polhemus also provided useful suggestions for this paper and the translocation efforts.

Literature Cited

Banko, P.C., R.E. David, J.D. Jacobi & W.E. Banko. 2001. Conservation status and recovery strategies for endemic Hawaiian birds. *Studies in Avian Biology* **22**: 359–376.

Caughley, G. 1980. *Analysis of vertebrate populations*. John Wiley & Sons, New York. 234 p.

Englund, R.A. 1998. Response of the Orangeblack Hawaiian Damselfly (*Megalagrion xanthomelas*), a Candidate Threatened Species, to increases in stream flow. *Bishop Museum Occasional Papers* **56**: 19-24.

———. 1999. The impacts of introduced poeciliid fish and Odonata on endemic *Megalagrion* (Odonata) damselflies on O'ahu Island, Hawai'i. *Journal of Insect Conservation* **3**: 225–243.

———. 2001. Long-term monitoring of one of the most restricted insect populations in the United States, *Megalagrion xanthomelas* (Selys-Longchamps), at Tripler Army Medical Center, O'ahu, Hawai'i (Zygoptera: Coenagrionidae). *Odonatologica* **30**: 255–263.

Evenhuis, N.L., D. Polhemus, S. Swift, K. Arakaki & D. Preston. 1995. A study of the biology of the orangeblack Hawaiian damselfly (*Megalagrion xanthomelas*), with special reference to conservation of the population at Tripler Army Medical Center, O'ahu. *Bishop Museum Technical Report* **8**, 81 p.

Johnson, C.G. 2001. Conservation biology of the orangeblack Hawaiian damselfly, *Megalagrion xanthomelas*, (Odonata: Coenagrionidae) on O'ahu. Unpublished Master's Thesis, University of Hawai'i at Mānoa, Honolulu. 92 p.

Pangelinan, A.A. 1997. Demography and life history of the orangeblack Hawaiian damselfly (*Megalagrion xanthomelas*) (Sélys-Longchamps, 1876) on O'ahu, Hawai'i. Unpublished Masters Thesis, University of Guam. 44 p.

Perkins, R.C.L. 1899. *Neuroptera. Fauna Hawaiiensis* **2**(3): 63–77.

———. 1913. Introduction. *Fauna Hawaiiensis* **1**(6): xv–ccxxviii.

Polhemus, D.A. 1993. Damsels in distress: a review of the conservation status of Hawaiian *Megalagrion* damselflies (Odonata: Coenagrionidae). *Aquatic Conservation* **3**: 343–349.

———. 1995. A survey of the aquatic insect faunas of selected Hawaiian streams. *Hawai'i Biological Survey Report for the Commission on Water Resource Management, Department of Land and Natural Resources, State of Hawai'i.* Department of Natural Sciences, Bishop Museum. 128 p.

———. 1996. The Orangeblack Hawaiian Damselfly, *Megalagrion xanthomelas* (Odonata: Coenagrionidae): clarifying the current range of a threatened species. *Bishop Museum Occasional*

Papers **45**: 30–53.

———. **& A. Asquith**. 1996. *Hawaiian damselflies: a field identification guide*. Bishop Museum Press, Honolulu, Hawai'i. 122 p.

———., **H. Oppenheimer, F. Starr & K. Martz**. 1999. Notable rediscoveries of *Megalagrion* species on Maui (Odonata: Coenagrionidae). *Bishop Museum Occasional Papers* **59**: 27–29.

Robichaux, R, S. Bergfeld, M. Bruegmann, J. Canfield. P. Moriyasu, T. Rubenstein, T. Tunison & F. Warshauer. 2000. Reintroducing Hawaii's silverswords. *Endangered Species Bulletin* **25**: 22–23.

Williams, F.X. 1937. Biological Studies in Hawaiian water-loving insects. Part 1. Coleoptera or beetles. Part 2. Odonata or dragonflies. *Proceedings of the Hawaiian Entomological Society* **9**: 235–345.

Biology of Hawaiian Streams and Estuaries. Edited by N.L. Evenhuis & J.M. Fitzsimons. *Bishop Museum Bulletin in Cultural and Environmental Studies* 3: 277–291 (2007).

Contaminants in the Watershed: Implications for Hawaiian Stream Biota

ANNE M.D. BRASHER
U.S. Geological Survey, 2329 West Orton Circle, Salt Lake City, UT 84119, USA; email: abrasher@usgs.gov

REUBEN H. WOLFF
U.S. Geological Survey, 677 Ala Moana Blvd., Suite 415, Honolulu, Hawai'i 96813, USA

Abstract

In aquatic systems throughout the world, urban and agricultural activities have resulted in elevated concentrations of organochlorine pesticides and metals. At sites on two Hawaiian Islands, O'ahu and Kaua'i, streambed sediment and fishes were analyzed for organic and metal contaminants; benthic invertebrates were collected for analysis of community composition and structure; and physical habitat characteristics were recorded. Sites were selected to represent a gradient of development from high-intensity urban to undeveloped (forested) areas. Many of the sites had both urban and agricultural development in the watershed. Concentrations of many organic and metal contaminants at the developed sites exceeded criteria for the protection of fish and benthic aquatic invertebrates, and the wildlife that consume them, and indicate potential deleterious effects. Streams in developed watersheds had elevated concentrations of organochlorine pesticides used for termite control (dieldrin and chlordane) and DDT. Concentrations of metals including arsenic, cadmium, lead, and zinc in sediment from developed areas were substantially higher than concentrations measured in undeveloped areas. Differences in macroinvertebrate species composition, diversity, and abundance were associated with elevated concentrations of contaminants at the developed sites compared with the undeveloped sites. The persistence of these contaminants, their tendency to accumulate in sediment and biota, and their potentially harmful effects are of concern for the management and protection of stream biota in Hawaiian watersheds.

Introduction

With increasing population and subsequent development, many watersheds in the Hawaiian Islands have undergone substantial land-use change that has resulted in degradation of both water and habitat quality (Brasher, 2003; Anthony *et al.*, 2004). Anthropogenic influences, both urban and agricultural, can adversely impact stream ecosystems. Effects such as stream channelization for flood control or roadways, increases in sedimentation from construction and farming, contaminants from agricultural, urban, and industrial activities transported in storm-water runoff, and diversions used to redirect stream water to agriculture and other off-stream uses can all affect stream quality (Oki & Brasher, 2003).

As is typical throughout the tropics, modifications of the stream ecosystems have been most severe at the lower elevations where urban growth is greatest (Resh *et al.*, 1992; Pringle & Ramirez, 1998). Urbanization is typically accompanied by stream channelization and the removal of riparian canopy cover that results in higher water temperatures and increased daily temperature fluctuations, as well as increased siltation and removal of larger substrate such as boulders (Brasher, 2003; Brasher *et al.*, 2004). Habitat alteration can have substantial impacts on aquatic communities that reside in the streams, resulting in changes in species composition, abundance, and diversity. Distinct invertebrate assemblages have been observed to occur with different land uses in the watershed and consequently can be used to indicate habitat quality (Brasher *et al.*, 2004).

Figure 1. Organochlorine pesticides and metals are transported from agricultural and urban sources through soil erosion, runoff, and precipitation. Once in the stream, they accumulate in streambed sediments and in tissues of aquatic invertebrates and fish.

Physical changes including decreased habitat heterogeneity and increased abiotic variability are compounded by the presence of organic and inorganic (metal) contaminants (Oki & Brasher, 2003). Urban land use is commonly associated with a variety of organochlorine, organophosphate, and metal contaminants in streams. These constituents are applied on land as pesticides, herbicides, or fertilizer, in both urban and agricultural areas, or may have industrial uses. Whether present in the water column or adhered to the sediment, they can have potentially negative effects on the benthic invertebrates that live in the streams (Nowell *et al.*, 1999; Oki & Brasher, 2003; Cain *et al.*, 2004).

Organochlorine pesticides are chlorine-containing compounds that are present in the environment as a result of human activities. Organochlorine pesticides were heavily used from the mid-1940s to the mid-1980s. The persistence of organochlorine pesticides, their tendency to accumulate in soil, sediment, and biota, and their harmful effects on wildlife resulted in restriction or banning of their use in the United States in the 1970s through mid-1980s (Nowell *et al.*, 1999). Despite use restrictions, these compounds continue to be detected in sediment and fish samples.

Metals (for example arsenic, copper, lead, mercury, and zinc) can occur naturally in the environment but may increase to elevated levels as a result of human activities. Urban sources for metals include vehicular traffic, batteries, paint, road surfacing, leaded gasoline, and wood preservatives. Many metals have been shown to be toxic to aquatic biota (Hare, 1992). As with organochlorine pesticides, leaded gasoline and lead-based paints were phased out in the 1970s, but lead persists in soils and continues to enter Hawaiian streams with sediment in runoff (DeCarlo & Anthony, 2002).

Organochlorine pesticides, metals, and trace elements enter the aquatic environment from a variety of sources including the atmosphere, industrial and municipal effluents, and soil erosion and other nonpoint-source runoff (Fig. 1). Once in the system, these compounds (which have low solubility) are mostly associated with bottom sediments that can be ingested by benthic organisms. These organisms are then eaten by fish and birds, transferring the contaminants to higher trophic levels in aquatic and terrestrial food chains.

Table 1. List of sites samples.

Site	Code	Island	Pesticides Sediment	Pesticides Fish	Metals Sediment	Metals Inverts
Ala Wai Canal	ALWI	O'ahu		X		
Hanakāpī'ai Stream	HNKP	Kaua'i				X
Hulē'ia Stream	HULA	Kaua'i				X
Kalauao Stream	KALA	O'ahu	X	X		X
Kalihi Stream (lower)	LKLI	O'ahu	X	X	X	
Kalihi Stream (upper)	UKLI	O'ahu	X	X	X	
Kaluanui Stream	KALU	O'ahu		X	X	X
Kāne'ohe Stream	KANE	O'ahu	X	X	X	X
Kapa'a Stream	MKPA	Kaua'i				X
Kaukonahua Stream	KKON	O'ahu		X	X	
Kaukonahua Stream (North Fork)	NFKK	O'ahu			X	
Kaukonahua Stream (South Fork)	SFKK	O'ahu		X	X	
Kea'ahala Stream	KEAH	O'ahu		X		
Kīpapa Stream	KIPA	O'ahu	X	X	X	
Kōloa Stream	KOLO	O'ahu		X	X	
Lawa'i Stream (lower)	LLWI	Kaua'i				X
Lawa'i Stream (upper)	ULWI	Kaua'i				X
Limahuli Stream	LMAH	Kaua'i				X
Luluku Stream	LULU	O'ahu	X	X	X	X
Makaleha Stream	UKPA	Kaua'i				X
Mānoa Stream	MANO	O'ahu	X	X	X	X
Maunawili Stream	MWIL	O'ahu	X	X	X	
Nawiliwili Stream	NWIL	Kaua'i	X	X	X	X
Nu'uanu Stream (lower)	NUUA	O'ahu	X	X	X	X
Nu'uanu Stream (upper)	NUUJ	O'ahu		X		
'Ōpa'eula Stream (lower)	LOPA	O'ahu		X	X	
'Ōpa'eula Stream (upper)	UOPA	O'ahu		X	X	
Poamoho Stream	POAM	O'ahu	X	X	X	
Pū'ali Stream	PUAL	Kaua'i		X		X
Punalu'u Stream (above dam)	PUNA	O'ahu				X
Punalu'u Stream (below dam)	PUNB	O'ahu				X
Wahiawā Reservoir	WAHI	O'ahu		X	X	
Waiāhole Stream	WHOL	O'ahu				X
Waiakeakua Stream	WKEA	O'ahu	X	X	X	X
Waiawa Stream	WAIW	O'ahu	X	X	X	X
Waihe'e Stream	WHEE	O'ahu	X	X	X	X
Waikakalaua Stream	WKAK	O'ahu	X	X	X	X
Waikele Stream	WKEL	O'ahu	X	X	X	X
Waimānalo Stream	WAIM	O'ahu	X	X	X	X
Waimano Stream	WANO	O'ahu	X	X	X	

Although organisms used in stream-quality monitoring programs include algae, invertebrates, and fish (Lenat & Barbour, 1994; Barbour *et al.*, 1999), benthic macroinvertebrates are by far the most commonly used group of organisms for this purpose (Rosenberg & Resh, 1993). Benthic macroinvertebrates offer many advantages in biomonitoring: (1) they are ubiquitous and, consequently, can be affected by environmental perturbation in a variety of aquatic systems and habitats, (2) the large number of species offers a wide spectrum of responses to environmental stressors, (3) their basic sedentary nature allows effective spatial analyses of pollutants or disturbance effects, and (4) they have relatively long life cycles, which allows elucidation of temporal changes caused by perturbation (Rosenberg & Resh, 1993).

Table 2. Organochlorine compounds and codes used in this paper.

Compound	Code	Compound	Code
Aldrin	ALD	delta-HCH	DHC
cis-Chlordane	CCH	gamma-HCH (Lindane)	LIN
trans-Chlordane	TCH	Heptachlor	HEP
Dacthal (DCPA)	DCP	Heptachlor epoxide	HEE
o,p'-DDD	ODD	Hexachlorobenzene	HEX
p,p'-DDD	PDD	o,p'-Methoxychlor	OME
o,p'-DDE	ODE	p,p'-Methoxychlor	PME
p,p'-DDE	PDE	Mirex	MIR
o,p'-DDT	ODT	cis-Nonachlor	CNO
p,p'-DDT	PDT	trans-Nonachlor	TNO
Dieldrin	DIE	Oxychlordane	OXY
Endrin	END	PCB	PCB
alpha-HCH	AHC	Pentachloroanisole	PEN
beta-HCH	BHC	Toxaphene	TOX

In this paper we discuss streams that are located in watersheds on two Hawaiian islands, O'ahu and Kaua'i, and represent a gradient of anthropogenic development from high-intensity urban to undeveloped (forested) areas. We will describe the results of our recent studies assessing organochlorine contaminants, trace elements, and metals in streambed sediment and fish from selected streams on these two islands. Because development frequently results in the co-occurrence of water-quality degradation (including organic and inorganic contaminants) and physical habitat alteration, we will also describe how urbanization affects physical habitat characteristics. Finally, we will discuss implications of stream-quality alteration, associated with anthropogenic activities in the watershed, to the benthic macroinvertebrate communities that live in these streams.

Methods

Site selection

Sites were selected to represent a gradient of development, from high-intensity urban to forested reference (Table 1). Sites with both urban and agricultural activities in the watershed are categorized as "mixed" land use. Streambed sediment and fish from 31 sites were analyzed for organic and metal contaminants. Benthic invertebrates from 24 sites were collected for analysis of community composition and structure, and physical habitat characteristics were recorded at each of these sites. For this overview paper, results from selected sites are presented.

Fish sampling

Fish were collected by using an electrofisher, throw-net, or seine. Species selection was based on availability. Poeciliidae (guppies and mollies) were the most commonly collected fish. Other fish included those that people might eat (bass and tilapia), aquarium introductions (such as cichlids), and one sample of the native eleotrid (*Eleotris sandwicensis*). At 6 sites, more than one species was collected (to compare among species). At four sites, samples were collected in two different years (to compare between years). Composite samples of approximately 100 grams of whole fish of the same species (for organochlorine compounds) or 10 grams (for metals analysis) were obtained at each site. For smaller species, composites consisted of approximately 100 individuals. For larger species, five to eight individuals were composited.

Sediment sampling

Sediment samples were collected from undisturbed depositional zones along a 100-meter reach at each site. Sampling was confined to the upper 2 cm of bed sediment that represents contaminants most recently contributed to the stream (Shelton & Capel, 1994). Subsamples from along the reach were composited and wet-sieved in the field, using a 2-mm stainless steel sieve for the organics and a 63- m sieve for metals. Samples were frozen on site and shipped to the laboratory for analysis.

Organic and metal analysis

Methods for processing sediment and fish followed nationally consistent procedures (Shelton & Capel, 1994; Crawford & Luoma, 1993), and all samples were analyzed at the U.S. Geological Survey National Water-Quality Laboratory in Denver, Colorado. Thirty-two organochlorine compounds and 43 metals were analyzed in streambed sediment. Twenty-eight organochlorine compounds were analyzed in fish tissue (Table 2). A list of constituents analyzed is available at: http://ca.water.usgs.gov/pnsp/pest.rep/bs-t.html.

Guidelines

Guidelines have been established for certain contaminants to help determine what concentrations of chemicals are likely to be associated with adverse biological effects. Concentrations measured in this study were compared with the Canadian Sediment Quality Guidelines (CSQG) for the protection of aquatic life (http://www.ec.gc.ca/ceqg-rcqe). Two assessment values have been calculated for the CSQG. The lower value of ISQG (interim sediment quality guideline) represents the concentration below which adverse effects to aquatic biota are rarely expected to occur. The upper value or PEL (probable effect level) defines the level above which adverse effects to aquatic biota are expected to occur frequently. These guidelines are based on chronic (long-term) effects of contaminants to aquatic organisms. Limitations of guidelines are that they have been established for only a limited number of pesticides and metals; the toxicity of mixtures and breakdown products is not generally considered; and sublethal effects, such as endocrine disruption, typically have not been assessed.

Benthic invertebrate sampling

Invertebrate samples were collected at each site by using a modified Surber-sampler (with a 425-mm mesh net) for quantitative riffle samples and a D-frame net (with a 210-mm mesh net) for qualitative multi-habitat samples (Cuffney *et al.*, 1993). All sampling was conducted during base-flow conditions. Quantitative samples provide estimates of organism relative abundance to allow comparison between sites. Qualitative samples were collected from all available habitats within the reach at each site to provide a comprehensive species list.

Invertebrate samples were identified and enumerated at the U.S. Geological Survey National Water-Quality Laboratory Biological Unit (Denver, Colorado) or by EcoAnalysts, Inc. in Moscow, Idaho. Verification of problematic taxa and routine quality-assurance checks on taxonomic identification were done by nationally recognized experts. Data reported for the quantitative samples include both species occurrence and density using numeric (300-fixed count) and time (total sorting time) criteria (Moulton *et al.*, 2000). Qualitative samples were analyzed only for species occurrence, by using a timed visual-sort method (Moulton *et al.*, 2000).

Habitat measurements

Habitat characteristics were determined at multiple spatial scales: basin, segment, reach, transect, and point (as described in: Fitzpatrick *et al.*, 1998; Brasher *et al.*, 2004). Basin and segment characteristics (watershed-scale features) such as land use, drainage area, and gradient were determined by using Geographic Information Systems (GIS) and topographic maps. Reach length at each site was determined as the linear distance equal to 20 times the average wetted channel width, with a minimum length of 100 meters. Within each reach, 11 equally spaced transects were established across the stream perpendicular to the direction of flow. Physical measurements of bank and riparian features and instream characteristics were made at each transect. Bank and riparian features included measurements such as bank angle, amount of erosion, and canopy cover. Instream habitat measurements included features such as aspect, wetted perimeter, depth, velocity, and substrate size. Point (microhabitat) measurements of depth and velocity were also made at each location where a quantitative invertebrate sample was collected. Water temperature (°C) was continuously recorded, using temperature loggers, at 30-minute intervals for approximately a 15-month period at 10 of the sampling sites.

Results and Discussion

Traditionally, water-quality monitoring has focused on water-column sampling. There are a number of reasons why it is important to analyze streambed sediment and fish for contaminants as well. Different types of compounds, including the organochlorine pesticides and metals, are more readily detected in sediment and tissue than in water because of their hydrophobic characteristics (they have low solubility in water but preferentially distribute into soil organic carbon and fatty tissues). Sediment serves as both a source and a removal mechanism for contaminants to and from the stream, and as a means of contaminant transfer downstream. Sediment also provides habitat for benthic biota and can be ingested by them. Aquatic biota are important in the food web around the stream, and some organisms such as fish are consumed by people and birds. In addition, fish and sediment can provide an integrated picture of contaminants over time.

The organochlorine pesticides most frequently detected at elevated concentrations in Hawaiian stream sediment and fish were dieldrin and chlordane (insecticides used for termite control) and DDT (an insecticide used in both agricultural and urban settings) and its derivative DDE (Fig. 2). Chlordane and aldrin (which breaks down into dieldrin) were used heavily in urban settings until discontinued in the mid-1980s (Takahashi, 1982). Despite the fact that they have not been used in almost 20 years, chlordane and dieldrin were prevalent in sediment and fish from urban watersheds, and concentrations consistently exceeded aquatic-life guidelines.

DDT was used to control agricultural pests in Hawai'i and also was sprayed widely for mosquito control in urban areas (Robert Boesch, Hawai'i Department of Agriculture, oral communication). Although DDT was banned in 1972, DDT and its derivatives (DDD and DDE) were prevalent in streambed sediment and fish from urban, agricultural, and mixed (urban and agricultural) watersheds. As with the organochlorine pesticides, the continued persistence of the DDT compounds in the environment likely reflects slow breakdown of the compounds, persistence in soils, and continual delivery to streams.

Following a trend that has been observed across the country (Schmitt *et al.*, 1990, 1999; Nowell *et al.*, 1999; Wong *et al.*, 2000), concentrations of organochlorine compounds are generally decreasing in Hawaiian streams (Brasher & Wolff, 2004). As part of the National Pesticide Monitoring Program, the U.S. Fish and Wildlife Service periodically analyzed organochlorine pesticides in fish from streams across the country, including two streams on O'ahu; Waikele Stream and Mānoa Stream. In Waikele Stream, dieldrin, chlordane, and total DDT concentrations have decreased substantially since the 1970s. In Mānoa Stream, total DDT concentrations have decreased substantially but dieldrin and chlordane do not appear to have decreased much, if at all (Brasher & Wolff, 2004).

Concentrations of a number of trace elements and metals were elevated in streambed sediment samples (Fig. 3) at concentrations exceeding aquatic-life guidelines. These include arsenic, lead, and zinc, which are associated with anthropogenic activities, and others such as chromium and copper that naturally occur at high concentrations in Hawaiian rocks and soils (Anthony *et al.*, 2004). Concentrations of arsenic were highest in streams from agricultural and mixed (agricultural and urban) watersheds. Arsenic is a common impurity in fertilizer, and this may be the source of arsenic in these areas (Anthony *et al.*, 2004). Concentrations of metals associated with anthropogenic activities typically exceeded guidelines in developed (urban and agricultural) watersheds while those that naturally occur in Hawaiian rocks and soils (such as chromium) were elevated in forested areas as well.

Concentrations of organic contaminants tended to be higher in fish tissue than in associated streambed sediment (Brasher & Wolff, 2004). Concentrations of a number of the compounds analyzed for in Hawaiian streams were among the highest in the nation (Brasher & Wolff, 2004; Anthony *et al.*, 2004). Replicate samples of sediment and fish collected over 2 years, and among species at a site over 1 year substantiate these findings. Two or three different fish species were sampled at six different sites for organochlorine compounds. Three examples are shown in Fig. 4. In general, if a compound was detected in one of the species, it was detected in the other (Fig. 4). Concentrations between species for a given compound were very similar.

The same species were sampled during two different years at four different sites; three examples are shown in Fig. 5. In general, if a compound was detected in 1998 it was also detected in 2000

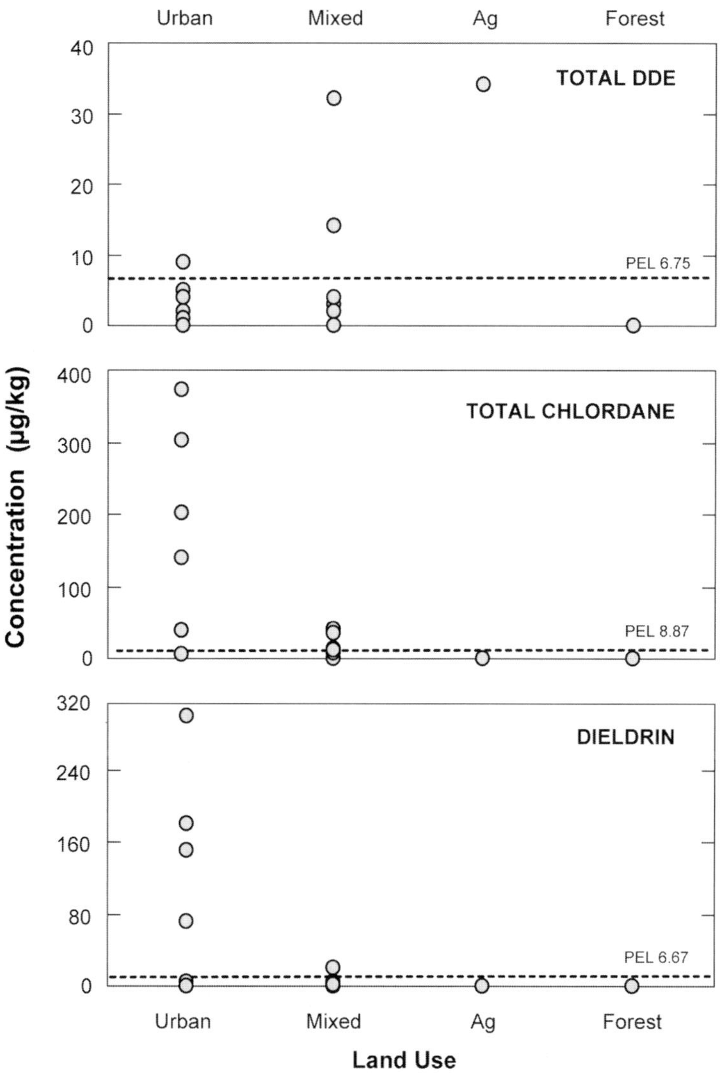

Figure 2. DDE (a breakdown product of DDT), chlordane, and dieldrin concentrations in sediment. Sites grouped by land use (Ag= agriculture). Sediment values are dry weight, concentrations in micrograms per kilogram (µg/kg). Canadian Sediment Quality Guidelines; PEL= probable effect level.

(Fig. 5). At two urban sites (KANE* and NUUA), concentrations of organochlorine compounds were very similar between years, and at a site in a forested watershed (WHEE), no organic compounds were detected either year.

A Principal Components Analysis (PCA) of organic contaminants and metals analyzed in sediment at 17 sites (where both organics and metals were analyzed) shows that the sites group together based on the type of contaminants that occur there, and that this grouping is associated with land use in the watershed (Fig. 6). Sites located in urban areas are characterized by the pesticides chlordane and dieldrin, PCBs, lead, cadmium, and zinc. Sites in agricultural and mixed (agricultural and urban) watersheds had elevated concentrations of DDT, and a number of metals including arsenic,

* see Table 1 for explanation of codes used for sites referred to in this paper.

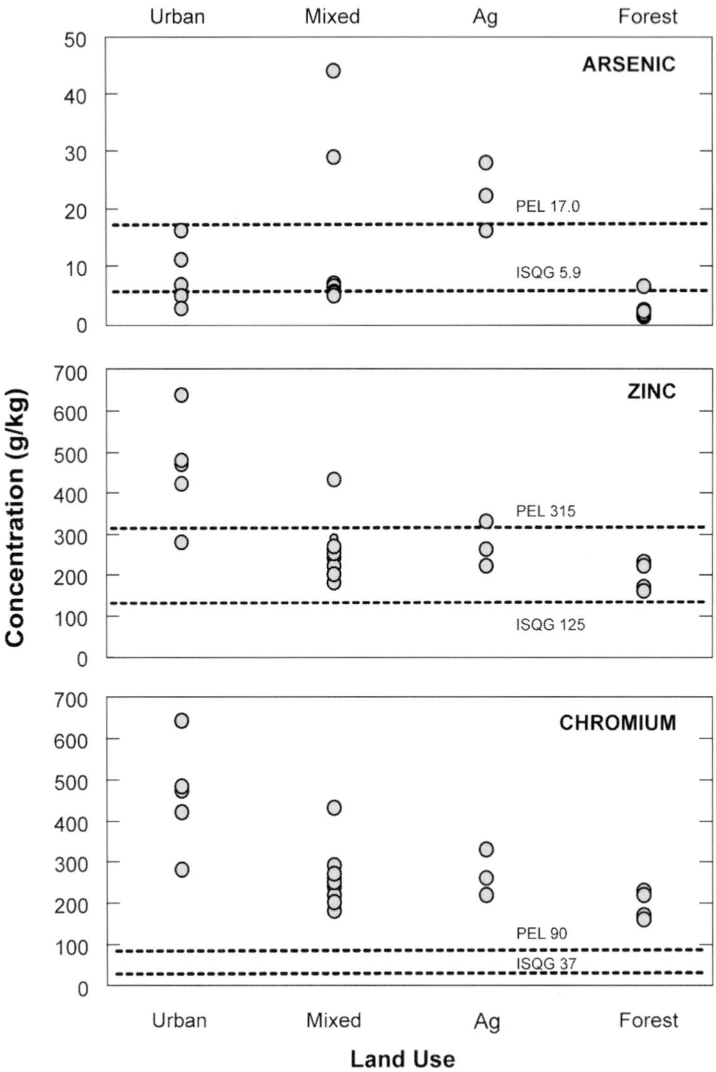

Figure 3. Arsenic, zinc, and chromium concentrations in sediment. Sites grouped by land use (Ag= agriculture). Sediment values are dry weight, concentrations in grams per kilogram (g/kg). Canadian Sediment Quality Guidelines; PEL= probable effect level, ISQG= interim sediment quality guidelines.

copper and mercury. Sites in forested areas had no detectable organic contaminants.

Many organochlorine pesticides and metals are known animal carcinogens and are potential human carcinogens. The adverse effects of DDT on reproduction in birds have been well documented, and other organochlorine pesticides have been linked to endocrine disruption, an alteration of natural hormonal activity (Kavlock *et al.*, 1996). These effects, combined with a slow rate of breakdown, make many organic and metal contaminants a long-term environmental concern. Endangered native waterbirds, such as the Hawaiian Stilt (*Himantopus mexicanus knudseni*), that consume fish or aquatic invertebrates may be at risk because of the elevated concentrations of these compounds in many Hawaiian streams.

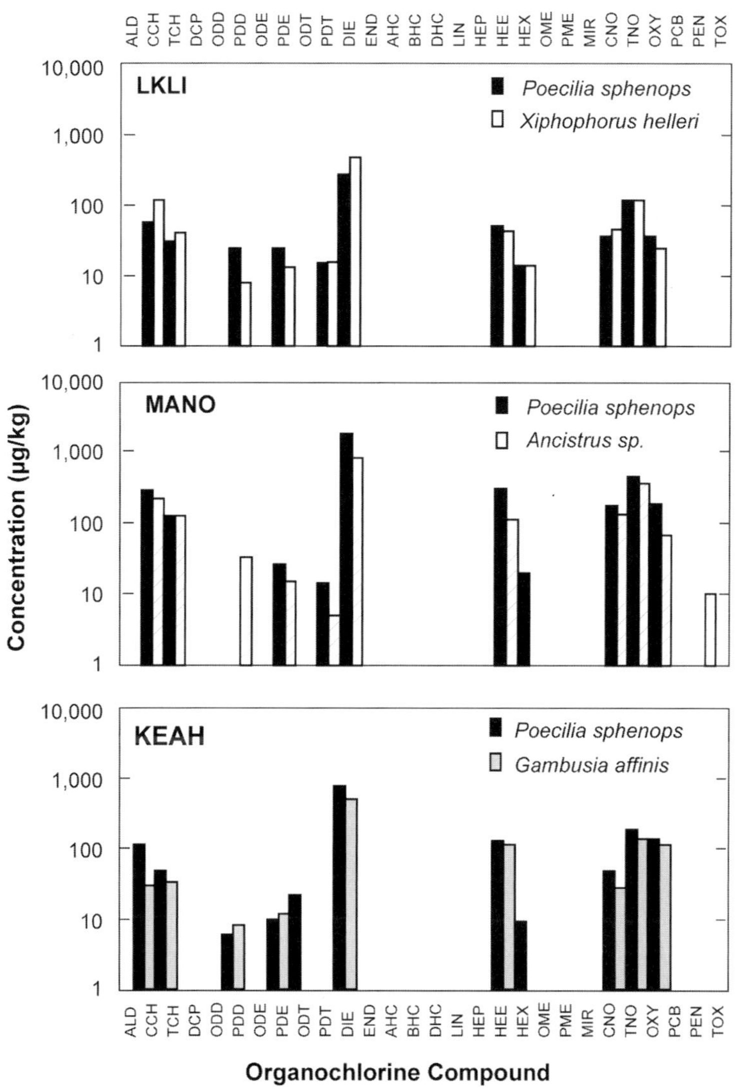

Figure 4. Comparison of organochlorine compound concentrations (Y axis is log scale) in different fish species collected at the same site. Tissue values are wet weight, concentrations in micrograms per kilogram (µg/kg). Site acronyms in Table 1. Acronyms for organochlorine compounds in Table 2.

Development is associated not only with changes in water quality but also with changes in physical habitat characteristics (Brasher, 2003; Brasher *et al.*, 2004). A PCA of physical habitat characteristics at the sampling sites indicates that the sites group together based on the type of anthropogenic activity in the watershed (Fig. 7). Sites in the lower left quadrant tended to be forested sites, with large substrate, natural channels, and low levels of silt. Sites in the lower right quadrant were urban sites with cement-lined channels, high levels of solar radiation, and silt. Sites in the upper right quadrant were agricultural sites, which tend to be channelized, with small substrate, but with more riparian cover than urban sites. Mixed sites, in the upper left quadrant, were similar to agricultural

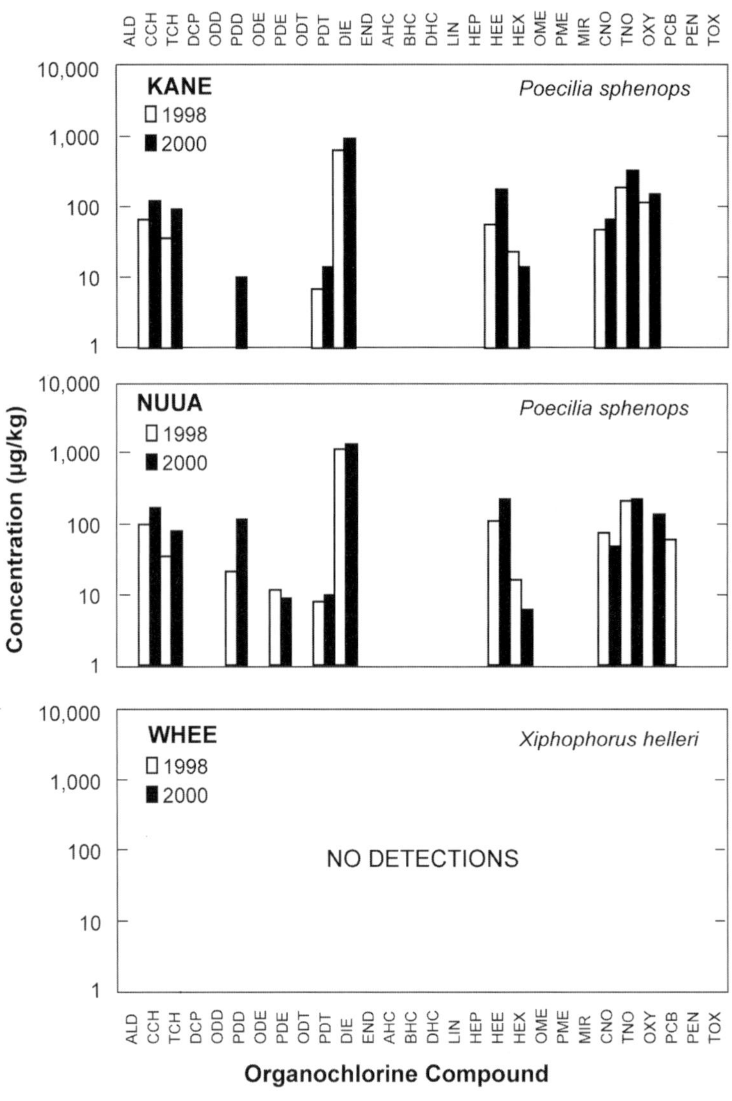

Figure 5. Comparison of organochlorine compound concentrations (Y axis is log scale) in fish collected in two different years. Tissue values are wet weight, concentrations in micrograms per kilogram (µg/kg). Site acronyms in Table 1. Acronyms for organochlorine compounds in Table 2.

sites in terms of higher levels of riparian cover (and consequently lower levels of solar radiation), to forested sites in terms of larger substrates, and they tended to lack the channel modification of urban and agricultural sites.

Overall, habitat characteristics of streams in urban and mixed (urban and agriculture) watersheds were markedly different from those in forested watersheds (Brasher *et al.*, 2004). Artificially straightened reaches with concrete-lined, flat-bottomed channels and reinforced banks are common in urban areas of Hawai'i. Such modifications are often accompanied by the removal of riparian vegetative cover and a reduction in substrate heterogeneity (removal of large boulders). The end result is a wide, shallow, unshaded, and generally homogeneous stream reach; a stark contrast to the steep,

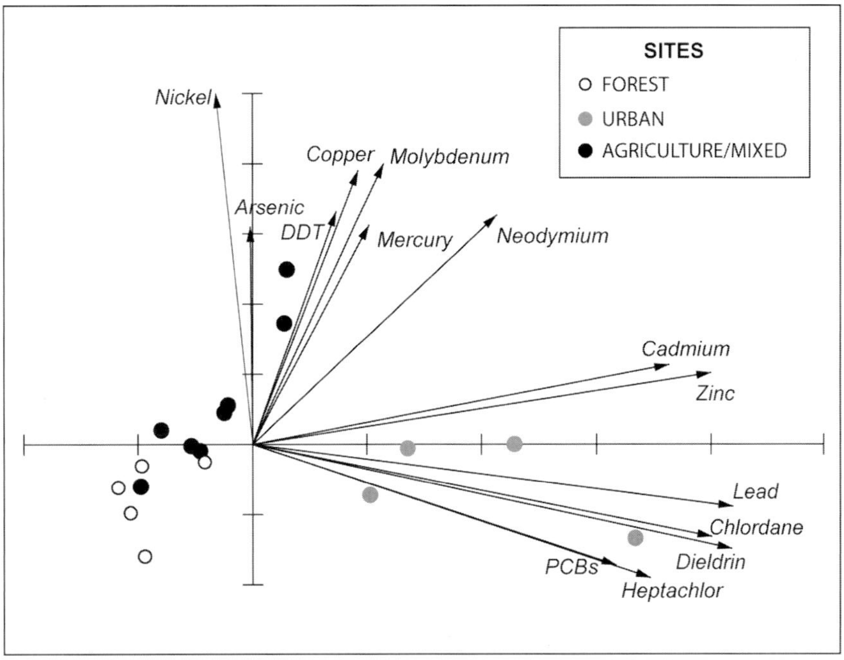

Figure 6. Principal Components Analysis of organochlorine pesticides and metals.

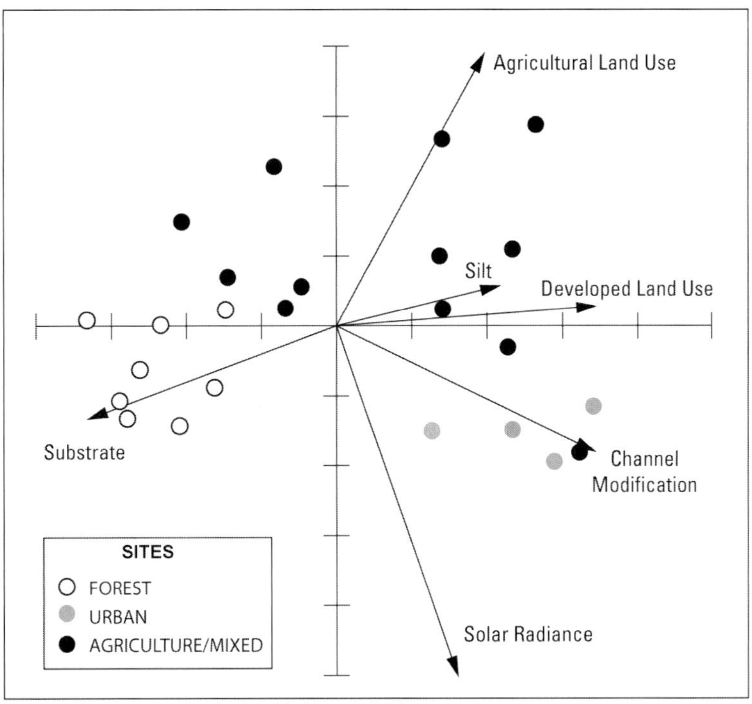

Figure 7. Principal Components Analysis of habitat variables.

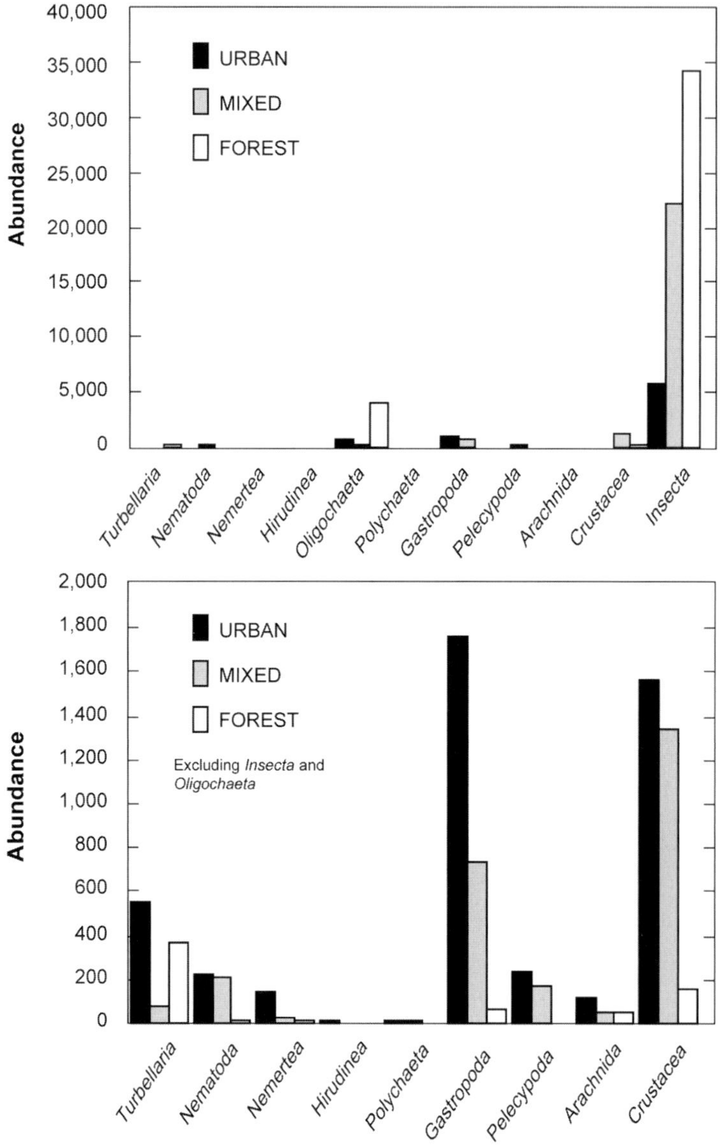

Figure 8. Invertebrate abundance (number of individuals) for selected classes. The top graph includes Insecta and Oligochaeta. Because of the high abundance in these two classes, the bottom figure excludes them, allowing for comparison among less abundant classes.

heavily vegetated, boulder-strewn reaches typical of more pristine streams in forested watersheds of Hawai'i (Brasher *et al.*, 2004).

One effect of physical habitat alteration associated with channelization is a large increase in mean daily temperature and maximum temperature, as well as substantially greater temperature fluctuations (Brasher, 2003). Continuous water temperature probes located at 10 sites representing a gradient of watershed development indicated that streams at urban sites, which tended to be shallow and have reduced riparian cover, had much higher temperatures than forested sites (Brasher *et al.*, 2004).

Benthic invertebrate community structure likely reflects both the physical and chemical quality of the stream. Species composition and abundance were very different in developed versus undeveloped watersheds (Fig. 8). Most notably, insects accounted for a much larger proportion of the benthic invertebrate abundance in undeveloped (forested) watersheds than in developed (urban, and mixed urban and agricultural) watersheds. Although oligochaetes were associated with forested watersheds in the analysis, the abundance of oligochaetes was the result of a single forested site that periodically goes dry and consequently, has physical habitat characteristics more similar to developed watersheds (such as high levels of silt) where oligochaetes typically occur (Brasher *et al.*, 2004).

In an analysis of ten of the sites on O'ahu, insects were determined to typically make up more than 80% of the invertebrate species in forested watersheds; introduced Trichoptera (caddisflies) were the dominant insect (Brasher, 2003). In contrast, insects typically made up only about half of the invertebrates at urban sites, with Diptera (true flies) more abundant than caddisflies (Brasher, 2003). A similar pattern has been documented in New Zealand, where a more pristine site was dominated by mayflies and caddisflies and a more developed site was dominated by true flies (Death, 2000).

Urban sites had the lowest abundance of insects and high numbers of gastropods and crustaceans (Fig. 8). No native gastropods were collected during this study. The most abundant gastropods were the pan-tropical thiarid snails, the introduced clam (*Corbicula fluminea*), and a limpet (*Ferrissia sharpi*). Gastropods occurred almost exclusively at channelized sites (Brasher *et al.*, 2004). As has been reported in Japan (Karr & Chu, 1997), the most common gastropods present in developed areas of Hawaii are nonnative and indicators of degraded conditions. Crustaceans collected from streams in forested watersheds tended to be either the native atyid shrimp (*Atyoida bisulcata*) or the widespread introduced Tahitian prawn (*Macrobrachium lar*). The dominant crustaceans at developed sites were the introduced crayfish (*Procambarus clarkii*), and the introduced atyid shrimp (*Neocaridina denticulata*).

Conclusions

Elevated concentrations of pesticides used for termite control (dieldrin and chlordane) and DDT were measured in sediment and fish from streams in developed watersheds. These concentrations often greatly exceeded guidelines for the protection of aquatic species. Metals associated with anthropogenic activities (such as lead, zinc, and arsenic) were measured in sediment from developed watersheds at concentrations that exceeded aquatic-life guidelines. Concentrations of some metals (such as chromium and copper) are naturally high in Hawaiian soil and rocks and were measured at concentrations exceeding guidelines even in forested areas.

Water-quality degradation and physical habitat alteration tend to co-occur in developed watersheds. Ongoing analysis is being conducted to further elucidate the relative contribution of these two factors to stream quality and to the integrity of benthic macroinvertebrate communities in Hawaiian streams. In addition, an understanding of associations between anthropogenic activities and the resulting introduction of organic and metal contaminants into aquatic systems will allow land-management practices to be designed to reduce the loading of contaminants to streams and nearshore waters.

Acknowledgments

Corene Luton contributed to data collection and analysis for this project. Steve Anthony provided helpful comments on data interpretation and presentation. Comments on an earlier draft by Sue Thiros, Larry Spangler, David Susong, Mike Fitzsimons, and two anonymous reviewers are appreciated. We were assisted in the field by numerous volunteers (too many to list by name) including community members, students, and State and Federal agencies. Funding for this research was provided by the U.S. Geological Survey National Water-Quality Assessment Program and the Hawaii State Department of Health. We thank all of the private landowners who generously allowed us access to their property.

Literature Cited

Anthony, S.S., C.D. Hunt, Jr., A.M.D. Brasher, L.D. Miller & M.S. Tomlinson. 2004. Water quality on the island of Oahu, Hawaii, 1999–2001. *United States Geological Survey Circular* **1239**: 37 p.

Barbour, M.T., J. Gerritsen, B.D. Snyder & J.B. Stribling. 1999. *Rapid bioassessment protocols for use in streams and wadeable rivers: periphyton, benthic macroinvertebrates and fish, second edition.* Environmental Protection Agency **841-B-99-002**. United States Environmental Protection Agency, Office of Water, Washington, D.C.

Brasher, A.M.D. 2003. Impacts of human disturbance on biotic communities in Hawaiian streams. *BioScience* **53**: 1052–1060.

———. **& R.H. Wolff.** 2004. Relations between land use and organochlorine pesticides, PCBs, and semi-volatile organic compounds in streambed sediments and fish on the island of Oahu, Hawaii. *Archives of Environmental Contamination and Toxicology* **46**: 385–398.

———., **R.H. Wolff & C.D. Luton.** 2004. Associations among land use, habitat characteristics, and invertebrate community structure in nine streams on the island of Oahu, Hawaii. *United States Geological Survey Water Resources Investigations Report* **03-4256**.

Cain, D.J., S.N. Luoma & W.G. Wallace. 2004. Linking metal bioaccumulation of aquatic insects to their distribution patterns in a mining-impacted river. *Environmental Toxicology and Chemistry* **23**: 1463–1473.

Crawford, J.K. & S.N. Luoma. 1993. Guidelines for studies of contaminants in biological tissues for the National Water-Quality Assessment Program. *United States Geological Survey Open-File Report* **94-458**, 69 p.

Cuffney, T.F., M.E. Gurtz & M.R. Meador. 1993. Methods for collecting benthic invertebrate samples as part of the National Water-Quality Assessment Program. *United States Geological Survey Open-File Report* **93-406**, 66 p.

Death, R.G. 2000. The effect of land-use on species-area relationships in benthic stream invertebrates. *Verhandlungen der Internationalen Vereinigung für Limnologie* **27**: 2519–2522.

DeCarlo, E.H. & S.S. Anthony. 2002. Spatial and temporal variability of trace element concentrations in an urban subtropical watershed, Honolulu, Hawaii. *Applied Geochemistry* **17**: 475–492.

Fitzpatrick, F.A., I.R. Waite, P.J. D'Arconte, M.R. Meador, M.A. Maupin & M.E. Gurtz. 1998. Revised methods for characterizing stream habitat in the National Water-Quality Assessment Program. *United States Geological Survey Open-File Report* **98-4052**, 67 p.

Hare, L. 1992. Aquatic insects and trace metals: bioavailability, bioaccumulation, and toxicity. *Critical Reviews in Toxicology* **22**: 327–369.

Karr, J.R. & E.W. Chu. 1997. *Biological monitoring and assessment: Using multimetric indexes effectively.* Seattle, University of Washington, United States Environmental Protection Agency **235-R97-001**, 149 p.

Kavlock, R.J., G.P. Daston, C. DeRosa, P. Fenner-Crisp, L.E. Gray, S. Kaattari, G. Lucier, M. Luster, M.J. Mac, C. Maczka, R. Miller, J. Moore, R. Rolland, G. Scott, D.M. Sheehan, T. Sinks & H.A. Tilson. 1996. Research needs for the risk assessment of health and environmental effects of endocrine disruptors: a report of the United States Environmental Protection Agency-sponsored workshop. *Environmental Health Perspectives* **104**: 715–739.

Lenat, D.R. & M.T. Barbour. 1994. Using benthic macroinvertebrate community structure for rapid, cost-effective, water quality monitoring: Rapid bioassessment, p. 187–215. *In* Loeb, S.L. & A. Spacie (eds.), *Biological Monitoring of Aquatic Systems.* Lewis Publishers, CRC Press, Inc., Boca Raton.

Moulton, S.R., II, J.L. Carter, S.A. Gotheer, T.F. Cuffney & T.M. Short. 2000. Methods of analysis by the U.S. Geological Survey National Water-Quality Laboratory–Processing, taxonomy, and quality control of benthic macroinvertebrate samples. *United States Geological Survey Open-File Report* **00-212**, 49 p.

Nowell, L.H., P.D. Capel & P.D. Dileanis. 1999. *Pesticides in stream sediment and aquatic biota–Distribution, trends, and governing factors.* Pesticides in the Hydrologic System series.

Lewis Publishers, Boca Raton, Florida.

Oki, D.S. & A.M.D. Brasher. 2003. Environmental setting and implications for water quality and aquatic biota, Oahu, Hawaii. *United States Geological Survey Water-Resources Investigations Report* **03-4156**, 98 p.

Pringle, C.M. & A. Ramirez. 1998. Use of both benthic and drift sampling techniques to assess tropical stream invertebrate communities along an altitudinal gradient, Costa Rica. *Freshwater Biology* **39**: 359–373.

Resh, V.H., J.R. Barnes, B. Benis-Steger & D.A. Craig. 1992. Life history features of some macroinvertebrates in a French Polynesian stream. *Studies on Neotropical Fauna and Environment* **27**: 145–153.

Rosenberg, D.M.& V.H. Resh. 1993. Introduction to freshwater biomonitoring and benthic macroinvertebrates, p. 1–9. *In*: Rosenberg, D.M. & V.H. Resh (eds), *Freshwater biomonitoring and benthic macroinvertebrates*. Chapman and Hall, New York.

Schmitt, C.J., J.L. Zajicek, T.W. May & D.F. Cowman. 1999. Organochlorine residues and elemental contaminants in US freshwater fish, 1976-1986: National Contaminant Biomonitoring Program. *Reviews in Environmental Contamination and Toxicology* **162**: 43–104.

———., **J.L. Zajicek & P.H. Peterman**. 1990. National pesticide monitoring program: residues of organochlorine chemicals in United States freshwater fish, 1976–1984. *Archives of Environmental Contamination and Toxicology* **19**: 748–781.

Shelton, L.R. & P.D. Capel. 1994. Guidelines for collecting and processing samples of streambed sediment for analysis of trace elements and organic compounds for the National Water-Quality Assessment Program. *United States Geological Survey Open-File Report* **94-458**, 20 p.

Takahashi, W. 1982. *Pesticide usage patterns in Hawaii, 1977*. Honolulu, University of Hawaii, Pesticide Hazard Assessment Project, Pacific Biomedical Research Center: 92 p.

Wong, C.S., P.D. Capel & L.H. Nowell. 2000. Organochlorine pesticides and PCBs in stream sediment and aquatic biota–initial results from the National Water-Quality Assessment Program, 1992-1995. *United States Geological Survey Water-Resources Investigations Report* **00-4053**, 88 p.

Biology of Hawaiian Streams and Estuaries. Edited by N.L. Evenhuis & J.M. Fitzsimons. *Bishop Museum Bulletin in Cultural and Environmental Studies* 3: 293–304 (2007).

Setting Instream Flow Standards for Hawaiian Streams—the Role of Science

EDWIN T. SAKODA
State of Hawaii, Commission on Water Resource Management, P.O. Box 621, Honolulu, Hawai'i, 96809, USA; email: edwin.t.sakoda@hawaii.gov

Abstract

The State of Hawaii, Commission on Water Resource Management (Commission) is responsible for protecting and managing Hawaii's ground-water and surface water resources. In its surface water management role, the Commission regulates all stream channel alterations, surface water diversions, and amendments to the interim instream flow standards. This paper tracks the Commission's progress towards setting instream flow standards for Hawaiian streams and the role of science in the process.

Introduction

The State Water Code (Water Code), Chapter 174C, Hawaii Revised Statutes (HRS), was adopted by the legislature of the State of Hawaii in 1987. The Commission, established under the Water Code, is responsible for protecting and managing Hawaii's ground-water and surface water resources. In its surface water management role, the Commission regulates all stream channel alterations, surface water diversions, and amendments to the interim instream flow standards. Additionally, the Commission has established an instream use protection program designed to protect, enhance, and reestablish, where practicable, beneficial instream uses of water in the State. This paper briefly reviews the history of the Water Code and follows the Commission's progress towards setting instream flow standards for Hawaiian streams, and the role of science in the process.

Brief History of the Water Code and Instream Flow Standards

1978 Constitutional Convention

The 1978 Constitutional Convention and 7 November 1978 election added Article XI, Section 7, to the Constitution of the State of Hawaii:

> *Section 7. The State has an obligation to protect, control and regulate the use of Hawaii's water resources for the benefit of its people.*
>
> *The legislature shall provide for a water resources agency which, as provided by law, shall set overall water conservation, quality and use policies; define beneficial and reasonable uses; protect ground and surface water resources, watersheds and natural stream environments; establish criteria for water use priorities while assuring appurtenant rights and existing correlative and riparian uses and establish procedures for regulating all uses of Hawaii's water resources.*

Section 7 reflected the concern of the people of Hawai'i to protect, control, and regulate its water resources, and authorized the legislature to set up a water resources agency.

Hawai'i Instream Use Protection Act of 1982

In 1982, the legislature passed Chapter 176D, PROTECTION OF INSTREAM USES OF WATER, or the Hawai'i Instream Use Protection Act of 1982. The intent of the legislature in passing the Act was that the State (for this Act, the Board of Land and Natural Resources) "develop instream flow standards and instream flow programs to protect and enhance, where practicable, beneficial instream uses of water". Although the Act only applied to windward O'ahu districts, and was programmed to expire upon the enactment of a state water code, it contained definitions (continuous flowing water, instream flow standard, instream use, stream channel, stream system) and provided guidelines toward the development of instream flow standards.

Windward O'ahu Interim Instream Flow Standard — 30 July 1987

On 30 July 1987, the Board of Land and Natural Resources adopted an interim instream flow standard for Windward O'ahu pursuant to Chapter 176D, HRS. The Department of Land and Natural Resources then began work on interim standards for Kaua'i and East Maui streams, by holding public information meetings on Kaua'i and Maui in October 1987. Also in 1987, Hawaii's Legislature adopted the Water Code that allowed for the establishment of new interim and permanent instream flow standards (Sec. 174C-71, HRS).

Section 8, Act 45, Session Laws of Hawai'i 1987 provided that Chapter 176D, HRS, be repealed two years from the effective date of the Water Code (i.e., on 1 July 1989). As such, the existing interim standard for windward O'ahu streams would cease to have the force and effect of law. Under the Water Code, the Commission would need to reconsider and readopt an interim instream flow standard for windward O'ahu before 1 July 1989, when Chapter 176D is repealed.

An interim instream flow standard, defined by Chapter 176D, HRS, is a temporary flow standard of general applicability that will identify a quantity of water to be set aside to protect instream uses, such as fish and wildlife habitats and recreational and aesthetic values, until permanent instream flow standards can be established on a stream-by-stream basis.

The interim standard, adopted on 30 July 1987 by the Board of Land and Natural Resources, reads as follows:

> *A. With respect to gaged streams, the standard be set at 100% of the median flow computed after existing diversions have been deducted.*
> *B. With respect to ungaged streams, no further diversion shall be allowed.*
> *C. These interim instream flow standards may be modified on a case-by-case, stream-by-stream basis by individual application, when additional and more specific data become available for each such individual stream.*

1987 State Water Code

With the adoption of the Water Code in 1987, the responsibility to develop instream flow standards passed from the Board of Land and Natural Resources to the newly formed Commission. Part VI of the Water Code, entitled Instream Uses of Water, outlines the Commission's responsibilities regarding statewide protection of instream uses, including the setting of instream flow standards.

PART VI. INSTREAM USES OF WATER

§174C-71 Protection of instream uses. *The commission shall establish and administer a statewide instream use protection program. In carrying out this part, the commission shall cooperate with the United States government or any of its agencies, other state agencies, and the county governments and any of their agencies. In the performance of its duties the commission shall:*

(1) Establish instream flow standards on a stream-by-stream basis whenever necessary to protect the public interest in waters of the State;

- *(A) The commission, on its own motion, may determine that the public interest in the waters of the State requires the establishment of an instream flow standard for streams;*
- *(B) In acting upon the establishment of instream flow standards, the commission shall set forth in writing its conclusion that the public interest does or does not require, as is appropriate, an instream flow standard to be set for the stream, the reasons therefore, and the findings supporting the reasons;*
- *(C) Each instream flow standard shall describe the flows necessary to protect the public inter-*

est in the particular stream. Flows shall be expressed in terms of variable flows of water necessary to protect adequately fishery, wildlife, recreational, aesthetic, scenic, or other beneficial instream uses in the stream in light of existing and potential water developments including the economic impact of restriction of such use;

(D) *Establishment or modification of an instream flow standard shall be initiated by the commission by providing notice of its intention to set an instream flow standard in a newspaper of general circulation published in the vicinity of the stream in question, to the mayor of the appropriate county, and to persons who have previously requested such notice in writing;*

(E) *After giving notice of its intention to set an instream flow standard, the commission or other agencies in participation with the commission shall investigate the stream. During the process of this investigation, the commission shall consult with and consider the recommendations of the department of health, the aquatic biologist of the department of land and natural resources, the natural area reserves system commission, the University of Hawai'i cooperative fishery unit, the United States Fish and Wildlife Service, the mayor of the county in which the stream is located, and other agencies having interest in or information on the stream, and may consult with and consider the recommendations of persons having interest in or information on the stream. In formulating the proposed standard, the commission shall weigh the importance of the present or potential uses of water from the stream for noninstream purposes, including the economic impact of restriction of such uses. In order to avoid or minimize the impact on existing uses of preserving, enhancing, or restoring instream values, the commission shall consider physical solutions, including water exchanges, modifications of project operations, changes in points of diversion, changes in time and rate of diversion, uses of water from alternative sources, or any other solution;*

(F) *Before adoption of an instream flow standard or modification of an established instream flow standard, the commission shall give notice and hold a hearing on its proposed standard or modification;*

(2) *Establish interim instream flow standards;*

(A) *Any person with the proper standing may petition the commission to adopt an interim instream flow standard for streams in order to protect the public interest pending the establishment of a permanent instream flow standard;*

(B) *Any interim instream flow standard adopted under this section shall terminate upon the establishment of a permanent instream flow standard for the stream on which the interim standards were adopted;*

(C) *A petition to adopt an interim instream flow standard under this section shall set forth data and information concerning the need to protect and conserve beneficial instream uses of water and any other relevant and reasonable information required by the commission;*

(D) *In considering a petition to adopt an interim instream flow standard, the commission shall weigh the importance of the present or potential instream values with the importance of the present or potential uses of water for noninstream purposes, including the economic impact of restricting such uses;*

(E) *The commission shall grant or reject a petition to adopt an interim instream flow standard under this section within one hundred eighty days of the date the petition is filed. The one hundred eighty days may be extended a maximum of one hundred eighty days at the request of the petitioner and subject to the approval of the commission;*

(F) *Interim instream flow standards may be adopted on a stream-by-stream basis or may consist of a general instream flow standard applicable to all streams within a specified area;*

(3) *Protect stream channels from alteration whenever practicable to provide for fishery, wildlife, recreational, aesthetic, scenic, and other beneficial instream uses;*

(A) *The commission shall require persons to obtain a permit from the commission prior to undertaking a stream channel alteration; provided that routine streambed and drainageway maintenance activities and maintenance of existing facilities are exempt from obtaining a permit;*

(B) Projects which have commenced construction or projects reviewed and approved by the appropriate federal, state, or county agency prior to July 1, 1987, shall not be affected by this part;

(C) The commission shall establish guidelines for processing and considering applications for stream channel alterations consistent with section 174C-93;

(D) The commission shall require filing fees by users to accompany each application for stream channel alteration;

(4) Establish an instream flow program to protect, enhance, and reestablish, where practicable, beneficial instream uses of water. The commission shall conduct investigations and collect instream flow data including fishing, wildlife, aesthetic, recreational, water quality, and ecological information and basic streamflow characteristics necessary for determining instream flow requirements. The commission shall implement its instream flow standards when disposing of water from state watersheds, including that removed by wells or tunnels where they may affect stream flow, and when regulating use of lands and waters within the state conservation district, including water development. [L 1987, c 45, pt of §2; am L 1988, c 276, §2]

Definition of Instream Use

Instream use is defined as "beneficial uses of stream water for significant purposes which are located in the stream and which are achieved by leaving the water in the stream. Instream uses include, but are not limited to: (1) Maintenance of fish and wildlife habitats; (2) Outdoor recreational activities; (3) Maintenance of ecosystems such as estuaries, wetlands, and stream vegetation; (4) Aesthetic values such as waterfalls and scenic waterways; (5) Navigation; (6) Instream hydropower generation; (7) Maintenance of water quality; (8) The conveyance of irrigation and domestic water supplies to downstream points of diversion; and (9) The protection of traditional and customary Hawaiian rights".

Definition of Instream Flow Standard

Instream flow standard is defined as "a quantity or flow of water or depth of water which is required to be present at a specific location in a stream system at certain specified times of the year to protect fishery, wildlife, recreational, aesthetic, scenic, and other beneficial instream uses".

Setting an Instream Flow Standard

The Water Code distinguishes between an instream flow standard and an interim instream flow standard. An Instream Flow Standard is to be established by the Commission, on its own motion, on a stream-by-stream basis. Each Instream Flow Standard needs to describe the flow necessary to protect the public interest in the particular stream. Flows are to be expressed in terms of variable flows of water necessary to adequately protect fishery, wildlife, recreational, aesthetic, scenic, or other beneficial instream uses in the stream. In investigating a stream to set an Instream Flow Standard, the Commission shall consult with and consider the recommendations of the department of health, the aquatic biologist of the department of land and natural resources, the natural area reserves system commission, the University of Hawai'i cooperative fishery unit, the United States Fish and Wildlife Service, and other agencies having interest or information on the stream. Finally, prior to setting an Instream Flow Standard, the Commission shall give notice and hold a hearing on its proposed standard or modification.

Setting an Interim Instream Flow Standard

Any person with proper standing may petition the Commission to establishing an Interim Instream Flow Standard. The Interim Instream Flow Standard may be adopted on a stream-by-stream basis or may consist of a general instream flow standard applicable to all streams within a specified area. A petition to adopt an Interim Instream Flow Standard requires information and data, but there is no requirement to consult with and consider recommendations from any agency. Finally, any Interim Instream Flow Standard adopted shall terminate upon the establishment of a permanent Instream Flow Standard.

East Maui, Kaua'i, Hawai'i and Moloka'i Interim Instream Flow Standards – 15 June 1988
The Interim Instream Flow Standards for all streams on East Maui, Kaua'i, Hawai'i, and Moloka'i were adopted by the Commission on 15 June 1988. Following is the Interim Instream Flow Standard for East Maui streams from the Hawaii Administrative Rules (HAR) §13-169-44 Interim instream flow standard for East Maui. The standards for Kaua'i, Hawai'i, and Moloka'i are identical except for the area names and section numbers in the rules.

> *§13-169-44 Interim instream flow standard for East Maui. The Interim Instream Flow Standard for all streams on East Maui, as adopted by the commission on water resource management on June 15, 1988, shall be that amount of water flowing in each stream on the effective date of this standard, and as that flow may naturally vary throughout the year and from year to year without further amounts of water being diverted offstream through new of expanded diversions, and under the stream conditions existing on the effective date of the standard, except as may be modified by the following conditions:*
> *(1) Based upon additional information or a compelling public need, a person may petition the commission on water resource management to amend the standard to allow future diversion, restoration, or other utilization of any streamflow.*
> *(2) The commission reserves its authority to modify the standard or establish new standards, including area-wide or stream-by-stream standards, based upon supplemental or additional information.*
> *(3) In any proceeding to enforce the instream flow standard, the commission, its delegated hearing officer, or a judicial officer may abate the enforcement proceeding if, under the circumstances and weighing the importance of the present or potential instream values with the importance of the present or potential uses of the stream's water for non-instream purposes (including the economic impact of restricting such uses), the enforcement of the instream flow standard would:*
> *(A) Create a substantial hardship on a use existing on the effective date of this standard; or*
> *(B) Impermissibly burden a right, title, or interest arising under law.*
> *(4) Projects under construction or projects that have secured all discretionary permits required by appropriate federal, state, or county agencies prior to July 1, 1987 shall not be affected by the standard.*
> *[Eff. Oct. 8, 1988] (Auth: HRS Sec. 174C-8) (Imp: HRS Sec. 174C-2, 174C-3, 174C-5, 174C-71)*

Act 276, Session Laws of Hawaii – 13 June 1988
The 1988 Legislature amended the Water Code by adding Section 174C-31, subsection (c) (4), requiring the Commission to: "Identify rivers or streams, or a portion of a river or stream, which appropriately may be placed within a wild and scenic rivers system, to be preserved and protected as part of the public trust. For the purposes of this paragraph, the term "wild and scenic rivers" means rivers or streams, or a portion of a river or stream of high natural quality or that possess significant scenic value, including but not limited to, rivers or streams which are within the natural area reserves system. The commission shall report its findings to the legislature twenty days prior to the convening of each regular legislative session." The 1988 Legislature also amended Section 174C-71(1)(E), by adding "the aquatic biologist of the department of land and natural resources, the natural area reserves system commission, the University of Hawai'i cooperative fishery unit" to the list of agencies consulted when setting an instream flow standard.

West Maui and Leeward O'ahu Interim Instream Flow Standards – 19 October 1988
The Interim Instream Flow Standards for West Maui and Leeward O'ahu streams were adopted by the Commission on 19 October 1988. The standards for West Maui and Leeward O'ahu are identical to the standards for East Maui, Kaua'i, Hawai'i, and Moloka'i except for the area names and section numbers in the rules.

Windward O'ahu Interim Instream Flow Standard – 19 April 1989
The Interim Instream Flow Standard for Windward O'ahu streams was adopted by the Commission on April 19, 1989. The standard adopted by the Commission replaced the original standard adopted by the Board of Land and Natural Resources on July 30, 1987. The new standard incorporated the language used for the other areas of the state.

Hawaii Stream Assessment – October 1988 to December 1990
The Commission initiated the Hawaii Stream Assessment through a cooperative agreement with the National Park Service's State and Local Rivers and Trails Conservation Assistance Program. This program was established in response to the National Wild and Scenic Rivers Act, that encouraged the National Park Service to assist states to consider needs and opportunities for establishing state and local wild and scenic, and recreational river areas [Public Law 90-542, Section 11(a)]. The primary task of the Hawaii Stream Assessment was to identify streams appropriate for protection. It made no attempt to assess then-existing or potential offstream use. It was presented from the conservation point of view. The Hawaii Stream Assessment was to be used as a reference document that included a physical inventory of Hawaii's 376 perennial streams and working maps, as assessment of resources associated with those streams, and a database. The Hawaii Stream Assessment was to help policy-makers, resource managers, developers, scientists and the interested public to locate published information for a particular stream, identify and prioritize areas where information is needed, understand stream resources within a statewide context, make management decisions based on data, develop general stream resource protection guidelines, and identify specific streams appropriate for protection and enhancement.

Senate Concurrent Resolution No. 130—April–May 1992
Senate Concurrent Resolution No. 130, Requesting Adoption of an Environmental Protection System for Hawaii's Streams, was passed by the Senate in April 1992 and by the House of Representatives in May 1992. The Resolution requested that the Commission "finalize, adopt, and put into place the stream protection system, including designation of streams to be protected pursuant to state or federal law".

The Stream Protection and Management Task Force—May 1993 to April 1994
The Stream Protection and Management (SPAM) Task Force was formed in part due to the request by the Legislature in Resolution No. 130, and in part due to the earlier Act 276 described above. The Commission, in November 1992, held a briefing to address Act 276 and resolution 130. The Commission reflected on the strengths and weaknesses of the past and existing surface water program and directed staff to develop a surface water management strategy that included a stream categorization system and a mechanism to include community-based planning. The Task Force was a multi-interest group, professionally facilitated, and chaired by a water commissioner. The Task Force gathered information to more clearly understand the various issues related to surface water, attended public meetings to gather more information, and began facilitated deliberations. The Task Force Report of April 1994, entitled Stream Protection and Management in Hawai'i: Recommendations and Suggestions, listed consensus agreements reached by the members and constituted their recommendations to the Commission. The Report also included a list of suggestions provided by the members, representing the opinions of the individual members, not the Task Force as a whole. The Commission, on May 18, 1994, accepted the Report and directed staff to hold public forums to discuss the Report in conjunction with the draft staff recommendations.

Stream Protection and Management in Hawai'i, Draft Staff Recommendations—May 1994
Draft staff recommendations, based on the work of the SPAM Task Force, were presented to the Commission on 18 May 1994. The Commission directed staff to hold public forums to discuss the Task Force Report in conjunction with the draft staff recommendations, and to submit final recommendations to the Commission along with draft rules and an implementation plan "as soon as possible".

Commission Staff Reorganization—July 2002

In July 2002, the Commission staff was reorganized. The Stream Protection and Management Branch was formed with two sections, the Surface Water Regulation Section and the Instream Use Protection Section. The Surface Water Regulation Section continued to regulate surface water through Stream Channel Alteration Permits, Stream Diversion Works Permits, and amendments to the Interim Instream Flow Standards. The Instream Use Protection Section was added to more adequately address the setting of Instream Flow Standards and improve the overall stream management objectives. Key objectives include: 1) data compilation, including developing and promoting a standardized watershed coding system and a surface water information management system; 2) developing a surface water monitoring program; 3) developing a stream protection program – categorizing streams; 4) setting an instream flow standard methodology; and 5) increasing public outreach and education.

Pristine Stream Policy Recommendations and Proposed Actions – February 2004

The Commission staff, at a Commission meeting in February 2004, proposed that the Commission adopt a pristine stream policy to protect streams of high natural quality, to identify streams for which the Commission would require a higher level of scrutiny before allowing any modifications or diversions, to continue the on-going process of developing a methodology of setting instream flow standards, and to inform and educate the public of this policy. Comments from interested persons during the meeting included statements that the Commission should direct its attention to establishing instream flow standards, that the Commission should seek more public input, and that input should be sought at the beginning of the process instead of at the end. Based on the testimony, the proposal was deferred and staff was directed to work with the community and other stakeholders by holding public meetings to discuss the matter. The Stream Policy Working Group, composed of many of the interested persons who attended the February 2004 Commission meeting, as well as other interested persons, began meeting a week after the Commission meeting, and met a total of five times, ending in July 2004. By the final meeting in July, it was the consensus of the group that the Commission staff should work on a methodology to set interim instream flow standards, rather than pursue a pristine stream policy. The Stream Policy Working Group will be recalled on an as-needed basis to provide input as staff develops the methodology.

Discussion

The Role of Science in the Commission's History

Now that we have briefly reviewed the history of the Commission with regard to setting instream flow standards, let us go back and see the role science has taken in the process.

1978 Constitutional Convention

The addition of Article XI, Section 7, to the Constitution provided for "a water resources agency", but there was no science involved.

Hawai'i Instream Use Protection Act of 1982

Though this Act was specifically addressed to Windward O'ahu streams, it was the first step towards setting instream flow standards. Of particular interest, from the "role of science" perspective, were references to "describe the flows necessary to protect the public interest in the particular stream," and to express flows "in terms of variable flows of water necessary to protect adequately fishery, wildlife, recreational, aesthetic, scenic, or other beneficial instream uses". The Act also provided for consultation with and consideration of recommendations of the Department of Health, the United States Fish and Wildlife Service, and other agencies having interest in or information on the stream.

Windward O'ahu Interim Instream Flow Standard of July 30, 1987

This pre-Water Code standard was adopted by the Board of Land and Natural Resources for

Windward O'ahu streams under the provisions of the previously mentioned Hawai'i Instream Use Protection Act. For gaged streams, the standard was set at "100% of the median flow computed after existing diversions have been deducted". For ungaged streams, no further diversions were allowed. There was a provision for modification of the standard on a case-by-case basis, when additional and more specific data become available. Science was used, via the gaging of streams, in setting the standard for the gaged streams, and in the provision for modifying the standard based on additional data.

1987 State Water Code

From the "role of science" perspective, the Water Code added new definitions (hydrologic unit, impoundment, interim instream flow standard, noninstream use, stream, stream diversion, stream reach, watercourse, etc.) and expanded some of the earlier definitions (channel alteration, stream system, etc.).

It authorized the Commission to: carry out topographic surveys, research, and investigations into all aspects of water use and water quality; maintain an advisory staff of experts; plan and coordinate programs for the development, conservation, protection, control, and regulation of water resources based on the best available information; catalog and maintain an inventory of all water uses and water resources; and determine appurtenant water rights, including quantification of the amount of water entitled to by that right.

Legislative intent regarding Interim Instream Flow Standard

Conference Committee Report No. 119, RE: H.B. No. 35, H.D. 1, S.D. 2, C.D.1, dated 27 April 1987, gives us some insight into the legislature's thinking (perhaps we could call this non-scientific or "wishful" thinking) regarding the ease (time-wise and science-wise) in the setting of Instream Flow Standard and Interim Instream Flow Standard. The Report, that recommended the final passage of the Water Code, included the following:

> *The Commission is directed to implement instream flow standards when disposing of water from state watersheds and when regulating use of lands and waters within conservation districts.*
>
> *To the fullest extent possible, it is the intent of the Legislature that interim instream flow standards be established prior to either new or expanded diversions of water from a stream. Protection of our streams is an important part of the water code.*

It is your Committee's recommendation that the interim instream flow standards be undertaken by a joint Department of Land and Natural Resources and appropriate federal agencies such as the U.S. Fish and Wildlife Service and U.S. Geological Survey, in order to take advantage of in-house, least-cost expertise.

Section (13) indicates that at least the conference committee, if not the entire Legislature, believed setting interim standards would be a relatively simple (science-wise) and quick (time-wise) task, and that it could be done with "in-house, least-cost expertise". The legislature gave the Commission the following deadlines to set interim instream flow standards: 1) Windward O'ahu by July 31, 1987; 2) East Maui and Kaua'i by December 31, 1987; 3) Hawai'i and Moloka'i by July 1, 1988; and 4) West Maui and Leeward O'ahu by December 31, 1988. However, it can also be concluded that the legislature regarded the protection of streams as an important part of the Water Code, and that the Water Code contemplated using the expertise of various state and federal agencies to assist in setting interim and permanent instream flow standards.

Setting Interim Instream Flow Standards for Various Areas of the State—15 June 1988 to 19 April 1989

The newly formed Commission, working with the deadlines set by the legislature to set interim instream flow standards, reached consensus in defining the interim instream flow standard for all streams statewide to be "that amount of water flowing in each stream on the effective date of this standard, and as that flow may naturally vary throughout the year and from year to year without further amounts of water being diverted offstream through new or expanded diversions, and under the stream conditions existing on the effective date of the standard, except as may be modified by the following…". The interim standard was not based on science, but on the requirements of the Water Code, deadlines set by the legislature, comments received at six public meetings held across the state, and several redrafts of the language at the Commission meeting on June 15, 1988.

Act 276, Session Laws of Hawai'i—13 June 1988

Act 276 contributed to the role of science in setting instream flow standards by adding to the Water Code "the aquatic biologist of the department of land and natural resources, the natural area reserves system commission, the University of Hawai'i cooperative fishery unit" to the list of agencies to be consulted when setting an instream flow standard.

Hawai'i Stream Assessment – December 1990

The Hawai'i Stream Assessment was undertaken as an effort to develop a broad-based collection of existing information on Hawaii's rivers and streams to help make water protection and management decisions. The inventory and assessment was of a general nature and was not intended to take the place of subsequent reviews and studies. However, it continues to serve as a baseline source of information for anyone interested in Hawaii's streams.

A list of 376 perennial streams on the five major islands (Hawai'i, Maui, O'ahu, Kaua'i, and Moloka'i) was compiled and assessed in four resource categories: 1) aquatic resources, 2) riparian resources, 3) cultural resources, and 4) recreational resources. Additional areas of concern were addressed, including gaging, water quality, water supply, dams and reservoirs, hydroelectric power, and channelization. A qualitative ranking system was applied, whereby streams were ranked as outstanding, substantial, moderate, or limited.

The Division of Aquatic Resources has expanded the aquatic resources portion originally contained in the Hawai'i Stream Assessment and has developed and maintained a large database, known as the Stream Biological Database, of freshwater stream and estuary biological resources throughout the State.

The Stream Protection and Management Task Force – May 1993 to April 1994

The Stream Protection and Management Task Force members (there were eight members including the chair) represented three generalized interests: stream biology, stream economy, and the cultural aspects of stream life and stream use. "Stream biology" was represented by: 1) the State Department of Land and Natural Resources, Division of Aquatic Resources; 2) the U.S. Fish and Wildlife Service; and 3) the Sierra Club Legal Defense Fund. Consensus recommendations related to science included: 1) verification of perennial streams by the Commission, Division of Aquatic Resources or the U.S. Fish and Wildlife Service; 2) recommendation that the Commission convene various working conferences, one of which was to insure implementation of water quality standards (including biological standards); 3) providing adequate funding for biological studies; and 4) endorsing the idea of holding an annual science conference and an annual stream conference for the public.

Specific Examples of the Role of Science in Setting Instream Flow Standards

Kuhiwa Well Contested Case Hearing—May–June 1991

The Kuhiwa Well Contested Case Hearing was the first contested case hearing before the Commission. Maui Pineapple Company, Ltd. (Maui Pine) applied for a pump installation permit to use water from the Kuhiwa Well to augment their water supply during dry periods. The central issue raised by the Hana Community Association was the possible effect of well pumpage on the flow of streams and springs in the area. Two opposing models of ground-water behavior were presented to the Commission by expert hydrologic witnesses. The data and analyses at that time were insufficient to determine which of the models better described the conditions in the region. Both models suggested that pumping the well would have some impact on the streams and springs in the area, but neither model could conclude whether the impact would be large enough to be detectable. The

Commission concluded that the only method available to fully predict in advance actual streamflow depletion from pumping the Kuhiwa Well was through construction of a ground-water model. It also concluded that since there was insufficient data available, the time and cost to obtain the data would be prohibitive. Three approaches to monitoring for impacts of pumping on stream flows were proposed to the Commission: a single-gage or observation point flow measurement approach; a paired-gage flow measurement approach; and a biological monitoring approach. In its final decision and order, the Commission allowed Maui Pine to install the well and use the water with the condition that a biological and hydrological monitoring system be established to provide information that would guide future Commission decisions and actions. If monitoring determined that there was detectable and not "insubstantial" reduction of instream flows, Maui Pine would be required to stop pumping and to obtain an amendment to the interim instream flow standard. Science, in this case, was not used to determine an instream flow standard. It was used to determine that amending the interim instream flow would not be required unless there was detectable and not "insubstantial" reduction of stream flows.

Waiāhole Contested Case Hearing–25 January 1995 to the Present

The Waiāhole Ditch Contested Case Hearing involves three different types of applications to use water from the Waiāhole Ditch system on O'ahu: 1) water use permit applications resulting from the designation of the windward O'ahu aquifer systems as ground-water management areas in May 1992; 2) applications to restore water to windward O'ahu streams by amending the interim instream flow standard for windward O'ahu; 3) petitions for reservations of windward O'ahu ground water. Protection of traditional and customary Hawaiian rights and principles of the Public Trust Doctrine are also implicated.

The interim instream flow standard for windward O'ahu streams is linked with the water use permit applications because the base flows of windward O'ahu streams and the development tunnels of the Waiāhole Ditch system develop water from the same high level aquifers in dike intruded lavas.

The Commission, on 25 January 1995, ordered that the contested case hearing be held. The formal portion of the hearing began on 9 November 1995 and continued to 21 August 1996, during which time there were fifty-two days of hearings, testimony from 161 witnesses, and 567 exhibits introduced into evidence. Closing arguments took three days, from 18–20 September 1996. The Commission issued its Findings of Fact, Conclusions of Law, and Decision and Order on 24 December 1997 (D&O I). The D&O I was appealed to the Hawaii Supreme Court. The Supreme Court issued its decision on 22 August 2000 (Waiāhole I), and remanded certain matters back to the Commission. The Commission held further hearings and issued its second Waiāhole Decision on 28 December 2001 (D&O II). The D&O II was also appealed to the Supreme Court. The Supreme Court issued its second decision on 21 June 2004 (Waiāhole II). The Commission is scheduled to hold hearings on the matter in early April 2005.

Many of the 161 witnesses that provided testimony during the hearing were scientists. Both scientists and nonscientists among the many participants in the proceedings were treated to (or some would say, "forced to endure") an abundance of information about general stream ecology; native Hawaiian gobies; alien, introduced, exotic fish species; parasites; native Hawaiian damselflies; Hawaiian and exotic plants; stream restoration in general; instream flow protection and techniques used elsewhere; and estuarine and marine ecology. Although there were many studies generated by the Waiāhole case, one of the biggest problems, from the stream restoration perspective, was the lack of baseline data. No one had done any studies on the abundance of native species in Waiāhole and other windward O'ahu streams prior to the construction of the Waiāhole ditch in 1916. No one had done any studies prior to the partial restoration of flow to Waiāhole Stream in December 1994. As a consequence of the lack of baseline data, the partial restoration of flow to Waiāhole Stream could only be characterized as having a positive effect on the native fish species in the stream, and that additional flows would be expected to increase the native biota habitat. It also became apparent that there was much more to be done to increase our knowledge regarding native fish species. There was testimony that "aquatic experts have insufficient knowledge of the ecosystem context for native

Hawaiian fish species to define the quantitative population improvements resulting from stream flow restoration;" "no data is currently available to tell what flow characteristics are necessary to promote movement of larval fishes from the ocean into freshwater streams and what current velocities are necessary to limit or prevent the occurrence of exotic fishes in such streams;" and "no one in the scientific community can conclusively state the amount of water that is necessary to positively impact *'o'opu* recruitment".

Waiāhole Contested Case Hearing—Requirement to Fund Studies

The Commission, in its D&O I, required the permittees using water from the Waiāhole Ditch to help fund studies and monitoring activities resulting from the order. This was a precedent-setting step patterned in part after the Mono Lake Decision, September 1994. Funding was to be based on the amount of water used and on a pro rata basis. The Commission was to establish a committee to recommend a reasonable amount for the funding, and coordinate and set up the mechanism for the collection, accounting, and distribution of the funds.

The requirement was appealed to the Supreme Court by the permittees who maintained that the Water Code requires the Commission to fund any studies. The Supreme Court concluded, in its 22 August 2000 Waiāhole I decision, that "the Commission has the general authority to condition the permits upon compliance with the instant funding requirement, which more properly falls under the category of a regulatory fee, rather than a land development exaction. Under the standard applicable to such fees, we hold that, as a general matter, the funding requirement does not constitute an illegal tax". The Commission, in May 2002, based upon recommendations by a Commission-appointed funding committee, determined that $0.025 per 1,000 gallons of water used, is a reasonable rate for the permittees.

Petitions to Amend the Interim Instream Flow Standard (IIFS) for 27 East Maui streams—May 2001 to the Present

In May 2001, the Commission received petitions to amend the interim instream flow standards for 27 East Maui streams. In March 2002, the Commission approved a cooperative agreement between the U.S. Geological Survey and the Commission for a study entitled Water Resource Investigations for Northeast Maui Streams. The study period for the $600,000 study runs for the period 1 October 2002 to 30 September 2005, coinciding with the Federal fiscal year. The U.S. Geological Survey will contribute 50%, or $300,000 to the program. The Commission's share of the matching program will be 12.5 %, or $75,000 ($25,000 per year), as it will be for the Land Division of the Department of Land and Natural Resources, the Maui Department of Water Supply, and Alexander & Baldwin Inc., respectively, for a total of $300,000 to match the U.S. Geological Survey's contribution.

Related to the U.S. Geological Survey's studies in East Maui, and in an effort to increase communication and foster greater collaboration among Hawaii's stream scientists, a working group composed primarily of aquatic biologists, and initially facilitated by staff of the Division of Aquatic Resources, was convened to provide a forum for the open exchange of ideas and suggestions related to instream flow standards and Hawaiian stream biology based upon scientifically defensible arguments, and to provide decision support to the Commission. The working group meets on an as-needed basis and is composed of Commission staff and scientists representing the U.S. Geological Survey, the University of Hawai'i Zoology Department, the Bishop Museum, the University of Nebraska Lincoln, and the Hawaii Division of Aquatic Resources.

Base Flows Representative of Natural Conditions for Streams Affected by the Waiāhole Ditch–U.S. Geological Survey, April 2004 to January 2006

The Commission has entered into an agreement with the U.S. Geological Survey to study streams affected by the Waiāhole Ditch system. The $120,000 project ($60,000 each by the U.S. Geological Survey and the Commission) involves data analysis, report writing, and publication of a U.S. Geological Survey Science Investigations Report that provides estimates of base flows representative of natural conditions for streams in windward O'ahu currently affected by the Waiāhole Ditch and Tunnel System. The project is part of the Commission's efforts to establish instream flow standards for the Waiāhole Ditch contested case hearing. The State's portion of the agreement is a combination of funding from the Division of Aquatic Resources and from the Waiāhole water users fund established through the contested case hearing.

Proposed GIS-Based Hydrologic Modeling Project

The Commission and the Division of Aquatic Resources are working on a project with Dr. James Parham at the University of Nebraska, as part of establishing a methodology of setting instream flow standards. This project seeks to calibrate a distributed rainfall-runoff model (like TOPMODEL) on a group of watersheds and streams with long-term discharge records. The calibrated model will then be tested against another group of streams with long-term discharge records to determine the applicability of using the model on streams with short or non-existent discharge records to predict pre-diversion flow conditions. This effort seeks to quantify the accuracy of the model to predict an unknown hydrograph and to determine the most sensitive variables (topography, land cover, soils, etc.) to developing the rainfall-runoff relationship. Ultimately, if successful, the rainfall-runoff model would be used together with GIS-based models of native fish habitat and instream distributions to allow pre-diversion conditions to be compared with current conditions.

Conclusion

The Commission must establish instream flow standards for Hawaii's streams. With the help of science, we must investigate the ecology of Hawaiian streams and determine the correlation between flow levels and instream values as the first steps in the methodology for determining instream flow standards. It takes lots of time and money to do such work. The following is our dilemma. The Commission could continue allowing water to be taken from streams until adequate scientific information becomes available. Conceivably, a stream could be incrementally drained without having enough information to determine its instream flow needs. On the other hand, some suggest that no water be taken from any streams until sufficient scientific information on instream requirements become available. The Supreme Court, in its 22 August 2000 Waiāhole I decision, recognized the Commission's dilemma and commented:

> *This dilemma offers no simple solution. At the present time, we hold only that the Commission's inability to designate more definitive instream flow standards neither allows the prolonged deferral of the question of instream use protection nor necessarily precludes present and future allocations for offstream purposes. Accordingly, the Commission must apply, in its own words, "a methodology that recognizes the preliminary and incomplete nature of existing evidence," and, indeed, incorporates elements of uncertainty and risk as part of its analysis. Such a methodology, by its nature, must rely as much on policy considerations as on hard scientific "facts".*
>
> *In furtherance of its trust obligations, the Commission may make reasonable precautionary presumptions or allowances in the public interest. The Commission may still act when public benefits and risks are not capable of exact quantification. At all times, however, the Commission should not hide behind scientific uncertainty, but should confront it as systematically and judiciously as possible — considering every offstream use in view of the cumulative potential harm to instream uses and values and the need for meaningful studies of stream flow requirements. We do not expect this to be an easy task. Yet it is nothing novel to the administrative function or the legal process in general. And it is no more and no less than what the people of this state created the Commission to do".*

We appreciate the scientific community and all the work on Hawaiian streams that has been done in the past and continues to be done today. We look forward to working with all of you as we continue the Commission's work towards setting instream flow standards for our streams. The Commission and staff, I believe even more than the Supreme Court, recognize that it is not an easy task. But that is what we have been asked to do by the people of Hawai'i, and by the Supreme Court, and we ask for, and need, your help to accomplish it.

Biology of Hawaiian Streams and Estuaries. Edited by N.L. Evenhuis & J.M. Fitzsimons. *Bishop Museum Bulletin in Cultural and Environmental Studies* 3: 305–313 (2007).

The Point Quadrat Method: a Rapid Assessment of Hawaiian Streams

GLENN R. HIGASHI
Hawaii Department of Land & Natural Resources, Division of Aquatic Resources
1151 Punchbowl Street, Room 330, Honolulu, Hawai'i 96813, USA; email: glenn.r.higashi@hawaii.gov

ROBERT T. NISHIMOTO
Hawaii Department of Land & Natural Resources, Division of Aquatic Resources
75 Aupuni Street, Room 204, Hilo, Hawai'i 96720, USA;
email: robert.t.nishimoto@hawaii.gov

Abstract

This paper describes the technical guidelines for the collection of data on the abundance and distribution of aquatic animals in Hawaiian streams by using the point quadrat method. This methodology is a standardized visual survey technique involving snorkeling, and it is well suited for the physical and ecological characteristics of Hawai'i streams. The small, steep, dynamic nature of Hawaiian streams with their unique aquatic species are easily observed with this methodology. The in-stream distribution by elevation, behavior, and amphidromous life cycles are easily observed using this technique.

Introduction

The Division of Aquatic Resources (DAR), Hawaii Department of Land and Natural Resources is responsible for managing, conserving, and protecting Hawaii's native stream biota and their habitats. There are 376 perennial streams and a greater number of non-perennial streams and drainage basins statewide. Ideally, surveying the entire length of all these streams would provide baseline data on which to make good management decisions. However, because of limited financial resources, the use of the point quadrat method provides a good characterization of the major stream habitat types and an inventory of native populations with sufficient detail and speed. It must be remembered that this survey methodology, unlike long term monitoring, represents only a snapshot in time (Baker & Foster, 1992). This method is a refinement of previous sampling methodology (modified quadrat) used by DAR biologists to obtain information on goby populations (Baker & Foster, 1992). In 1989, DAR began using the point quadrat methodology developed by DAR field biologists and technicians.

DAR stream surveys using the point quadrat methodology provide data on: species presence, in-stream distribution by discrete location or elevation, relative density and abundance. This method also provides a characterization of habitat types and animal distribution within the stream reaches.

Material and Methods

Personal field gear

The point quadrat methodology requires underwater observation. Minimally, a dive mask and snorkel are all that is necessary, but the addition of other field gear will help to make the surveys more enjoyable, comfortable, and safe. Gear requirements include a full length two-piece wet suit with hood and gloves. The custom-fit wet suit ("Farmer John" style resembling bib overalls plus jacket) protects the observer from cuts, scrapes, and bruises, and from becoming chilled in water with temperatures ranging from about 15–25 °C. The suit should fit comfortably when walking. A separate jacket is best because it can be removed to prevent overheating when hiking into the survey

Figure 1. Stream width measurements.

area. Elbow and knee pads provide further protection and may help reduce slipping on wet rocks. A hood, gloves, and hooded long-sleeved dive skin shirt (lycra/polyolefin) are required in cold water to prevent hypothermia. The gloves are also useful to protect hands from rocks with sharp edges and to provide a firmer grip on branches or rocks during hiking and climbing during the survey. Felt-soled flyfishing boots or Japanese spiked *tabis* are superior to hard-soled footwear that slips easily on wet, algae-covered rocks. A rock climbing or kayak helmet with face plate or goggles may be necessary for head and eye protection from falling debris and rock slides when working at high stream elevations in narrow steep-walled valleys or near high waterfalls.

Other field gear

A meshed-back pack is necessary to carry other field gear and will allow the biologist to have both hands free for measuring, writing, hiking, or climbing. A plastic slate with binder clip and rubber bands to secure the survey book when underwater and a No. 2 lead pencil with which to write is a requirement. A folding fiberglass ruler is ideal for measuring the point quadrat site. A USGS quadrangle map of the stream, an altimeter, thermometer, and GPS (global positioning system) unit are all kept in a Pelican® dry box and shared between observers for recording elevation, water temperature, and latitude/longitude position readings at each survey site.

Survey Procedure

In the point quadrat method, visual counts are made at discrete points in the stream by a stationary observer. The observer enters the water quietly, moves upstream to the observation site, and remains motionless until the fishes and other animals have resumed their "normal" behavior. Site selection can be nonrandom (selected by the observer) or random (selected by use of a table of random numbers). The stream width measurement is taken at the bankfull wetted edge level (the edgeheight of the water level on the bank under maximum normal flow conditions or the top elevation of the current active channel where the water touches the stream bank) of the left bank (10%) and ends at the wetted edge bankfullbank full level on the right bank (100%) (Fig. 1). The nonrandom point quadrat method will take the observer to sites thought to contain concentrations of native fish and invertebrates, and may be biased toward the selection of pools and runs. The random approach will take the observer to sites that may or may not be occupied by stream animals, is unbiased in the selection of habitat type, and provides a truer representation of the composition of different habitat types along the stream length. The use of nonrandom or random site selection in the point quadrat method will depend on the kinds of biological questions being asked, such as the presence, absence, instream distribution, relative density of aquatic species, or the characterization of the different habitat types making up a stream reach. The point quadrat method can also involve the use of both random and nonrandom selection of sites during a stream survey. For example, the use of the random numbers may result in the selection of distance across the stream being on a boulder that is out of the water or on a dry section of the streambed. This would provide a null observation with no water. A nonrandom observation could be made to the left or right of the boulder or in another area in close proximity to the random site. The site information on elevation, water temperature, and GPS readings will be the same as the random site observation, and the site number will also be the same except followed by the letter (n), e.g., random site number is 3r and the nonrandom site number is 3n.

Visual counts are accompanied by noting the size of the observation area with measurements taken with the folding fiberglass ruler to include length, width, and depth of the point quadrat site. The observation area should not be larger than 3 feet by 3 feet, and will depend on the field of view. Obstructions such as boulders or fallen logs, and water clarity-depth, turbid water, or bubbles may decrease the area of observation. Previous survey tests (Baker & Foster, 1992) have demonstrated that in viewing larger areas, the observer might miss small fish in the distance and other animals hiding under or in ledges and crevices. Native Hawaiian stream gobies especially are cryptic, and responsive to rapid movement, and thus difficult to detect and observe.

The survey duration should range from 3 to 7 minutes and may be for a shorter duration depending on the site size, depth, water clarity, and laminar flow. A survey team of 3 observers is optimum with a total of 30 observations per stream survey section.

A quadrangle map is used for locating survey area and sites, landmarks, trails, and as a reference for altimeter elevation readings and DAR coding of stream sections (Fig. 2).

The GPS reading is also recorded when available and is important in determining the position on the stream.

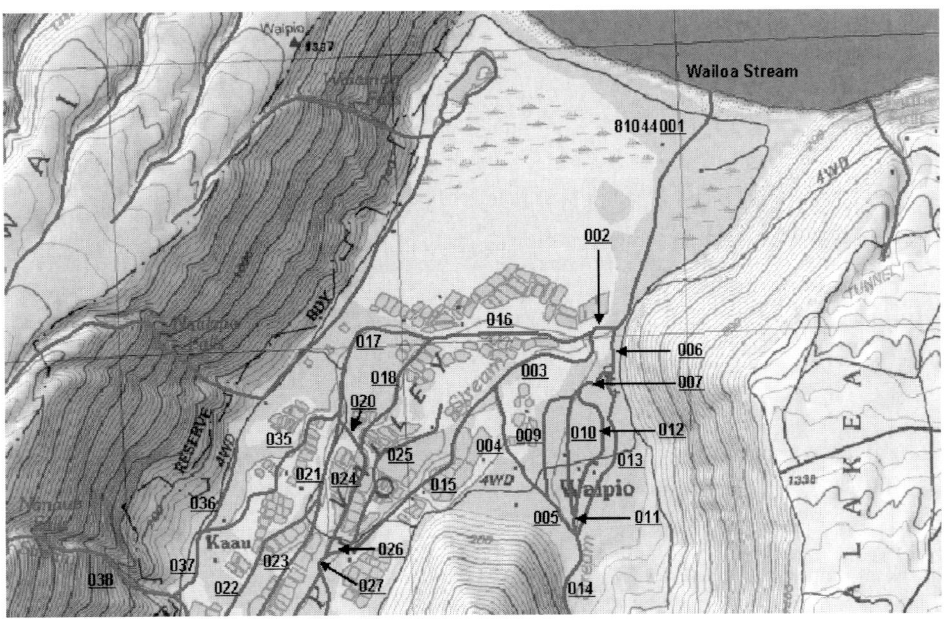

Figure 2. USGS quadrangle map with tributary and stream sections coded.

Table 1. Aquatic habitats of Hawaiian streams [after Baker (1992)]

Habitat Type	Depth (m)	Current (m/sec)	Turbulence
Pool	variable	< 0.2	no
Side Pool	< 0.5 usually	nil usually	no
Plunge Pool	< 2.0 usually	< 0.20	yes
Run	variable	0.20 – 0.75+	no
Riffle	< 0.5	> 0.75	yes
Cascade	2.0 usually	> 2.0	much

Upon arrival at the site, observers take the following measurements by using the equipment in the Pelican™ dry box. Elevation will be taken in feet with the altimeter, temperature in centigrade with the digital thermometer, and the GPS unit will provide latitude/longitude readings in decimal degree units. The GPS readings may not always be obtainable especially under a heavy vegetative canopy, in narrow steep-walled valleys, or next to high cliffs. The next observer takes the Pelican™ box and walks along the stream bank to his or her site while, keeping out of the water as much as possible so as not to disturb the bottom and make the water turbid for the observer downstream. After completing his measurements, the next observer takes the box to his site.

At the site, the observer waits for the fish to go back to their normal behavior before beginning the count. The observer notes the species (fish, crustacean, mollusk, etc.) present in the survey site, their number, size, and sex if identifiable.

DAR's aquatic habitat classification is based upon the five habitats delineated by Baker & Foster in 1992. The classification suggested here is an extension and modification of those habitats based on our stream experience on all five of the major Hawaiian Islands (Hawai'i, Maui, Moloka'i, O'ahu, and Kaua'i). The objective is to enable surveyors to recognize important differences in the habitat types efficiently without requiring them to invest a large amount of time evaluating a large number of details. The habitat types are based on a combination of depth, current speed, and turbulence (Table 1). A definitive description of the habitat types is provided below:

Pool: Area within streams that have reduced current velocity (<20 cm/sec), and which often has water deeper than that of surrounding areas.

Side pool: Small, shallow, slow current habitats very similar to shallow pools. The main difference is that it is situated well off the main stream channel.

Plunge pool: Pool that is situated at the base of a waterfall.

Run: An area of swiftly flowing water, without surface agitation or waves, and in which the slope of the surface water is roughly parallel to the overall gradient of the stream reach (Helm *et al.*, 1985).

Riffle: Shallow rapids where the water flows swiftly over completely or partially submerged obstructions to produce surface agitation, but standing waves are absent (Helm *et al.*, 1985).

Cascade: Extremely swift, turbulent habitats which are generated by a steep slope.

No water: A "No water" classification is included because when using a random survey the site selection could possibly be on top of a boulder sticking out of the water or dry stream bed.

Dirty water: Is included to accommodate situations where visibility is limited due to rain or other causes making the water cloudy or dirty.

Substrate types are noted by percent coverage of the point quadrat site and recorded according to the definitions of Wentworth's (1922) scale of particle classification (Cummins, 1962; Minshall, 1984; Helm *et al.*, 1985) (Table 2). The detritus classification is not part of the Wentworth scale, but was added for sites where there was no strong flow and leaf litter, branches, and other vegetative matter covered the bottom.

Comments are included to note any important facts about the survey site, e.g., located just below a tributary on the left, first site or last site of survey, etc. This information will also be included in the survey database.

Random number and site number are recorded to provide additional information about the distance up stream from the previous site and the site location across the width of the stream. This information can also be useful in locating the site when a GPS coordinate is unattainable due to vegetative canopy or in steep walled areas.

Table 2. The classification of mineral substrates by particle size, according to the Wentworth Scale [After Cummins (1962), Marshall (1982)]

Size Category	Particle Diameter (mm)	Particle Diameter (in)	Reference
Boulder	>256	>10	head-size and larger
Cobble	64–256	2.5–10.0	fist-size
Gravel	2–64	0.08–2.50	thumb-size
Sand	0.062 –2.000	0.002–0.080	sand-size
Silt	>0.062	>0.002	smaller than pin head

Databasing stream survey information

The DAR stream survey form (Fig. 3) contains the following general information on the stream: name of stream, DAR code, observer, date, temperature, elevation, and start and stop time for the site observations. The DAR code is a straight eight-digit number (00000000) that is used to identify streams for incorporation of the point quadrat survey data into the Division's freshwater Access database. The spatially nested hierarchy is used for all the survey types in the database and is based on island chain, island, hydrographic unit, watersheds, stream segments and sites (Fig. 4).

- The first digit identifies the ISLAND (Island) **0**0000000
  2 Kaua'i
  3 O'ahu
  4 Moloka'i
  6 Maui
  8 Hawai'i

- The second digit identifies the HYDROGRAPHIC UNIT (Hgu_code) 0**0**000000 within each island. These units are regional drainage areas established in the 1970s by U.S. Geological Survey (USGS) and Hawaii Division of Water & Land Development (DOWALD).

- The third through the fifth digits identify the WATERSHEDS 00**000**000. Expansion to 3-digits to incorporate intermittent streams and drainage basins will be included at a later date. The coding of the intermittent streams and drainage basins will probably start from the last number where the perennial streams left off for that island and continue clockwise around the island.

- The sixth through the eighth digits identifies the stream's individual SEGMENTS (Segments) 00000**000**.

Discussion

Hawaiian streams are characterized by being small and rugged in nature, with irregular bottoms, and not easily accessible. Hawaiian fishes and crustaceans quickly move into holes and crevices between rocks and boulders or dive into the loose sand and gravel when disturbed. Because of the behavior of stream animals and the irregularity of the bottom topography, collections by seining or backpack electroshocking are not likely to provide reliable assessments of species presence, distribution, density, and relative abundance. However, visual surveys are the one method that has proven to be reliable, efficient, and effective in the assessment of all Hawaiian stream animals. The point quadrat method, in particular, is most effective.

STREAM SURVEY FORM

SURVEY TYPE: RANDOM NONRANDOM OBSERVER: ____________

HSA CODE: ____-____-______.______ TEMPERATURE: ____________ (°C)

NAME: ____________________ ELEVATION: ____________ (ft.)

DATE: ____/____/____ (Mo. Day Yr.) TIME: ________ (START) AM PM TO ________ (STOP) AM PM

SUBSTRATE TYPE (%):

SEDIMENT ________ SAND ________ GRAVEL ________

COBBLE ________ BOULDERS ________ BEDROCK ________

DETRITUS ________

HABITAT TYPE: SURVEY AREA (l" x w" x h"): ____________

POOL SIDE POOL PLUNGE POOL RUN RIFFLE CASCADE

NO WATER DIRTY WATER

SPP: SIZE ♂ ♀ ?

COMMENTS:

Random No.: ____ , ____ Site No.: ____

DIVISION OF AQUATIC RESOURCES
GRH 2/02

DURA Rite WATERPROOF

G-212

J. L. DARLING CORP.
www.RiteintheRain.com
(253) 922-5000

Figure 3. DAR stream survey form.

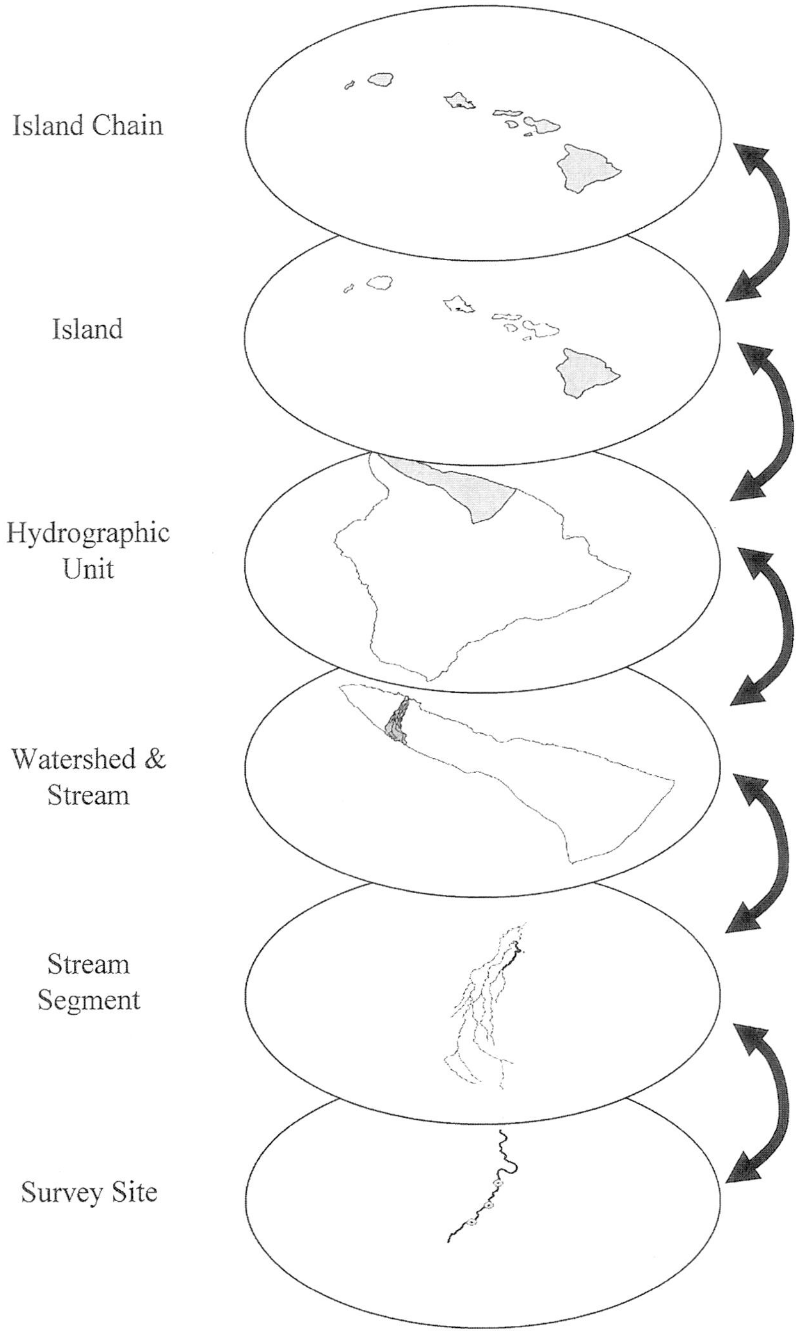

Figure 4.The spatially nested hierarchy used in the Division of Aquatic Resources Survey Database.

The choice between random and nonrandom point quadrat methods depends on the kinds of biological questions being asked. Because island streams have different types of habitats according to elevation or distance inland, it is convenient to place randomly selected sites with lower, middle, and upper stream reaches. This type of sampling (selecting a large area non-randomly and selecting sampling sites within it randomly) is referred to as "stratified random point quadrat method" (Fitzsimons *et al.*, 2005). This method has become the DAR standard for surveying Hawaiian streams.

The point quadrat method, however, should not be used to provide population density estimates but only to provide species presence, distribution, relative abundance, and density at that period in time during which the survey was taken.

The point quadrat survey data play an important part in the management of aquatic resources by DAR by providing baseline data on native biota, and it is a method to quickly assess the aquatic fauna, flora, habitat types, and species/habitat relationships within a stream ecosystem. The incorporation of these data into DAR's freshwater Access database along with other survey types-larval trapping, drift net sampling, etc. and the integration of this database into a Geographic Information System (GIS) isare providing a powerful management tool for setting permanent instream flow standards for all Hawaiian streams. The protection of native stream biota and their habitat is essential to maintain healthy streams, a neglected but essential ecosystem connecting the forest and the ocean.

Acknowledgments

We are grateful to J. Michael Fitzsimons for comments on the manuscript. J.M. Fitzsimons, J.E. Parham, and D.G.K. Kuamo'o provided input into the development and fine tuning of this stream survey method. This report is supported in part by funds from the Division of Aquatic Resources, Department of Land and Natural Resources, State of Hawaii (Dingell-Johnson/Wallop-Breaux Sport Fish Restoration Program.

Literature Cited

Baker, J.A. & S.A. Foster. 1992. *Estimating density and abundance of endemic fishes in Hawaiian streams.* Division of Aquatic Resources, Department of Land and Natural Resources, State of Hawaii. 50 p.

Cummins, K.W. 1962. An evaluation of some techniques for the collection and analysis of benthic samples with special emphasis on lotic waters. *American. Midland Naturalist* **67**: 477–504.

Fitzsimons, J.M. & R.T. Nishimoto. 1991. Behavior of gobioid fishes from Hawaiian freshwaters, p. 106–124. *In:* Devick, W.S. (ed.), New Directions in Research, Management, and Conservation of Hawaiian Freshwater Stream Ecosystems: *Proceedings of the 1990 Symposium on Stream Biology and Fisheries Management*, Division of Aquatic Resources, Departmentt. of Land and Natural Resources, State of Hawaii, Honolulu.

Fitzsimons, J.M. & R.T. Nishimoto. 1999. Application of fish behavior to stream monitoring on tropical Pacific Islands, p. 797–804. *In*: Seret, B. & J.-Y. Sire (eds.), *Proceedings of the 5th Indo-Pacif. Fish Conference, Noumea, 1997.* B. Société Française d'Ichtyologie, Paris.

———, **R.T. Nishimoto, & J.E. Parham**. 2005. Methods for analyzing stream ecosystems on oceanic islands of the Tropical Pacific, p. 103–136. *In:* Mueller Dombois, D., K. Bridges & C. Daehler (eds.), *Biodiversity Assessment of Tropical Islands Ecosystems: PABITRA Manual for Interactive Ecology and Management.*

Helm, W.T., P. Brouha, M. Aceituna, C. Armour, P. Bisson, J. Hall, G. Holton, & M. Shaw. 1985. Aquatic habitat inventory, glossary of stream habitat terms. *Western Division, American Fisheries Society*. 33 p.

Higashi, G.R. & M.N. Yamamoto. 1993. Rediscovery of "extinct" *Lentipes concolor* (Pisces: Gobiidae) on the island of O'ahu, Hawai'i. *Pacific Science* **47(2)**: 115–117.

Kuamo'o, G.K., G.R. Higashi & J.E. Parham. 2005. Structure of the Division of Aquatic Resources surveys database and use with a geographical information system. *Proceedings of the 2005 Symposium on the Biology of Hawaiian Streams and Estuaries.* State of Hawaii, Division of Aquatic Resources, Hilo.

Minshall, G.W. 1984. Aquatic insect-substratum relationships, p. 358–400. *In*: Resh, V.H. & Rosenberg, D.M. (eds.), *The ecology of aquatic insects.* Praeger Scientific. 625 p.

Nishimoto, R.T. & J.M. Fitzsimons. 1999. Behavioral determinants of the instream distribution of native Hawaiian steam fishes, p. 813–818. *In*: Seret, B. & J.-Y. Sire (eds.), *Proceedings of the 5th Indo-Pacif. Fish Conference, Noumea, 1997.* B. Société Française d'Ichtyologie, Paris.

Parham, J.E. 2002. Spatial models of Hawaiian streams and stream fish habitats. Ph.D. Dissertation Louisiana: Louisiana State University. 144 p.

Rosgen, D. & H.L. Silvey. 1998. *Field guide for stream classification.* Wildland Hydrology Consultants, Pagosa Springs. 193 p.

Wentworth, C.K. 1922. A scale of grade and class terms for clastic sediments. *Journal of Geology* **30**: 377–392.

Yamamoto, M.N. & A.W. Tagawa. 2000. *Hawai'i's native & exotic freshwater animals.* Mutual Publishing, Honolulu. 200 pp.

Biology of Hawaiian Streams and Estuaries. Edited by N.L. Evenhuis & J.M. Fitzsimons. *Bishop Museum Bulletin in Cultural and Environmental Studies* 3: 315–322 (2007).

Structure of the Division of Aquatic Resources Survey Database and Use with a Geographic Information System

Darrell G.K. Kuamo'o
Hawaii Department of Land & Natural Resources, Division of Aquatic Resources, 75 Aupuni Street, Room 204, Hilo, Hawai'i 96720 USA; email: darrell.gk.kuamoo@hawaii.gov

Glenn R. Higashi
Hawaii Department of Land & Natural Resources, Division of Aquatic Resources 1151 Punchbowl Street, Room 330, Honolulu, Hawai'i 96813 USA; email: glenn.r.higashi@hawaii.gov

James E. Parham
School of Natural Resources Sciences, 311 Biochemistry Hall, University of Nebraska, Lincoln, Nebraska 68583-0759 USA; email: jparham2@unl.edu

Abstract

Stream management and protection from overuse are becoming serious areas of conflict in Hawai'i. Improvements in the ability to coordinate, synthesize, and communicate information concerning stream animals in over 350 perennial streams is critical to the Division of Aquatic Resources' public trust responsibilities. One of the Division's major concerns is the protection and management of the unique amphidromous fauna inhabiting streams. The native Hawaiian stream animals are distributed throughout the streams in response to elevation and habitat type. A relational database was redeveloped to aid in the storage, retrieval, and analysis of the information collected from years of stream surveys throughout the state of Hawaii. In addition to redeveloping the database system, the Division of Aquatic Resources coded all stream segments to allow integration with a Geographical Information System. Combining survey observation information from the database with data layers showing stream characteristics such as waterfalls, intermittent and channelized sections, diversions, etc. provides insight into important factors affecting the distribution of native and exotic animals within the stream. This paper reports on the current iteration of the Division of Aquatic Resources Survey Database. Additionally, we describe how the data can be integrated with a Geographical Information System program and provide an example of how the database information may be used to answer important questions about the distribution and habitat use of Hawaiian stream animals.

Introduction

The State of Hawaii's Department of Land and Natural Resources (DLNR) has a leading role in watershed ownership and management responsibility. Essentially all of the State's streams arise in forest reserves or other state-owned areas. These fresh waters are important in providing essential habitat for Hawaii's aquatic flora and fauna. As part of the DLNR, the Division of Aquatic Resources (DAR) is tasked with upholding the public trust by ensuring that Hawaii's aquatic resources are managed for the benefit of current and future generations. One aspect of this public trust is the protection of Hawaii's native aquatic animals and the assurance of their long-term population viability. As the development of Hawai'i increases and diversifies, so does demand for fresh water. Stream management and protection from overuse are becoming serious areas of conflict in Hawai'i, and improvements in the ability of DAR to coordinate, synthesize, and communicate information concerning stream animals in over 350 perennial streams is critical to DAR's public trust responsibilities.

One major concern for DAR is the protection and management of the unique amphidromous fauna inhabiting the streams. The amphidromous animals include fishes, crustaceans, and mollusks.

Amphidromy is a type of diadromy where the adult animals live and reproduce in fresh waters, while the larvae develop in the ocean (McDowall, 1997). Critical to the existence of amphidromous animals are suitable instream habitat, a pathway between instream habitat and the ocean during migratory events, and suitable conditions in the ocean for the developing larvae (Fitzsimons *et al.*, 2002). Current management focuses mostly on the protection of critical instream habitats and the maintenance of the migratory pathway, as these two areas are presumed to be most susceptible to disruption by human activities.

Recognizing the need to protect and manage this valuable and limited resource, the State Commission on Water Resources Management (CWRM) initiated the Hawai'i Stream Assessment (HSA, 1990) through a cooperative agreement with the National Park Service in 1988. This Program, established in response to the National Wild and Scenic Rivers Act, assisted states in determining the needs and opportunities for establishing state and local wild, scenic, and recreational rivers areas (16 USC 1271-1287, Public Law 90-542, Section 11(a), 1968).

The primary task of the HSA was to identify streams appropriate for protection based on a broad-based collection of existing information on Hawaii's rivers and streams. The results of the HSA effort provided a reference to help policy makers, resource managers, developers, scientists, and the interested public to:

- locate published information for a particular stream;
- identify and prioritize areas where information is needed;
- understand stream resources within a statewide context;
- make management decisions based on available data;
- develop general stream resource protection guidelines; and,
- identify specific streams appropriate for protection and enhancement.

After the completion of the HSA effort, DAR and many cooperators developed appropriate biological survey methodologies (Baker & Foster, 1992; Nishimoto & Kuamo'o, 1997; Fitzsimons *et al.*, 2005; Parham, 2005), surveyed many streams, and supported a wide range of research to better understand the life history dynamics of the stream animals. With the increased efforts and the baseline information from the HSA, issues of data storage, timely reporting, and the distribution of results became important.

This information is critical to DAR's ability to comment on the many surface water development proposals that will divert water from or otherwise modify stream habitats and, therefore, potentially affect stream animals.

The need for managing, conserving, and protecting Hawaii's native stream biota and their habitats led to the development of a database system to store and retrieve stream data in 1991. The initial design of the database was in FoxBASE+ on a Macintosh platform (Fox Software, 1989). This design was abandoned because it required too much programming experience, and thus did not support DAR's need for an easy to use database. The next generation of the DAR database was developed in the relational database in ACIUS 4th Dimension (ACIUS, 1987). The data was transferred to this software in part to ACQUIRE the capabilities of 4th Dimension to import/export modified FoxBASE data.

In 1998, the stream database was transferred yet again, this time to Filemaker Pro 4.0 (Claris, 1998). This conversion was done in part because Filemaker Pro 4.0 was upgraded to include relational database functions, was in more universal usage than 4th Dimension, required less storage space, and provided more flexibility in data searches.

In 2001, the database was redeveloped in Microsoft Access (Microsoft Corporation, 2000) on a Windows platform. The redesign of the database was prompted by two major needs. First, the functionality of the database needed improving to speed data entry, management, analysis, and to allow internet access to the data by the public. Secondly, the design of the database needed to be spatially explicit to allow data integration into a Geographic Information System (GIS) that was not easily accomplished on a Macintosh system.

The goal of this paper is threefold. First is to provide an update on the current iteration of the DAR Survey Database. Second is to show how the data can be integrated with a GIS program, and third is to provide an example of how the database may be used to answer important questions about the distribution and habitat use of Hawaiian stream animals.

Results

Current Database Design

The current iteration of the DAR Survey Database was designed to fit a spatially nested hierarchical structure. At the largest spatial extent, the hierarchy begins with the Hawaiian island chain. Nested within the island chain, at increasingly smaller spatial extents, are the individual islands, hydrographic units within an island, watersheds and their streams within a hydrographic unit, stream segments within a stream, survey locations within a stream segment, and individual observations within a survey location (Fig. 1). This basic structure provided the ability to incorporate different survey designs within a single database and allowed for the integration of data collected from multiple different research projects.

At the largest spatial extent, the island chain level provides little direct application to general users of the DAR Survey Database but will allow this database and its structure to be distributed to other Pacific Islands with minimal change. Given Hawaii's leadership role in the Pacific island community and the ability of the database to incorporate a wide range of survey types, use of this database will aid other island groups with their aquatic resource management. Additionally, the island chain level will allow future large scale comparisons to be made to better understand the general patterns in species distributions and overarching commonalities in effective resource protection (Fitzsimons *et al.*, 2002; McDowall, 2003).

At the next smaller spatial extent, the island level provides the ability to compare and contrast the habitat use and distributions of aquatic organisms among islands. The islands that make up the Hawaiian Island chain vary greatly in age (Grigg, 1988) and with the differences in age is a corresponding change in typical stream morphology and available instream habitats (Parham, 2002).

The hydrographic unit is the next level within the nested hierarchy. These units are regional drainage areas established in the 1970s in part by the U.S. Geological Survey (USGS) and the State of Hawaii Division of Water and Land Development. There were 27 hydrographic units on the main Hawaiian Islands, and these are groups of spatially contiguous watersheds that share similar environmental conditions. At this level, differences in general conditions of groups of streams can be addressed.

The watershed is the next level within the nested hierarchy. The watershed is the runoff basin for a single stream. As a result, the watershed and the stream are contained at the same level in the database and made practically interchangeable in a structural sense. The level is called the watershed level as the stream is part of the overall watershed. The definition and use of the watershed label is undergoing changes in Hawai'i. In the past, the watersheds were only defined for perennial streams, and areas without perennial streams may have been grouped into watershed regions. Currently all watersheds, as defined as the upstream runoff basin terminating at the ocean, are being coded for all watersheds of non-perennial streams in the state. To ensure consistency with older version of the data, all codes for both the old coding systems and newly created coding system have been retained. Comparisons at this level include the obvious stream by stream comparison, but also comparisons of streams with varying levels of modification within the watershed. Collections of larval fishes either emigrating out of or immigrating into the stream near the stream mouth can be linked at this level, as the sample is a representation of potential upstream populations. Additionally, marine or estuarine surveys in the area extending out into the ocean from the watershed edges can be linked to the watershed level. This will allow comparisons of species populations in areas with differing freshwater contributions.

The stream segment is the next level within the nested hierarchy. Stream segments were defined

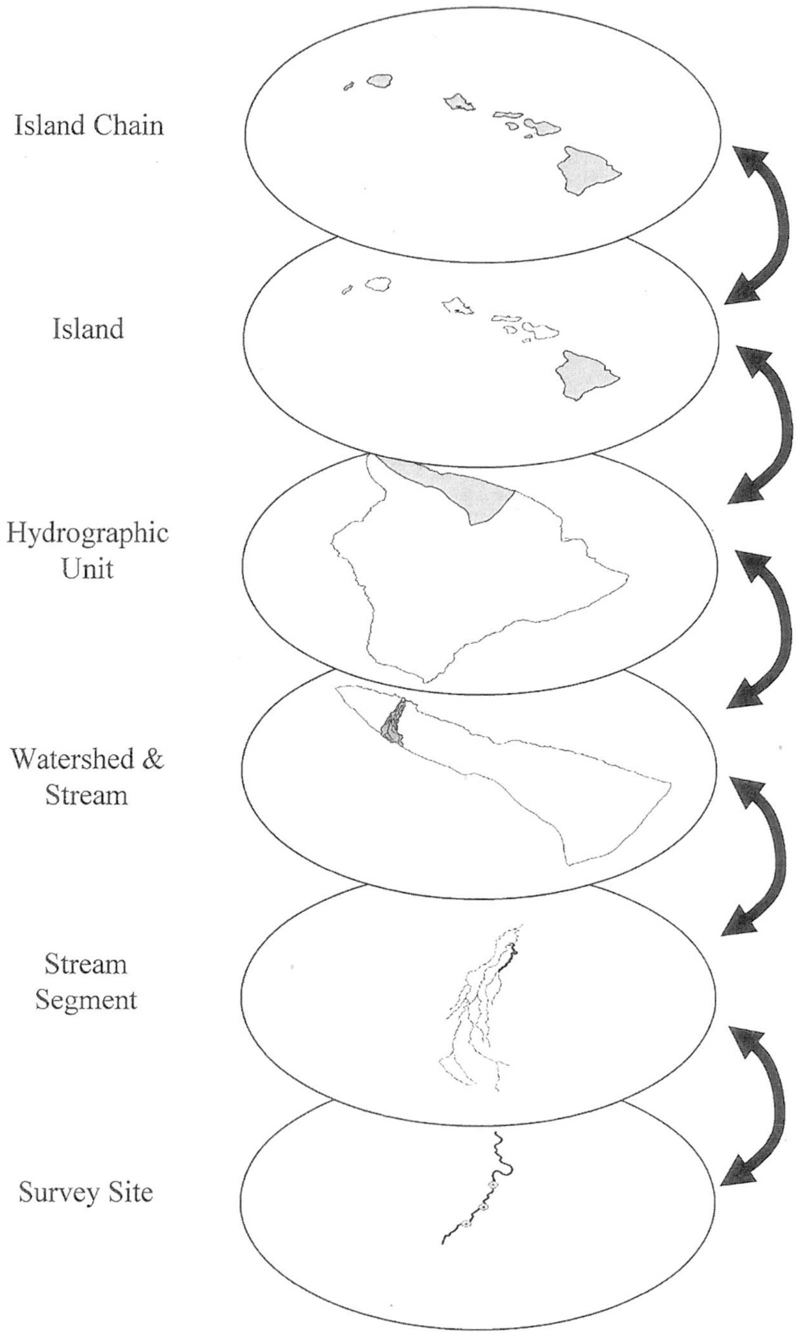

Figure 1. The spatially nested hierarchy used in the Division of Aquatic Resources Survey Database.

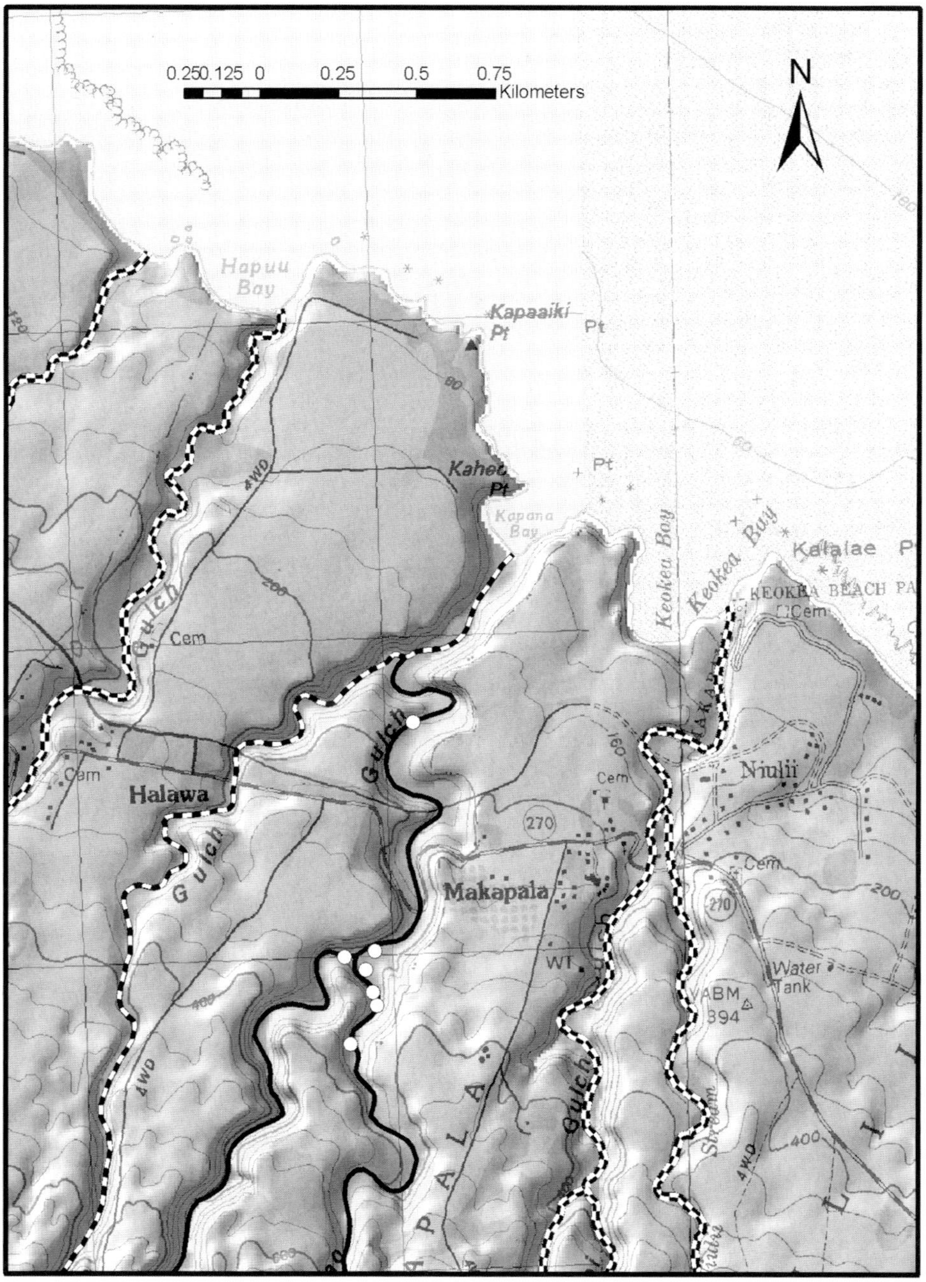

Figure 2. 'A'amakāō Stream in Kohala displayed with hillshading and digital raster graphics added. Nonsurveyed stream segments (dotted line), stream segments surveyed but no *Lentipes concolor* observed (dashed line), segments with *Lentipes concolor* observed (solid line), and the individual survey points are displayed.

as any linear, undivided segment of stream. This definition results in a stream segment being coded as a unique segment where the downstream node terminates at either the ocean or the next downstream segment, and the upstream segment terminates at either the end of the stream channel or at a junction with another segment. In the simplest case, a stream could contain just one stream segment if the stream contain no tributaries. The current stream network derived from USGS digital line graphs is based on their 7.5-minute topographic maps. Fields containing the DAR coding as well as their corresponding names were added to the attribute table. The stream segment level within the database has undergone the most changes as the coding system is completely updated to include all stream segments within the stream coverage. This level is likely to continue to undergo changes as higher resolution flow networks will allow systematic coding of all intermittent and perennial channels with a stream system. As in the watershed level, past coding systems were retained along with newer coding systems to allow for backward compatibility within the database system. Comparison developed at the stream segment level includes positional comparison (distance from the ocean, upstream area, etc.) of species distributions within a stream and quantification of extent of segments surveyed. This level is also appropriate for linking to general reconnaissance surveys where long sections of streams are surveyed without specific sampling locations.

The next level within the spatial hierarchy is the site level. The majority of surveys link to the database at this level. The site level requires higher positional accuracy than the stream segment level, and typically has a GPS coordinate location recorded with the site. Given the difficultly of obtaining satellite fixes in some of the deep gorges containing streams in Hawai'i, it is critical that stream segment information is recorded in the field. Sites usually can be relatively accurately georeferenced after the surveys if the stream segment, elevation, and a few sites during the sampling effort received GPS locations in the field. It is important to understand that a GPS location alone is not always enough to locate the site within the appropriate stream segment. Horizontal error in the GPS reading and errors in the digitizing of the stream segments may result in a situation where the point location is closer to the incorrect stream segment than to the correct one. This is especially common near tributary junctions.

The site is the smallest spatial extent within the database and can be used in a number of ways. The site shows the location that individual species were observed and allows for repeated sampling to see if there are changes over time. The site also holds information on site attributes like water depth or habitat type to allow comparison of habitat use vs. habitat availability.

Numerous additional tables exist or could be added to the database, but these tables will primarily hold information on the species observed in the surveys. These tables can be linked to the overall nested spatial hierarchy of tables at the appropriate level of spatial resolution. Currently, the data date back the 1960s and includes information on perennial and intermittent streams, lakes, reservoirs, ponds, ditches, and diversions.

Linking the Database to GIS

As a result of the design of the DAR Survey Database around a spatially nested hierarchy, linking the data to a GIS is relatively straightforward. A GIS in its simplest description is a spatial database, where each piece of information can be located on a map. While the DAR Survey Database could be fully replicated in a geodatabase, the intention was not to require the use of a GIS in data entry, storage, or access, but to allow full integration with a GIS when needed.

As a concurrent effort to the redesign and development of the DAR Survey Database, the spatial data layers for the streams of Hawai'i were updated to include all of the necessary location fields to allow seamless integration with the DAR Survey Database. Each of the levels from island to stream segment had attribute fields with data added to the stream coverage, so that a unique location field in both the database and the stream coverage existed for each level. To add the database information into a GIS, the location field of the table must be included with any attribute information in a query. The resulting database file can be linked to the comparable location field in the GIS coverage, and then the results can be displayed in map form within the GIS.

By integrating the database information with the GIS, we can relate information observed in a relatively small location with conditions not only nearby, upstream, or downstream of the sampling location, but to the watershed, region, or island as well. This provides the resource manager a mechanism to aid in determining the positive or negative effects of changes to the environment on the stream ecosystem.

An example of database and GIS integration

In order to display the data from the DAR Survey Database in the GIS, database queries are needed to be developed to extract the relevant information. In order for these database queries to be linked to the GIS, the query must contain a related location field to join the information. In the DAR Survey Database, the location fields are generally codes. The codes remain static within the database, but as new information on a particular area is acquired, the data can be updated without disrupting the database structure. These codes occur only once in the database so there is no chance of the information being duplicated.

In the example query for this paper, the first database query was developed to find all the tributaries in the state, followed by a query to list all tributaries that had at least one point quadrat survey. Finally, from the tributaries that were surveyed, a query was developed that lists the tributaries that recorded the presence of *Lentipes concolor*.

In this example, the first query resulted in 4,988 segments contained within all Hawaiian streams. From the 4,988 perennial segments, 282 (5.6%) had at least one point quadrat survey. Of these, *Lentipes concolor* was present in 132. The total number of point quadrat surveys as of 17 June 2004 included 6,574 individual surveys (Sakuda, 1989–1994, Devick, 1995–2004). Of the point quadrat surveys 1,538 or 23% showed the presence of *Lentipes concolor*.

After the appropriate queries were created and exported from the database, the next step was to import and display the information in the GIS. The perennial stream coverage, named darstreams, can be obtained from the Department of Business and Economic Development and Tourism (DBEDT), Office of Planning (OP) GIS dataserver at http://www.hawaii.gov/dbedt/gis/download.htm. The attribute table for the darstream coverage includes all location fields necessary for linking to the database at all levels of the spatially nested hierarchy. In the example, the join was based on NDAR_CODE field from the darstreams coverage and the DAR_CODE from the database table. Additionally, the individual survey points can be displayed by using the coordinate fields Lat DD and Long DD. After developing the queries and linking the appropriate location fields, the data can be displayed in the GIS (Fig. 2).

Conclusions

The Division of Aquatic Resources has redesigned a relational database to make survey information available to interested parties. The information is available over the internet at http://www.hawaii.gov/dlnr/dar/streams/stream_data.htm. The database allows for efficient integration with GIS software packages and supports a wide range of survey methodologies. The example showed the ability to take the information from the database and link it in several different ways to the GIS stream coverage. The important concept is to use the location fields of the spatially nested hierarchy and be able to correctly link the database to the GIS. In addition to the location fields, the attributes fields are important as they allow the display of many different types of information. These attributes are limited only by the information collected by the surveyors.

Acknowledgments

We thank Dr. Robert Nishimoto, Dr. J. Michael Fitzsimons for all their encouragement and support, and especially Ms. Marjorie Hale-Fishman for her invaluable assistance in designing the DAR Survey Database. This report is supported in part by funds from the Division of Aquatic Resources, Department of Land and Natural Resources, State of Hawaii (Dingell-Johnson/Wallop-Breaux Sport Fish Restoration Program).

Literature Cited

ACIUS. 1987. *4th Dimension*, Cupertino, California.

Baker, J.A. & S.A. Foster. 1992. *Estimating density and abundance of endemic fishes in Hawaiian streams*. Hawaii State Department of Land and Natural Resources, Division of Aquatic Resources, Honolulu. 50 p.

Claris. 1998. *Filemaker Pro 4.0*. Santa Clara, California.

Devick, W.S. 1995–2004. *Freshwater fisheries research and surveys*. United States Fish & Wildlife Sportfish Restoration Project **F-14-R** numbers **19–28**.

Fitzsimons, J.M., J.E. Parham & R.T. Nishimoto. 2002. Similarities in behavioral ecology among amphidromous and catadromous fishes on oceanic islands of Hawai'i and Guam. *Environmental Biology of Fishes* **65**: 123–129.

———., **R.T. Nishimoto & J.E. Parham**. 2004. Stream ecosystems, p. 103–136. *In*: Mueller-Dombois, D. *et al.* (eds.), *Biodiversity Assessment of Tropical Island Ecosystems*. University of Hawai'i Press, Honolulu.

Fox Software. 1989. *FoxBASE+*. Tychy, Poland.

Grigg, R.W. 1988. Paleoceanography of coral reefs in the Hawaiian-Emperor chain. *Science* **240**: 1737–1743.

Hawai'i Stream Assessment. 1990. *A preliminary appraisal of Hawai'i stream resources*. Report R84. Prepared for the Commission of Water Resources Management. Hawai'i Cooperative Park Service Unit, Honolulu. 294 p.

McDowall, R.M. 1997. Is there such a thing as amphidromy? *Micronesica* **30**: 3–14.

———. 2003. Hawaiian biogeography and the islands' freshwater fish fauna. *Journal of Biogeography* **30**: 703–710.

Microsoft Corporation. 2000. *Microsoft Access*. Redmond, Washington.

Nishimoto, R.T. & D.G.K. Kuamo'o. 1997. Recruitment of goby postlarvae into Hakalau Stream Hawai'i Island. *Micronesica* **30**: 41–49.

Parham, J.E. 2002. Spatial models of Hawaiian streams and stream fish habitats. Ph.D. Dissertation. Louisiana State University, Baton Rouge. 144 p.

———. 2005. Survey techniques for freshwater streams on oceanic islands: Important design considerations for the PABITRA project. *Pacific Science* **59**: 283–292.

Sakuda, H.M. 1984–1994. *Freshwater fisheries research and surveys*. United States Fish & Wildlife Sportfish Restoration Project **F-14-R** numbers **13-18**.

Biology of Hawaiian Streams and Estuaries. Edited by N.L. Evenhuis & J.M. Fitzsimons. *Bishop Museum Bulletin in Cultural and Environmental Studies* 3: 323–326 (2007).

A Template for Conducting Ecosystem-Based Instream Flow Studies

ALLAN LOCKE

Alberta Department of Sustainable Resource Development, Fish & Wildlife Division, Box 1420, Provincial Building, Cochrane, Alberta, Canada, T4C 1B4; email: Allan.Locke@gov.ab.ca

Abstract

Since the development of the first quantitative instream flow techniques in the 1970s and 1980s much has been learned about ecological processes of rivers. Though much remains to be learned, many technical tools or methods have been developed to incorporate that understanding into developing and applying instream flow prescriptions. However, most studies to date have focused on only one or a few of the several elements that affect biological processes and ecosystem needs of rivers. In a recent project by the Instream Flow Council, a template was developed for conducting instream flow studies and improving strategies for riverine management. That construct draws on existing facts and knowledge to emphasize that effective instream flow management should integrate 5 riverine components (hydrology, geomorphology, biology, water quality and connectivity), public education and involvement, and legal/institutional elements. This report will illustrate the general characteristics of these various components, explain why it is important to integrate them all, and provide an example of how they might be addressed to improve the quantification of instream flow needs and protection of riverine resources.

Introduction

The following is the Executive Summary from the recent publication (Annear *et al.*, 2004), *Instream Flows for Riverine Resource Stewardship, Revised Edition*, published by the Instream Flow Council. The book was a collaborative effort with sixteen contributing authors: Peter Aarrestad, Connecticut Department of Environmental Protection; Tom Annear, Wyoming Game and Fish Department; Hal Beecher, Washington Department of Fish and Wildlife; Ian Chisholm, Minnesota Department of Natural Resources; Chuck Coomer, Georgia Department of Natural Resources; Christopher Estes, Alaska Department of Fish and Game; Joel Hunt, Manitoba Water Stewardship; Rick Jacobson, Connecticut Department of Environmental Protection; Gerrit Jobsis, South Carolina Department of Natural Resources; John Kauffman, Virginia Department of Game and Inland Fisheries; Allan Locke, Alberta Department of Sustainable Resource Development; John Marshall, Ohio Department of Natural Resources; Kevin Mayes, Texas Parks and Wildlife Department; Gary Smith, California Department of Fish and Game; Clair Stalnaker, and Rod Wentworth, Vermont Department of Fish and Wildlife.

Executive Summary
Instream Flows for Riverine Resource Stewardship, Revised Edition

Federal, state, provincial, tribal, and local governments are charged with stewardship responsibilities to wisely manage the quantity and quality of waters within their jurisdictions for current and future generations. Fundamental to that charge is ensuring that sufficient water is retained in rivers and lake systems at all times of year to sustain fishery and aquatic wildlife resources and ecological processes.

In the United States, the authority of state fishery and wildlife agency stewardship is derived, in part, from the Public Trust Doctrine. The basic tenets of the doctrine, as described in this book, intertwine principles of common and statutory law, including property law. When combined with other laws, the doctrine can be a very powerful tool for protecting and restoring instream flows in some situations. At present, the Public Trust Doctrine forms no recognized basis of law in Canada. Instead, Canadian stewardship responsibilities are broadly based on the provinces's need to act in the public interest. Although the Public Trust Doctrine is not formally incorporated in statute, we argue that the principles of the doctrine may be included in current Canadian law.

Instream Flows for Riverine Resource Stewardship presents the collective views and recommendations of the Instream Flow Council's (IFC) state and provincial fishery and wildlife agency members regarding appropriate instream flow strategies for managing, maintaining, or restoring riverine fishery and aquatic wildlife resources and processes. We also identify eight components that should serve as guidelines in establishing or improving existing agency instream flow programs and in developing prescriptions to be addressed in each instream flow assessment. These components include hydrology, biology, geomorphology, water quality, and connectivity as well as legal, institutional, and public involvement.

Although our work is similar to that of other works published on instream flow methods, it extends beyond a mere listing of methods and a description of their strengths and weaknesses; it incorporates ideas, policies, and recommendations that the IFC believes should, at a minimum, be addressed in all instream flow assessments.

In overseeing the management of game, fishery, and wildlife programs, resource managers routinely set goals and objectives, monitor progress, and evaluate performance. The IFC recommends that this approach be expanded to include the management of instream flow programs and that program development and evaluation encompass the concepts recommended herein. When developing goals and objectives for riverine management, it is important that resource managers seek strategies that explicitly address public demands and help them fulfill their legal responsibilities to maintain and restore healthy aquatic ecosystems.

The IFC promotes the goal of maintaining the ecological integrity of unregulated rivers and restoring regulated rivers to the ecological conditions that more nearly approximate their natural form and function. To move toward this goal in any increment, instream flow practitioners should address the eight ecosystem components in developing an instream flow program and incorporate them as appropriate.

Prior to the 1980s, many of the instream flows that were provided for water projects were limited to a flat-line "minimum" flow because water developers and managers had little or no appreciation for the importance of natural flow variability. In most of these cases, water managers ignored the recommendations of the early instream flow practitioners who noted the potential shortcomings and negative riverine effects that could result from a flat-line minimum instream flow versus maintaining or restoring variable flows that more nearly resemble the seasonal flow patterns and processes that sustain natural ecological functions.

Since the 1980s, laws and regulations have been developed in many states and some provinces to begin addressing these water management needs. These statutory and regulatory changes, combined with better understanding of riverine processes and enlightened attitudes, have improved the opportunity for instream flow practitioners to quantify and establish variable flow regimes for sustaining viable fisheries and riverine processes. Although opportunities now exist for protecting instream flows in most jurisdictions, the legal and institutional opportunities for reserving water for riverine purposes in most states and provinces are still more restrictive than those that exist for out-of-channel users and uses.

We provide guidelines for quantifying flow regimes and developing recommendations for replacing formerly assigned minimum flows and insufficient flow caps with more appropriate variable flows. Before initiating plans that will modify the flow of streams and rivers, water developers and managers should ask what the maximum amount of water is that can be removed at any given time without adversely altering the river system and its natural functions and processes rather than the minimum flow needed to sustain the resource.

The natural flow paradigm (preservation of the natural flow variability and ecological function of river systems) is axiomatic to ecological integrity of river systems. Managers establishing instream flows must recognize the importance of inter- and intra-annual flow variability in riverine systems because different flow levels enable critical ecological processes that cannot occur otherwise. For example, it was formerly accepted, and in some cases is still believed, that higher flows represent "excess" water in rivers and that flood water can be removed without harm and, perhaps, even benefit the ecological function of the river. However, as initially recognized by instream flow practitioners as early as the 1970s, seasonal high flows are critical components of river ecology. This is especially true at the terrestrial/aquatic interface where high flows deposit sediment, shape channels, rejuvenate and maintain riparian vegetation and habitats, improve water quality, expand and enrich food webs, maintain the valley, and provide access to spawning and rearing sites in the floodplain. The same can be said about the importance of natural periods of low flow (drought). Likewise, there is a growing body of evidence that supports the maintenance of natural processes of ice formation and breakup.

Most traditional state and provincial fishery and wildlife instream flow programs have limited, if any, staff primarily trained as fishery biologists. Although many agencies provide some training in instream flow assessment methods, trained staff often spends limited time on instream flow issues. Effective instream flow programs entail more than an occasional instream flow study or periodic mitigation negotiation with water development interests or consultants for a planned water project. Such programs require well-trained specialists who are capable of integrating the five riverine components into complex legal and institutional procedures while also ensuring effective public involvement. This is a daunting challenge for state and provincial fishery managers and one that makes their responsibilities uniquely different from those of instream flow practitioners whose assignment is less broad.

In developing an instream flow prescription to enhance or restore a degraded stream, it is prudent to involve all stakeholders, formally identify the problems to be addressed, solicit technical expertise, and devote attention to study design. Perhaps the most critical aspect of developing an instream flow prescription is to routinely and formally document the rationale used to decide a particular course of action. The strategy should address the riverine components individually and collectively, whether or not inclusion of a particular component in the study design is warranted. This approach drives the practitioner to consider all factors in developing prescriptions, documents the considerations for the record, and provides the basis for adjusting recommendations as new information or opportunities become available.

There is no universally accepted method, or combination of methods, that is appropriate for establishing instream flows on all rivers or streams. Selection of a method or adaptation of methods is dependent on the water body and potential modification under consideration. Moreover, selected methods should only be applied in accordance with the guidelines recommended in this document to ensure a solid scientific basis for establishing an instream flow prescription. Only when used in conjunction with other techniques can a specific tool afford adequate instream flow protection for all of a river's needs.

In some situations, scientifically sound demonstrations of tangible harm and benefit resulting from a range of experimentally controlled instream flow levels over extended periods of time may be required. Adaptive management may be a useful tool in some, but not all, of these situations. It is most appropriate when financial investment is significant, values for riverine resources are high, risk to all parties is considerable, and the time frame for the project allows prolonged monitoring. Further, binding commitments among stakeholders must be established at the start of studies to ensure that adequate resources (water and money) are available to fulfill testing needs for the full range of potential mitigation strategies, and that safeguards are provided to avoid irrecoverable impacts to riverine resources. As we emphasize throughout this work, studies that focus on only a limited number of components, such as the response of fish populations, should be avoided because they are often confounded by interrelations with riverine components other than streamflow. Consequently, managers must be critical of efforts to ascribe perceived short-term changes, or lack thereof, to a single factor such as streamflow alone. Measurable targets or conditions and defined decision points are essential.

Of the many instream flow quantification methods and variations developed over the last 30 years, we assess 34 of those most commonly used. In some cases, a broad class of methods is addressed collectively within a single review (e.g., flushing flow methods and biological response correlations). Although not exhaustive, tool evaluation is sufficiently broad to allow practitioners to identify an appropriate methodology for most study designs that may be encountered. Evaluation addresses 14 categories, ranging from the method's purpose to the IFC's critical opinion.

Understanding the underlying mechanism(s) responsible for the biological, physical, and chemical outcomes evident in river systems must underpin the instream flow prescriptions if they are to succeed. However, much remains to be learned about the role and interrelations of factors governing riverine resources and processes. We do not presume to imply that this document is the definitive resource for all instream flow study needs. The science of instream flow management is a relatively young and evolving discipline and much additional research is needed. Although we do not provide a summary of additional research needs, we note that research is being conducted in many settings throughout the United States, Canada, and elsewhere. Clearly, there is a need for research that better identifies how to build a flow prescription that addresses the eight ecosystem components in concert. We urge instream flow practitioners and water managers to remain open to the application of new assessment tools as they continue to be developed and accepted by the instream flow community.

Literature Cited

Annear, T., I. Chisholm, H. Beecher, A. Locke, P. Aarestad, C. Coomer, C. Estes, J. Hunt, R. Jacobson, G. Jobsis, J. Kauffman, J. Marshall, K. Mayes, G. Smith, C. Styalnaker & R. Wentworth. 2004. *Instream flows for riverine resource stewardship.* Revised edition. Instream Flow Council, Cheyenne, Wyoming. 268 p.

Biology of Hawaiian Streams and Estuaries. Edited by N.L. Evenhuis & J.M. Fitzsimons. *Bishop Museum Bulletin in Cultural and Environmental Studies* 3: 327–330 (2007).

Establishment of an Integrated Instream Flow Program in Hawai'i Consistent with Public Trust Doctrine

WILLIAM S. DEVICK
94-374 Makalu Loop, Mililani Town, Hawai'i 96789, USA; e-mail: billdevick@hawaii.rr.com

Introduction

Hawai'i is the water state, renowned for its spectacular waterfalls and surrounding blue sea. Our hundreds of freshwater streams by contrast barely register. To the callously uninformed, streams are at best environmental annoyances that catch waste from the land, pollute the ocean, kill coral reefs, and at times inconveniently inundate structures built on flood plains. Unless these waters are captured or diverted for beneficial uses such as irrigation, the water flowing out to the sea is wasted. Streams flowing through urbanized areas must of course be bulldozed, reconfigured, and filled with concrete to move water through more rapidly and contain storm flows within artificial banks.

The select but small group attending the Symposium on the Biology of Hawaiian Streams and Estuaries knows better. In biological reality, Hawaii's streams are *mauka-makai* lifelines that inextricably tie the land and sea in a unified system that sustains a native biota unique to the islands. Properly managed streams literally feed the sea, not just with water. They replenish groundwater and wetlands. They form estuaries, green-blue bridges, where fresh and salt waters mix. These transition zones serve as nurseries for a wide variety of marine organisms and are especially susceptible to upset from human interference with naturally functioning ecosystems.

The original Polynesian immigrants to Hawai'i developed a complex society inseparably linked with stream resources and formed deep emotional attachment to stream biota. Postlarval *'o'opu*, known collectively as *hinana*, were harvested *en masse* and considered a delicacy. Certain of the inland *'o'opu* were reserved as food for the *ali'i*. Streams, their watersheds, their biota, diversions for taro with return of water to the streams, fishponds, and fronting waters of the ocean were intelligently managed with a unit known as an *ahupua'a*. Some present-day local communities have retained an intense appreciation of these resources and understand their contribution to the cultural heritage of native Hawaiians.

The ascendance of westernized customs in Hawai'i highlighted exploitation of resources and transformed local practices definable as sustainable management to maximization of economic gain. The *ahupua'a* ideal, which was roughly analogous to ecosystem-based management, was subverted by many actions but most dramatically by diversion of water from streams for industrial scale agriculture, especially sugar cane cultivation. With the loss of stream water, the native stream biota were devastated, taro culture declined, fishponds deteriorated, and biological productivity of nearshore marine waters was likely reduced. By the 1980s, it became blatantly apparent that unless positive action was taken, the native stream biota would inevitably dip to threatened or endangered status.

Discussion

Creation of a State Water Code in 1987 and formation of the State Commission on Water Resource Management (CWRM), reinforced by a growing appreciation of streams in the conservation community, provided a vehicle for halting the slide. For the first time there was an assertion of State authority over surface waters, which previously had been viewed as privately owned and therefore beyond the scope of state regulation. With the new assertion of authority came an unavoidable responsibility to protect the surviving biological resources despite the increasing demands of the ever-growing human population. Although far from perfect, the resultant process has tended to meet an implicit objective of "no net loss of habitat" occupied by native biotic populations recognized as significant and has encouraged rapid improvement in understanding fundamental elements of the occurrence, distribution, behavior, habitat requirements, and other components of endemic and indigenous species and of non-human threats which persist, in particular invasions by non-native species.

There is an inherent presumption that the State Water Codes should accommodate the requirements of Public Trust Doctrine (PTD), which legal authorities have emphasized is expressed in the State Constitution. PTD seems to be an expression of the original meaning of "conservation" in which natural resources should be managed in a manner that ensures their perpetuation for use and enjoyment by future generations. Government is supposed to manage these resources in a responsible manner, weighing human needs or wants against the ability of the environment to accommodate those demands.

That can lead to a conflict between perceived "public interest" and the expectations of PTD. They should be identical. But given economic and political pressures, combined with a lack of understanding at all levels of the consequences of decisions that are made, short term interests that maximized use have historically prevailed in the decision-making process at the expense of long term sustainability of not only the targeted resource but also of the natural system to which they are indivisibly linked. Ideally, recognition of the priority that should be given to PTD will in the future shift the fulcrum in balancing those decisions from the side of exploitation to protection.

The past half century has seen a dramatic shift in attitudes towards how living natural resources should be managed. Originally these living resources were presumed to be renewable, with appropriate management, and the highest good was to increase their availability to man. Accordingly, habitats were being altered for human convenience. If we build on a flood plain, channelize the stream to prevent flooding. As we produce more and more waste, just dump it in a stream or in the ocean. While we are at it bring in lots of alien species that intentionally or accidentally become permanent residents, reducing native species biodiversity in the process. Virtually all of living resource problems we are facing—from declining fish populations, loss of native species and habitats to alien species invasions and ultimately global warming—have as their root cause excessive human exploitation.

That is hardly a revelation. But only recently has it been generally recognized that living resource management can not be done in isolation from what is happening in the rest of the world. The mantra now is "ecosystem-based management", which can be quite elusive in practice. Although in practice we still tend to be mired in single-species mindsets, it is imperative that we strive to develop the tools to construct an effective management framework that approaches an ecosystem-based ideal.

At the heart of successful ecosystem-based management must be good science. At the biological level we must do our best to understand the requirements of species, their relationships to other species, and the impact of perturbations associated with landscape changes and other impacts of human activity. This is open-ended and can not be done cheaply, but as understanding increases so will better identification of key questions that need to be addressed. If good science is set aside as an economic expedient, it is certain that the messes accumulated in the last half century will continue to pile up.

The group represented in the symposium has made enormous progress in developing the good, hard science needed to support the stream-related decision-making process under the Water Code in Hawai'i. But we are far from knowing all that needs to be understood, and what is known must be integrated in a strong administrative structure to be meaningful. The Instream Flow Council in an exceptionally high quality publication (Annear *et al.*, 2004) on instream flow issues, programs and methodologies recognizes the following:

> "Managing instream flows to protect public natural resources within the legal and institutional bounds of water allocation is a highly complex job. It is a daunting and potentially confusing maze that state and provincial managers must negotiate to fulfill their responsibilities. Few, if any, fishery biologists possess the skill to integrate their own biological knowledge with knowledge of hydrologic and geomorphologic concepts, legal and administrative processes, and public involvement. Clearly,

> agency managers need to expand their horizons. To fulfill fishery and wildlife management responsibilities for present and future generations, state and provincial fishery and wildlife agencies must make water management a top priority, hire and retain qualified and trained staff, and participate fully in water management decisions. Although Hawai'i has a nascent structure to meet these expectations, it has to be formalized and strengthened to meet the constitutional imperatives of PTD".

At present, stream management and regulation in Hawai'i ostensibly falls under the Department of Land and Natural Resources (DLNR), which has basic regulatory authority through CWRM and the Division of Aquatic Resources (DAR), which serves as an informal information source to CWRM about stream biota and has some related rule-making authority. Peculiarly, water quality regulation falls under the Department of Health (DOH), although there is a connection in that the DOH Director is a CWRM member. Given the many pressures, differing mandates, and limited funding available to agencies, this system is too diffuse and dependent upon the personal predilections of staff or higher authority at any point in time. A system that unifies these responsibilities within a single institutional authority is needed.

This can be done by formalizing the presently *de facto* operations and integrating them into a genuine instream flow program. There is a proposal for establishment of a stream, estuarine, and alien species center in Hilo, taking advantage of facilities that already exist. There is already an informal agreement to link the center with the Hawai'i Institute of Marine Biology. The center would have to be recognized within the DLNR organization as a bridge between DAR and CWRM. Because the center would be primarily a science-based operation, it could incorporate water quality considerations in its contributions to CWRM.

To reiterate, the Hawaii State Constitution, Article X, Section 1 states the following:

> "For the benefit of present and future generations, the state and its political subdivisions shall conserve and protect Hawaii's natural beauty and all natural resources, including land, water, air, minerals and energy sources, and shall promote the development and utilization of these resources in a manner consistent with their conservation and in furtherance of the self-sufficiency of the state. All public natural resources are held in trust by the state for the benefit of the people".

This is a clear directive that in combination with the Water Code should impel the state to establish a focused and comprehensive instream flow program. More than the foundation, PTD provides the glue to bond the pieces of the program together permanently. In the broader sense it mandates all government entities to look further into the future, and account for their actions accordingly, to ensure perpetuation of the resources for which they are responsible. And their actions should not be taken in isolation. Much greater emphasis on public outreach and education will be needed. PTD may be a step ahead of much of our political leadership, but that too may change as more people realize that it is not just an unattainable ideal, but it instead provides a realistic framework for effective resource management.

The Instream Flow Council (IFC) was formed in 1998 following the conclusion of a federally supported National Instream Flow Program Assessment in 1995 that brought together instream flow coordinators or their equivalents from each of the 50 states and the federal instream flow coordinator from each of the seven regions of the U.S. Fish and Wildlife Service. Now comprised of state and Canadian fish and wildlife agencies (Hawai'i is a charter member), the IFC is a nonprofit organization whose mission is to improve the effectiveness of instream flow programs for conserving aquatic resources. It has issued 46 policy statements, of which their statement on Public Trust Doctrine warrants special notice here:

> "Laws, regulations, and/or policies affecting fishery and wildlife resources and the habitats upon which they depend should be based on the state or province's legal stewardship responsibilities to manage those resources for the benefit and enjoyment of present and future generations".

Literature Cited

Annear, T., I. Chisholm, H. Beecher, A. Locke, P. Aarestad, C. Coomer, C. Estes, J. Hunt, R. Jacobson, G. Jobsis, J. Kauffman, J. Marshall, K. Mayes, G. Smith, C. Styalnaker & R. Wentworth. 2004. *Instream Flows for Riverine Resource Stewardship.* Revised edition. Instream Flow Council, Cheyenne, Wyoming. 268 p.

Biology of Hawaiian Streams and Estuaries. Edited by N.L. Evenhuis & J.M. Fitzsimons. *Bishop Museum Bulletin in Cultural and Environmental Studies* 3: 331–333 (2007).

Directions for Future Research in Hawaiian Streams and Estuaries: Results of Group Discussion at the Closing of the Symposium on Hawaiian Streams and Estuaries 26 & 27 April 2005 in Hilo, Hawai'i

Lori K. Benson McRae
Department of Biology, University of Tampa, 401 W. Kennedy Blvd., Tampa, Florida 33606-1490 USA; email: lbenson@ut.edu

Mark G. McRae
Louisiana State University, Museum of Natural Science, 119 Foster Hall, Baton Rouge, Louisiana 70803 USA; email: mmcrae@paws.lsu.edu

Introduction

On 26 and 27 April 2005, a Symposium on Hawaiian Streams and Estuaries was sponsored by the Division of Aquatic Resources (Department of Land and Natural Resources, State of Hawaii) and the United States Fish and Wildlife Service Sport Fish Restoration Program. This Symposium (co-organized by the Louisiana State University Museum of Natural Science) had several goals. These included bringing together researchers who had been working in Hawaiian streams and estuaries, as well as biologists who study similar systems in other parts of the World. This forum allowed updates and reports on research projects and facilitated collaboration between these scientists. Other goals included educating local people on the status of freshwater and estuarine organisms and describing the nature and scope of research on aquatic plants and animals. In addition to talks by experts, the conference incorporated presentations by local naturalists and students and thus provided unique opportunities for the scientists to learn from residents whose lives and livelihood have been associated with streams and estuaries for many generations. These latter presentations provided historical and cultural perspectives that otherwise would have been missing from the conference. Finally, symposium participants were charged with providing ideas and directions for future research on Hawaiian streams and estuaries, and a discussion section was scheduled at the end of the symposium. This final discussion section was moderated by M.G. McRae and recorded by L.K. Benson McRae. Participants were provided with a list of possible research directions and talking points as a springboard for discussion. These initial ideas and directions were co-authored by J.M. Fitzsimons & M.G. McRae from LSU and by R.T. Nishimoto & G.R. Higashi of the Hawaii Division of Aquatic Resources. The list below includes these items as well as additional future directions provided by conference participants.

Directions for Future Research in Hawaiian Streams and Estuaries

1. Hawaiian Stream Surveys:
   a. Increase frequency of statewide surveys of streams, especially at the lower elevations where data are limited.
   b. Continue building the GIS-correlated stream database and make it web accessible to potential stream users and researchers.
   c. Improve the integrity of spatial models of Hawaiian streams and native stream fish habitats.
   d. Establish flow requirements for streams and estuaries in Hawai'i.
   e. Maintain early life history studies of amphidromous fishes in the context of identifying source and sink populations of adults.
   f. Investigate the ecological importance of flash floods.

2. Synthesize all information that has been compiled by different agencies and research groups pertaining to hydrologic models.
3. Continue studies on Hawaiian stream organisms.
    a. Complete life history studies for the principal species of fishes and aquatic invertebrates.
    b. Develop a field identification key for nonindigenous species reported for Hawaiian streams.
    c. Continue studies contrasting the two species of *Kuhlia* as a background for management recommendations.
    d. Support studies on the distribution, systematics, ecology, and behavior of damselflies, dipterans, and other invertebrates whose life cycles are tied to streams and estuaries.
    e. Continue to explore the use of aquatic insects as indicator species for aquatic environments.
    f. Pursue studies of freshwater and euryhaline algae and other aquatic plants to identify their significance in community composition and structure in streams and estuaries.
    g. Expand investigations of the genetics of stream/estuarine organisms and the use of naturally occurring stable isotopes to amplify details of the marine ecology of amphidromous fishes and invertebrates.
    h. Determine the role of anatomy and biomechanical capabilities of stream and estuarine species in maintaining faunal composition and the distribution of plants and animals in stream/estuarine ecosystems.
    i. Extend surveys to determine *Lentipes* distributions.
    j. Investigate the role of increased stream temperatures on goby climbing effectiveness.
    k. Perform additional diet research on understudied species of gobies, and expand *Sicyopterus* feeding studies to contrast preference vs. availability of algae.
    l. Accumulate information on traditional use and knowledge of native gobies by Hawaiians.
4. Support research on invasive aquatic species.
    a. Develop procedures for removing alien aquatic species and assess their effectiveness.
    b. Establish a program to educate the public about alien species and set up a hotline for reporting.
    c. Develop creative volunteer programs to assist with alien control.
    d. Assess the impact of nonindigenous species on native fishes, crustaceans, and mollusks, especially at the stream mouths where the interaction between amphidromous native stream animals and coastal recreational fish species are temporally sympatric.
    e. Apply stock enhancement technology to analyze the impact of alien mullet (*Valamugil*) on native mullet (*Mugil*).
    f. Determine if movements of native amphidromous and coastal species are transmitting parasites into pristine native habitats.
    g. Assess parasite implications of the continued import of established exotics.
    h. Evaluate links of parasite introductions with the aquarium trade. Determine whether the import of poeciliids should be banned.
    i. Investigate the pathology of parasites over time on the reproductive success, etc. of native stream organisms.
    j. Study the effects of exotics on native insects.
5. Extend research into larger watershed scale studies, with emphasis on the links between terrestrial/freshwater species.
    a. Explore the biological links between terrestrial vegetation and aquatic plants and animals.
    b. Ascertain the influence of terrestrial inputs on water chemistry in stream watersheds and the implications for coastal marine fisheries.
    c. Determine the effects of exotic trees, ungulates, etc. on stream quality.
    d. Study the role of nutrient input from woody debris and riparian vegetation in streams and estuaries.
    e. Institutionalize the PABITRA methodology for stream assessment studies. (PABITRA = Pacific-Asia Biological Transect Network; http://www.botany.hawaii.edu/pabitra/)

   f. Investigate the role of forested/nonforested habitats in cloud formation, rainfall, and water flow in streams.
   g. Study how vegetation, agriculture, and land use changes affect groundwater discharge.
   h. Develop predictive models about the correlation between water quality and cover.
   i. Revisit zoning/land use planning issues regarding building and development near streams and rivers and determine whether the currently mandated buffer zone is being enforced and is adequate.
   j. Expand *ahupua'a* research to studies of management plans— learn from past management approaches used by traditional Hawaiian cultures.
6. Extend toxicology research to address effects of contaminants on native fish fauna.
   a. Broaden contaminant studies to include water quality issues with other native biota.
   b. Determine how toxins in the marine environment affect growth and survival of amphidromous larvae and investigate the subsequent implications for population sizes of adults.
7. Establish and coordinate restoration efforts.
   a. Develop procedures for restoring streams, estuaries, and anchialine ponds.
   b. Initiate a program of replacing exotics with native forest/riparian vegetation.
   c. Encourage aquaculture of native species.
   d. Restore goby populations to harvestable levels.
   e. Examine natural vs. original vs. pristine conditions for developing attainable restoration goals.
   f. Determine historic flow regimes and determine how they are linked to the presence/absence of stream animals.
   g. Facilitate better public involvement in management and restoration.
8. Investigate the response of native stream animals to climate and other environmental changes.
9. Determine priority catchments for biodiversity conservation in the Hawaiian Islands.
10. Prepare formal reports that summarize past studies on the biology, conservation, and management of native Hawaiian stream fishes and other aquatic animals.
11. Periodically organize symposia and publish the proceedings to set directions for future projects.
12. Develop partnerships with professionals on other Pacific Islands facing similar threats to stream habitats.
13. Strengthen public education regarding the significance of streams and estuaries in the Hawaiian Islands.

Closing Remarks

In summary, there was a general consensus that better communication and cooperation between research groups, universities, and government agencies needs to exist. This conference led to numerous discussions and brainstorming sessions that hopefully will facilitate a continuation of these dialogues; conference organizers were reminded to provide everyone involved with a list of participants and contact information. It was noted that another potential venue for this type of meeting would be in an aquatic session of the Hawai'i Conservation Conference, hosted by the Hawai'i Conservation Alliance and held each summer in Honolulu. In addition, conference attendees were reminded of the responsibility they have to report research findings and recommendations to the appropriate management organizations. Researchers also were encouraged to facilitate building a common website for reporting research findings, listing government documents and scientific publications, and describing upcoming projects. Finally, educational activities and continued participation of local students of all ages were regarded as being of utmost importance, with the hope that future conferences on the state's streams and estuaries will showcase more research being done in Hawai'i by Hawaiians.